Understanding the Prefrontal Cortex

OXFORD PSYCHOLOGY SERIES

Editors

Mark D'Esposito Daniel Schacter
Jon Driver Anne Treisman
Trevor Robbins Lawrence Weiskrantz

1. The Neuropsychology of Anxiety *J. A. Gray*
2. Elements of Episodic Memory *E. Tulving*
3. Conditioning and Associative Learning *N. J. Mackintosh*
4. Visual Masking *B. G. Breitmeyer*
5. The Musical Mind *J. A. Sloboda*
6. Elements of Psychophysical Theory *J.-C. Falmagne*
7. Animal Intelligence *L. Weiskrantz*
8. Response Times *R. D. Luce*
9. Mental Representations *A. Paivio*
10. Memory, Imprinting, and the Brain *G. Horn*
11. Working Memory *A. Baddeley*
12. Blindsight *L. Weiskrantz*
13. Profile Analysis *D. M. Green*
14. Spatial Vision *R. L. DeValois and K. K. DeValois*
15. The Neural and Behavioural Organization of Goal-Directed Movements *M. Jeannerod*
16. Visual Pattern Analyzers *N. V. S. Graham*
17. Cognitive Foundations of Musical Pitch *C. L. Krumhansl*
18. Perceptual and Associative Learning *G. Hall*
19. Implicit Learning and Tacit Knowledge *A. S. Reber*
20. Neuromotor Mechanisms in Human Communication *D. Kimura*
21. The Frontal Lobes and Voluntary Action *R. Passingham*
22. Classification and Cognition *W. K. Estes*
23. Vowel Perception and Production *B. S. Rosner and J. B. Pickering*
24. Visual Stress *A. Wilkins*
25. Electrophysiology of Mind *Edited by M. D. Rugg and M. G. H. Coles*
26. Attention and Memory *N. Cowan*
27. The Visual Brain in Action *A. D. Milner and M. A. Goodale*
28. Perceptual Consequences of Cochlear Damage *B. C. J. Moore*
29. Binocular Vision and Stereopsis *I. P. Howard and B. J. Rogers*
30. The Measurement of Sensation *D. Laming*
31. Conditioned Taste Aversion *J. Bures, F. Bermúdez-Rattoni, and T. Yamamoto*
32. The Developing Visual Brain *J. Atkinson*
33. The Neuropsychology of Anxiety, 2e *J. A. Gray and N. McNaughton*
34. Looking Down on Human Intelligence *I. J. Deary*
35. From Conditioning to Conscious Recollection *H. Eichenbaum and N. J. Cohen*
36. Understanding Figurative Language *S. Glucksberg*
37. Active Vision *J. M. Findlay and I. D. Gilchrist*
38. The Science of False Memory *C. J. Brainerd and V. F. Reyna*
39. The Case for Mental Imagery *S. M. Kosslyn, W. L. Thompson, and G. Ganis*
40. Seeing Black and White *A. Gilchrist*
41. Visual Masking, 2e *B. Breitmeyer and H. Öğmen*
42. Motor Cognition *M. Jeannerod*
43. The Visual Brain in Action *A. D. Milner and M. A. Goodale*
44. The Continuity of Mind *M. Spivey*
45. Working Memory, Thought, and Action *A. Baddeley*
46. What Is Special about the Human Brain? *R. Passingham*
47. Visual Reflections *M. McCloskey*
48. Principles of Visual Attention *C. Bundesen and T. Habekost*
49. Major Issues in Cognitive Aging *T. A. Salthouse*
50. Perceiving in Depth *Ian P. Howard*
51. The Neurobiology of the Prefrontal Cortex: Anatomy, Evolution, and the Origin of Insight *Richard E. Passingham and Steven P. Wise*
52. The Evolution of Memory Systems: Ancestors, Anatomy, and Adaptations *Elisabeth A. Murray, Steven P. Wise, and Kim S. Graham*
53. Understanding the Prefrontal Cortex: Selective advantage, connectivity, and neural operations *Richard E. Passingham*

Understanding the Prefrontal Cortex

Selective Advantage, Connectivity, and Neural Operations

RICHARD E. PASSINGHAM
Emeritus Professor of Cognitive Neuroscience
Department of Experimental Psychology
Oxford University

OXFORD
UNIVERSITY PRESS

Great Clarendon Street, Oxford, OX2 6DP,
United Kingdom

Oxford University Press is a department of the University of Oxford.
It furthers the University's objective of excellence in research, scholarship,
and education by publishing worldwide. Oxford is a registered trade mark of
Oxford University Press in the UK and in certain other countries

© Oxford University Press 2021

The moral rights of the author have been asserted

First Edition published in 2021

Impression: 3

All rights reserved. No part of this publication may be reproduced, stored in
a retrieval system, or transmitted, in any form or by any means, without the
prior permission in writing of Oxford University Press, or as expressly permitted
by law, by licence or under terms agreed with the appropriate reprographics
rights organization. Enquiries concerning reproduction outside the scope of the
above should be sent to the Rights Department, Oxford University Press, at the
address above

You must not circulate this work in any other form
and you must impose this same condition on any acquirer

Published in the United States of America by Oxford University Press
198 Madison Avenue, New York, NY 10016, United States of America

British Library Cataloguing in Publication Data

Data available

Library of Congress Control Number: 2020945296

ISBN 978–0–19–884457–0

DOI: 10.1093/oso/9780198844570.001.0001

Printed and bound by
CPI Group (UK) Ltd, Croydon, CR0 4YY

Oxford University Press makes no representation, express or implied, that the
drug dosages in this book are correct. Readers must therefore always check
the product information and clinical procedures with the most up-to-date
published product information and data sheets provided by the manufacturers
and the most recent codes of conduct and safety regulations. The authors and
the publishers do not accept responsibility or legal liability for any errors in the
text or for the misuse or misapplication of material in this work. Except where
otherwise stated, drug dosages and recommendations are for the non-pregnant
adult who is not breast-feeding

Links to third party websites are provided by Oxford in good faith and
for information only. Oxford disclaims any responsibility for the materials
contained in any third party website referenced in this work.

In memory of Alan Cowey (1935–2012) and Lawrence Weiskrantz (1926–2018).

'If you want to understand function, study structure'
Francis Crick (*What Mad Pursuit*, 1988)

Note on the Cover

The picture on the front cover shows that, compared to the macaque monkey brain, it is the prefrontal cortex and other association areas (in yellow and orange) that have expanded most in the human brain.

Preface

Steve Wise and I published *The Neurobiology of the Prefrontal Cortex* in 2012. This date might seem recent to some, but scientific monographs have a shelf life of five years or less. This is the typical timespan over which many scientists consider such publications to be relevant. As in other scientific disciplines, neuroscience moves on apace, and ideas that once seemed novel and exciting soon become standard knowledge.

I decided that it was time to write a second edition. But, as I began revising the book, changes accumulated so rapidly that a new book took shape. That is why it has a new title. It retains some of the organizational structure of the original book, as well as some of the figures; and I have also made use of some of the material in the text. So, a clear palimpsest of the original remains. But the main burden of this book is totally new, and most of it has been written without reference to the original.

There are several reasons why a new book is required. First, I have changed my mind on many issues. In the original book, we attempted to frame our proposals as clearly as possible, but the danger of clarity is that errors can become obvious. Good scientists embrace the prospect that they might be wrong. Marvin Minsky, for example, cultivated 'the habit of not wanting to be right for very long.' As he put it: 'If I still believe something after five years, I doubt it.'

A second reason is that the original book dealt mainly with functional neuroanatomy, and we described functions in general psychological terms. We argued that anatomical inputs constrain the function that each area performs and that the outputs convey the result to other areas. But we did not take things further by providing an explicit proposal concerning the *transformation* that each area performs from its inputs to its outputs. Chapters 3–7 discuss each area in turn, and they end with a proposal concerning the transformation that that area performs.

A focus on input–output transforms means that book must now consider what can be learned from computational neuroscience. It is this field of research that helps us to understand the transforms that are computed by cortical areas. Beginning with Zipser and Anderson (1988), neuroscientists have produced computational models that use neurophysiological data to suggest how populations of neurons with particular properties could perform specific functions. For example, Rolls (2016) has considered how attractor networks can account for the operations of the cerebral cortex. It is also increasingly common in functional brain-imaging studies to look for activations that are related to specific parameters in a computational model (e.g. Boorman et al., 2016).

I am not an expert in this field and am not therefore competent to present detailed computational models in this book. The transforms are therefore presented without fully worked-out proposals as to how they are actually implemented in the brain.

A third reason for a new book is that new methods have clarified the evolution of prefrontal areas. For example, imaging methods have enabled anthropologists to construct virtual endocasts from fossil crania (e.g., Long et al., 2015). These studies have augmented the methods of comparative neuroanatomy with a direct examination of ancestral species. Studies of the pattern of anatomical connections have provided a method for identifying which prefrontal areas are homologous in different species (Mars et al., 2017). And the development of phylogenetic statistics (Smaers & Rohlf, 2016) has also been a major advance.

There is a final reason why a new book was necessary. The literature on functional brain imaging has exploded in recent years, as Figure P.1 illustrates. The purple curve shows the number of publications on the prefrontal cortex stemming from research on humans, with the vast majority involving functional brain imaging.

In the original book, we based our proposals on the animal literature. We then included a single chapter in which we argued that the literature on the human brain from functional imaging was consistent with these proposals. But the imaging literature has now developed to the point where it is important to consider what new it has to tell us on its own. Only studies of the human brain itself can help us to understand capacities such as language that are unique to humans.

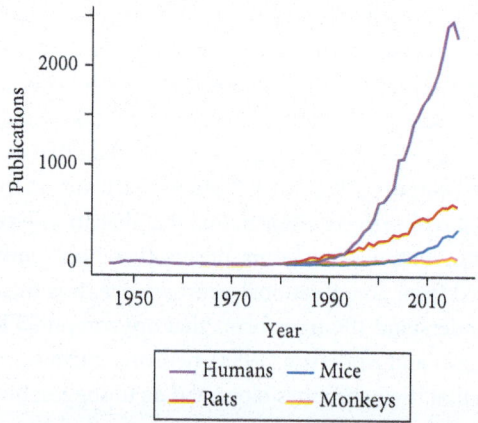

Figure P.1 The annual number of papers on the prefrontal cortex of humans (purple), rats (red), mice (blue), and monkeys (orange) since the late 1940s

Reproduced from *eNeuro*, 5 (4), Mark Laubach, Linda M. Amarante, Kyra Swanson, and Samantha R. White, What, if anything, is rodent prefrontal cortex? Figure 1a, Doi: https://doi.org/10.1523/ENEURO.0315-18.2018 Copyright (c) 2018, The Authors. Licensed under CC BY 4.0.

Much of the evidence comes from functional imaging. But the extent of the imaging literature means that I cannot hope to do more than mention a selection of the published results. Rather than attempt a comprehensive review, I have chosen to highlight the results that best illuminate the topic. I therefore plead guilty to cherry-picking throughout this book. All that I can say is that I have done my best to avoid picking the rotten ones.

One final word of warning. This book conveys my understanding, as of now. There is always a danger that in trying to explain things, one skates over points that are inconvenient for the account. I hope that I have described the experimental evidence in enough detail that others can come to different conclusions. After all, that is how science progresses.

References

Boorman, E.D., Rajendran, V.G., O'Reilly, J.X., & Behrens, T.E. (2016) Two anatomically and computationally distinct learning signals predict changes to stimulus-outcome associations in hippocampus. *Neuron*, 89, 1343–54.

Laubach, M., Amarante, L.M., Swanson, K., & White, S.R. (2018) What, if anything, is rodent prefrontal cortex? *eNeuro*, 5 315–18.

Long, A., Bloch, J.I., & Silcox, M.T. (2015) Quantification of neocortical ratios in stem primates. *Am J Phys Anthropol*, 157, 363–73.

Mars, R.B., Passingham, R.E., Neubert, F.X., Verhagen, L., & Sallet, J. (2017) Evolutionary specializations of human association cortex. In Preuss, T.M., Kaas, J. (eds) *Evolution of Nervous Systems*. Elsevier, New York.

Rolls, E.T. (2016) *Cerebral Cortex: Principles of Operation*. Oxford University Press, Oxford.

Smaers, J.B. & Rohlf, F.J. (2016) Testing species' deviation from allometric predictions using the phylogenetic regression. *Evolution*, 70, 1145–49.

Zipser, D. & Andersen, R.A. (1988) A back-propagation programmed network that simulates response properties of a subset of posterior parietal neurons. *Nature*, 331, 679–84.

Acknowledgements

I could not have written this book without the help of Steve Wise. He read all the chapters, commented and made extensive revisions throughout the text. He contributed especially to Chapter 2 because he knows much more than I do about the evolution of the non-human primates, and in particular about the fossil record. I value in particular the critical comments that he made on Chapters 10 and 11: disagreement can be very much more helpful than agreement. Steve also helped me by producing many of the figures and telling me how to produce the others. I cannot say how grateful I am for all the work he has put in.

I confess to having stolen the idea of dividing the book into parts or sections from the book on 'The Evolution of Memory Systems' by Betsy Murray, Steve Wise, and Kim Graham (2017). Their book won a prize; so, it must be a good idea.

The fundamental way of thinking about the brain on which this book is based came out of a conversation with Klaas Stephan. It was written up a long time ago in a paper with Klaas and Rolf Kötter on 'The anatomical basis of functional localization in the cortex'. The idea has since been further developed by Matthew Rushworth, Heidi Johansen-Berg, and Rogier Mars. The motivation for this book was to take it yet further, by changing the way in which we think about the function of brain areas.

I have other colleagues to thank for sharing ideas and demolishing some of my more outrageous speculations. In particular, Eleanor Maguire, Celia Heyes, and Hakwan Lau have patiently answered my emails and questions, and put up with my excitable nature.

Because I am long retired, it is hard to find the new literature. I am grateful in particular to Earl Miller and Matthew Rushworth for sending me some of their more recent papers. I have also been lucky enough to benefit from comments from Nils Kolling, Laurence Hunt, Jill O'Reilly, and their graduate students and postdocs who have read the whole book as a summer project during lockdown over COVID-19. They have had the advantage of viewing the manuscript with young eyes.

I am indebted to Larry Weiskrantz and Alan Cowey who supported me throughout my career, and this book is dedicated to them. When I worked with them as a postdoctoral fellow, they allowed me to work on the prefrontal cortex although the grant was on vision. Those were the days.

I could not have published a book which is so lavishly illustrated had not the authors of the relevant papers taken the trouble to send me the originals of their figures. If the text is in danger of becoming dull, it always pays to add in a colour figure to spice things up.

Research is a collaborative project. So, I need to thank my graduate students, postdoctoral fellows, and others who actually did the work. In various ways all have contributed to the ideas in this book, whether they worked on prefrontal cortex or not. I have benefited from the contributions of Katie Alcock, Sara Bengtsson, Pierre Burbaud, Tony Canavan, Jim Colebatch, Marie-Pierre Deiber, Julie Grezes, Katie Hadland, Harri Jenkins, Louise Johns, Markus Jueptner, Mike Krams, Hakwan Lau, Rogier Mars, Phil Nixon, Narender Ramnani, James Rowe, Matthew Rushworth, Katsuyuki Sakai, Nat Schluter, Jeroen Smaers, David Thaler, and Ivan Toni. In retirement it is their stimulating company that I miss.

I would also like to thank the editors at Oxford University Press, Martin Baum and Charlotte Holloway, for encouraging a new book and guiding me through the process. Apart from dealing with the manuscript, Charlotte put in an enormous amount of work to seek the permissions for all the figures.

Finally, I would like to thank my wife Clare. I had promised her that I wouldn't write another book; and she has forgiven me, I think.

Contents

List of Figures   xvii
List of Tables   xxiii
Style, Scope, and Terminology   xxv
Abbreviations   xxvii

PART I: FOUNDATIONS

1. Introduction   3
2. Evolution of the Prefrontal Cortex in Non-human Primates   34

PART II: SUBAREAS OF THE PREFRONTAL CORTEX

3. Medial Prefrontal Cortex: Self-Generated Actions   71
4. Orbital Prefrontal Cortex: Evaluating Resources   118
5. Caudal Prefrontal Cortex: Searching for Objects   153
6. Dorsal Prefrontal Cortex: Planning Sequences   191
7. Ventral Prefrontal Cortex: Associating Objects   236

PART III: THE PREFRONTAL CORTEX WITHIN THE SYSTEM AS A WHOLE

8. Prefrontal Cortex: Abstract Rules and Attentional Performance   287

PART IV: THE HUMAN PREFRONTAL CORTEX

9. Evolution of the Prefrontal Cortex in the Hominins   333
10. Human Prefrontal Cortex: Reasoning, Imagination, and Planning   372
11. Human Prefrontal Cortex: Language, Culture, and Social Rules   420

Index   469

List of Figures

P.1	The annual number of papers on the prefrontal cortex of humans (purple), rats (red), mice (blue), and monkeys (orange) since the late 1940s	x
1.1	Connectional fingerprints for the PF areas 9 and 14	6
1.2	Connectional clusters in the PF cortex	7
1.3	(A & B) Connectional fingerprints of the SMA and the rostral ventral premotor cortex (PMvr)	9
1.4	Functional fingerprints showing selectivity for visually guided versus memory-guided movement sequences in the premotor cortex (lateral area 6), the primary motor cortex (area 4), and the supplementary motor area (medial area 6)	10
1.5	Functional fingerprints for the cortex in the anterior cingulate sulcus and the right anterior insular cortex (Ia)	11
1.6	Comparison of anatomical and functional borders	13
1.7	(A) Area 9 (part of the granular medial PF cortex) in human subjects; (B) Area 9 in macaques. The letters refer to the different sulci. (C) The functional connectivity fingerprints in humans (light blue) and macaques (purple) with the overlap shown in dark blue. (D) The summed absolute differences between the functional coupling scores	18
1.8	Effect of a lesion on neuronal activity	21
1.9	Map of the macaque cortex	24
1.10	Maps of the human cortex (A) and macaque cortex (B)	26
1.11	The five regions of the PF cortex used in this book	28
2.1	The PF cortex and other frontal areas in macaque monkeys (A), rats (B), and (C) bushbabies (*galagos*)	35
2.2	Cortical map of tree shrew brains	44
2.3	Effect of granular PF lesions in tree shrews	45
2.4	Phylogenetic reconstruction of primate social systems	47
2.5	Virtual brain endocasts of stem and crown primates	53
2.6	(A) Encephalization: Brain–body mass relationship in *Rooneyia* and other fossil primates. (B) Corticalization	54
2.7	Brain–body mass relationship for modern strepsirrhines, modern anthropoids, fossil anthropoids (red letters), and a fossil pongid (blue letters)	56

xviii LIST OF FIGURES

2.8	Circular cladogram showing a stack of encephalization quotients (EQs) for a series of species in each mammalian lineage, ordered by EQ value	57
2.9	Reconstructed phylogeny of encephalization quotients (EQs) for primates	58
2.10	Virtual brain endocasts of extinct catarrhines	59
3.1	Medial PF cortex in macaque monkeys (left) and humans (right), indicated by shading	72
3.2	Homologies among agranular parts of the medial PF cortex in rodents and anthropoids	74
3.3	Selected connections of the medial PF cortex in macaque monkeys	75
3.4	Reversal impairment for choices between two actions	83
3.5	(A) Areas that were specifically activated when monkeys watched videos of other monkeys interacting. (B) Flat map of areas that showed a difference in grey matter depending on the size of the group	90
3.6	Overlap between medial (default) network and social network in macaques	93
3.7	Medial network as visualized by fMRI	94
3.8	Cell activity encoding choices at feedback time for populations of cells in the polar PF cortex (area 10) (orange) and the orbital PF cortex (area 11) (green)	103
3.9	Pictures (p) for 'object-in-scenes' task, shown for two trials or runs, with 20 pictures per run	104
3.10	The effect of polar PF lesions in macaque monkeys	104
4.1	The orbital PF cortex in monkeys (left) and humans (right) indicated by shading	119
4.2	Selected connections of the orbital PF cortex	120
4.3	Total selection frequency for the category of nonpreferred foods (garlic, lemons, and monkey chow) for each experimental group across the pre-surgery, post-surgery, and shuffled testing phases	123
4.4	The effects of satiation on the choice between objects that were associated with particular foods	127
4.5	Effect of three lesions on choices in the devaluation task	129
4.6	The effect of amygdala lesions on the encoding of reward in the orbital PF cortex	132
4.7	Action reversal and object reversal tasks	135
4.8	Impairment in object reversal set in monkeys after lesions of the central and medial sectors of the orbital PF cortex	136
4.9	Errors made on the object reversal task, before and after the first correct choice	138
4.10	Performance on probabilistic reversals after lesions of the orbital PF cortex	140
4.11	Performance on a probabilistic reversal learning task: the 3-arm bandit task	142
4.12	Data for object reversal learning for different groups of lesioned macaques	144

5.1	The caudal PF cortex in the macaque monkeys (left) and humans (right)	154
5.2	Selected connections of the caudal PF cortex	156
5.3	Layout of the FEF circuit	162
5.4	Topographic maps in the frontal eye fields and posterior parietal cortex in one subject	165
5.5	Intrinsically defined dorsal and ventral attention systems and the overlap between them	168
5.6	A common version of the oculomotor delayed response task	172
5.7	Attention versus memory coding in the PF cortex	174
5.8	Performance on the oculomotor delayed response task for one lesioned monkey	176
5.9	Corrective saccades after a frank error	177
5.10	Change in population vectors on an oculomotor delayed response task	179
5.11	An example of a search array on which are superimposed the scan paths of one patient	180
6.1	The dorsal PF cortex in macaque monkeys (left) and humans (right)	192
6.2	Selected connections of the dorsal PF area 46	193
6.3	Areas 9/46 and 46 in the human brain as identified on the basis of the similarity of their connectional fingerprints to the same areas in a macaque monkey	197
6.4	(A) The anterior area of the dorsal PF cortex (green) that co-activated with the anterior cingulate cortex. (B) The posterior area of the dorsal PF cortex (red) that co-activated with the parietal cortex	198
6.5	Testing procedure for the classic delayed response task in a Wisconsin general testing apparatus (WGTA)	201
6.6	The top section shows the matching task and the bottom section the recall task that were used in an fMRI experiment	204
6.7	The course of the BOLD signal during the tasks that were illustrated in Figure 6.6	205
6.8	(A) The data for sites at which the spiking activity was informative. (B) The data for sites at which the spiking activity was not informative	214
6.9	(A) The task used in an imaging experiment to test for the effect of distraction in memory. (B) The relation between sustained activation during the delay and the accuracy of performance on trials on which there were no distractors and trials on which there were distractors	217
6.10	(A) The activation in the dorsal PF cortex on a task in which human subjects decide of their own accord which finger to move. (B) The plot of the degree of activation as a function of the equipotentiality index	220

6.11	(A) Task in which the monkeys used a handle to make sequences of four movements. (B) Activity of cells that encoded sequences with a particular abstract structure	222
6.12	Visual maze task	223
6.13	Population analysis of cells in the dorsal PF cortex during planning	224
7.1	The ventral PF cortex in macaque monkeys (left) and humans (right)	237
7.2	Cross section through the ventral limb of the arcuate sulcus on an MRI scan in a macaque monkey	238
7.3	Areas in the human ventral PF cortex that correspond with areas in the macaque monkey, as demonstrated by the functional fingerprint based on resting state covariance	239
7.4	Selected connections of the ventral PF cortex	241
7.5	Effect of lesions of the ventral PF cortex on simultaneous matching	245
7.6	Performance across trials within problems on a series of visuo-spatial problems before and after ventral PF surgery	254
7.7	Stimuli used in a categorization task	257
7.8	(A) A cell in the ventral PF cortex that encoded the category 'dog'. (B) A cell in the ventral PF cortex that encoded the category 'cat'	258
7.9	A cell in the ventral PF cortex that encoded arbitrary categories, demarcated by the lightly stippled vertical lines in Figure 7.7	260
7.10	Single cells reflecting both shape-shape associations and motion direction categories	263
7.11	Activations for shifting between categories, shown on inflated surface reconstructions of the macaque monkey brain (left) and human brain (right)	270
7.12	(A) The effect of a TMS pulse over the FEF on the activation in the MT complex when motion is relevant. (B) The effect of a TMS pulse over the FEF on the activation in the fusiform face area when the shape of the face is relevant	273
8.1	The central neocortical hub or core as shown by graph theory	289
8.2	The organization of the feedforward (blue) and feedback (red) connections in the neocortex	291
8.3	Organization of basal ganglia and cerebellar outputs to the cerebral cortex	293
8.4	The overlap of the projections to the striatum from different PF areas is shown in orange. The projections from the inferior parietal area PG are shown in green	295
8.5	PF cell encoding the abstract matching rule	300
8.6	The location of rule selective and generalist cells in the PF cortex that coded for the general rule 'greater than' or 'lesser than'	301
8.7	Development of a learning set for visual discrimination problems, in a selection of mammalian species	303

8.8	Histograms showing mean percent error in trials 2–11 at each performance test on discrimination learning set	305
8.9	Strategy score (performance on repeat compared with change trials) before (black) and after (white) lesions	307
8.10	Population coding for abstract rules	308
8.11	Timing of the development of selectivity for visual objects, behavioral goals, and actions	310
9.1	Encephalization quotients (EQs) for fossil hominins, modern humans, and modern chimpanzees	335
9.2	CT scans of skulls of a chimpanzee, *Homo heidelbergensis*, *Homo neanderthalensis*, and *Homo sapiens*	336
9.3	Granular PF cortex as a percentage of the frontal lobe, plotted versus function of cerebral extent, in modern primates	338
9.4	Log-log plot of the estimated volume of the granular PF cortex as a function of non-PF cortex in the frontal lobe using the data from Smaers et al. (2011)	339
9.5	Myeloarchitectonics in the brains of humans, chimpanzees, and macaque monkeys	343
9.6	Regressions of the volume of the frontal pole cortex (area 10) as a function of brain volume in selected primates	344
9.7	Expansion of the area 10 in humans	345
9.8	(A) Relative expansion of cortical regions from macaque to human brains, shown on the human brain. (B) Relative expansion of cortical regions from chimpanzee to human brain, shown on the chimpanzee brain	347
9.9	Log-log plot of volume of the PF cortex as a function of an estimate of the other association areas (see text)	348
9.10	Degree of expansion and closeness centrality	351
9.11	Areas of the brain that have changed most in their connectivity as estimated from visualizing the major tracts using diffusion weighted imaging (DWI)	352
9.12	Arcuate fasciculus in the human and macaque monkey brain as visualized by diffusion weighted imaging	354
9.13	Postnatal cortical surface expansion	358
10.1	The multiple-demand system as visualized on the parcellation of the human brain by Glasser et al.	376
10.2	A typical problem on the Raven's Progressive Matrices	377
10.3	The relation between fluid intelligence and the volume of damage to the frontal, parietal, or temporal lobe	379
10.4	Computerized version of the 'Tower of London'	384
10.5	Areas that are activated when subjects plan	387

10.6	Activations in the left inferior caudal PF cortex (areas 44 and 45B) and the STS during deductive reasoning	389
10.7	The accuracy of decoding using a multivoxel pattern analysis for memories that were of events that occurred either 2 years ago or 10 years ago	392
10.8	The area in yellow shows the source of the theta oscillations as measured by MEG while the subjects retrieved personal memories from the past	394
10.9	Indices relating the activity for hits and misses	399
10.10	ERPs recorded on the Libet task for the conditions in which the subjects timed their intention to move (W) or the actual movement itself (M)	403
10.11	Brain regions with a significant difference between the prediction of the subjective ratings and skin conductance reactivity	408
11.1	(A) Areas that were activated during signing and speaking. (B) Areas that were only activated during signing	424
11.2	The arcuate fasciculus in the macaque monkeys and human brain	426
11.3	(A) The areas that were activated both when observing and when imitating. (B) The areas that were activated when imitating was contrasted with observing alone in green, and when observation was contrasted with imitation in red	428
11.4	Activations for speaking, imitation, and motor inhibition in the left and right hemisphere	429
11.5	Horizontal sections showing the areas that were under-activated in the brains of affected members of the KE family	430
11.6	Comparing sentences and small clauses	433
11.7	Areas of the neocortex which showed activity that was specific to naming pictures (blue), naming by definition (red), or that showed activity for naming in both conditions (pink)	436
11.8	Setting up the current task	442
11.9	Semantic task used by Vandenberghe et al. (1996)	443
11.10	Activations while expert stone knappers make Oldowan and Acheulean tools	446
11.11	The activations resulting from metanalyses of fMRI data for metacognition (yellow and red) and mentalizing (green and blue)	449
11.12	The overlap for the lesions that included the medial and orbital PF cortex in 19 patients who had had a medial meningioma removed at surgery (vmpfc)	451

List of Tables

1.1	Prefrontal areas in human and macaque monkey brains, with area numbers in parentheses, where applicable.	25
1.2	Groups of PF areas and a compact abbreviation for each.	29
2.1	PF areas with homologues (+) in mammals, strepsirrhine (prosimian) primates, anthropoids, and humans.	46
9.1	Remapping factors.	340
9.2	Remapping factors.	341
9.3	Remapping factors.	349
9.4	Factors that might account for the expansion of the PF cortex in hominins.	360
10.1	Remapping factors.	393
11.1	Handedness in chimpanzees.	421
11.2	Handedness in human subjects.	421

Style, Scope, and Terminology

I have thought it important to tell readers how the researchers obtained the results that are summarized in this book. Accordingly, I have avoided simply stating findings and conclusions as facts, followed by a long series of references in brackets. This practice is perfectly reasonable in a paper or review article, especially in view of the word limits imposed by many publishers. But the point of a textbook or monograph is not to provide references, however convenient that might be. The aim of this book is to further understanding, and this means spelling out in detail why the results of a particular experiment lead to certain conclusions and not to others.

Because each section describes a series of experiments in detail, to help the reader I have added summaries at the end of each section. These provide the provisional conclusions that I take to follow from these experiments. The final interpretation of the results is left to the end of the chapter,

For convenience, I have adopted certain several semantic conventions:

- The word *animal*, when unmodified, refers to a non-human animal.
- The word *monkey*, when unmodified, refers to macaque monkeys.
- I have used the phrase *granular prefrontal cortex*, although its architecture is not truly granular in the same sense as the granular cortex of the primary sensory areas.
- I use the term *lesion* to cover all procedures that prevent a cortical area from functioning normally, including various forms of temporary disruption.
- The term *cell* is used to refer specifically to neurones throughout.
- I use the term *activity* to describe the rate of neuronal action potentials, commonly known as firing rate, discharge rate, or modulation; but I use the term *activation* to describe results from functional brain-imaging experiments because the BOLD signal is a vascular one.
- I have resisted using the term *volunteers* to describe people who take part in fMRI experiments. I doubt that readers will think that these people have been dragged kicking and screaming into the scanner. I know that the label *participants* is now *de rigueur*, but the word *subject* has the advantage that it can be used for all species, humans included.
- I will quite rightly draw the ire of anatomists by phrases that suggest that anatomical connections run from or to anatomical sulci such as the intraparietal sulcus. Of course, they run from the cortical tissue in the sulcus, not from the sulcus itself. I also realize that fMRI activations are in the cortex of the

sulcus, not the sulcus itself. But it becomes cumbersome to labour this point throughout, and I have not done so.
- I use the term *association cortex* when it would be more accurate to use the anatomical term homotypical cortex which is defined on the basis of cytoarchitecture. The term association cortex is an outdated functional term, but it has the advantage that many readers will be more familiar with it; and I don't want to frighten the horses.
- I use the term *posterior parietal cortex* to distinguish it from the somatosensory cortex in the parietal lobe. The association cortex of the parietal lobe lies posterior to the primary somatosensory cortex.
- I have tried to keep the abbreviations to a minimum. It is fine to use them in an anatomy paper but can put off readers of a book. So, I have confined the abbreviations to the ones that are most common. Thus, I use IPS for intraparietal sulcus and STS for the superior temporal sulcus, but I spell out inferior frontal sulcus in full. Similarly, though I used SMA for the supplementary motor area, I spell out dorsal and ventral premotor cortex in full, rather than using PFd and PMv. I toyed with using abbreviations for the different prefrontal subareas, PFm, PFo, PFc, PFd, and PFv, but decided that I was more interested in sales. To help the reader I define any abbreviations that I use the first time that I mention them in a chapter. A list of all the abbreviations, together with their definitions follows on the next page.

Abbreviations

2-DG	2-deoxyglucose
ACC	anterior cingulate cortex
AIP	anterior intraparietal area
Area F5	rostral part of ventral premotor cortex
Area X	part of a bird brain
ASL	American sign language
BOLD	blood oxygen-level dependent [signal]
CMA	cingulate motor area
CMAr	rostral cingulate motor area
CoCoMac	database of corticocortical connections in macaques
DA	delayed alternation task
DR	delayed response task
DREADDS	designer receptors exclusively activated by designer drugs
DWI	diffusion weighted imaging
ECoG	electrocorticography
EEG	electroencephalography
EQ	encephalization quotient
ERP	event-related potential
FEF	frontal eye field
FEFd	dorsal frontal eye field
FEFv	ventral frontal eye field
FFA	fusiform face area
fMRI	functional magnetic resonance imaging
FoxP2	forkhead box protein P2
GABA	gamma-aminobutryic acid, an inhibitory neurotransmitter
gACC	gyral part of the anterior cingulate cortex
Hz	Hertz
Ia	agranular insular cortex
IFJ	inferior frontal junction
IPS	intraparietal sulcus
IQ	intelligence quotient
Ka	thousand years ago
LIP	lateral intraparietal area
M1	primary motor area
Ma	million years ago
MD	mediodorsal nucleus of the thalamus
MEG	magnetoencephalography
MEP	motor evoked potential
MIP	medial intraparietal area

MNI	Montréal Neurological Institute
MST	middle superior temporal area
MT	middle temporal area
MVPA	multivoxel pattern analysis
N2pc	ERP component linked to selective attention
NA	nucleus accumbens
NMDA	N-methyl-D-aspartate, an excitatory neurotransmitter
ODR	oculomotor delayed response task
PET	positron emission tomography
PF	prefrontal cortex
preSMA	presupplementary motor area
ROC	receiver operating characteristic
RSA	representational similarity analysis
rTMS	repetitive transcranial magnetic stimulation
S1	primary somatosensory cortex
S2	secondary somatosensory cortex
sACC	anterior cingulate sulcus
SEF	supplementary eye field
SEM	standard error of the mean
SMA	supplementary motor area
SOA	stimulus-onset asynchrony
STP	superior temporal polysensory area
STS	superior temporal sulcus
TMS	transcranial magnetic stimulation
TPJ	temporal-parietal junction
Tpt	temporo-parietal area, transition between the lobes
V1	primary (striate) visual area (area 17)
V2	second visual area (area 18), an extrastriate area
V3	a low-order area in the extrastriate visual cortex
V4	a low-order area in the extrastriate visual cortex
WGTA	Wisconsin general testing apparatus

PART I
FOUNDATIONS

1
Introduction

This chapter has two purposes. The first is to explain why the book is subtitled 'selective advantage, connectivity, and neural operations'. The second is to tell the reader why the second part of the book describes neurophysiological and neuropsychological experiments that were carried out on macaque monkeys. After all, many readers would have started out with the assumption that the way to understand the prefrontal (PF) cortex was to use functional brain imaging with human subjects.

Selective Advantage

The reason why the words 'selective advantage' appear in the subtitle is that, as with any other biological system, a complete understanding of the PF cortex requires answers to four questions. Tinbergen (1951) listed them as follows:

- How did it evolve (phylogeny)?
- How does it promote fitness (selective advantage)?
- How did it develop (ontogeny)?
- How does it work (physiology)?

The literature on the PF cortex is mainly devoted to the fourth question, though it tackles the third one as well. But it rarely addresses the first two. Yet, they hold an important key to understanding the human PF cortex. The reason is that characteristics that are unique to humans evolved by co-opting and elaborating mechanisms that existed in ancestral primates. We cannot properly understand the human PF cortex if we do not appreciate that the fundamental organization of the human PF cortex is the same as in other anthropoid primates, the monkeys and apes.

Chapter 2 explains that some parts of the PF cortex first appeared in early primates and others came along during the evolution of the anthropoids, the monkeys and apes. If we take into account the life and times of the ancestral species in which particular PF areas first appeared, we can obtain clues about what these areas did then and the selective advantage that they conferred.

It truly matters that humans are primates. Unlike other mammals, primates forage by reaching and grasping food with their hands. This involves the coordination of hand and eye. Specialized areas evolved to support this, including the anterior parietal area (AIP) and the ventral premotor cortex. If either the inferior parietal cortex (Stepniewska et al., 2009) or the ventral premotor cortex (Graziano et al., 2002) is stimulated electrically in non-human primates, the hand reaches towards the mouth irrespective of its initial starting position. Shadmehr and Wise (2005) provide a full description of the complex mechanisms that are involved in reaching for a target.

However, as Chapters 6 and 7 demonstrate, it is the PF cortex that generates the goal or target for searching with the eye and reaching with the hand. The PF cortex sends connections not only to the frontal eye field (FEF) but also to the premotor areas that are specialized for the movements of the hand and arm, though not the leg. It is these mechanisms that have been co-opted and elaborated to generate new goals in humans. These goals are as different as choosing the item that completes a series on tests of non-verbal intelligence (IQ) or generating the verbs that are appropriate given a particular noun.

Unfortunately, the tasks that comparative psychology and psychometrics have devised to study intelligence were devised without consideration of the ecological conditions in which human and other primates evolved. Yet Chapter 2 argues that it is critical to understand that primates originally evolved to forage in the fine-branch niche, and that later many old-world monkeys and apes came to the ground to forage. However, the fruit and leaves were patchy; heat stress meant that the periods for foraging were short; and because there were periods in which there was a shortage of particular foods, there was a premium on behavioural flexibility and the ability to solve new problems rapidly.

This was particularly true during the evolution of the hominins (Chapter 9). This means that to understand how humans came to have unique capacities, it is important to appreciate the selection pressures that encouraged them. For example, we cannot fully understand how it is that humans can reflect on the intentions of others without realizing that the survival of our ancestors depended critically on cooperation.

However, we cannot study the brains of ancestral hominins or the primates from which they evolved. And unfortunately, as a later section explains, we have very little evidence on the brains of chimpanzees, our closest ancestor. Therefore, much of this book is devoted to trying to understand the PF cortex of macaque monkeys.

Of course, humans are not descended from either chimpanzees or macaque monkeys. The last common ancestor of macaques and chimpanzees lived between 25 and 30 million years ago (Stevens et al., 2013). Since that time, the lines leading to modern macaques and chimpanzees have themselves undergone further change. Nonetheless, studies of macaque monkeys can provide a clue, however indirect, that allows us to infer the PF mechanisms that the common ancestors of monkeys and humans were likely to have possessed.

Summary

The human PF cortex cannot be understood without recognizing that it evolved by co-opting and elaborating mechanisms that existed in the brains of our primate ancestors. Its fundamental organization is the same as in monkeys and apes.

Connectivity

The word 'connectivity' in the subtitle means anatomical connectivity. As Devlin and Poldrak (2007) have pointed out, many regard anatomy as tedious. The word anatomy can conjure up the memory of classes in which brains were cut up and a host of labels had to be learned so as to distinguish between the different areas.

This sort of anatomy is indeed tedious. But understanding how the brain is wired up is not, because it provides clues as to mechanisms. Working out the structure of deoxyribonucleic acid (DNA) proved crucial to understanding how genetic material is copied from parents to offspring. No wonder Francis Crick (1988) wrote 'if you want to understand function, study structure'.

When in later life Crick turned his attention to the brain, he looked first for evidence on the anatomical connections of the human brain. He was disturbed to discover how 'backward' human neuroanatomy was at the time (Crick & Jones, 1993). Fortunately, the situation has changed with the development of diffusion- weighted tractography (DWI) and the study of the covariance between the activations in different brain areas while the subjects are at rest.

Modern work on the connections between the areas of the neocortex began with the studies by Pandya and Kuypers (1969) and Jones and Powell (1970) on macaque monkeys. However, the methods have become more sensitive, and the picture has changed. For example, Markov et al. (2014) injected tracers into 29 of the 91 areas of the macaque neocortex. They found 1,615 paths between areas, and roughly a third of them had not been described before.

One way of visualizing the anatomical inputs and outputs of an area is to chart them on polar plots. These can be regarded as 'connectional fingerprints' (Passingham et al., 2002). The analogy is with the polar plots produced by Zilles et al. (2001) for the receptor densities in each cortical area. The circumference shows the areas that are connected with the featured region, and the radius plots the strength of each connection, ranging from one to three.

Passingham et al. (2002) plotted these using the data CoCoMac (cocomac.g-node.org/). This is a database for the cortical (Co) connections (Co) of macaque (Mac) monkeys. In the version available at the time, it contained data from 413 studies with 39,748 connectional entries. A second edition called CoCoMac 2 has been developed from the Donders Institute, and it includes many more connections.

By using the data from CoCoMac 1, Passingham et al. concluded that each cortical area has a unique set of inputs and outputs. They plotted the strength of the connections in polar coordinates for various PF and premotor areas. Figure 1.1 shows two examples for the PF areas 9 and 14.

After constructing connectional fingerprints, Passingham et al. (2002) then used multidimensional scaling to prove that each area had a *unique* pattern of inputs and outputs. As Figure 1.2A (next page) illustrates, the analysis showed that all the PF areas lie at different points in a 'connectional space'. This is evidence that no two areas have exactly the same overall pattern of connections. The distance between two areas in this space (Figure 1.2A) is a measure of the difference in the connectivity.

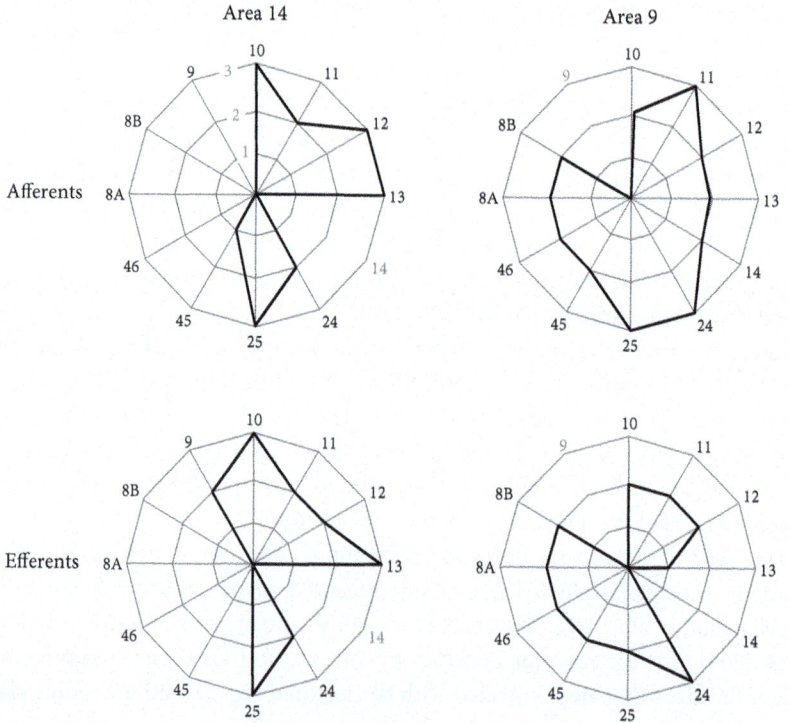

Figure 1.1 Connectional fingerprints for the PF areas 9 and 14
Each plot shows the PF areas on the circumference that are connected with the featured area, with the intensity of the projection along the radius: light (1), moderate (2), or heavy (3). Projections within an area are not included.

Reproduced from Passingham R.E., Stephan K.E., & Kotter R. The anatomical basis of functional localization in the cortex. *Nat Rev Neurosci*, 3 (8), 606-16, Figure 1, Doi: 10.1038/nrn893 Copyright © 2002, Springer Nature.

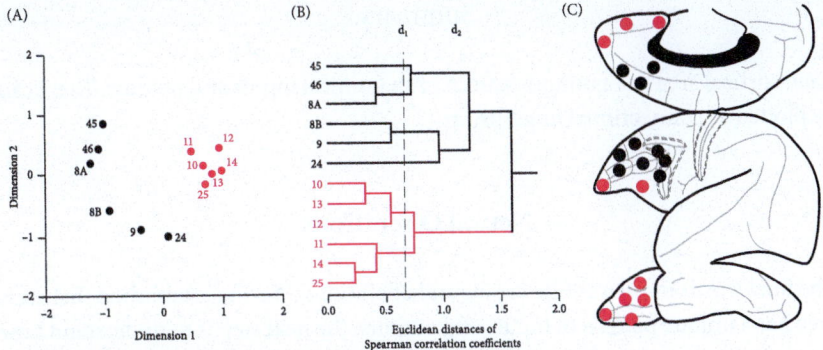

Figure 1.2 Connectional clusters in the PF cortex
(A) Plot of two connectional dimensions with the area denoted next to each point. (B) Hierarchical clustering along a connectional dimension, denoted as d_1. (C) Clusters colour coded to match (A) and (B).

Reproduced from Passingham R.E., Stephan K.E., & Kotter R. The anatomical basis of functional localization in the cortex. *Nat Rev Neurosci*, 3 (8), 606–16, Figure 2, Doi: 10.1038/nrn893 Copyright © 2002, Springer Nature.

Systems

Passingham et al. (2002) then used hierarchical cluster analysis to identify clusters of regions with similar intrinsic PF connections. Figure 1.2B shows five clusters within the PF cortex. Averbeck and Seo (2008) identified similar clusters, basing their analysis on the afferent connections to the PF cortex. In both analyses, a cluster of dorsal areas contrasts with a cluster of ventral areas (Figure 1.2C). The same conclusion was reached by Blumenfeld et al. (2014) who have produced open-source software so that users can plot the connections of any cortical area based on the CoCoMac database.

Considering the neocortex as a whole, Passingham et al. (2002) suggested that there are 'families' of areas or 'systems' that shared a similar, though not identical, pattern of connections. More recently, Markov et al. (2013) came to a similar conclusion based on an analysis of the whole neocortical connectome. For example, they identified a large-scale cortical network that includes the posterior parietal cortex and the PF cortex. Markov et al. regarded these two regions as part of a 'core' network, with other groups of areas such as the occipital and temporal components of the ventral visual system being situated on the periphery.

The terms 'systems' or 'networks' are sometimes used loosely in the functional magnetic resonance imaging (fMRI) literature to refer to the constellation of activations found in a particular experiment. However, these terms only have a meaning if they are restricted to those areas that can be shown to be closely connected anatomically.

Summary

Each cortical area has a unique pattern of anatomical inputs and outputs. These can be plotted as connectional fingerprints.

Neural Operations

The final words in the subtitle are 'neural operations'. Having established that each area has a unique pattern of inputs and outputs, the next step is to understand how an area processes the incoming information so as to transform it into the influence that it exerts on other areas via its outputs.

Much of the fMRI literature is devoted to identifying the 'function' of an area. Yet, the functional labels that are attached to areas come from psychology. Thus, an area might be said to be involved in 'attention', 'decision making', or 'conflict monitoring'. But these labels say nothing about the neural operation that an area performs.

Function

There are other problems with using functional labels. Many have been inherited from common sense psychology (James, 1890). Some are vague such as 'executive', one of the labels in BrainMap (Lancaster et al., 2012). Others such as 'working memory' are appropriate for systems, not single areas. And there is a danger that the functional labels are allocated on the basis of studying just one or two tasks.

To avoid this danger, Passingham et al. (2002) introduced the concept of 'functional fingerprints'. The idea was that the function of an area could be characterized by producing polar plots in which a series of tasks were shown round the circumference, and the degree of activation on each task plotted on the radius. However, at the time the fMRI literature was not rich enough to provide an illustration.

So instead, Passingham et al. plotted data from neurophysiology. They took five properties of cell activity: auditory or visual responses; proprioceptive or cutaneous responses; a muscle-like pattern of activity; a temporal correlation of activity with movements; and persistent delay-period activity.

Figure 1.3C (next page) shows functional fingerprints for two premotor areas, the supplementary motor area (SMA) and the ventral premotor cortex. It shows that they differ in the relative frequency of the various cell classes. For example, a higher proportion of the cells in the SMA have somatosensory responses. For comparison, the figure also shows that the two areas differ in their connectional fingerprints (Figure 1.3A and B).

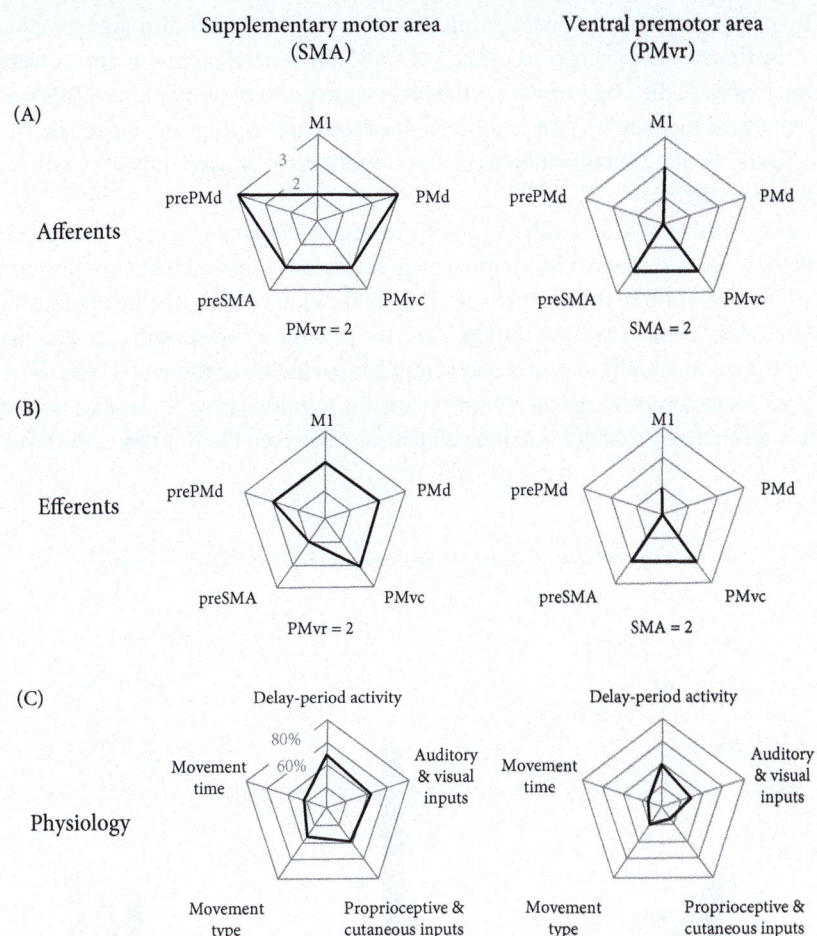

Figure 1.3 (A & B) Connectional fingerprints of the SMA and the rostral ventral premotor cortex (PMvr)

Abbreviations: M1, primary motor cortex (area 4); PMd, dorsal premotor cortex (caudal dorsal area 6); PMvc (caudal ventral area 6); pre-SMA, presupplementary motor area (rostral medial area 6); pre-PMd, rostral part of the dorsal premotor cortex (rostral dorsal area 6). The strength of connections between the areas depicted in the two columns is presented beneath each polar plot. C = Functional fingerprints for the same two areas.

A and B: Reproduced from Passingham, R.E. & Wise, S.P. *The Neurobiology of Prefrontal Cortex*, p. 16, Figure 1.9a and b Copyright © 2012, Oxford University Press.

C: Adapted from Passingham R.E., Stephan K.E., & Kotter R. 'The anatomical basis of functional localization in the cortex', *Nat Rev Neurosci*, 3 (8), 606–16, Doi: 10.1038/nrn893 Copyright © 2002, Springer Nature.

10 INTRODUCTION

By using multidimensional scaling, Passingham et al. were also able to show that the functional fingerprints of the SMA and the ventral premotor cortex were unique, as were the fingerprints for the other motor and premotor areas. This was shown by the fact that they lay at different locations in two-dimensional space. The conclusion is that functional fingerprints are unique to an area, just like connectional fingerprints.

To show that the differences can be one of degree, Passingham et al. (2002) also plotted the preference of cells for movement sequences that were either performed from memory or based on visual cues (Mushiake et al., 1991). The histograms in Figure 1.4 show the results for the SMA and the premotor cortex, with activity classified from 1 to 7. Cells in class 1 had complete specificity for the visual task; those in class 7 had complete specificity for the memory-guided task. A classification of 4 indicates activity that did not differ statistically between the two tasks. The SMA

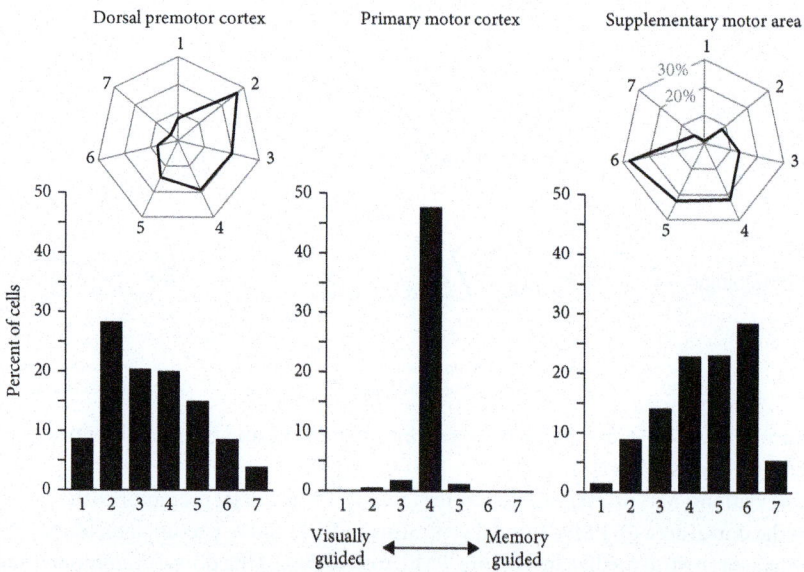

Figure 1.4 Functional fingerprints showing selectivity for visually guided versus memory-guided movement sequences in the premotor cortex (lateral area 6), the primary motor cortex (area 4), and the supplementary motor area (medial area 6) Cells in class 1 show complete specificity for visually guided sequences; cells in class 7 show complete specificity for memory-guided sequences. Cells in the other five classes show intermediate properties. The polar plots above the left and right bar graphs show the same data as the bar graphs.

Reproduced from Mushiake H., Inase M., & Tanji J. Neuronal activity in the primate premotor, supplementary, and precentral motor cortex during visually guided and internally determined sequential movements. *J Neurophysiol*, 66 (3), 705–18, Doi: 10.1152/jn.1991.66.3.705 Copyright © 1991, The American Physiological Society, with permission.

showed a preponderance of activity for the memory-guided task, with cells in the lateral premotor cortex showing the opposite bias.

Figure 1.4 shows these data in the form of functional fingerprints. The polar plots at the top of figure show the same data as the bar charts below them, with the cell class (1–7) plotted around the circumference and the proportion in each class along the radius.

Given the rapid rise in the number of studies using fMRI, functional fingerprints can now be plotted on the basis of activations. Figure 1.5 presents an example. Based on the BrainMap database, the plots compare the range of tasks over which fMRI experiments have revealed activations in the anterior insular cortex (Ia) and the cortex in the anterior cingulate sulcus (Sporns, 2014).

Unfortunately, the functional labels used in the BrainMap database are crude (Lancaster et al., 2012). Nonetheless, the plots in Figure 1.5 are adequate to show that these two areas have different functional fingerprints, even though the areas are interconnected (Hutchison et al., 2012). For example, the anterior insula cortex

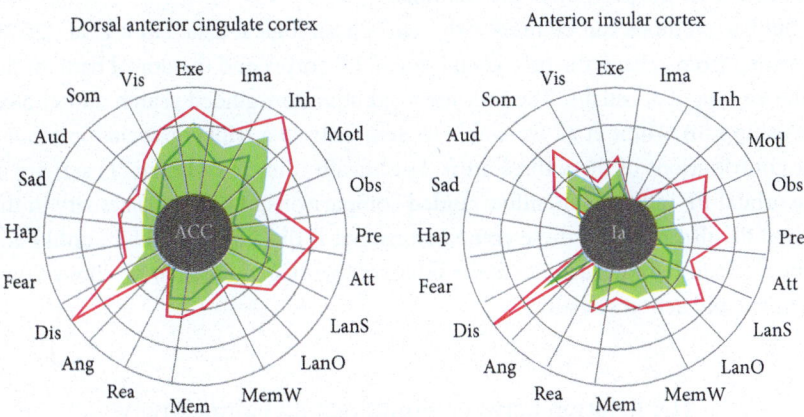

Figure 1.5 Functional fingerprints for the cortex in the anterior cingulate sulcus and the right anterior insular cortex (Ia)
The green shaded area gives the estimated involvement of the cortical area in each kind of task or emotion plotted on the circumference of each circle, based on fMRI activation studies. The magenta and dark green lines reflect the upper and lower bounds of these estimates. Abbreviations, in clockwise order beginning at 12 o'clock: Exe, executing movements; Ima, imagining movements; Inh, inhibiting movements; Motl, motor learning; Obs, observing movements; Pre, planning movements; Att, attention; LanS, semantic aspects of language; LanO, other (nonsemantic) aspects of language; MemW, working memory; Mem, memory; Rea, reasoning; Ang, anger; Dis, disgust; Hap, happiness; Sad, sadness; Aud, auditory perception; Som, somatosensory perception; Vis, visual perception.
Adapted from Sporns O. Contributions and challenges for network models in cognitive neuroscience. *Nat Neurosci*, 17, 652–60, https://doi.org/10.1038/nn.3690 Copyright © 2014, Springer Nature.

(Ia) has stronger somatosensory activations compared to most other features plotted for that area. And the anterior cingulate cortex (ACC) has stronger activations that relate to motor execution. These differences reflect the inputs and outputs of both areas. The Ia has strong somatosensory inputs whereas the ACC does not; and the ACC has direct outputs to the motor cortex whereas the Ia has none (Saleem et al., 2008).

Multivoxel pattern analysis (MVPA) and representational similarity analysis (RSA) can also be used to show that different areas encode different information. In an fMRI study, Woolgar et al. (2011) taught human subjects two incompatible sets of stimulus–response mappings. The subjects saw one of four stimuli and pressed one of four response keys. The colour of the background specified the appropriate set of mappings between each stimulus and a response key.

Woolgar et al. then used MVPA to study four aspects of cortical encoding: the mapping rule, the position of the stimulus, the response, and the colour of the background. Their analysis showed that, for example, an activation in the ventral PF cortex coded for the rule, and an activation in the medial parietal cortex (area 7m) coded for the position of the stimulus.

Similar methods can be used with neurophysiological data. Hunt et al. (2018) recorded from cells in the orbital and dorsal PF cortex and the dorsal bank of the anterior cingulate sulcus. The task involved attention-guided search and choice. By using RSA, Hunt et al. were able to demonstrate a triple dissociation among the contributions of these three areas. Specifically, they found that: (1) activity in the orbital PF reflected attention-guided comparisons of value; (2) activity in the dorsal PF also reflected these comparisons, but it did not store them, unlike the orbital PF; and (3) that activity in the sulcus of the ACC reflected both choice commitment and action selection.

The Relation between Function and Connectivity

The fundamental claim made by Passingham et al. (2002) was that localization of function depends on the unique pattern of connections of each area. However, they did not present any data to prove this.

Mars et al. (2018) have now done so. They analysed data from the Human Connectome Project. This enabled them to detect where there were changes in the pattern of connections as visualized by resting-state covariance, and to see if these changes matched the borders between areas as visualized by task-related fMRI activations.

Figure 1.6 (next page) shows the result. There was a remarkable degree of overlap between the borders as established by these two methods.

This demonstration would be even more convincing if the anatomical borders were determined by using DWI to visualize the connections. It was Johansen-Berg

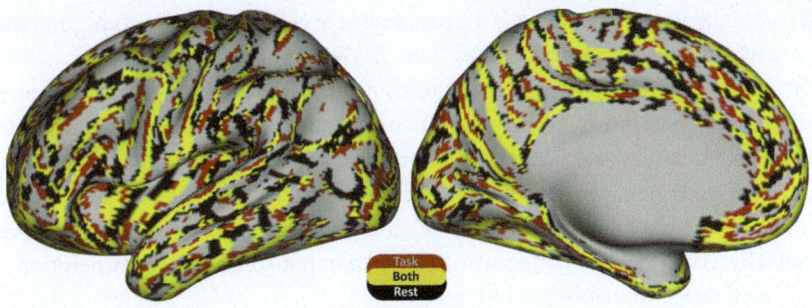

Figure 1.6 Comparison of anatomical and functional borders
Using data from the Human Connectome Project, Mars et al. based cortical borders on changes in connectivity profiles as determined using resting-state activation (black) or by similarities in fMRI activations during tasks (red).

Reproduced from Mars R.B., Passingham R.E., & Saad J. Connectivity and function in cognitive neuroscience: From areal fingerprints to abstract spaces. *T Cog Neurosci*, 22, 1026–37, Figure 2, https://doi.org/10.1016/j.tics.2018.08.009 Copyright © 2018, The Authors. Licensed under CC BY 4.0.

et al. (2004) who were the first to demonstrate that the borders between areas can be visualized by using DWI. They were able to detect the border between the pre-SMA and the SMA by finding a change in the overall pattern of connections.

To validate the method, it is critical that the borders as revealed by DWI should correspond to the borders as identified by a study of the cytoarchitecture of the same areas. Klein et al. (2007) used Broca's area as a test. It turned out that there was good agreement on the boundary between Broca's area 44 and 45 as determined by the two methods.

Beckmann et al. (2009) used DWI to detect the borders between nine subareas in the cingulate cortex. This allowed them to test whether the anatomical borders corresponded to functional border. To do this, they carried out a meta-analysis of 171 studies in which fMRI had been used to identify the functions of the different subareas. It proved possible to relate differences between areas as established by DWI to differences in functional specialization. Though this study did not plot functional fingerprints, it nonetheless confirmed that functional localization depends on connectional fingerprints.

Transformation

Though functional fingerprints can be shown to be unique to an area, they are simply a way of using polar plots to summarize the responsiveness of an area. They do not tell us about the flow of information from inputs to outputs. In other words, they do not tell us about the neural operations that occur in the area. This is constrained by the pattern of inputs. But it also depends on the intrinsic micro-circuitry of the area, that is the percentage of cells of particular types and the way in which they are interconnected (Douglas & Martin, 2004).

It is this circuitry that performs the *transformation* from inputs to outputs. Chapters 3–7 therefore make proposals concerning the way in which the inputs to different PF subareas are combined and transformed so as to generate the outputs. These proposals are given at the end of each of these chapters, and they are printed in italics. They are mainly framed at the computational and algorithmic level as defined by Marr (1982). In other words, they attempt to describe *what* operation is performed rather than *how* it is implemented.

To give an example, Chapter 5 deals with the frontal eye-field, an area that is involved in identifying the targets for saccadic eye movements. Neurophysiological studies have shown that some of cells respond to visual stimuli, others are visuo-motor, and yet others are motor (Schall, 1995). This suggests that these cell types are involved in the transformation from visual inputs to motor outputs. However, it is quite another enterprise to present a precise computational account of how the intrinsic micro-circuitry implements that transformation (Heinzle et al., 2007).

The proposals at the end of Chapters 3–7 might be thought to imply that the transformation that each area performs is static. We know, of course, that this is not the case. For example, in an electroencephalogram (EEG) study Stephani et al. (2020) have demonstrated that the responses of the primary somatosensory cortex (SI) are variable, depending on the excitability of the system, and in particular oscillatory activity in the alpha band.

Clearly, therefore, the proposals are over-simplified. Research is needed on the extent to which the transformation that an area performs is dynamic, depending on the internal state of the area, the specific inputs at the time, and the current task or rule. Chapter 11 describes an experiment which shows that the PF cortex sends outputs to different areas depending on the task that the subject is going to perform (Sakai & Passingham, 2006).

Summary

Localization of function depends on the fact that each area has a unique pattern of inputs and outputs. The challenge is to understand the transformation that each area performs between these inputs and outputs.

Macaque Monkeys

Understanding the transformation that each PF area performs requires a specification of the connections and an account of the neurophysiological activity in that area.

Connectivity

Some readers might think that DWI and resting-state covariance are wholly adequate for specifying the connections and that fMRI can replace the methods that have been developed for studying function in animals. So, it is necessary to explain why Chapters 3–8 depend mainly, though not entirely, on experiments that have been carried out with macaque monkeys.

It is true that DWI and resting-state covariance provide information about connections, but the problem is that these methods have limitations. DWI depends on the fact that the MRI signal can be sensitized to the diffusion of water along axon bundles (Basser & Ozarlsan, 2009). Resting-state covariance measures the degree to which the activation of different areas follows the same time course. Together, these methods have been used to produce a 'connectome' of the neocortex (Bullmore & Sporns, 2009; Behrens & Sporns, 2012).

However, DWI has the limitation that it can produce false-positive results (Maier-Hein et al., 2017) and though methods have been developed to prune these (Roberts et al., 2017), it is not clear to what extent they are successful. Also, as currently used, DWI cannot trace fibres into the cortical grey matter (Reveley et al., 2015). To overcome this problem, researchers seed the white matter under a cortical area. However, there is a danger of seeing fibres of passage by doing this, and thereby producing misleading findings.

Resting-state covariance or 'functional connectivity' is assumed to be related to anatomical connectivity (Bullmore & Sporns, 2009; O'Reilly et al., 2013). However, it has a major limitation, in that there is no way of knowing whether the connections are mono- or poly-synaptic.

The advantage of using tracers such as fluorescent diamidino yellow is that we can be sure that the connections that are identified are mono-synaptic. Diamidino yellow is an example of a retrograde tracer, meaning that if it is injected into area B it travels from the terminals of the cells in area B to the cell bodies in area A.

Of course, if the aim is to characterize a system, tracers can be used to visualize poly-synaptic connections. For example, viruses have been used to study the projections of the basal ganglia and cerebellum to the neocortex via the thalamus (Bostan & Strick, 2018). There are also magnesium-enhanced tracers that are paramagnetic and therefore detectable by MRI, and these have been used to visualize circuits (Murayama et al., 2006). However, they cannot be used for the human brain because it would not be ethical to inject tracers of any sort into human subjects. For this reason, the experiments have been conducted with macaque monkeys.

Activity versus Activation

Understanding how the system processes or transforms information requires knowledge of the cellular mechanisms. But the BOLD signal recorded during fMRI is a vascular one. It depends indirectly on the fact that when cells in an area become active there is an increase in the flow of the arterial blood bringing oxygen and glucose.

The typical spatial resolution of the signal is in the order of millimetres, and this is adequate for visualizing activations that are located within neocortical areas. With higher magnetic strength and improved coils, it is now possible to achieve spatial resolutions that are close to the size of the functional patches that have been demonstrated in macaque monkeys. For example, there are patches in the inferior temporal cortex that respond to object-like stimuli, and these are roughly 0.4mm in size (Tanaka, 1997; Wang et al., 1998). Similarly, in the PF cortex there are functional patches for spatial and motion direction that are roughly 0.7mm in size (Masse et al., 2017).

However, to understand *how* areas perform their functions, it is necessary to record from the cells or neurones themselves, and neuronal cells measure in the order of microns.

It is true that recordings can be taken in patients, either with arrays of electrodes on the cortex (Forseth et al., 2018) or by implanting microelectrodes during surgery (Mukamel & Fried, 2012). But this can only be done when there is clinical justification. Furthermore, if recordings are taken in patients during surgery, it is only possible to record over relatively short periods, whereas recordings can be taken over months when animals such as macaque monkeys are used.

Macaques as a Model

There are several advantages of using macaque monkeys. They are highly adaptable animals. They live in a very wide range of habitats from Afghanistan to India, Indonesia, Japan, and China. This means that they adapt well to laboratory conditions. They are also able to learn complex cognitive tasks, even if it can take months to train them.

However, there are two important reservations. First, macaques cannot be treated as if they are 'representative' of monkeys. There are seventy-nine genera of primates, and they are remarkably diverse (Preuss, 2000). Just consider, for example, the golden lion tamarin (*Leontopithecus rosalia*), the hamadryas baboon (*Papio hamadryas*), and the red-faced spider monkey (*Ateles paniscus*).

Next, humans are, of course, much more closely related to common chimpanzees (*Pan troglodytes*) and bonobos (*Pan paniscus*) (Staes et al., 2019). But it has not been felt to be ethically acceptable to carry out anatomical tracing or cell recording experiments in chimpanzees.

However, fortunately, there is some brain imaging data on chimpanzees using positron emission tomography (PET), and this is mentioned in Chapters 3 and 11.

PET measures the influx of arterial blood bringing oxygen and glucose when the cells in an area increase in their activity. The positron emitting label 2-18F-fluoro-2-deoxyglucose has the advantage that it has a long half-life. This means that the chimpanzee can be tested on a task outside the scanner, and then placed in the scanner while anaesthetized. The region uptake of the labelled tracer that is associated with performance of the task will still be evident in the PET images.

It is still, of course, possible to carry out observational studies of chimpanzees in the wild or captivity. But experimental procedures of any sort have now been banned on chimpanzees in the USA. So, in this book it has only been possible to present data on chimpanzees that were acquired before this ban took effect.

Homology

Though humans are not descended from macaque monkeys, it is still possible to establish whether different neocortical areas are homologous in the two groups. The suggestion that an area is homologous in humans and macaques simply means that it can be traced to a common ancestry.

One way of establishing homology is to plot connectional fingerprints. Sallet et al. (2013) used resting-state covariance to identify the different areas in the dorsal PF cortex in both macaque monkeys and humans. The method works by identifying the area in the macaque brain that shows the least difference in its connectional fingerprint from a particular area in the human brain.

Figure 1.7 (next page) illustrates the technique for delineating area 9, part of the dorsomedial PF cortex. Sallet et al. first used DWI to establish the borders between the different PF areas in the human brain. Then they compared the connectional fingerprints by using resting-state covariance. Figure 1.7C shows the human connectional fingerprint in light blue and the macaque fingerprint in purple. The overlap is shown in dark blue.

The histograms in Figure 1.7D show the difference between the fingerprints for the different areas. It will be seen that the area with the least difference is area 9. Figure 1.7A shows the lateral and medial views of this area in the human brain, and Figure 1.7B in the macaque monkey brain.

Summary

Much of this book describes research on macaque monkeys. The reason is that to understand the transformations that are performed by the different PF areas, it is necessary to chart the inputs and outputs and to record the activity of cells in the area. Macaque monkeys have been widely used to chart the anatomical

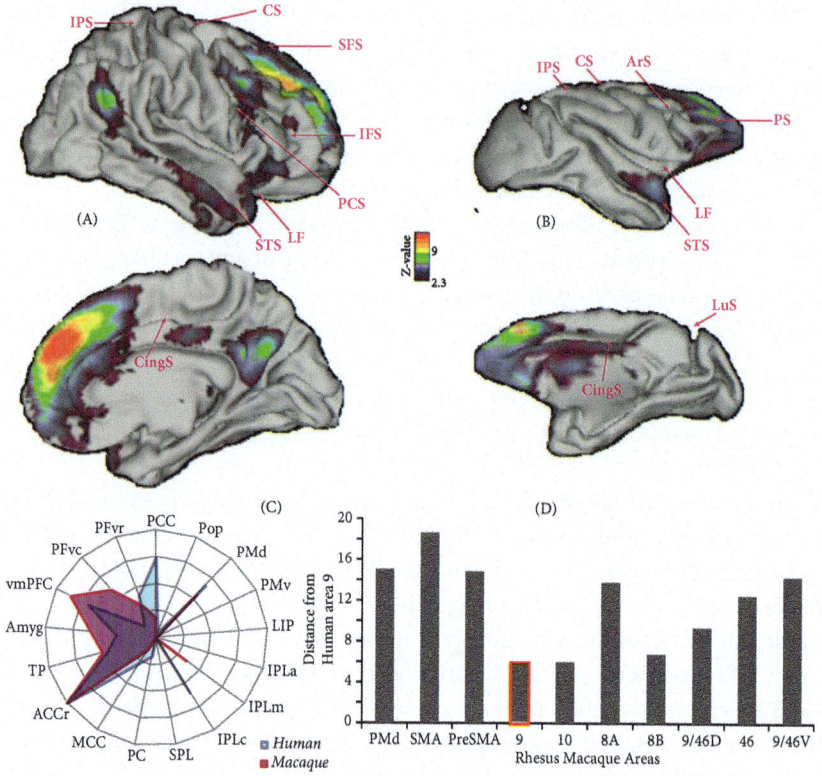

Figure 1.7 (A) Area 9 (part of the granular medial PF cortex) in human subjects; (B) Area 9 in macaques. The letters refer to the different sulci. (C) The functional connectivity fingerprints in humans (light blue) and macaques (purple) with the overlap shown in dark blue. (D) The summed absolute differences between the functional coupling scores

Reproduced from Jérôme Sallet, Rogier B. Mars, MaryAnn P. Noonan, Franz-Xaver Neubert, Saad Jbabdi, Jill X. O'Reilly, Nicola Filippini, Adam G. Thomas, & Matthew F. Rushworth. The organization of dorsal frontal cortex in humans and macaques. *J Neurosci*, 33 (30), 12255–74, Figure 7, Doi: https://doi.org/10.1523/JNEUROSCI.5108-12.2013 Copyright © 2013, The Society for Neuroscience, with permission.

connections and to record the activity of cells while the animals perform complex laboratory tasks. The homologies between human and macaque PF areas can be established by plotting connectional fingerprints.

Brain Lesions

Given that fMRI enables us to study human brains, the reader might think that there is no need now for studies of the effects of lesions. After all, fMRI has two great advantages over lesion studies. First, the peaks of the activations lie in the grey and not the white matter. Second, the peaks can, and often do, lie within specific areas.

However, it is a commonplace that an activation in an area does not tell us whether that area is critical for performance of the task, that is whether the subject can perform the task its absence. Unfortunately, fMRI is a correlational technique, and so it cannot provide direct and conclusive evidence concerning causes. And this is true even when analytic methods such as structural equation modelling (McIntosh & Gonzalez-Lima, 1994) or dynamic causal modelling (Friston et al., 2003) are used.

The advantage of studying the effect of a lesion in area A is that it tells us the contribution that is unique to that area. The behavioural impairment caused by the lesion is an indication that areas B, C, and D do not have the connectivity that can support that behaviour. Because they do not have the identical pattern of inputs and outputs, they are not able to perform the identical transformation.

But there are two problems with studying the effects of lesions in patients. The first is that the lesion will almost always be spread over several areas. The second is that, whether these result from strokes, tumours, or surgical resection, the lesion will include the underlying white matter, that is the fibres of passage that underlie that area. For example, left hemisphere strokes can cause apraxia, but it is clear from structural scanning that the underlying fibres of passage are compromised, and not just the grey matter (Pizzamiglio et al., 2019).

Lesions in Macaque Monkeys

There are two advantages of placing lesions in the brains of macaque monkeys. One is that it is possible to ensure that the lesion is confined to a single area by aspiration of the tissue at open surgery. The technique is the same as for human neurosurgery, and this precision can be achieved by using an operating microscope.

The second advantage is that it is possible to leave the white matter intact by injecting an excitotoxin such as kainic or ibotenic acid. Excitotoxins cause the cells in the area to die due to epileptic firing. However, the underlying fibres of passage are little affected because excitotoxins act at synapses.

There is an alternative, and this is to induce a temporary lesion. This can be done by inhibiting cell activity by injecting muscimol; this is an agonist for gamma-aminobutyric acid (GABA), an inhibitory neurotransmitter. It is also possible to temporarily disrupt the activity of cells in the human brain by applying repetitive transcranial magnetic stimulation (rTMS) over the scalp. But it is not yet possible to reach deep structures in this way in the human brain.

However, Folloni et al. (2019) have explored a promising new method that overcomes this limitation. They have applied focused ultrasound in macaque monkeys.

Folloni et al. were able to target two deep structures with this method, the anterior cingulate cortex and the amygdala. The stimulation of each area decreased the degree to which it showed resting-state connectivity with other areas.

The Relation between Lesions and Cell Activity

One reason why it is essential to consider the effects of lesions is that the interpretation of the activity of the cells in an area is not straightforward. This area will be connected with other areas in the system. Recording alone will not establish whether the cell activity in area B relates to the transformation that is carried out by that area. There is an alternative possibility, and this is that that activity is derived from areas A or C with which area B is interconnected.

The point can be illustrated by taking the cell activity shown in Figure 1.4. The figure shows cell activity in the SMA and the lateral premotor cortex. Many cells in both areas discharge whether movements are visually cued or self-guided. So, it might seem that both areas can support movement however they are guided.

But experiments have shown that the results of placing lesions in the two areas are not the same. Lesions of the SMA severely disrupted self-generated movements; these are defined as movements that are not prompted by visual cues (Thaler et al., 1995). By contrast, lesions of the lateral premotor cortex did not have this effect; they affected visually cued movements instead (Thaler et al., 1995).

Figure 1.8 (next page) is a modification of Figure 1.4, and it illustrates how the results from lesions and neurophysiology can be reconciled. To simplify the explanation, the central histogram (class 4) has been omitted. As in Figure 1.14, classes 1, 2, and 3 refer to cell activity that encoded visually cued sequences, whereas classes 5, 6, and 7 reflect activity that encoded memory-guided sequences.

In an intact brain, cells in the lateral premotor cortex were found to show increased activity even when a macaque monkey performs a sequence from memory (classes 5, 6, and 7) (Mushiake et al., 1991). Thus, one might expect the lateral premotor cortex to be able to take over control of memory-guided sequences after the removal or inactivation of the SMA. But suppose that the lateral premotor cortex gets the relevant input from the SMA (as indicated by the arrow Figure 1.8A). Then, in lesioned monkeys, the cells in the premotor cortex would no longer have the same properties as they do in intact monkeys (Figure 1.8B). Accordingly, in the absence of the SMA the lateral premotor cortex would no longer be able to support memory-guided sequences.

This example shows that, when considering neurophysiological data, it is essential to distinguish cell activity that is *intrinsic* to the area from which the recordings are taken and cell activity that is *derived* from other areas within the connected system.

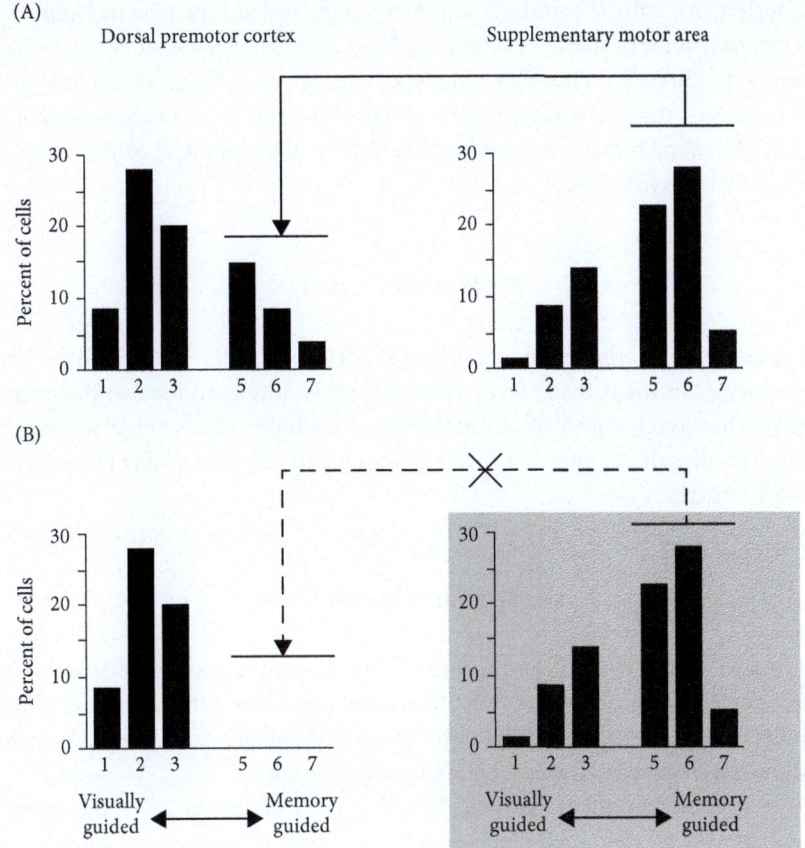

Figure 1.8 Effect of a lesion on neuronal activity
This figure has the same format as Figure 1.4, except that I have omitted class 4. As in Figure 1.4, classes 1–7 refer to the degree of selectivity for either a visually-guided sequence or the same sequence performed from memory. Class 1 showed complete specificity for the visually guided sequence, and class 7 did so for the memory-guided sequence. (A) The situation in the intact brain. (B) The situation as imagined for a brain with an SMA lesion (shaded area).

Chapter 7 describes a recent experiment by Brincat et al. (2018). They recorded in multiple cortical areas, including visual areas and the PF cortex, while monkeys performed a task in which they had to categorize visual stimuli. The authors suggested that their findings cast doubt on the classical view of localization of function, because there was a similarity in the information that was encoded in related areas. Brincat et al. interpreted this finding as support for models of graded functional specialization.

But they made their recordings in intact brains, and as explained in Figure 1.8 this means that the properties of connected areas depend on each other, even when connected indirectly via multiple synapses. Given this fact, the wealth of information processed in a cortical area often obscures the specific contributions that it makes. The moral is that cell recordings do not reveal their unique contribution in the way that lesions can do.

Prefrontal Cortex

For humans and other primates, the PF cortex excludes the premotor and motor areas. In the text the term 'frontal lobe' is only used for two purposes. One is when referring to the frontal lobe as a whole. The other is when discussing studies of patients in which the lesion extends beyond the boundaries of the PF cortex.

Cytoarchitecture

In primates, much of the PF cortex consists of areas with a 'granular' cytoarchitecture. The neocortex comes in cytoarchitectonic types that differ according to the number and density of cell bodies in the internal granular layer, layer 4. Granular areas have a conspicuous layer 4; agranular areas do not.

Two systems have been used for labelling the different areas of the cortex. Brodmann (1909) examined the human brain as well as the brains of some other primates. He simply numbered his areas sequentially from the top of the brain to the bottom, as he encountered them in a horizontal series of brain sections. Area 1 is just the first area he encountered, the one at the top of the human brain. Von Economo (1929) also studied the human brain, but he used a lettering system instead to label the areas. The same lettering system was then adopted by von Bonin and Bailey when they studied the brains of the macaque monkey (Von Bonin & Bailey, 1947) and chimpanzee (Bailey et al., 1950).

Unfortunately, there are discrepancies between the maps that different authors have produced. One reason is that, in addition to the granular and agranular areas of the PF cortex, there are intermediate types of cortex. Unlike agranular areas, in which layer 4 is almost completely absent, and granular areas, which have a conspicuous layer 4, other areas have an intermediate or 'dysgranular' cytoarchitecture.

This type of cortex has a thin and sometimes discontinuous internal granular layer. For example, the rostral premotor area FC in the map of the macaque monkey by von Bonin and Bailey (1947) has this property. Area FC is intermediate between the agranular premotor area FB and the granular PF area FD.

Mackey and Petrides (2010) examined the ventral premotor cortex as well as the inferior part of the caudal PF cortex (areas 44 and 45). Via a quantitative analysis, they were able to show that layer 4 becomes thicker and more conspicuous as one progresses in a caudal to rostral direction along the ventromedial surface of the human and macaque cortex.

For more than a century, neuroanatomists have recognized that the bulk of the primate PF cortex as consisting of 'homotypical' or 'eulaminate' areas. The name 'homotypical' alludes to something homogenous and common, and these areas have the typical distribution of the neocortical layers. The general term 'association cortex' is often used to refer to these areas.

The 'heterotypical' areas include the primary visual cortex (area 17) in which layer 4 has several sublayers and is very thick. They also include the motor cortex in which layer 4 is absent.

This book uses the cytoarchitectural divisions of the macaque monkey brain as identified by (Petrides & Pandya, 2007) and included in a recent monograph by Pandya et al. (2015). They are shown in Figure 1.9 (next page).

As will be seen from Figure 1.9, a mixture of labels is used, some numbers and some letters as, for example, for the inferior temporal area (TE).

Granular and Agranular Prefrontal Areas

The motor and premotor areas of the frontal lobe differ from the PF cortex in that they send direct projections to the spinal cord (Murray & Coulter, 1981; Dum & Strick, 1991; He et al., 1993). By contrast, the granular PF cortex has no direct corticospinal projections.

Granular areas make up the bulk of the primate PF cortex, especially in the brains of the anthropoids, including humans. But as Chapter 2 explains, many neuroscientists use the term PF cortex for particular cortical areas in rats (*Rattus norvegicus*) and mice (*Mus musculus*), among other mammalian species. These agranular areas include the anterior cingulate, prelimbic, and infralimbic areas on the medial surface of the hemisphere, along with the anterior insular cortex (Ia) on the orbital surface.

To borrow a term from Preuss and Robert (2014), there is nothing inherently wrong with 'rebranding' these areas in non-primate mammals as being prefrontal, even though they are agranular. These agranular PF areas are like the granular PF cortex in that they lack direct projections to the spinal cord.

However, including these agranular areas as prefrontal raises two questions. One is what the granular PF cortex in primates adds to the functions of the agranular PF areas in other mammals. The second is what the granular PF areas in primates add to their own agranular PF areas.

Table 1.1 (page 25) provides a list of granular and agranular PF areas.

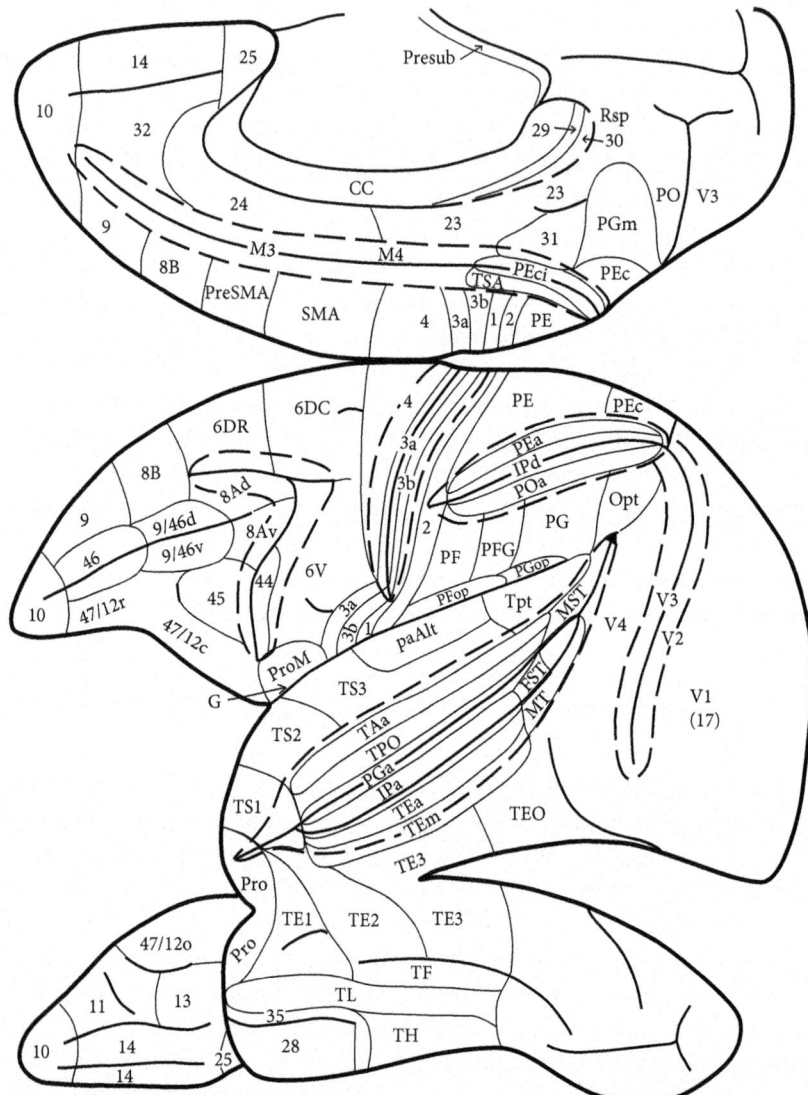

Figure 1.9 Map of the macaque cortex

Rostral is to the left, with a medial view at the top (ventral up), a lateral view in the middle (dorsal up), and a ventral view at the bottom (lateral up). Abbreviations: CC, corpus callosum; G, gustatory cortex; Rsp, retrosplenial cortex; Pro, proisocortex, which is a variant of neocortex; preSMA, pre-supplementary motor area; SMA, supplementary motor area. Subdivisions of areas are often designated as dorsal (d or D), rostral (r or R), ventral (v or V), orbital (o), opercular (op); medial (m), caudal (c), or anterior (a). Note that, in this figure, PF does not stand for prefrontal; instead, it refers to field F in the parietal (P) lobe.

Reproduced from Petrides, M. & Pandya, D.N. Efferent association pathways from the rostral prefrontal cortex in the macaque monkey. *J Neurosci*, 27 (43), 11573–86, Figure 2, Doi: 10.1523/JNEUROSCI.2419-07.2007 © 2007, The Society for Neuroscience, with permission.

Table 1.1 Prefrontal areas in human and macaque monkey brains, with area numbers in parentheses, where applicable.

Type	Area	Humans	Macaque monkeys
Granular PF areas	Caudal PF cortex	Caudal lateral PF (FEF, 8, 45B, 44)	Arcuate cortex (FEF, 8, 45B, 44, inferior frontal junction)[a]
	Dorsal PF cortex	Middle frontal gyrus (46, 9/46)	Periprincipal cortex (46, 9/46)
	Ventral PF cortex	Inferior frontal gyrus (45A, 47)	Inferior convexity (45A, 12)
	Orbital PF cortex	Orbital areas 11, rostral 13, rostral 14	Orbital areas 11, rostral 13, rostral 14
	Dorsomedial PF cortex	Superior and medial frontal gyrus (area 9)	Superior and medial frontal gyrus (area 9)
	Polar PF cortex	Frontal pole (area 10)	Frontal pole (area 10)
Agranular PF areas	Medial agranular (24, 25, 32)	Anterior cingulate[b] (24), infralimbic (25), prelimbic (32)	Anterior cingulate[b] (24), infralimbic (25), prelimbic (32)
	Lateral agranular	Caudal orbital (13a, 14c), agranular insular areas	Caudal orbital (13a, 14c), agranular insular areas

Abbreviations: frontal eye field (FEF); prefrontal cortex (PF).
[a]Broca's area 44 is dysgranular, and the inferior frontal junction (IFJ) forms the dorsal division of this area (Amunts & Zilles, 2012).
[b]The ACC excludes the rostral cingulate motor area, even though it is also part of area 24.

Figure 1.10A (next page) illustrates the granular PF areas for the human brain and Figure 1.10B for the macaque monkey brain (Petrides & Pandya, 1999). The lateral views are shown at the top, the medial views in the middle, and the orbital views at the bottom of the figures.

The maps shown in Figures 1.9 and 1.10 are those that are used in this book, because they were based on anatomical criteria alone. Furthermore, the divisions shown are those that are illustrated in *The Atlas of the Human Brain* that Petrides (2018) has produced to allow those using fMRI to localize their activations. This atlas is in MNI space, meaning that it uses the template produced by the Montreal Neurological Institute so that activations could be reported in a common space.

The atlas has been used throughout this book to correct the allocation of activations to particular areas where they were inaccurate in the published papers. It is not uncommon to find that the peaks are not where the authors had claimed. The reason is that many cognitive psychologists who use fMRI have had no training in neuroanatomy. I hope that the authors will forgive my presumption in using my own judgement as to where the peaks were most likely to be localized.

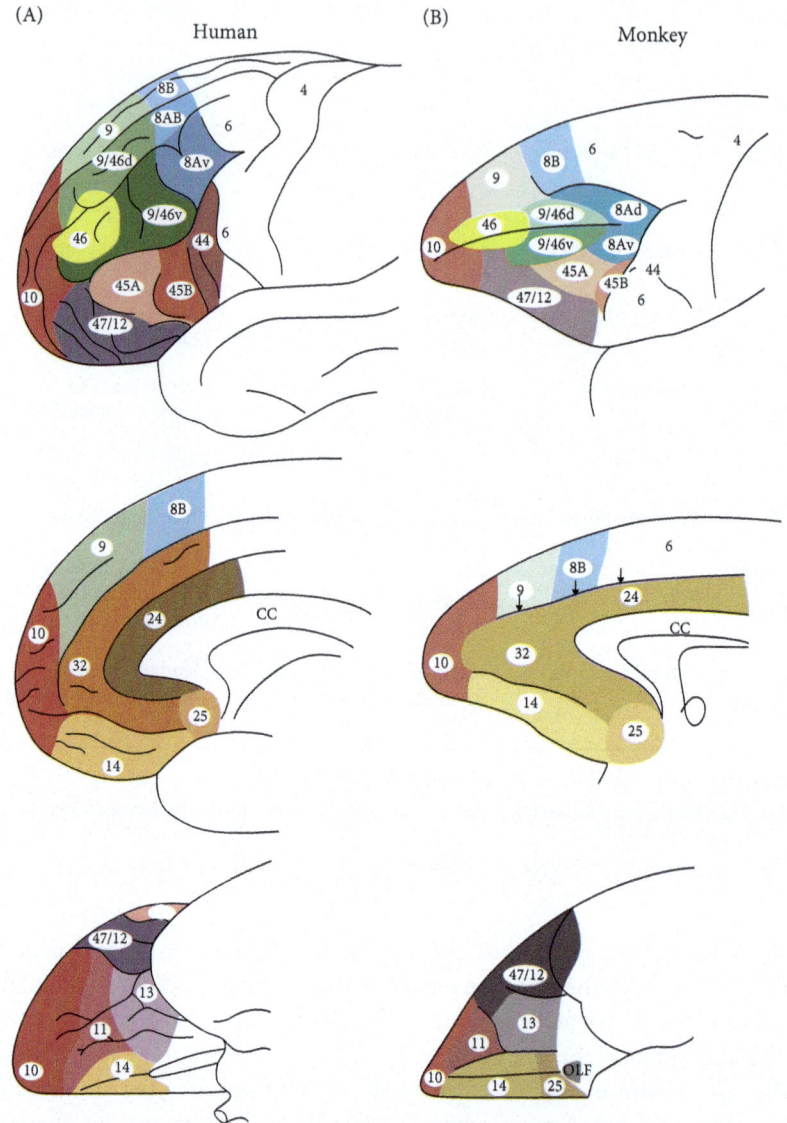

Figure 1.10 Maps of the human cortex (A) and macaque cortex (B)
Rostral is to the left in all cases. In the lateral (top) and medial (middle) views, dorsal is up; and in the ventral view (bottom), lateral is up.

Reproduced from Petrides M. & Pandya D.N. Dorsolateral prefrontal cortex: Comparative cytoarchitectonic analysis in the human and the macaque brain and corticocortical connection patterns. *Eur J Neurosci*, 11 (3), 1011–36, Figure 3, Doi: 10.1046/j.1460-9568.1999.00518.x © 1999, European Neuroscience Association.

Glasser et al. (2016) have produced a multimodal map of the human brain, with area boundaries detected in an observer-independent fashion. The data came from 1,200 subjects scanned as part of the Human Connectome Project (Van Essen et al., 2012). Glasser et al. (2016) generated the divisions on the basis of cortical thickness; the degree of myelination; fMRI activations during task performance; and resting-state covariance. They checked some, but not all, of the borders with light microscopy. Their human map includes 183 areas, which contrasts with 137 areas that Saleem and Logothetis (2006) distinguished in an atlas of the macaque monkey cortex. Readers can find a digital form of this macaque map in Reveley et al. (2017). The multimodal map of the human brain is only referred to in this book when the authors of papers used it as their template for localizing activations.

Since the manuscript was finalized, Amunts et al. (2020) have published a complete 3-D probabilistic atlas of the human brain, the so-called Julich-Brain. This resource became available too late for it to be used for localizing the activations from fMRI studies.

Five Subareas of the Prefrontal Cortex

Five major subareas of the PF cortex are referred to throughout this book. The scheme closely resembles the one advanced by Price and Drevets (2010). They recognized five divisions of the PF cortex—medial, orbital, caudal, dorsal, and ventral—based on the similarity of their connections and their proximity within the primate PF cortex. The five subareas are illustrated in (Figure 1.11 and Table 1.2, pages 28 and 29) for the macaque monkey brain.

Two conventions that I have used need to be noted. First, the medial polar PF cortex is taken to be part of the medial PF cortex. Second, Broca's area (44 and 45B) together with the inferior frontal junction (IFJ) are taken to be part of the inferior caudal PF cortex. The term 'Broca's area' is not used in this book as an anatomical term.

Summary

The PF is unlike the premotor or motor cortex in that it has no direct projections to the spinal cord. However, there are agranular area on the medial surface and anterior insula which also lack these projections, and these are included in this book as being part of the PF cortex. The challenge is to discover what the granular PF cortex does that the agranular PF areas are not able to do. Within the granular PF cortex, the book distinguishes five subareas, the medial PF cortex, orbital PF cortex, caudal PF cortex, dorsal PF cortex, and ventral PF cortex.

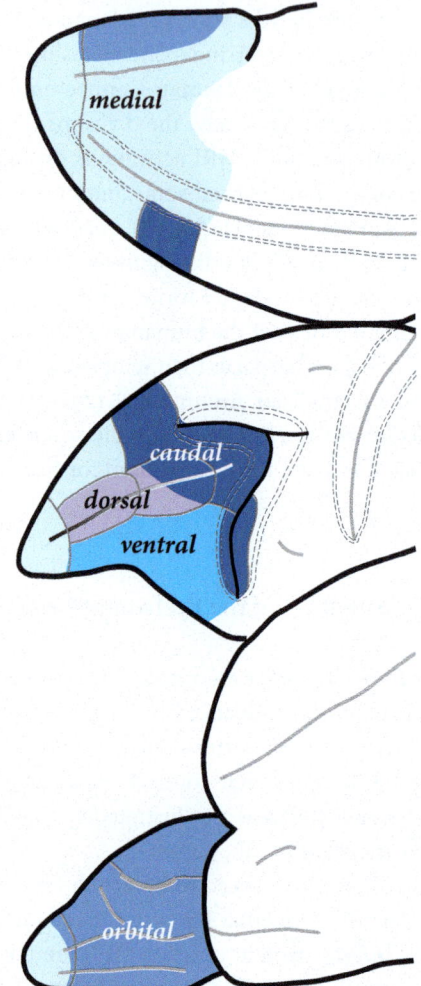

Figure 1.11 The five regions of the PF cortex used in this book
The middle figure shows the dorsal or periprincipalis PF cortex (grey), ventral PF cortex (blue) and the caudal PF cortex (dark blue); the top figure shows the medial PF cortex (light blue); the bottom figure shows the orbital PF cortex.

Reproduced from Passingham, R.E. & Wise, S.P. *The Neurobiology of Prefrontal Cortex*, p. 15, Figure 1.8 Copyright © 2012, Oxford University Press.

Organization of the Book

This book is organized into four sections.

In Section I, the introduction presents foundational material that is essential for following the argument of the book. Chapter 2 takes up the first theme of the book. It reviews the evolution of the PF cortex in non-human primates and the selection pressures that led to an increase in its size.

Table 1.2 Groups of PF areas and a compact abbreviation for each.

Region	Areas
Medial PF cortex	Areas 9 and 10, ACC, PL, IL
Orbital PF cortex	Areas 11, orbital 12, 13, 14, Ia
Caudal PF cortex	Areas 8, FEF, 45B, 44, inferior frontal junction
Dorsal PF cortex	Areas 46, 9/46
Ventral PF cortex	Areas 45A, 47/12

Abbreviations: anterior cingulate cortex (ACC) (area 24); frontal eye field (FEF); agranular insular cortex (Ia); infralimbic cortex (IL) (area 25); prefrontal cortex (PF); prelimbic cortex (PL) (area 32).

In Section II, Chapters 3–7 discuss the five subareas of the PF cortex. Though the emphasis is on studies with macaque monkeys, imaging studies with human subjects are discussed where the results support the argument. The chapters start with the second theme of the book, connectivity, by describing the main inputs and outputs of the area in question. They end by taking up the third theme of the book, neural operations. The chapters each make a specific proposal about the transformation that the area in question performs from input to output.

In Section III, Chapter 8 considers the PF cortex as a whole. In doing so, it draws conclusions about the fundamental functions of the PF cortex as established in macaque monkeys.

The final Section IV is devoted to the human prefrontal cortex. Chapter 9 documents the expansion of the PF cortex during the evolution of the hominins. Chapters 10 and 11 then suggest that to understand the human PF cortex we need to realize that it has co-opted and elaborated mechanisms that were likely to have been present in ancestral anthropoid primates. These chapters also discuss the ways in which the human brain is specialized and attempt to explain the extraordinary intellectual gap between humans and other primates.

References

Amunts, K., Mohlberg, H., Bludau, S., & Zilles, K. (2020) Julich-Brain: A 3D probabilistic atlas of the human brain's cytoarchitecture. *Science*, 369, 988–92.

Amunts, K. & Zilles, K. (2012) Architecture and organizational principles of Broca's region. *Trends Cogn Sci*, 16, 418–26.

Averbeck, B.B. & Seo, M. (2008) The statistical neuroanatomy of frontal networks in the macaque. *PLoS Comput Biol*, 4, e1000050.

Bailey, P., von Bonin, G., & McCullogh, W.S. (1950) *The Isocortex of the Chimpanzee*. University of Illinois Press, Urbana.

Basser, P.J. & Ozarlsan, E. (2009) Introduction to Diffusion MRI. In Johansen-Berg, H., Behrens, T. (eds) *Diffusion MRI*. Academic Press, Oxford, pp. 3–10.

Beckmann, M., Johansen-Berg, H., & Rushworth, M.F. (2009) Connectivity-based parcellation of human cingulate cortex and its relation to functional specialization. *J Neurosci*, 29, 1175–90.

Behrens, T.E. & Sporns, O. (2012) Human connectomics. *Curr Opin Neurobiol*, 22, 144–53.

Blumenfeld, R.S., Bliss, D.P., Perez, F., & D'Esposito, M. (2014) CoCoTools: Open-Source software for building connectomes using the CoCoMac anatomical database. *J Cogn Neurosci*, 26, 722–45.

Bostan, A.C. & Strick, P.L. (2018) The basal ganglia and the cerebellum: Nodes in an integrated network. *Nat Rev Neurosci*, 19, 338–50.

Brincat, S.L., Siegel, M., von Nicolai, C., & Miller, E.K. (2018) Gradual progression from sensory to task-related processing in cerebral cortex. *Proc Natl Acad Sci USA*, 115, E7202–11.

Brodmann, K. (1909) *Vergleichende Lokalisationlehre der Grosshirnrinde*. Barth, Leipzig.

Bullmore, E. & Sporns, O. (2009) Complex brain networks: Graph theoretical analysis of structural and functional systems. *Nat Rev Neurosci*, 10, 186–98.

Crick, F. & Jones, E. (1993) Backwardness of human neuroanatomy. *Nature*, 361, 109–10.

Crick, F.H.C. (1988) *What Mad Pursuit*. Weidenfeld and Nicholson, London.

Devlin, J.T. & Poldrack, R.A. (2007) In praise of tedious anatomy. *Neuroimage*, 37, 1033–1041.

Douglas, R.J. & Martin, K.A. (2004) Neuronal circuits of the neocortex. *Annu Rev Neurosci*, 27, 419–51.

Dum, R.P. & Strick, P.L. (1991) The origin of corticospinal projections from the premotor areas in the frontal lobe. *J Neurosci*, 11, 667–89.

Folloni, D., Verhagen, L., Mars, R.B., Fouragnan, E., Constans, C., Aubry, J.F., Rushworth, M.F.S., & Sallet, J. (2019) Manipulation of subcortical and deep cortical activity in the primate brain using transcranial focused ultrasound stimulation. *Neuron*, 101, 1109–16, e1105.

Forseth, K.J., Kadipasaoglu, C.M., Conner, C.R., Hickok, G., Knight, R.T., & Tandon, N. (2018) A lexical semantic hub for heteromodal naming in middle fusiform gyrus. *Brain*, 141, 2112–26.

Friston, K., Harrison, L., & Penny, W. (2003) Dynamic causal modelling. *Neuroimage*, 19, 1273–302.

Glasser, M.F., Coalson, T.S., Robinson, E.C., Hacker, C.D., Harwell, J., Yacoub, E., Ugurbil, K., Andersson, J., Beckmann, C.F., Jenkinson, M., Smith, S.M., & Van Essen, D.C. (2016) A multi-modal parcellation of human cerebral cortex. *Nature*, 536, 171–8.

Graziano, M.S., Taylor, C.S., & Moore, T. (2002) Complex movements evoked by microstimulation of precentral cortex. *Neuron*, 34, 841–51.

He, S.-Q., Dum, R.P., & Strick, P.L. (1993) Topographnic organization of corticospinal projections from the frontal lobe: Motor areas on the lateral surface of the hemisphere. *J Neurosci*, 13, 952–80.

Heinzle, J., Hepp, K., & Martin, K.A. (2007) A microcircuit model of the frontal eye fields. *J Neurosci*, 27, 9341–53.

Hunt, L.T., Malalasekera, W.M.N., de Berker, A.O., Miranda, B., Farmer, S.F., Behrens, T.E.J., & Kennerley, S.W. (2018) Triple dissociation of attention and decision computations across prefrontal cortex. *Nat Neurosci*, 21, 1471–81.

Hutchison, R.M., Womelsdorf, T., Gati, J.S., Leung, L.S., Menon, R.S., & Everling, S. (2012) Resting-state connectivity identifies distinct functional networks in macaque cingulate cortex. *Cereb Cortex*, 22, 1294–308.

James, W. (1890) *The Principles of Psychology*. Holt, New York.
Johansen-Berg, H., Behrens, T.E., Robson, M.D., Drobnjak, I., Rushworth, M.F., Brady, J.M., Smith, S.M., Higham, D.J., & Matthews, P.M. (2004) Changes in connectivity profiles define functionally distinct regions in human medial frontal cortex. *Proc Natl Acad Sci USA*, 101, 13335–40.
Jones, E.G. & Powell, T.P. (1970) An anatomical study of converging sensory pathways within the cerebral cortex of the monkey. *Brain*, 93, 793–820.
Klein, J.C., Behrens, T.E., Robson, M.D., Mackay, C.E., Higham, D.J., & Johansen-Berg, H. (2007) Connectivity-based parcellation of human cortex using diffusion MRI: Establishing reproducibility, validity and observer independence in BA 44/45 and SMA/pre-SMA. *Neuroimage*, 34, 204–11.
Lancaster, J.L., Laird, A.R., Eickhoff, S.B., Martinez, M.J., Fox, P.M., & Fox, P.T. (2012) Automated regional behavioral analysis for human brain images. *Front Neuroinform*, 6, 23.
Mackey, S. & Petrides, M. (2010) Quantitative demonstration of comparable architectonic areas within the ventromedial and lateral orbital frontal cortex in the human and the macaque monkey brains. *Eur J Neurosci*, 32, 1940–50.
Maier-Hein, K.H., Neher, P.F., Houde, J.C., Cote, M.A., Garyfallidis, E., Zhong, J., Chamberland, M., Yeh, F.C., Lin, Y.C., Ji, Q., Reddick, W.E., Glass, J.O., Chen, D.Q., Feng, Y., Gao, C., Wu, Y., Ma, J., Renjie, H., Li, Q., Westin, C.F., Deslauriers-Gauthier, S., Gonzalez, J.O.O., Paquette, M., St-Jean, S., Girard, G., Rheault, F., Sidhu, J., Tax, C.M.W., Guo, F., Mesri, H.Y., David, S., Froeling, M., Heemskerk, A.M., Leemans, A., Bore, A., Pinsard, B., Bedetti, C., Desrosiers, M., Brambati, S., Doyon, J., Sarica, A., Vasta, R., Cerasa, A., Quattrone, A., Yeatman, J., Khan, A.R., Hodges, W., Alexander, S., Romascano, D., Barakovic, M., Auria, A., Esteban, O., Lemkaddem, A., Thiran, J.P., Cetingul, H.E., Odry, B.L., Mailhe, B., Nadar, M.S., Pizzagalli, F., Prasad, G., Villalon-Reina, J.E., Galvis, J., Thompson, P.M., Requejo, F.S., Laguna, P.L., Lacerda, L.M., Barrett, R., Dell'Acqua, F., Catani, M., Petit, L., Caruyer, E., Daducci, A., Dyrby, T.B., Holland-Letz, T., Hilgetag, C.C., Stieltjes, B., & Descoteaux, M. (2017) The challenge of mapping the human connectome based on diffusion tractography. *Nature Communications*, 8, 1349.
Markov, N.T., Ercsey-Ravasz, M., Van Essen, D.C., Knoblauch, K., Toroczkai, Z., & Kennedy, H. (2013) Cortical high-density counterstream architectures. *Science*, 342, 1238406.
Markov, N.T., Ercsey-Ravasz, M.M., Ribeiro Gomes, A.R., Lamy, C., Magrou, L., Vezoli, J., Misery, P., Falchier, A., Quilodran, R., Gariel, M.A., Sallet, J., Gamanut, R., Huissoud, C., Clavagnier, S., Giroud, P., Sappey-Marinier, D., Barone, P., Dehay, C., Toroczkai, Z., Knoblauch, K., Van Essen, D.C., & Kennedy, H. (2014) A weighted and delete directed interareal connectivity matrix for macaque cerebral cortex. *Cereb Cortex*, 24, 17–36.
Marr, D. (1982) *Vision*. MIT Press, Cambridge.
Mars, R.B., Passingham, R.E., & Jbabdi, S. (2018) Connectivity fingerprints: From areal descriptions to abstract spaces. *Trends Cogn Sci*, 22, 1026–37.
Masse, N.Y., Hodnefield, J.M., & Freedman, D.J. (2017) Mnemonic encoding and cortical organization in parietal and prefrontal cortices. *J Neurosci*, 37, 6098–112.
McIntosh, A.R. & Gonzalez-Lima, F. (1994) Network interactions among limbic cortices, basal forebrain, and cerebellum differentiate a tone conditioned as a Pavlovian excitor or inhibitor: Fluorodeoxyglucose mapping and covariance structural modeling. *J Neurophysiol*, 72, 1717–33.
Mukamel, R. & Fried, I. (2012) Human intracranial recordings and cognitive neuroscience. *Annu Rev Psychol*, 63, 511–37.
Murayama, Y., Weber, B., Saleem, K.S., Augath, M., & Logothetis, N.K. (2006) Tracing neural circuits in vivo with Mn-enhanced MRI. *Magn Reson Imaging*, 24, 349–58.

Murray, E.A. & Coulter, J.D. (1981) Organization of corticospinal neurons in the monkey. *J Comp Neurol*, 195, 339–65.

Mushiake, H., Inase, M., & Tanji, J. (1991) Neuronal activity in the primate premotor, supplementary, and precentral motor cortex during visually guided and internally determined sequential movements. *J Neurophysiol*, 66, 705–18.

O'Reilly, J.X., Croxson, P.L., Jbabdi, S., Sallet, J., Noonan, M.P., Mars, R.B., Browning, P.G., Wilson, C.R., Mitchell, A.S., Miller, K.L., Rushworth, M.F., & Baxter, M.G. (2013) Causal effect of disconnection lesions on interhemispheric functional connectivity in rhesus monkeys. *Proc Natl Acad Sci USA*, 110, 13982–7.

Pandya, D.M., Seltzer, B., Petrides, M., & Cipolloni, P.B. (2015) *Cerebral Cortex: Architecture, Connections and the Dual Origin Concept*. Oxford University Press, Oxford.

Pandya, D.N. & Kuypers, H.G. (1969) Cortico-cortical connections in the rhesus monkey. *Brain Res*, 13, 13–36.

Passingham, R.E., Stephan, K.E., & Kotter, R. (2002) The anatomical basis of functional localization in the cortex. *Nat Rev Neurosci*, 3, 606–16.

Petrides, M. (2018) *Atlas of the Morphology of the Human Cerebral Cortex on the Average MNI Brain*. Academic Press, New York.

Petrides, M. & Pandya, D.N. (1999) Dorsolateral prefrontal cortex: Comparative cytoarchitectonic analysis in the human and the macaque brain and corticocortical connection patterns. *Eur J Neurosci*, 11, 1011–36.

Petrides, M. & Pandya, D.N. (2007) Efferent association pathways from the rostral prefrontal cortex in the macaque monkey. *J Neurosci*, 27, 11573–86.

Pizzamiglio, G., Zhang, Z., Kolasinski, J., Riddoch, J.M., Passingham, R.E., Mantini, D., & Rounis, E. (2019) A role for the action observation network in apraxia after stroke. *Front Hum Neurosci*, 13, 422.

Preuss, T.M. (2000) Taking the measure of diversity: Comparative alternatives to the model-animal paradigm in cortical neuroscience. *Brain Behav Evol*, 55, 287–99.

Preuss, T.M. & Robert, J.S. (2014) Animal models of the human brain: Repairing the paradigm. In Gazzaniga, M. (ed.) *The Cognitive Neurosciences*. MIT Press, Cambrdige M.A.

Price, J.L. & Drevets, W.C. (2010) Neurocircuitry of mood disorders. *Neuropsychopharmacology*, 35, 192–216.

Reveley, C., Gruslys, A., Ye, F.Q., Glen, D., Samaha, J., Russ, B.E., Saad, Z., Seth, A.K., Leopold, D.A., & Saleem, K.S. (2017) Three-dimensional digital template atlas of the macaque brain. *Cereb Cortex*, 27, 4463–77.

Reveley, C., Seth, A.K., Pierpaoli, C., Silva, A.C., Yu, D., Saunders, R.C., Leopold, D.A., & Ye, F.Q. (2015) Superficial white matter fiber systems impede detection of long-range cortical connections in diffusion MR tractography. *Proc Natl Acad Sci USA*, 112, E2820–8.

Roberts, J.A., Perry, A., Roberts, G., Mitchell, P.B., & Breakspear, M. (2017) Consistency-based thresholding of the human connectome. *Neuroimage*, 145, 118–29.

Sakai, K. & Passingham, R.E. (2006) Prefrontal set activity predicts rule-specific neural processing during subsequent cognitive performance. *J Neurosci*, 26, 1211–18.

Saleem, K.S., Kondo, H., & Price, J.L. (2008) Complementary circuits connecting the orbital and medial prefrontal networks with the temporal, insular, and opercular cortex in the macaque monkey. *J Comp Neurol*, 506, 659–93.

Saleem, K.S. & Logothetis, N.K. (2006) *A Combined MRI and Histology Atlas of the Rhresus Monkey*. Academic Press, Amsterdam.

Sallet, J., Mars, R.B., Noonan, M.P., Neubert, F.X., Jbabdi, S., O'Reilly, J.X., Filippini, N., Thomas, A.G., & Rushworth, M.F. (2013) The organization of dorsal frontal cortex in humans and macaques. *J Neurosci*, 33, 12255–74.

Schall, J.D. (1995) Neural basis of saccade target selection. *Rev Neurosci*, 6, 63–85.
Shadmehr, R. & Wise, S. (2005) *The Computational Neurobiology of Reaching and Pointing*. MIT press, Cambridge.
Sporns, O. (2014) Contributions and challenges for network models in cognitive neuroscience. *Nat Neurosci*, 17, 652–60.
Staes, N., Smaers, J.B., Kunkle, A.E., Hopkins, W.D., Bradley, B.J., & Sherwood, C.C. (2019) Evolutionary divergence of neuroanatomical organization and related genes in chimpanzees and bonobos. *Cortex*, 118, 154–64.
Stephani, T., Waterstraat, G., Curio, G., Villringer, A., Nikulin, V.V. (2020) Temporal signatures of criticality in human cortical excitability as probed by early somatosensory responses. *J Neurosci*, 40, 6572–83.
Stepniewska, I., Cerkevich, C.M., Fang, P.C., & Kaas, J.H. (2009) Organization of the posterior parietal cortex in galagos: II. Ipsilateral cortical connections of physiologically identified zones within anterior sensorimotor region. *J Comp Neurol*, 517, 783–807.
Stevens, N.J., Seiffert, E.R., O'Connor, P.M., Roberts, E.M., Schmitz, M.D., Krause, C., Gorscak, E., Ngasala, S., Hieronymus, T.L., & Temu, J. (2013) Palaeontological evidence for an Oligocene divergence between old world monkeys and apes. *Nature*, 497, 611–14.
Tanaka, K. (1997) Mechanisms of visual object recognition. *Curr Opin Neurobiol*, 7, 523–9.
Thaler, D., Chen, Y.-C., Nixon, P.D., Stern, C., & Passingham, R.E. (1995) The functions of the medial premotor cortex (SMA). I. Simple learned movements. *Exper Brain Res*, 102, 445–60.
Tinbergen, N. (1951) *The Study of Instinct*. Oxford University Press, Oxford.
Van Essen, D.C., Ugurbil, K., Auerbach, E., Barch, D., Behrens, T.E., Bucholz, R., Chang, A., Chen, L., Corbetta, M., Curtiss, S.W., Della Penna, S., Feinberg, D., Glasser, M.F., Harel, N., Heath, A.C., Larson-Prior, L., Marcus, D., Michalareas, G., Moeller, S., Oostenveld, R., Petersen, S.E., Prior, F., Schlaggar, B.L., Smith, S.M., Snyder, A.Z., Xu, J., Yacoub, E., & Consortium, W.U.-M.H. (2012) The Human Connectome Project: A data acquisition perspective. *Neuroimage*, 62, 2222–31.
Von Bonin, G. & Bailey, P. (1947) *The Neocortex of Macaca Mulatta*. University of Illinois, Urbana.
von Economo, C. (1929) *The Cytoarchitectonics of the Human Cerebral Cortex*. Oxford University Press, London.
Wang, G., Tanifuji, M., & Tanaka, K. (1998) Functional architecture in monkey inferotemporal cortex revealed by *in vivo* optical imaging. *Neurosci Res*, 32, 33–46.
Woolgar, A., Thompson, R., Bor, D., & Duncan, J. (2011) Multi-voxel coding of stimuli, rules, and responses in human frontoparietal cortex. *Neuroimage*, 56, 744–52.
Zilles, K. & Palomero-Gallagher, N. (2001) Cyto-, myelo-, and receptor architectonics of the human parietal cortex. *Neuroimage*, 14, S8–20.

2
Evolution of the Prefrontal Cortex in Non-human Primates

Introduction

The previous chapter argues that we will not fully understand the prefrontal (PF) cortex if we fail to appreciate the selective advantages that it brings. In particular, we need to know what the granular PF cortex can do that other brain areas are unable to do.

As Chapter 1 also mentions, there are two groups of PF areas in primate brains. One consists of the agranular PF areas and these have homologues in the brains of other mammals. The other consists of the granular PF areas and these have no homologues in other mammals.

The granular PF cortex evolved in phases, but it is not possible to carry out cytoarchitectural studies on ancestral species. Nonetheless, by comparing modern species, Preuss and Goldman-Rakic (1991b) were able to show that bushbabies (*Galago*) have granular PF areas that rats lack. And they also demonstrated that macaque monkeys have granular PF areas that bushbabies lack.

Figure 2.1 (next page) illustrates the granular and agranular PF areas in macaque monkeys (A), rats (B), and bushbabies (C). Humans aside, most of our knowledge about the PF cortex comes from studies of these species, along with mice.

Figure 2.1 depicts the motor areas in gold and the agranular PF areas in green. Rats, bushbabies, and macaques all have areas of these two types. Motor areas (gold) send projections directly to the spinal cord whereas agranular PF areas (green) do not. The granular PF cortex appears in blue. The figure also distinguishes between the lightly myelinated granular PF cortex, shaded in light blue, and the moderately myelinated granular PF cortex, in dark blue.

Some of these areas evolved in the last common ancestor of tree shrews and primates; others first appeared in early primates; and yet others emerged during the evolution of the monkeys and apes. With the exception of a few areas shared with tree shrews, only primates have homologues of the areas depicted in blue.

The term 'strepsirhine' refers to busbabies, lemurs, and lorises. The term 'haplorhine' refers to monkeys and apes, together with small nocturnal primates called tarsiers (*Tarsiidae*). The monkeys, apes and humans collectively form the 'anthropoids'.

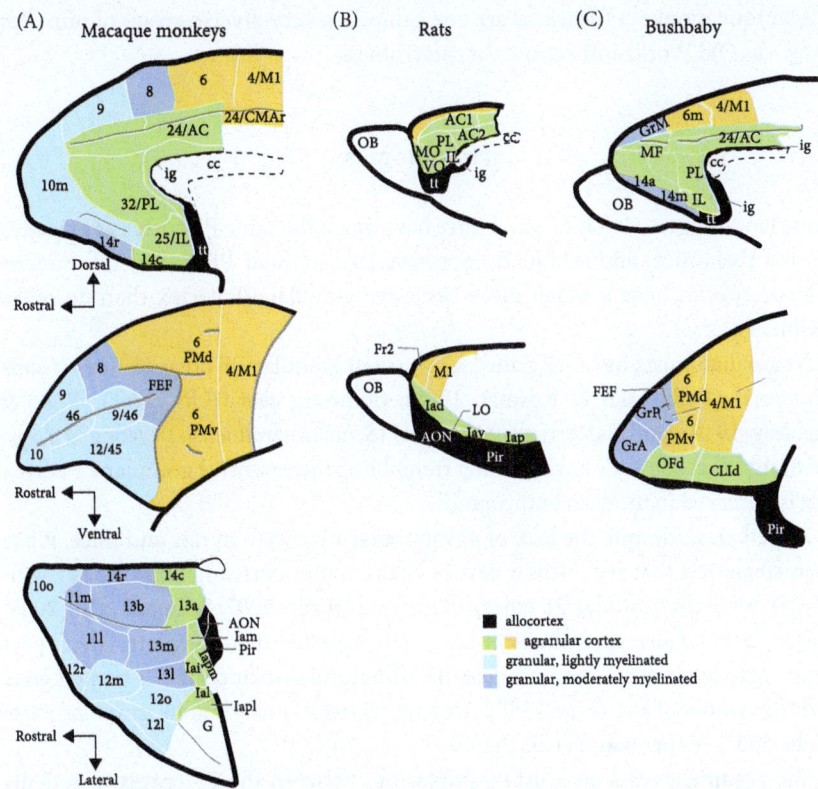

Figure 2.1 The PF cortex and other frontal areas in macaque monkeys (A), rats (B), and (C) bushbabies (*galagos*)

Top row: medial view; middle row: lateral view; bottom row: ventral view. Gold shading = motor and premotor areas, green shading = agranular PF areas, and blue shading = granular PF areas. Across species, the brain drawings are not to scale. Grey lines indicate sulci; white lines indicate boundaries between areas. Abbreviations for orientation: c, caudal; i, inferior; l, lateral; m, medial; o, orbital; p, posterior; r, rostral. Abbreviations for areas: AC, anterior cingulate cortex; AON, anterior olfactory nucleus; cc, corpus callosum; CLId, claustro-insular dysgranular area; FEF, frontal eye field; Fr2, medial agranular motor area; G, gustatory cortex; GrA, anterior granular area; GrP, posterior granular area; Ia, agranular insular cortex; ig, induseum griseum; IL, infralimbic cortex; LO, lateral orbitofrontal cortex; M1, primary motor cortex (area 4); MF, medial frontal cortex; MO, medial orbitofrontal cortex; OB, olfactory bulb; OFd, dorsal orbitofrontal cortex; Pir, piriform cortex; PL, prelimbic; tt, tenia tecta; PMd, dorsal premotor

Adapted from Wise, S.P., 'The evolution of the prefrontal cortex in early primates and anthropoids', in Krubitzer, L., Kaas, J. (eds) The Evolution of Nervous Systems, Second Edition, Volume 3, pp. 387-422, Figure 2, DOI: https://doi.org/10.1016/B978-0-12-804042-3.00092-0 Copyright © 2017, Elsevier Inc. All rights reserved.

Macaque monkeys (*Macaca*) are one genus in a very diverse group of primates known as Old World anthropoids or catarrhines.

Homologies

Some have suggested that rats and mice have much the same PF cortex as primates do. But rats, mice, and rabbits do not have any granular PF areas; and modern primate species have a much more extensive granular PF cortex than do other mammals.

Neuroanatomists have described a few small granular PF areas in dogs (*Canis familiaris*) (Rajkowska & Kosmal, 1988), domestic cats (*Felis catus*) (Rose & Woolsey, 1948), and Eastern grey squirrels (*Sciurus carolinensis*) (Wong & Kaas, 2008). But these species have nothing resembling the extensive granular PF cortex that is observed in modern anthropoids.

Nonetheless, despite the lack of any granular PF cortex in rats and mice, it has been suggested that the medial part of their frontal cortex is homologous with most or all of the granular PF cortex in primates (Kolb, 2007; Seamans et al., 2008; Carlen, 2017). Correspondingly, it has been suggested that the ventrolateral part of the rat frontal cortex, consisting of the orbital and insular cortex, is homologous with the whole of the orbital PF cortex in primates, including its granular parts (Kolb, 2007; Schoenbaum et al., 2009).

The argument rests on a list of similarities between the PF cortex of rodents and primates. The proposed similarities relate to neurophysiology, neuroanatomy, neuropharmacology, and neuropsychology. But as Preuss (1995) has stressed, the identification of homologues depends on traits that are *diagnostic* of a particular brain area, not on a list of similarities.

To illustrate the necessity for diagnostic traits, the agranular PF cortex in rats has cells that encode valuations (Burton et al., 2014) as the granular PF cortex does in monkeys. But there are cells in both the agranular and granular PF areas of monkeys that have these properties (Wallis & Kennerley, 2011). So, these similarities do not serve as diagnostic traits.

The same argument holds for reciprocal connections with the mediodorsal nucleus of the thalamus (Akert, 1964). On this basis, Uylings et al. (2003) held that the agranular PF cortex in rats is homologous with the granular PF cortex in primates. However, the mediodorsal nucleus of macaques' projects to virtually all of the frontal lobe, as well as to the parietal cortex (Giguere & Goldman-Rakic, 1988; Matelli et al., 1989). So, connections with the mediodorsal nucleus do not establish a homology either.

More recently, Heilbronner et al. (2016) have described corticofugal projections to the nucleus accumbens in rats and macaque monkeys. It is true that this nucleus, part of the striatum, receives inputs from the orbital PF cortex both in macaques and rats. However, both the granular and agranular parts of the orbital PF cortex send projections to the nucleus accumbens in macaques (Ferry et al., 2000). So, these connections in question are not specific to either the granular or agranular orbital PF cortex and do not establish a homology.

Another argument is that lesions of the PF cortex can have similar behavioural effects in rats and macaques (Brown & Bowman, 2002). For example, lesions of the medial PF cortex in rats (Kolb et al., 1974) and of the dorsal PF cortex (areas 46 and 9/46) in monkeys (Goldman et al., 1971) cause an impairment on the delayed response task. But the impairment in rats is mild: in one experiment, the rats relearned the task in 60 trials after a medial frontal cortex lesion (Kolb et al., 1974), and in another in 140 trials (Kolb et al., 1994). Lesions of the anterior cingulate cortex cause a similar mild impairment in macaques (Meunier et al., 1997). But the impairment is totally unlike the one that follows lesions of the dorsal PF cortex in monkeys: the monkeys fail to relearn the task at all in 1,000 trials (Goldman et al., 1971).

A final observation is that the functional connectivity of the medial PF cortex is not the same in rats and primates. Schaeffer et al. (2020) studied this using high-field functional magnetic resonance imaging (fMRI). The functional fingerprint for the rat medial PF cortex is like that of the premotor areas in the marmoset, not the granular PF cortex.

The idea that primates have new PF areas should not generate as much controversy as it sometimes does. For other parts of the cerebral cortex, neuroscientists readily accept the suggestion that new areas emerged during evolution. So, it is not clear why some are unwilling to apply this principle when applied to the PF cortex.

Take the visual cortex as an example. Many new visual areas developed during the evolution of the monkeys and apes; anthropoid primates have more than two dozen visual areas (Felleman & van Essen, 1991; Kaas, 2006a). One of these areas, called MT, for the middle temporal area, has sophisticated specializations for analysing visual motion across the fovea. A useful review of the evolution of this area and its thalamic connections can be found in Baldwin et al. (2017b).

Yet, rodents have many fewer visual areas (Rosa & Krubitzer, 1999; Lyon, 2007), and in addition they lack a fovea. So, it is simply not plausible that the smaller number of visual areas in rats replicate all of the functions of the more than two dozen visual areas in primates. For one thing, the fine-grained analysis of foveal motion would be especially unexpected in a species that lacks a fovea.

To take another example, consider the auditory cortex of echolocating bats (*Microchiroptera*). It is clear that these species have more auditory areas than rodents do (Kossl et al., 2015). Echolocating bats use a sonar-like system to detect the distance and velocity of their insect prey, and their auditory cortex has many specialized areas for processing the acoustic signals involved in echolocation, including echo delays and Doppler shifts (Suga et al., 1997; Fitzpatrick et al., 1998).

Yet rodents do not track their food items with echolocation. The idea that the auditory areas in rats have all of the same functions and properties as the specialized areas in echolocating bats is simply not plausible.

Given the widespread acceptance that echolocating bats have new auditory areas and that anthropoid primates have new visual areas, it is surprising there is resistance to the suggestion that primates have new PF areas and that these evolutionary innovations support novel functions.

New Areas

Part of the problem comes from using the word 'new'. It has been suggested that new areas appear by a process of differentiating out of earlier ones (Krubitzer & Huffman, 2000). If so, then it is reasonable to assume that some of these new areas change little over evolutionary time scales; and we can recognize these conserved areas as homologous when they occur in many mammalian species (Krubitzer & Seelke, 2012).

However, compared to these relatively conserved areas, other areas change more and therefore merit the designation 'new'. As they change, these areas develop a new connectional fingerprint and come to perform modified functions, thus providing an advantage over the ancestral condition. But neither their connections nor novel functions will be entirely without precedent. For example, these new areas will usually have connections with the earlier areas, and the connectional fingerprint will therefore have a family resemblance with the fingerprint of the original area.

Accordingly, it should come as no surprise that granular and agranular PF areas have many similarities, including projections to some of the same phylogenetically old brain structures (Heilbronner et al., 2016). But this does not mean that they should be treated as being the same.

Summary

Some have argued that agranular PF areas in rodents are homologous with the granular PF areas in primates. But similarities alone do not demonstrate homologies. The traits have to be diagnostic. The properties of an agranular area in rodents cannot be said to homologous with those of a granular area in primates if there is a corresponding agranular area in primates that has the same properties.

Diagnostic Traits

The argument does not depend on cytoarchitecture alone. There are four diagnostic traits that support the conclusion that rats and mice lack homologues of the granular PF areas found in primates. These features involve spatial arrangement, cortico-striatal projections, sensory inputs, and autonomic outputs. Taken one at a time, these might be said to be debatable. However, when taken together they show how sharply granular and agranular PF areas differ from each other.

Spatial Arrangement

A particularly useful aspect of the arrangement of the PF cortex involves its relationship with the allocortex. Allocortex has three layers, in contrast to the more complex laminar structure of neocortex. Typical examples of allocortex include the hippocampus and piriform cortex, but smaller allocortical areas such as the induseum griseum and the tenia tecta are also relevant for the spatial arrangement.

In Figure 2.1 allocortical areas within the frontal lobe are shaded in black. As this figure shows, some PF areas lie directly adjacent to allocortex and some do not. Specifically, the agranular PF areas (green) lie adjacent to the allocortex both in rodents and primates.

By contrast, the granular PF areas (blue in Figure 2.1) rarely, if ever, share a boundary with the allocortex. Instead, the granular PF areas lie next to the agranular areas (green) or to each other. An apparent exception in Figure 2.1C, where area 14 m seems to abut the tenia tecta, probably results from a failure to recognize a small agranular area separating the two. So, the spatial arrangement agrees with cytoarchitecture in designating the granular PF cortex as something different from the agranular PF cortex. Adjacency to allocortex serves as a diagnostic trait for homologues of agranular PF areas.

Cortico-Striatal Projections

The granular PF cortex has cortico-striatal projections that agranular areas do not have. As an earlier section mention, the nucleus accumbens is part of the ventral striatum. This differs from the dorsal striatum. In primates, the dorsal striatum has two parts: the caudate and putamen. The dorsal part of the head of the caudate nucleus appears to be new to primates.

A key observation that supports this view comes from the study of Heilbronner et al. (2016). They showed that areas 24, 25, and 32, the agranular parts of the medial PF cortex, send axons with terminals that fill the rostroventral aspect of the striatum. But they also provide an illustration that demonstrates a feature of macaques that rats do not share. This illustration indicates that the macaque anterior cingulate area (ACC, area 24) projects to both the dorsal and ventral caudate at an intermediate rostro-caudal level but that in the rostral caudate the dorsal part is almost devoid of projections. In macaque monkeys, the cortico-striatal connections from the medial, dorsal, and ventral PF cortex project to this 'empty' zone (Selemon & Goldman-Rakic, 1985; Choi et al., 2017), and they are all granular PF areas.

Choi et al. (2017) studied cortico-striatal connections in humans. They seeded the rostral head of the caudate and looked for areas that showed covariance in resting-state activation. Their results confirmed an anatomical analysis that they carried out on macaque monkeys using tracers. The cortical areas that showed covariance with the rostral head of the caudate nucleus included the medial, dorsal, and ventral parts of the granular PF cortex, as well as parts of the posterior parietal cortex.

The most parsimonious account for these findings is that the dorsal part of the rostral caudate nucleus is an evolutionary innovation of primates, which probably evolved in tandem with their new granular PF areas and also with new areas in the posterior parietal cortex. Although connections can change during evolution, it seems that the cortico-striatal projections of the agranular PF areas are conserved for the most part. New areas in the primate cortex project to new parts of the primate striatum.

Sensory Inputs

Connections that convey sensory inputs to the PF cortex can also serve as diagnostic traits. Agranular parts of the PF cortex receive relatively direct olfactory, gustatory, and visceral inputs in both rats (Ray & Price, 1992) and macaque monkeys (Ray & Price, 1993). The olfactory inputs come from the piriform cortex; gustatory and visceral sensory inputs arrive via relays in the brainstem and thalamus. These connections support the homology of agranular orbital PF cortex in primates and rodents.

However, in macaques the granular areas of the orbital PF cortex do not receive direct olfactory, gustatory, and visceral inputs. Instead, these sensory inputs arrive only indirectly from the agranular PF areas. In contrast to the agranular orbital PF cortex, the granular part has prominent cortico-cortical inputs from visual areas of the temporal lobe (Carmichael & Price, 1995).

Autonomic Outputs

A final diagnostic trait involves the ability to elicit autonomic outputs by cortical stimulation. In macaque monkeys, the medial PF cortex has a granular cytoarchitecture in areas 9 and 10 but an agranular architecture in the anterior cingulate, infralimbic, and prelimbic areas (areas 24, 25, and 32, respectively). Similarly, in macaque monkeys, areas 13 and 14 of the orbital PF cortex have both rostral granular and caudal agranular parts, with the fully granular area 11 located more rostrally and the fully agranular insular areas located more caudally (Carmichael & Price, 1994). Figure 2.1A shades these agranular areas in green.

As a diagnostic feature, outputs from the agranular frontal areas influence the autonomic nervous system more directly than those from the granular PF cortex. Specifically, in rats, rabbits, cats, dogs, monkeys, and humans, electrical stimulation of the agranular parts of the PF cortex elicits autonomic effects. These areas include the anterior cingulate, prelimbic, and infralimbic cortex, as well as the agranular orbital PF cortex (Kaada, 1960). Stimulation of these areas elicits changes in respiratory rate, blood pressure, pulse rate, pupil dilation, and piloerection. Furthermore, lesions of these agranular areas disrupt autonomic responses in rabbits (Powell et al., 1997), rats (Frysztak & Neafsey, 1994), and macaques (Rudebeck et al., 2014).

By contrast, electrical stimulation of the granular PF areas fails to produce autonomic changes, even when matched in frequency and intensity (Kaada et al., 1949). Thus, relatively direct autonomic outputs serve as a diagnostic trait for the agranular PF cortex.

Summary

The claim that the primate PF cortex has new areas does not depend on cytoarchitecture alone. There are four diagnostic traits that add support. These are separation from the allocortex, cortico-striatal inputs to the head of the caudate nucleus, lack of direct gustatory, olfactory, and visceral inputs, and no direct autonomic outputs.

The Origins of the Granular Prefrontal Cortex

There is continuing debate about the classification of the extinct, primate-like mammals called plesiadapiforms (Silcox et al., 2017). Some consider them to be stem primates; others do not. However, assuming that plesiadapiforms are indeed stem primates, one such species probably gave rise to the first true primates (*Euprimates*).

The founding euprimates specialized in a nocturnal life confined to a fine-branch, arboreal niche. They had modifications of their hand and foot structure, along with the development of finger- and toenails. Most notably, their eyes shifted into a forward orientation. The frontal orientation of the eyes of the euprimates increased the extent of the visual field available for depth perception via stereopsis, which provided advantages in visually guided grasping. It might also have helped with vision in dim light by summing inputs from the two eyes (Hughes, 1977; Allman, 2000).

Taken together, their visual and grasping innovations enabled early primates to move through the trees by leaping from one branch and grasping another. This leaping and grasping mode of locomotion depended on precise stereoscopic vision, exquisite visuomotor coordination, and the ability to grasp and hold onto branches.

The fossil record indicates that the grasping specializations developed before the visual ones. *Torrejonia*, the oldest known plesiadapiform, dates to ~62 million years ago (Ma). Fossil evidence concerning its shoulder, elbow, hip, knee, and ankle morphology reveals that this Palaeocene species had already adapted to an arboreal life (Chester et al., 2017). It could cling and climb on vertically oriented branches, so it probably had a leaping and grasping form of locomotion. The later plesiadapiform *Carpolestes*, dated to ~55 Ma, had similar grasping specializations but like *Torrejonia*, it lacked forward-facing eyes (Bloch & Boyer, 2002).

Other studies show that the specializations for leaping and grasping probably evolved in stages, beginning with manual grasping (Boyer et al., 2013b). Modifications for grasping with the foot emerged later (Boyer et al., 2013a), with the combination of grasping with forelimb and hindlimb and forward-facing eyes coming together in early euprimates.

New Prefrontal Areas

The common ancestor of strepsirrhines and anthropoids probably had homologues of both the caudal PF cortex and the granular orbital PF cortex. This conclusion is based partly on the architectonic and connectional evidence adduced by Preuss and Goldman-Rakic (1991b; a) and partly on more recent findings (Wong & Kaas, 2010).

Intracortical micro-stimulation of the caudal PF cortex evokes saccadic eye movements in bushbabies (Wu et al., 2000; Stepniewska et al., 2018); and stimulation of a part of area 8, the frontal eye field (FEF), does so in macaques (Bruce et al., 1985). By contrast, Baldwin et al. (2017a) stimulated the frontal cortex of tree shrews and could not evoke eye movements from any part of the frontal cortex. A previous study (Remple et al., 2006) reached the same conclusions. This finding means that the FEF probably emerged in the primate lineage after the divergence of tree shrews and primates (Schall et al., 2017).

Two related conclusions follow from this work. First, the common ancestor of strepsirrhines and tree shrews lacked a homologue of the FEF. Second, the FEF-like areas in cats (Orem & Schlag, 1973) and rodents (Hall & Lindholm, 1974) have resulted from convergent evolution. Therefore, they are not homologous with the FEF of bushbabies and macaques.

In Figure 2.2 (next page), the granular PF areas in tree shrews are marked in blue shading (Wong & Kaas, 2009). Because both tree shrews and bushbabies have a granular orbital PF cortex, the emergence of this area probably predates the divergence of tree shrews and primates. Chapter 4 describes how this area is involved in the evaluation of foods in relation to current needs.

Figure 2.2 also illustrates small dorsal and medial areas in tree shrews. Passingham (1978) removed these areas after training tree shrews on the delayed alternation task. On this task, a hidden food item alternates between two locations from trial to trial.

After the surgery, one of these animals failed to relearn the task in 2,000 trials; another failed in 1,200 trials; and a tree shrew with an incomplete lesion in one hemisphere relearned the task in 821 trials. By contrast, all the three intact tree shrews in the study improved rapidly in the corresponding test period, relearning the task in an average of only 77 trials (Figure 2.3, page 45).

The severity of this impairment contrasts with the mild impairment on related tasks after lesions of the medial PF cortex in rats (Kolb et al., 1974; 1994). But it resembles the severe impairment on the delayed alternation task after lesions of the dorsal PF cortex in macaques (Goldman et al., 1971).

By contrast to the granular orbital PF cortex and FEF, Preuss and Goldman-Rakic (1991b) noted that many granular PF areas in macaque brains have much less myelin (Figure 2.1A, light blue) than these phylogenetically older areas (dark blue). They also found little, if any, myelin-poor granular cortex in the PF cortex of bushbabies (Figure 2.1C). Because many granular PF areas are poorly myelinated in macaque and human brains, Preuss and Goldman-Rakic concluded that bushbabies lack homologues of these areas.

Along with a homologue of the FEF, Preuss and Goldman-Rakic (1991b) described areas on the lateral surface of bushbaby brains that might correspond to areas 8B and 45 in macaques. In macaques, area 8B receives visual information from the posterior parietal cortex (Petrides & Pandya, 1999). Area 45 is divided into a posterior 45B and an anterior 45A. Area 45B is connected with the FEF (Gerbella et al., 2010), and it is possible that it is this area that Preuss and Goldman-Rakic (1991b) were able to identify in bushbabies.

The conclusion is that strepsirrhine primates lack homologues of many areas that are found in anthropoids such as macaques. These innovations include the dorsal PF cortex (areas 46 and 9/46), the ventral PF cortex (areas 47/12 and 45A), and the granular parts of the medial PF cortex (areas 9 and 10).

Table 2.1 (page 46) summarizes these conclusions. In addition to their new PF areas, primates evolved new areas in the posterior parietal, premotor, and temporal cortex. These developments have been reviewed elsewhere (Kaas, 2006b; Preuss, 2007; Cooke et al., 2014).

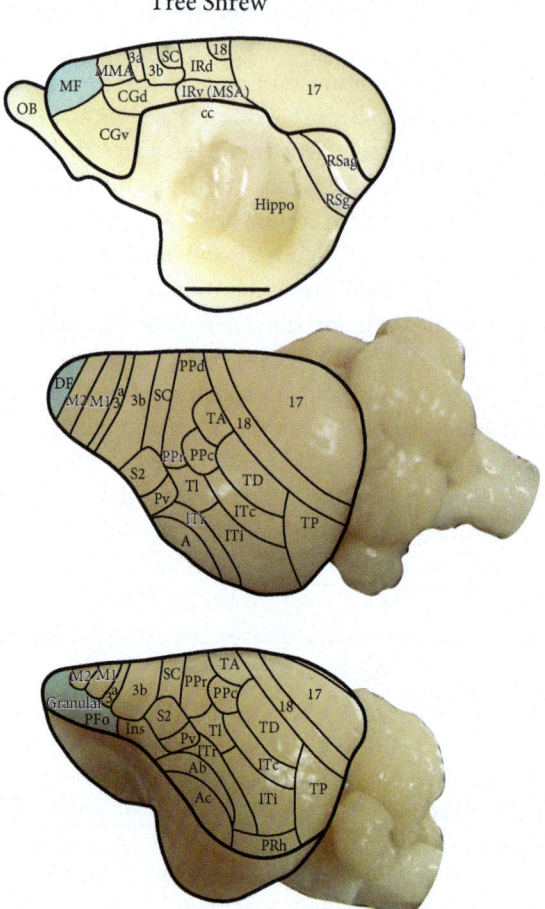

Figure 2.2 Cortical map of tree shrew brains
The areas shaded in blue have a conspicuous layer 4. Top: medial view; middle: dorsolateral view; bottom: ventrolateral view. Abbreviations for orientation in the brain: ag, agranular; c, caudal; d, dorsal; g, granular; i, inferior; l, lateral; m, medial; o, orbital; p, posterior; r, rostral. Abbreviations for areas: A, auditory area; Ab, auditory belt area; Ac, auditory core area; cc, corpus callosum; CG, cingulate gyrus; DF, dorsal frontal cortex; Hippo, hippocampus; Ins; insular cortex; IR, infraradiate area IT, inferior temporal cortex; M2, supplementary motor area; MF, medial frontal cortex; OB, olfactory bulb; Granular PFo, granular orbitofrontal cortex; PMv, ventral premotor cortex (area 6); PP, posterior parietal cortex; PRh, perirhinal cortex; Pv, posteroventral somatosensory area; RS, retrosplenial cortex; S2, second somatosensory area; SC, caudal somatosensory cortex; TA, anterior temporal area; TD, dorsal temporal areal TI, inferior temporal area; TP, posterior temporal area; VO, ventral orbitofrontal cortex.

Adapted from Wong, P. & Kaas, J.H. Architectonic subdivisions of neocortex in the tree shrew (*Tupaia belangeri*). *Anatom Rec*, 292 (7), 994–1027, Figures 14f and 13c, Doi: 10.1002/ar.20916 Copyright © 2009, Wiley-Liss, Inc. With kind permission from Jon Kaas.

Adapted from Wise, S.P. The evolution of the prefrontal cortex in early primates and anthropoids. In Krubitzer, L. & Kaas, J. (eds) *The Evolution of Nervous Systems, Second Edition, Volume 3*, pp. 387–422, Figure 2, Doi: https://doi.org/10.1016/B978-0-12-804042-3.00092-0 Copyright © 2017, Elsevier Inc. All rights reserved.

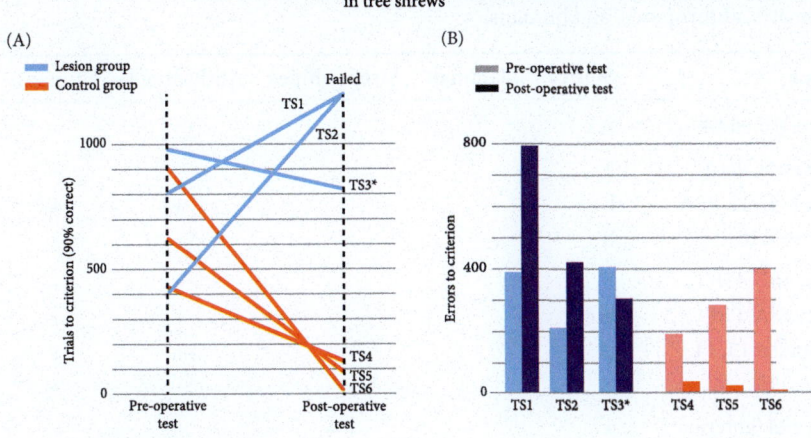

Figure 2.3 Effect of granular PF lesions in tree shrews
(A) Number of trials required to reach criterion performance of 90% correct choices on the spatial delayed alternation task: pre-operative versus post-operative scores and the corresponding test period for control animals. (B) Errors accumulated prior to reaching criterion performance: pre-operatively versus post-operatively.

Summary

By comparing the PF areas in tree shrews, bushbabies, and macaques it is possible to reconstruct the likely sequence in which the various areas evolved. The earliest granular PF to evolve were the FEF and the orbital PF cortex. This was followed by areas 8A and 45B. Finally, in anthropoid primates many new areas evolved, the dorsal PF cortex (areas 46 and 9/46), the ventral PF cortex (areas 45A and 47/12) and the granular parts of the medial PF cortex (areas 9 and 10).

Paleoecology of the Early Primates

To understand what advantages their granular PF areas conferred on stem primates, we need to understand something about how these extinct species lived. The stem primates and their closest relatives developed a suite of adaptations for finding, tracking, choosing, moving towards, reaching for, grasping, and feeding in the fine branches of angiosperm trees. These flowering plants or angiosperms emerged ~125 Ma and had diversified into many species of trees and shrubs by the time of the end-Cretaceous extinction (~66 Ma). The thinnest and most distal branches of these trees have the bulk of the nectar-bearing flowers, fruits, nuts, and

Table 2.1 PF areas with homologues (+) in mammals, strepsirrhine (prosimian) primates, anthropoids, and humans.

Area	Source	Mammals	Strepsirhines	Anthropoids	Humans
Lateral area 10	a				+
Areas 46, 9/46	b			+	+
Areas 47/12, 45A	b			+	+
Areas 9, 10	b			+	+
Areas 8B, 45B	b		+	+	+
Areas 8A, FEF	b, c, d		+	+	+
Granular 13, 14	b		+	+	+
Area 11	e			+	+
Area 24 (anterior cingulate)	e	+	+	+	+
Area 25 (infralimbic)	e	+	+	+	+
Area 32 (prelimbic)	e	+	+	+	+
Agranular 13, 14	e	+	+	+	+
Agranular insula	e	+	+	+	+

[a] Neubert (2014)
[b] Preuss and Goldman-Rakic (1991b)
[c] Wu et al. (2000)
[d] Preuss et al. (1993)
[e] Wise (2008)

seeds. They also have many of the youngest and most tender leaves, which have more protein and less fibre than mature leaves.

Indeed, the close relationship between primates and angiosperms has led some primatologists to suggest that primates evolved to exploit the adaptive radiation of angiosperms (Sussman, 1991). Silcox et al. (2015) concluded from tooth morphology in the fossil record that early primates were predominantly frugivorous.

Lehmann & Dunbar (2009) have emphasized the role of social groups, but early primates lived relatively solitary lives (Muller & Thalmann, 2000; DeCasien et al., 2017). Accordingly, social influences on brain evolution must have occurred later, long after the appearance of euprimates and for the most part after the strepsirrhine–haplorhine divergence. Monkeys and apes developed complex social systems partly as a defence against predators.

Figure 2.4 (next page) presents a phylogenetic analysis of primate social groups from DeCasien et al. (2017). In it, each purple symbol indicates a phylogenetically reconstructed solitary species. This reconstruction indicates that ancestral euprimates were solitary, as many of their strepsirrhine descendants are today.

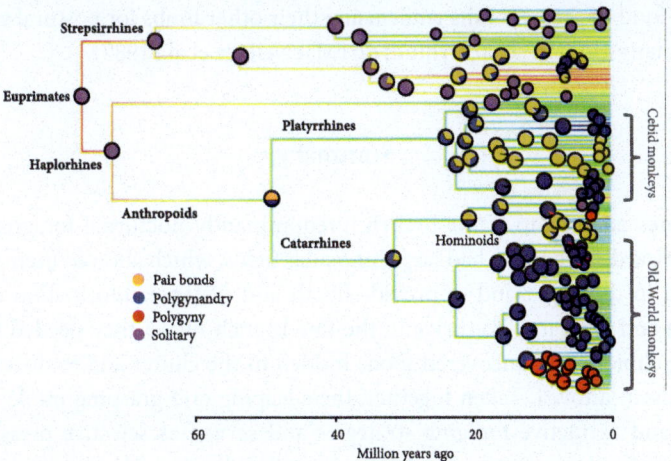

Figure 2.4 Phylogenetic reconstruction of primate social systems
Early euprimates, like most modern strepsirrhines led solitary lives (purple circles). Later, some lineages developed multi-male and multi-female groups, and others adopted pair-living or harems.

Reproduced from DeCasien, A.R., Williams, S.A., & Higham, J.P. Primate brain size is predicted by diet but not sociality. *Nat Ecol Evol*, 1 (5), 112, Doi: https://doi.org/10.1038/s41559-017-0112 Copyright © 2017, Springer Nature.

As Jenkins (1974) pointed out, euprimates adopted a life confined to the fine branches of trees rather than one that involved occasional entry into that niche, as many squirrel species do today. An entirely arboreal life presented early euprimates with several challenges in obtaining necessary resources. In a cluttered and mechanically unstable environment, their visual adaptations provided several advantages. They probably used vision to evaluate resources in relation to their current needs; and, of course, they also depended on vision for grasping food items and branches.

Stem primates lacked a fovea and most of them foraged nocturnally in dim light (Rose, 2006). As an earlier section mentions, their forward-facing eyes probably evolved for vision in dim light, in part by summation of inputs from the two eyes (Hughes, 1977; Allman, 2000). The improved stereopsis and depth perception that resulted from a large binocular field of vision also contributed to accuracy in reaching for and grasping food items and branches, and it could have permitted at least one eye to see around obstructions (Changizi, 2009).

Early euprimates also evolved a specialized mode of locomotion, which reflects life in the fine-branch niche. In euprimates, a hindlimb-dominated leaping and grasping form of locomotion generated less force on tree limbs. This adaptation decreased the rustling sounds that might attract predators, and it freed the forelimbs for other functions (Schmidt, 2010). The assumption is that the earliest euprimates brought food to their mouths while using their other limbs for postural stabilization, as many modern strepsirrhines do (MacNeilage et al., 1987).

Summary

Euprimates adapted to a fine-branch, predominantly nocturnal foraging niche. Their forward-facing eyes had large binocular fields, which allowed them to function well in dim light and to provide depth and distance information to guide reaching and grasping. To thrive in the fine-branch niche, they needed to make foraging choices that uncovered foods hidden in the clutter and to do so in dim light without a fovea. Taken together, their leaping and grasping mode of locomotion and extractive foraging strategies and served as selective pressures for enhanced visuomotor control and an improved ability to evaluate resources in relation to their current needs.

Paleoecology of the Anthropoids

After the divergence of strepsirrhines and haplorhines, the latter shifted from the predominantly nocturnal foraging habit of their euprimate ancestors to a diurnal life. However, tarsiers later reverted to foraging in dim-light conditions. Early haplorhines also developed a fovea, which led to improved visual acuity; and their retinas lost the reflective layer of cells called the tapetum lucidum, which is characteristic of nocturnal species (Fleagle, 1999; Heesy & Ross, 2004).

An extreme global cooling event of more than 2°C occurred ~34Ma, at the beginning of the Oligocene, after a long regime of steadily falling global temperatures. It seems likely that such dramatic climate changes made life difficult for anthropoids. There were probably widespread shortfalls in preferred food items.

After the platyrrhine–catarrhine split, catarrhines evolved the kind of three-colour (trichromatic) vision that humans have (De Valois & Jacobs, 1998; Williams et al., 2010). This 'routine' form of colour vision depends on the emergence of a novel gene, which codes for a third opsin pigment in retinal cone cells. Tarsiers developed a different form of trichromacy (Melin et al., 2013) based on genetic polymorphisms, and several platyrrhines have similar adaptations (Williams et al., 2010). One New World genus, howler monkeys (*Alouatta*) evolved the routine form of trichromacy convergently with catarrhines (Surridge & Mundy, 2002).

As anthropoids increased in size during the Miocene (23–5.3Ma), they required more energy from nutrients. Fossil evidence shows that early anthropoids weighed 100–300g, which is about the size of modern marmosets (Fleagle, 1999; Rose, 2006). Later anthropoids became larger, and like most modern anthropoids they weighed more than 1 kg, with many exceeding 10kg. Once larger anthropoid species appeared, they had adaptations in tooth morphology that indicate their diet consisted mainly of fruit (Williams et al., 2010) and tender leaves (Kirk & Simons, 2001; Dominy, 2004). Anthropoids have developed many specializations for extractive foraging. For example, in addition to exploiting ripe fruits, some species specialize in harvesting seeds from unripe fruits that have tough outer cortices.

Because of the dearth of resources, anthropoids needed to travel farther and diversify their feeding habits in order to obtain necessary resources. Larger animals have many advantages in foraging over long distances: greater energy reserves, more power for overcoming obstacles, and protection from predators that feed only on small animals. Some evidence suggests that folivory intensified at this time (Kirk & Simons, 2001; Dominy, 2004), which indicates a shift to fallback foods.

Dominy (2004) has proposed that trichromacy evolved in response to this shift to folivory. Colour discrimination among red and green hues can help foraging primates select the most tender and nutritious leaves, which have more red and less green than older, tougher leaves.

As larger animals, the anthropoids abandoned the leaping and grasping form of locomotion of early primates. Instead, they became arboreal quadrupeds, moving through the larger tree branches using all four limbs (Fleagle, 1999; Schmidt, 2010). Long, arboreal journeys make high demands on energy reserves (Janson, 1988), especially when elevation changes during movement, as it inevitably does. Accordingly, their large size and mode of movement required a high energy intake. These evolutionary changes meant that the ancestral anthropoids had to exploit a large home range, as most modern species do (Martin, 1981).

An especially important selective factor resulted from the fact that daylight foraging put these animals at risk for predation. So, the anthropoids needed to watch for predators during and between bouts of foraging. The fovea that anthropoids inherited from the haplorhine ancestors presumably helped them detect and identify predators at a distance, but the risk of predation remained much greater than for their nocturnal ancestors. The development of larger social groups mitigated predation threats, but the anthropoids and their descendants faced an influx of new predators and so always faced serious treats.

After the early Oligocene cooling had altered the ecosystem, new challenges emerged for both New and Old World anthropoids during the Miocene (23–5.3Ma). As Simons et al. (2007) put it, the fossil record shows that the 'catarrhines likely faced new demands . . . after the influx of relatively large-brained . . . competitors and predators into Afro-Arabia during the early Miocene, whereas platyrrhines likely experienced similarly strong selection pressures after their arrival on a landmass (South America) with new competitors, predators, and seasonal patterns'.

Competition with other diurnal animals also intensified. Modern anthropoid primates compete with fruit bats, squirrels, and frugivorous birds (Oates, 1987). Birds can fly directly to their food, and primates cannot. Ancestral anthropoids also had a limited amount of time to obtain their food because of heat stress in their tropical habitats. Accordingly, they probably foraged soon after awakening in the morning and just before their nightly rest period (Oates, 1987). So foraging periods probably remained restricted to the cooler parts of the day. Predators surely learned about these activity patterns.

As larger animals, crown anthropoids came to rely on the energy-rich products of angiosperm trees. However, in exploiting this resource, they faced several problems. One involved the patchy distribution of their preferred foods, dispersed throughout a large home range. They also had to cope with seasonal variations in plant life and other periodicities. Each kind of tree had a characteristic fruiting pattern, usually involving inconsistent fruiting intervals and dramatic variation during the year and from year to year (Chapman et al., 1999; Janmaat et al., 2006). Given that anthropoids cannot eat or digest much of the plant matter in the tropics, the volatility in preferred-food availability probably led to frequent and unpredictable food shortages.

Ancestral anthropoids surmounted the challenges of resource volatility to become a diverse and successful group, including more than 260 modern species of monkeys, not to mention the apes and humans. But abrupt climate change probably made life more demanding.

To understand how the anthropoids coped with these challenges, there are field observations that bear on the foraging strategies of modern anthropoids. They illustrate various ways that anthropoids cope with the problem of resource volatility.

The daily life of one anthropoid species illustrates the importance of the fovea. Struhsaker (1980) reported on a field study of red-tail monkeys (*Cercopithecus ascanius*), a catarrhine species. He found that they spend 21% of each day scanning their visual world and 17% of the day travelling from place to place based on what they have seen. Once they reach a fruitful location, foraging takes up 34% of their day. Taken together, the time spent looking around and acting on what they see amounts to more than 70% of their active time. By contrast, they spend about 10% of their time on social activities such as grooming.

There is evidence that anthropoids are able to distinguish which trees are most desirable, including species that look identical to all but the most expert anthropoid eyes (Zuberbuhler & Janmaat, 2010). They enter trees more frequently when they have previously eaten high-quality fruit there (Janmaat et al., 2006).

Monkeys of various species appear to monitor and remember a tree's previous food-production history in order to predict future food availability (Milton, 1988); for example, mangabeys (*Lophocebus albigena*) move fastest to trees with higher quantities of fruit (Zuberbuhler & Janmaat, 2010). Mangabeys also predict food availability based on the weather. Heat speeds the production and ripening of fruits, and mangabeys return more quickly to a previously productive tree in hot weather (Janmaat et al., 2006).

During periods of scarcity, anthropoids that depend on fruits and insects tend to increase their foraging time, spend less of their time resting, and compete with each other more (Kavanagh, 1978; Oates, 1987). The ability to adopt alternative foraging strategies allows anthropoids to overcome irregular shortages in preferred foods. Rather than continuing to try to exploit previous foods, they can explore new ones.

Summary

As anthropoids became larger animals during the Miocene, they encountered increased predation risks as they travelled long distances on energetically demanding foraging excursions in daylight. When anthropoids moved out of the fine-branch niche, they not only faced a higher risk of predation, but they also needed to obtain more nutrition to support their larger bodies. They also faced competition from other members of their social group, other primate species, and frugivorous birds. Given the volatility of the resources on which they depended, it paid to be flexible, exploring new foods when old ones were no longer available.

Upward Grade-Shifts

Fossil evidence shows that primate brains underwent several upward grade-shifts in size. One such shift occurred in stem euprimates and another in the early anthropoids. For anthropoids, some of these developments occurred after the platyrrhine–catarrhine split and again during the evolution of the great apes (*Pongidae*).

In biology, the term 'grade' refers to species or systems that have a comparable level of complexity or size. As applied to the brain, the concept of a grade has traditionally referred to the size of the brain or the neocortex in relation to the size of the body, but many other measures can also establish grades.

This can be illustrated by considering the neocortex. Passingham (1981) plotted the volume of the neocortex against the volume of the whole brain on a log–log plot, using data from primates, carnivores, ungulates, and rodents. The regression line for modern primates had a higher *y*-intercept than regressions for these other mammals. Accordingly, if we compare a primate brain with a similarly sized brain of a carnivore, an ungulate, or a rodent, the primate brain has a larger volume of neocortex. This observation means that primates have evolved a new grade of neocortex size compared to these other mammals.

The term 'corticalization' refers to the percentage of the brain taken up by cortex, and it usually refers specifically to the neocortex. So, another way of expressing the finding is to say that primates evolved a new grade of corticalization. So did cetaceans (Wright et al., 2017) which did so convergently.

The same concept applies to other brain measures. For example, Herculano-Houzel et al. (2015) measured the density of neurons in the neocortex in modern mammals. They found that primates also reached a new grade in this regard, with a higher density of neurons per unit area.

The size of the neocortex, and indeed the brain as a whole, depends on several factors. As Jerison (1973) pointed out, the size of the brain and the weight of the body have a close relationship to each other. This correlation occurs partly because of the embodied nature of the brain: sensory inputs and motor outputs both involve mappings between the brain and the body. But brain size also reflects an additional factor, which reveal something about how a species has adapted to its ecological circumstances. This is an adaptive factor.

To measure this adaptive factor, Jerison introduced the encephalization quotient (EQ). This term refers to the difference for the size of the brain for a species as compared with the average brain size for a mammal that is matched for body size. So, the EQ can point to a difference in grade, that reflects the specific adaptation of a species.

As Chapter 9 points out, it is specifically the homotypical or association areas that increase in size according to the demands of the ecological niche to which the species is adapted. The reason is that it is these areas that are critical for learning, and thus for ensuring that the species is capable of acquiring the knowledge and skills that are necessary for survival in its particular niche.

Euprimate Grade Shift

According to the fossil record, several grade-shifts in brain size occurred during the evolution of primates, many more than have been recognized previously. An early one occurred during the major evolutionary transition from stem primates to the first euprimates.

The top section of Figure 2.5 reproduces two virtual fossil endocasts from Long et al. (2015). Both come from stem primates, one for a genus called *Microsyops* and the other for *Ignacius*. The most obvious feature of the brains of stem primates is that they looked more like rodent brain than like modern primates' brains. They have the same convex curvature and lissencephalic surface, except for a shallow rhinal sulcus. The bottom section of Figure 2.5 shows a drawing of the brain of a laboratory rat for comparison. The visual cortex of these stem primates was small enough to expose part of the superior colliculus.

The middle section of Figure 2.5 illustrates a virtual endocast of the extinct euprimate *Rooneyia*. This could either be an early haplorrhine or a strepsirrhine (Ni

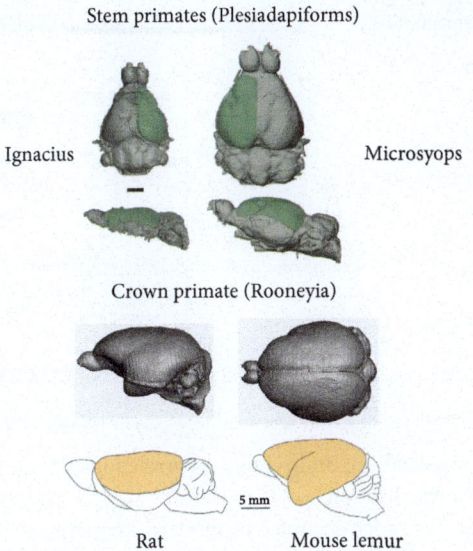

Figure 2.5 Virtual brain endocasts of Stem and Crown primates
(Top) Two extinct stem primates (plesiadapiforms): *Ignacius* and *Microsyops*, with the neocortex shaded green. (Middle) An extinct euprimate, *Rooneyia*. (Bottom) Drawings of the brains of a laboratory rat (left) and a mouse lemur (right) for comparison. Note that the scales differ among the different parts of this figure, so they say nothing about relative brain size.

(Top) Reproduced from Long, A., Bloch, J.I., & Silcox, M.T. Quantification of neocortical ratios in stem primates. *Am J Physic Anthropol*, 157 (3), 363–73, Figure 3a, b, and d, Doi: https://doi.org/10.1002/ajpa.22724 © 2015 Wiley Periodicals, Inc.

(Middle) Reproduced from Kirk, E.C., Daghighi, P., Macrini, T.E., Bhullar, B.S., & Rowe, T.B. Cranial anatomy of the Duchesnean primate Rooneyia viejaensis: New insights from high resolution computed tomography. *J Hum Evol*, 74, 82–95, Figure 1a, b, and d, Doi: 10.1016/j.jhevol.2014.03.007 Copyright © 2014 Elsevier Ltd. All rights reserved.

(Bottom) Reprinted from Andrew C. Halley & Leah Krubitzer. Not all cortical expansions are the same: The coevolution of the neocortex and the dorsal thalamus in mammals. *Curr Opin Neurobiol*, 56, 78–86, Figure 3a, https://doi.org/10.1016/j.conb.2018.12.003 © 2018, Elsevier Ltd. All rights reserved.

et al., 2013; Ramdarshan & Orliac, 2016). The brain of *Rooneyia* had the characteristic shape of modern primate brains, with a prominent temporal lobe and a lateral sulcus. For example, its brain strongly resembled that of modern strepsirrhines. The bottom section of Figure 2.5 shows the drawing of the brain of a mouse lemur (*Microcebus*) for comparison. *Rooneyia*'s brain differed from most modern primate brains in that only the lateral sulcus appeared as a prominent landmark on the lateral surface of the hemisphere.

Figure 2.6A (next page) displays the relationship between the size of the brain and body mass for modern strepsirrhines; the values all lie within the outlined

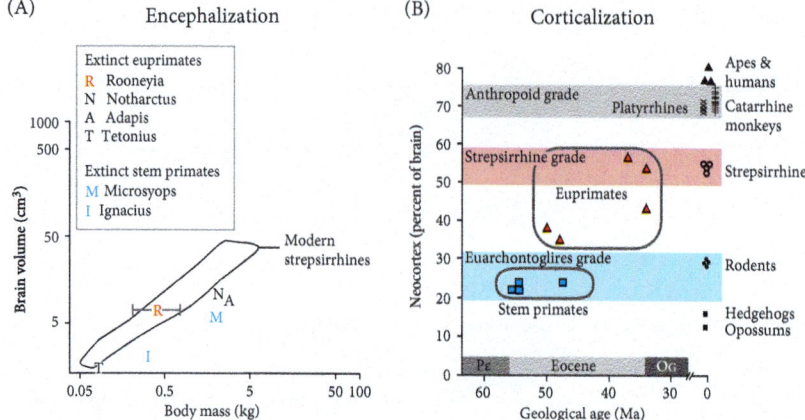

Figure 2.6 (A) Encephalization: Brain–body mass relationship in *Rooneyia* and other fossil primates. (B) Corticalization
(A) The outline shows the range occupied by modern strepsirrhine brains. (B) The percent of a brain consisting of neocortex in fossil primates. Values to the right come from selected modern mammals. Coloured shading indicates three grades of corticalization. Abbreviations: Pε, Palaeocene; OG, Oligocene

Drawing adapted from Wise, S.P. The evolution of the prefrontal cortex in early primates and anthropoids In Krubitzer, L., Kaas, J. (eds) *The Evolution of Nervous Systems, Second Edition, Volume 3*, pp. 387–422, Doi: https://doi.org/10.1016/B978-0-12-804042-3.00092-0 Copyright © 2017, Elsevier Inc. All rights reserved.

A: Data from Kirk, E.C., Daghighi, P., Macrini, T.E., Bhullar, B.S., & Rowe, T.B. Cranial anatomy of the Duchesnean primate Rooneyia viejaensis: New insights from high resolution computed tomography. *J Hum Evol*, 74, 82–95, Doi: 10.1016/j.jhevol.2014.03.007, 2014.

B: Data from Long, A., Bloch, J.I., & Silcox, M.T. Quantification of neocortical ratios in stem primates. *A J Phys Anthropol*, 157, 363–73, 2015.

area. The figure also includes the values for two extinct stem primates (in blue font) and four extinct euprimates (three in black font and one, *Rooneyia*, in red). Of the fossil primates, *Rooneyia* fell within the encephalization range typical of modern strepsirrhines. Harrington et al. (2016) have made virtual endocasts of *Notharctus* and *Adapis*, two other euprimates. The values of these fall just below the range of modern strepsirrhines.

These observations imply that the euprimate entry into the strepsirrhine grade of brain size occurred long after the divergence of euprimates from stem primates.

Figure 2.6B shows corticalization values in fossil primates (Long et al., 2015), with corresponding data for selected modern species to the right. The neocortex of stem primates took up ~20–25% of their brains, a little less than modern rodents and a little more than hedgehogs and opossums. Figure 2.6B suggests that euprimate brains expanded into the modern strepsirrhine range in the middle to late Eocene (~45–34 Ma) (Harrington et al., 2016). For the first time, the neocortex of a primate brain occupied more that 50% of the brain's volume.

Anthropoid Grade-Shift

During the evolution of crown anthropoids, their brains underwent another upward grade-shift in size, both in terms of encephalization and corticalization.

Figure 2.7 (next page) addresses encephalization. Taking body mass into account, as in Figure 2.7, most modern anthropoid brains (marked with plus signs) are larger than most modern strepsirrhine brains (marked with green triangles).

Of course, the study of modern anthropoids cannot tell us when this lineage stepped up in EQ, but fossil evidence can. Figure 2.7 presents data indicating that Miocene anthropoids had smaller brains than modern species. The fossil anthropoids are shown in red font. For example, both *Aegyptopithecus*, a catarrhine primate dated to ~30Ma (Seiffert, 2006), and *Simonsius* (*Parapithecus*) (Bush et al., 2004), which lived as much as 3 million years earlier, had brains that fell within the strepsirrhine range (Beard et al., 2016). Simons et al. (2007) commented that '*Aegyptopithecus* was ... at best strepsirrhine-like, and perhaps even non-primate-like, in its brain-to-body mass relationship'. Figure 2.7 includes data for three specimens of this species, each marked with a red 'A'.

An extinct platyrrhine from ~20 Ma, *Chilecebus*, also had a small brain (Sears et al., 2008) as did another stem platyrrhine, *Homunculus* (Kay et al., 2013). These species lived during the Miocene (from 23 Ma), long after the platyrrhine–catarrhine divergence in the Eocene (~46 Ma). These two Miocene platyrrhines and two others, *Tremacebus* and *Dolichocebus*, all had relatively low EQs, with at least *Homunculus* being in the strepsirrhine range (Kay et al., 2013; Halenar-Price & Tallman, 2019). The fact that both platyrrhine and catarrhine specimens of the Miocene had relatively small brains means that the upward grade-shift in brain size occurred convergently in these two taxa.

A later Old World catarrhine, *Victoriapithecus*, lived ~15 Ma (Gonzales et al., 2015). This anthropoid species had a brain only slightly larger than in *Aegyptopithecus*. In Figure 2.7, the values for its brain–body mass are designated with two red Vs, based on the range of estimated values for the size of the body. This observation suggests that at least part of the upward grade-shift in brain size occurred after ~15 Ma in catarrhines.

Gonzales et al. (2015) concluded that brain expansion occurred independently several times in catarrhines: in great apes (*Pongidae*), which diverged from other apes ~18–20 Ma, and separately in both cercopithecines and colobines, two lineages of Old World monkeys. The earliest endocast for an ape, *Proconsul*, dates to ~17–18 Ma. This specimen had a large endocranial volume, as well as an expanded frontal lobe and reduced olfactory bulbs (Gonzales et al., 2015). In Figure 2.7, its brain–body mass value is indicated with two blue Ps, which capture a range of values. Taken together, these findings demonstrate that at least some brain expansion occurred earlier and independently in the hominoid line than in Old World monkeys.

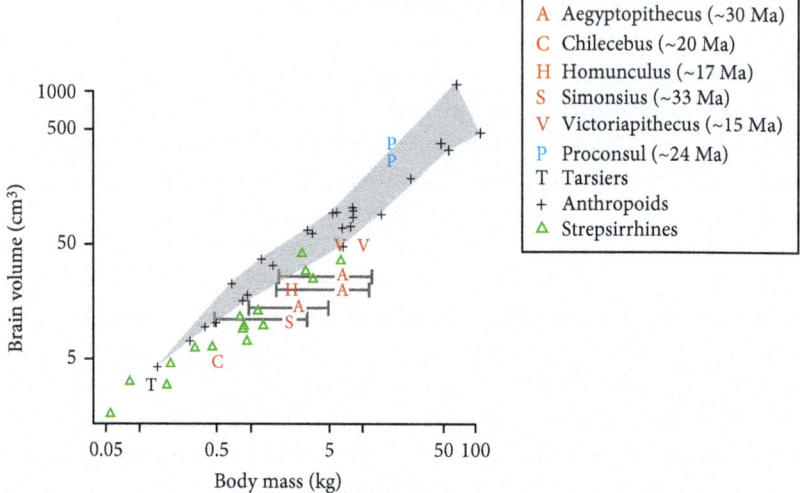

Figure 2.7 Brain–body mass relationship for modern strepsirrhines, modern anthropoids, fossil anthropoids (red letters), and a fossil pongid (blue letters)

Reproduced from Bush, E.C., Simons, E.L., & Allman, J.M. High-resolution computed tomography study of the cranium of a fossil anthropoid primate, Parapithecus grangeri: New insights into the evolutionary history of primate sensory systems, *Anat Rec. Part A, Disc Mol Cell Evol Biol*, 281a (1), 1083–7, Figure 1b, Doi: 10.1002/ar.a.20113 Copyright © 2004 Wiley-Liss, Inc.

Error bars reproduced from Simons, E.L., Seiffert, E.R., Ryan, T.M. & Attia, Y., 'A remarkable female cranium of the early Oligocene anthropoid Aegyptopithecus zeuxis (Catarrhini, Propliopithecidae)', Proceedings of the National Academy of Sciences, 104 (21), pp. 8731-8736, DOI: https://doi.org/10.1073/pnas.0703129104 Copyright © 2007, National Academy of Sciences.

The upward grade-shift in brain size, relative to mammals as a whole or relative to strepsirrhines, has produced high EQs in modern anthropoids. Figure 2.8 (next page) comes from Boddy et al. (2012), and it uses a colour code with each wedge consisting of a sorted stack of curved lines, one for each species in a given taxon. An EQ of 1, that is the EQ of an average mammal, lies just before the border between the navy blue and light blue.

This analysis shows that primates have many species with high EQs, as shown by the yellow and red segments on the primate wedge. Almost all of these species are anthropoids.

Cetaceans (whales and dolphins) and carnivores (such as cats and bears) have also developed large brains relative to body mass, but not to the same extent as in anthropoids. However, several cetacean species have attained a very high level of corticalization, matching the ~80% level of modern humans (Wright et al., 2017).

DeCasien et al. (2017) have performed a phylogenetic analysis on a large and improved database for primate EQ values. In the illustration of DeCasien et al. (2017), presented in Figure 2.9 (page 58), red lines indicate relatively low-EQ

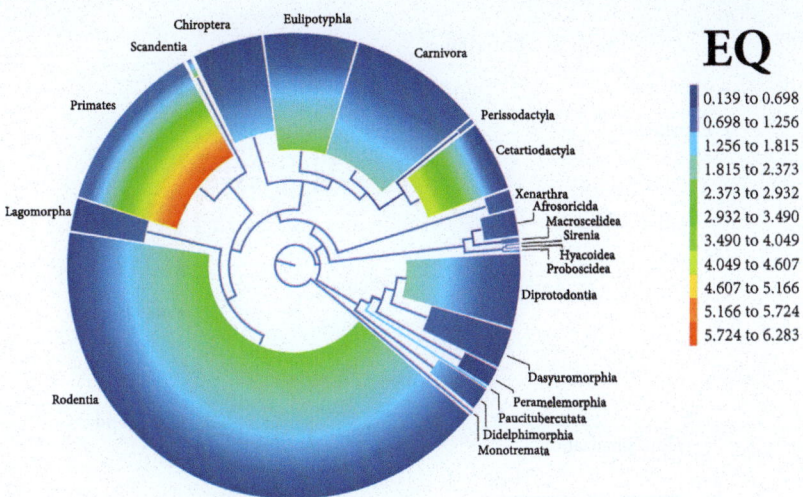

Figure 2.8 Circular cladogram showing a stack of encephalization quotients (EQs) for a series of species in each mammalian lineage, ordered by EQ value
Red indicates high EQ values; blue indicates low EQs.

Reproduced from Boddy, A.M., McGowen, M.R., Sherwood, C.C., Grossman, L.I., Goodman, M., & Wildman, D.E. Comparative analysis of encephalization in mammals reveals relaxed constraints on anthropoid primate and cetacean brain scaling, *J Evol Biol*, 25 (5), 981–94, Figure 2a, https://doi.org/10.1111/j.1420-9101.2012.02491.x © 2012 The Authors. Journal of Evolutionary Biology © 2012, European Society for Evolutionary Biology.

values and blue lines indicate high values. The highest EQs are for the great apes and cebus monkeys.

Recently, DeCasien et al. (2017) studied a large sample of primates, controlling for body size and phylogeny. They found that diet predicted brain size but that several different measures of social complexity did not. Powell et al. (2017) reached the same conclusion. Brain size correlates with diet, home-range size, and activity period rather than with measures of group size. These phylogenetic analyses point to the demands of foraging as providing the principal driving force for the upward grade-shift in anthropoid brain size, although this conclusion might not apply to platyrrhines (Halenar-Price & Tallman, 2019).

In light of the evidence from anthropoid paleoecology, it is likely that it was the combination of and unpredictable shortfalls in preferred foods and predation risks that drove the increase in brain size in anthropoids. However, Lehman and Dunbar (2009) measured the social complexity and social cohesion of the females in a number of Old World primates. They found that species with closely bonded grooming clans has a large proportion of neocortex in their brains. The implication is that social factors may have influenced the degree of corticalization in anthropoid brains.

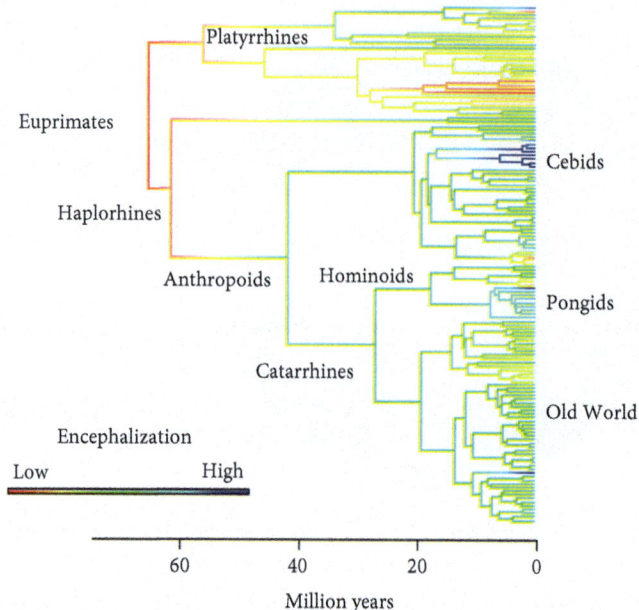

Figure 2.9 Reconstructed phylogeny of encephalization quotients (EQs) for primates Blue lines indicate high EQ values; red lines indicate low EQ values.

Adapted from DeCasien, A.R., Williams, S.A., & Higham, J.P. Primate brain size is predicted by diet but not sociality. *Nat Ecol Evol*, 1 (5), Article 112, Figure 1 (left), Doi: 10.1038/s41559-017-0112 Copyright © 2017, Springer Nature.

Frontal Expansion in Anthropoids

The evidence from endocasts bears on the size of the frontal lobe. Unfortunately, we do not know of any quantitative data on this point, but we can quote the subjective assessments of palaeontologists who have studied these fossils.

They suggested that the extinct anthropoids, *Chilecebus*, *Homunculus*, *Aegyptopithecus*, and *Victoriapithecus*, all had small frontal lobes. The previous section mentions that the first two were platyrrhines, the other two catarrhines. Accordingly, these observations date most of the expansion of the frontal lobes to sometime after the divergence of New World and Old World primates.

Figure 2.10A (next page) reproduces a virtual endocast of *Victoriapithecus*, dated ~15 Ma. Its frontal lobes were small, even though the rest of the brain appears to be similar to that of modern catarrhines, for example in its pattern of gyri. Nonetheless, Gonzales et al. (2015) identified both the arcuate sulcus and principal sulcus, features of the granular PF cortex which modern macaques have.

The fact that *Victoriapithecus* had relatively small frontal lobes means that the sulcal pattern preceded a least some of the frontal-lobe expansion that occurred in

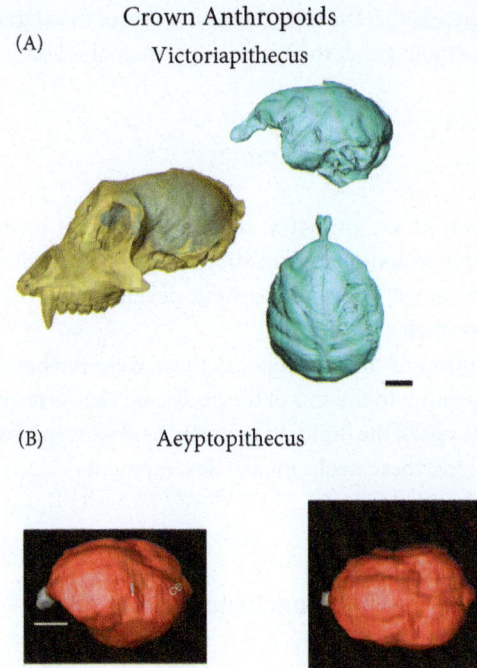

Figure 2.10 Virtual brain endocast of extinct catarrhines
(A) *Victoriapithecus*, an extinct catarrhine that had a small frontal lobe compared with modern macaques but was similar in having both the arcuate and principal sulci. (B) *Aegyptopithecus*, an extinct catarrhine without these sulci.

A: Reproduced from Gonzales, L.A., Benefit, B.R., McCrossin, M.L., & Spoor, F. Cerebral complexity preceded enlarged brain size and reduced olfactory bulbs in Old World monkeys. *Nat Commun*, 6, (7580), Figure 1a, b, and d, https://doi.org/10.1038/ncomms8580 Copyright © The Authors. Licensed under CC BY 4.0.

B: Reproduced from Simons, E.L., Seiffert, E.R., Ryan, T.M., & Attia, Y. A remarkable female cranium of the early Oligocene anthropoid Aegyptopithecus zeuxis (Catarrhini, Propliopithecidae), *Proc Nat Acad Scienc USA*, 104 (21), 8731–6, Figure 3b and e, Doi: https://doi.org/10.1073/pnas.0703129104 Copyright © 2007, The National Academy of Sciences.

catarrhine monkeys. Earlier catarrhines, such as *Aegyptopithecus* and *Simonsius*, lacked these sulci, as the bottom part of Figure 2.10B illustrates.

The expansion of visual cortex seems to have reached the range of modern anthropoids by the time of *Aegyptopithecus*, ~30 Ma (Radinsky, 1975; Simons et al., 2007). So, the relatively recent expansion of the homotypical cortex was probably the most significant factor in the upward grade-shift in later anthropoids.

A New World species, capuchin monkeys (*Cebus apella*), has both arcuate and principal sulci. But the fossil evidence shows that these are not homologous with the sulci in macaques; they arose convergently (Ni et al., 2019). Only the central sulcus had evolved by the time of the last common ancestor of New and Old World primates. The evidence is that the last common ancestor of catarrhines and platyrrhines had a brain without either the arcuate or principal sulcus.

Summary

During the evolution of the primates, there were several upward grade-shifts. Sometime during the evolutionary transition to euprimates, the brain expanded into the strepsirrhine range. These phylogenetic developments played a pivotal role in the success of euprimates.

During the evolution of the anthropoids there were further grade-shifts. The brain increased in relation to the size of the body, and the cortex increased in relation to the size of the rest of the brain. In parallel, there were increases in the frontal lobes. To some degree, these evolutionary developments occurred in parallel in New and Old World primates.

Conclusions

To understand the selective advantage that the PF cortex confers, we need to appreciate that it evolved in animals that were specialized for feeding by picking fruit and leaves. This involves the coordination of eye and hand which depends on parietal and premotor areas, as well as the FEF.

The first granular PF areas probably evolved in an ancestor common to both primates and tree shrews. The first euprimates exploited the fine branch niche. They developed a leaping and grasping form of locomotion, in which the legs produced the propulsion as well as clinging onto branches while the hands were used for foraging. Unlike a paw, the hand is capable of skilled movements, selecting small items from the tips of branches.

Because fruit and leaves are not edible until ripe, there was a need to choose between items within a patch. The situation was very different from that faced by ungulates feeding on the grasslands or carnivores chasing individual animals. To forage in a patch, it is necessary to search the display, looking for those items that are edible.

When monkeys and apes evolved, they also became diurnal, developing a fovea and some form of trichromacy. This enabled them to distinguish which fruit and leaves they could eat on the basis of their colour and shape. At the same time, the foods they ate were volatile, and there was a necessity for flexibility, exploiting current resources when available or exploring new ones in times of hardship.

The anthropoids were also at greater risk of predation as they came increasingly into the open. These selection pressures led to a further expansion of the brain and comparative evidence suggests that this was accompanied by the development of new, lightly myelinated, granular PF areas.

Studying fossil endocasts cannot tell us what the different regions of the PF cortex do, but studies of modern primates allows us to do this. The next section reviews the functions of five subareas of the PF cortex. The medial PF cortex supports the spontaneous generation of the search, based on the memory of where the animal has searched before (Chapter 3). The orbital PF cortex evaluates the foods in terms of the current needs of the animal (Chapter 4). The caudal PF cortex moves the eyes to the target (Chapter 5). The dorsal PF cortex plans the order of the search, specifying the targets that are currently desirable (Chapter 6). Finally, the ventral PF cortex chooses the food items that are suitable on the basis of their type or category (Chapter 7).

References

Akert, K. (1964) Comparative anatomy of frontal cortex and thalamofrontal connections. In Warren, J.M., Akert (eds) *The Frontal Granular Cortex*. McGraw-Hill, New York, pp. 372–96.

Allman, J. (2000) *Evolving Brains*. Freeman, New York.

Baldwin, M.K., Cooke, D.F., & Krubitzer, L. (2017a) Intracortical microstimulation maps of motor, somatosensory, and posterior parietal cortex in tree shrews (*Tupaia Belangeri*) reveal complex movement representations. *Cereb Cortex*, 27, 1439–56.

Baldwin, M.K.L., Balaram, P., & Kaas, J.H. (2017b) The evolution and functions of nuclei of the visual pulvinar in primates. *J Comp Neurol*, 525, 3207–26.

Beard, K.C., Coster, P.M., Salem, M.J., Chaimanee, Y., & Jaeger, J.J. (2016) A new species of Apidium (*Anthropoidea, Parapithecidae*) from the Sirt Basin, central Libya: First record of Oligocene primates from Libya. *J Hum Evol*, 90, 29–37.

Bloch, J.I. & Boyer, D.M. (2002) Grasping primate origins. *Science*, 298, 1606–10.

Boddy, A.M., McGowen, M.R., Sherwood, C.C., Grossman, L.I., Goodman, M., & Wildman, D.E. (2012) Comparative analysis of encephalization in mammals reveals relaxed constraints on anthropoid primate and cetacean brain scaling. *J Evol Biol*, 25, 981–94.

Boyer, D.M., Seiffert, E.R., Gladman, J.T. & Bloch, J.L. (2013a) Evolution and allometry of calcaneal elongation in living and extinct primates. *PLoS One*, 8, e6772.

Boyer, D.M., Yapuncich, G.S., Chester, S.G., Bloch, J.I., & Godinot, M. (2013b) Hands of early primates. *Am J Phys Anthropol*, 152 (Suppl 57), 33–78.

Brown, V.J. & Bowman, E.M. (2002) Rodent models of prefrontal cortical function. *Trends Neurosci*, 25, 340–43.

Bruce, C.J., Goldberg, M.E., Bushnell, M.C., & Stanton, G.B. (1985) Primate frontal eye fields. II. Physiological and anatomical correlates of electrically evoked eye movements. *J Neurophysiol*, 54, 714–34.

Burton, A.C., Kashtelyan, V., Bryden, D.W., & Roesch, M.R. (2014) Increased firing to cues that predict low-value reward in the medial orbitofrontal cortex. *Cereb Cortex*, 24, 3310–21.

Bush, E.C. & Allman, J.M. (2004) The scaling of frontal cortex in primates and carnivores. *Proc Natl Acad Sci USA*, 101, 3962–6.

Bush, E.C., Simons, E.L., & Allman, J.M. (2004) High-resolution computed tomography study of the cranium of a fossil anthropoid primate, Parapithecus grangeri: New insights into the evolutionary history of primate sensory systems. *Anat Rec A Discov Mol Cell Evol Biol*, 281, 1083–7.

Carlen, M. (2017) What constitutes the prefrontal cortex? *Science*, 358, 478–82.

Carmichael, S.T. & Price, J.L. (1994) Architeconic subdivision of the orbital and medial prefrontal cortex in the macaque monkey. *J Comp Neurol*, 346, 366–402.

Carmichael, S.T. & Price, J.L. (1995) Limbic connections of the orbital and medial prefrontal cortex in macaque monkeys. *J Comp Neurol*, 363, 615–41.

Changizi, M. (2009) *The Visual Revolution*. Barbella, Dallas.

Chapman, C.A., Wrangham, R.W., Chapman, L.J., Kennard, D.K., & Zanne, A.E. (1999) Fruit and flower phenology at two sites in Kibale National Park, Uganda. *J Tropic Ecol*, 15, 189–211.

Chester, S.G.B., Williamson, T.E., Bloch, J.I., Silcox, M.T., & Sargis, E.J. (2017) Oldest skeleton of a plesiadapiform provides additional evidence for an exclusively arboreal radiation of stem primates in the Palaeocene. *R Soc Open Sci*, 4, 170329.

Choi, E.Y., Tanimura, Y., Vage, P.R., Yates, E.H., & Haber, S.N. (2017) Convergence of prefrontal and parietal anatomical projections in a connectional hub in the striatum. *Neuroimage*, 146, 821–32.

Cooke, D.F., Goldring, A., Recanzone, G.H., & Krubitzer, L. (2014) The evolution of. parietal areas associated with visuomanual behavior: From grasping to tool use. In Chalupa, L.M., Werner, J. (eds) *The Visual Neurosciences*. MIP Press, Cambridge, MA.

DeCasien, A.R., Williams, S.A., & Higham, J.P. (2017) Primate brain size is predicted by diet but not sociality. *Nat Ecol Evol*, 1, 112.

deValois, R.L. & Jacobs, G.H. (1998) Primate color vision. *Science*, 162, 533–40.

Dominy, N.J. (2004) Fruits, fingers, and fermentation: The sensory cues available to foraging primates. *Integr Comp Biol*, 44, 295–303.

Dunbar, R.I. (2009) The social brain hypothesis and its implications for social evolution. *Ann Hum Biol*, 36, 562–72.

Felleman, D.J. & van Essen, D.C. (1991) Distributed hierachical processing in primate cerebral cortex. *Cer Cort*, 1, 1–47.

Ferry, A.T., Ongur, D., An, X., & Price, J.L. (2000) Prefrontal cortical projections to the striatum in macaque monkeys: Evidence for an organization related to prefrontal networks. *J Comp Neurol*, 425, 447–70.

Fitzpatrick, D.C., Suga, N., & Olsen, J.F. (1998) Distribution of response types across entire hemispheres of the mustached bat's auditory cortex. *J Comp Neurol*, 391, 353–65.

Fleagle, J.G. (1999) *Primate Adaptation and Evolution*. Academic Press, San Diego.

Frysztak, R.J. & Neafsey, E.J. (1994) The effect of medial frontal cortex lesions on cardiovascular conditioned emotional responses in the rat. *Brain Res*, 643, 181–93.

Gerbella, M., Belmalih, A., Borra, E., Rozzi, S., & Luppino, G. (2010) Cortical connections of the macaque caudal ventrolateral prefrontal areas 45A and 45B. *Cereb Cortex*, 20, 141–68.

Giguere, M. & Goldman-Rakic, P.S. (1988) Mediodorsal nucleus: Areal, laminar and tangential distribution of afferents and efferents in the frontal lobe of the rhesus monkey. *J Comp Neurol*, 277, 195–213.

Goldman, P.S., Rosvold, H.E., Vest, B., & Galkin, T.W. (1971) Analysis of the delayed-alternation deficit produced by dorsolateral prefrontal lesions in the rhesus monkey. *J Comp Physiol Psychol*, 77, 212–20.

Gonzales, L.A., Benefit, B.R., McCrossin, M.L., & Spoor, F. (2015) Cerebral complexity preceded enlarged brain size and reduced olfactory bulbs in Old World monkeys. *Nat Commun*, 6, 7580.

Halenar-Price, L. & Tallman, M. (2019) Investigating the effect of endocranial volume on cranial shape in platyrrhines and the relevance of this relationship to interpretations of the fossil record. *Am J Phys Anthropol*, 169, 12–30.

Hall, R.D. & Lindholm, E.P. (1974) Organization of motor and somatosensory neocortex in the abbino rat. *Brain Res*, 66, 23–38.

Harrington, A.R., Silcox, M.T., Yapuncich, G.S., Boyer, D.M., & Bloch, J.I. (2016) First virtual endocasts of adapiform primates. *J Hum Evol*, 99, 52–78.

Heesy, C.P. & Ross, C.F. (2004) Mosaic evolution of activity patterns, diet, and color vision in haplorhine primates. In Ross, C.F., Kay, R.F. (eds) *Anthropoid Origins*. Plenum, New York, pp. 665–698.

Heilbronner, S.R., Rodriguez-Romaguera, J., Quirk, G.J., Groenewegen, H.J., & Haber, S.N. (2016) Circuit-based corticostriatal homologies between rat and primate. *Biol Psychiatry*, 80, 509–21.

Herculano-Houzel, S., Catania, K., Manger, P.R., & Kaas, J.H. (2015) Mammalian brains are made of these: A dataset of the numbers and densities of neuronal and nonneuronal cells in the brain of Glires, primates, Scandentia, Eulipotyphlans, Afrotherians and artiodactyls, and their relationship with body mass. *Brain Behav Evol*, 86, 145–63.

Hughes, A. (1977) The topography of vision in mammals of contrasting lifestyle: Comparative optics and retinal organization. In Crescitelli, F. (ed.) *The Visual System of Vertebrates*. Springer, New York, pp. 615–697.

Janmaat, K.R., Byrne, R.W., & Zuberbuhler, K. (2006) Primates take weather into account when searching for fruits. *Curr Biol*, 16, 1232–7.

Janson, C.H. (1988) Food competition in brown capuchin monkeys (*Cebus apella*): Quantitative effects of group size and tree productivity. *Beh*, 105, 53–76.

Jenkins, F.A. (1974) Tree shrew locomotion and the origin of primate arborealism. In Jenkins, F.A. (ed.) *Primate Locomotion*. Academic Press, New York, pp. 85–115.

Jerison, H. (1973) *Evolution of the Brain and Intelligence*. Academic Press, New York.

Kaada, B.R. (1960) Cingulate, posterior orbital, anterior insular and temporal pole cortex. In Field, J. (ed.) *Handbook of Physiology*. American Physiological Society, Bethesda, pp. 1345–72.

Kaada, B.R., Pribram, K.H., & Epstein, J.A. (1949) Respiratory and vascular responses in monkeys from temporal pole, insula, orbital surface and cingulate gyrus; a preliminary report. *J Neurophysiol*, 12, 347–56.

Kaas, J. (2006a) The evolution of visual cortex in primates. In Kremers, J. (ed.) *The Structure, Function and Evolution of Primate Visual Systems*. Wiley, New York.

Kaas, J.H. (2006b) Evolution of the neocortex. *Curr Biol*, 16, R910–14.

Kavanagh, M. (1978) The diet and feeding behaviour of Cercopithecus aethiops tantalus. *Folia Primatol* (Basel), 30, 30–63.

Kay, R.F., Perry, J.M.G., Malinzak, M., Allen, K.L., Kirk, E.C., Pavcan, J.M., & Fleagle, J.G. (2013) Paleobiology of Santacrucian primates. In Vizcaino, S.F., Kay, R.F., Bargo, M.S. (eds) *Miocene paleobiology in Patagonia: High Latitude Paleocommunities of the Santa Cruz Formation*. Cambridge University Press, Cambridge.

Kirk, E.C., Daghighi, P., Macrini, T.E., Bhullar, B.S., & Rowe, T.B. (2014) Cranial anatomy of the Duchesnean primate *Rooneyia viejaensis*: New insights from high resolution computed tomography. *J Hum Evol*, 74, 82–95.

Kirk, E.C. & Simons, E.L. (2001) Diets of fossil primates from the Fayum Depression of Egypt: A quantitative analysis of molar shearing. *J Hum Evol*, 40, 203–229.

Kolb, B. (2007) Do all mammals have a prefrontal cortex? In Kaas, J., Krubitzer, L. (eds) *Evolution of Nervous Systems: Mammals*. Academic Press, New York, pp. 443–50.

Kolb, B., Buhrmann, K., McDonald, R., & Sutherland, R.J. (1994) Dissociation of the medial prefrontal, posterior parietal, and posterior temporal cortex for spatial navigation and recognition memory in the rat. *Cereb Cortex*, 4, 664–80.

Kolb, B., Nonneman, A.J., & Singh, R.K. (1974) Double dissociation of spatial impairments and perseveration following selective prefrontal lesions in rats. *J Comp Physiol Psychol*, 87, 772–80.

Kossl, M., Hechavarria, J., Voss, C., Schaefer, M., & Vater, M. (2015) Bat auditory cortex – model for general mammalian auditory computation or special design solution for active time perception? *Eur J Neurosci*, 41, 518–32.

Krubitzer, L. & Huffman, K.J. (2000) Arealization of the neocortex in mammals: Genetic and epigenetic contributions to the phenotype. *Brain Behav Evol*, 55, 322–35.

Krubitzer, L.A. & Seelke, A.M. (2012) Cortical evolution in mammals: The bane and beauty of phenotypic variability. *Proc Natl Acad Sci USA*, 109 Suppl 1, 10647–54.

Lehmann, J. & Dunbar, R.I. (2009) Network cohesion, group size and neocortex size in female-bonded Old World primates. *Proc Biol Sci*, 276, 4417–22.

Long, A., Bloch, J.I., & Silcox, M.T. (2015) Quantification of neocortical ratios in stem primates. *Am J Phys Anthropol*, 157, 363–73.

Lyon, D.C. (2007) The evolution of visual cortex and visual systems. In Kaas, J. (ed.) *Evolution of Nervous Systems*. Elsevier, New York, pp. 267–306.

MacNeilage, P., Studdert-Kennedy, M., & Lindblom, B. (1987) Primate handedness reconsidered. *Beh Brain Sci*, 10, 247–303.

Martin, R.D. (1981) Relative brain size and basal metabolic rate in terrestrial vertebrates. *Nature*, 293, 57–60.

Matelli, M., Luppino, G., Rogassi, L., & Rizzolatti, G. (1989) Thalamic input to inferior area 6 and area 4 in the Macaque monkey. *J Comp Neurol*, 280, 468–88.

Melin, A.D., Matsushita, Y., Moritz, G.L., Dominy, N.J., & Kawamura, S. (2013) Inferred L/M cone opsin polymorphism of ancestral tarsiers sheds dim light on the origin of anthropoid primates. *Proc Roy Soc B* 280, 20130189.

Meunier, M., Bachevalier, J., & Mishkin, M. (1997) Effects of orbital frontal and anterior cingulate lesions on object and spatial memory in rhesus monkeys. *Neuropsychol*, 35, 999–1015.

Milton, K. (1988) Foraging behaviour and the evolution of primate intelligence. In Byrne, R.W., Whiten, A. (eds) *Machiavellian Intelligence: Social Expertise and the Evolution of Intellect in Monkeys, Apes and Humans*. Oxford University Press, Oxford, pp. 285–305.

Muller, A.E. & Thalmann, U. (2000) Origin and evolution of primate social organisation: A reconstruction. *Biol Rev Camb Philos Soc*, 75, 405–35.

Neubert, F.X., Mars, R.B., Thomas, A.G., Sallet, J., & Rushworth, M.F. (2014) Comparison of human ventral frontal cortex areas for cognitive control and language with areas in monkey frontal cortex. *Neuron*, 81, 700–13.

Ni, X., Flynn, J.J., Wyss, A.R., & Zhang, C. (2019) Cranial endocast of a stem platyrrhine primate and ancestral brain conditions in anthropoids. *Sci Adv*, 5, eaav7913.

Ni, X., Gebo, D.L., Dagosto, M., Meng, J., Tafforeau, P., Flynn, J.J., & Beard, K.C. (2013) The oldest known primate skeleton and early haplorhine evolution. *Nature*, 498, 60–4.

Oates, J.F. (1987) Food distribution and foraging behavior. In Smuts, B.B., Wrangham, R.W., Struhsaker, T.T. (eds) *Primate Societies*. University of Chicago Press, Chicago, pp. 197–209.

Orem, J. & Schlag, J. (1973) Relations between thalamic and corticofrontal sites of oculomotor control in the cat. *Brain Res*, 60, 503–7.

Passingham, R. (1978) The functions of prefrontal cortex in the tree shrew (Tupaia belangeri). *Brain Res*, 145, 147–52.

Passingham, R.E. (1981) Primate specialization in brain and intelligence. *Symp Zool Soc Lond*, 46, 361–8.

Petrides, M. & Pandya, D.N. (1999) Dorsolateral prefrontal cortex: Comparative cytoarchitectonic analysis in the human and the macaque brain and corticocortical connection patterns. *Eur J Neurosci*, 11, 1011–36.

Powell, D.A., Chachich, M., Murphy, V., McLaughlin, J., Tebbutt, D., & Buchanan, S.L. (1997) Amygdala-prefrontal interactions and conditioned bradycardia in the rabbit. *Behav Neurosci*, 111, 1056–74.

Powell, L.E., Isler, K., & Barton, R.A. (2017) Re-evaluating the link between brain size and behavioural ecology in primates. *Proc Biol Sci*, 284, 1765.

Preuss, T.M. (1995) Do rats have prefrontal cortex? The Rose-Woolsey-Akert program reconsidered. *J Cogn Neurosci*, 7, 1–24.

Preuss, T.M. (2007) Primate brain evolution in phylogenetic context. In Kaas, J., Preuss, T.M. (eds) *The Evolution of Nervous Systems*. Elsevier, New York, pp. 3–34.

Preuss, T.M., Beck, P.D., & Kaas, J.H. (1993) Areal, modular, and connectional organization of visual cortex in a prosimian primate, the slow loris (*Nycticebus coucang*). *Brain Behav Evol*, 42, 321–35.

Preuss, T.M. & Goldman-Rakic, P.S. (1991a) Ipsilateral cortical connections of granular frontal cortex in the strepsirhine primate Galago, with comparative comments on anthropoid primates. *J Comp Neurol*, 310, 507–49.

Preuss, T.M. & Goldman-Rakic, P.S. (1991b) Myelo- and cytoarchitecture of the granular frontal cortex and surrounding regions in the strepsirhine primate Galago and the anthropoid primate Macaca. *J Comp Neurol*, 310, 429–74.

Radinsky, L. (1975) Primate brain evolution. *Amer Sci*, 63, 656–63.

Rajkowska, G. & Kosmal, A. (1988) Intrinsic connections and cytoarchitectonic data of the frontal association cortex in the dog. *Acta Neurobiol Exp* (Wars), 48, 169–92.

Ramdarshan, A. & Orliac, M.J. (2016) Endocranial morphology of Microchoerus erinaceus (*Euprimates, Tarsiiformes*) and early evolution of the Euprimates brain. *Am J Phys Anthropol*, 159, 5–16.

Ray, J.P. & Price, J.L. (1992) The organization of the thalamocortical connections of the mediodorsal thalamic nucleus in the rat, related to the ventral forebrain-prefrontal cortex topography. *J Comp Neurol*, 323, 167–97.

Ray, J.P. & Price, J.L. (1993) The organization of projections from the mediodorsal nucleus of the thalamus to orbital and medial prefrontal cortex in macaque monkeys. *J Comp Neurol*, 337, 1–31.

Remple, M.S., Reed, J.L., Stepniewska, I., & Kaas, J.H. (2006) Organization of frontoparietal cortex in the tree shrew (Tupaia belangeri). I. Architecture, microelectrode maps, and corticospinal connections. *J Comp Neurol*, 497, 133–54.

Rosa, M.G. & Krubitzer, L.A. (1999) The evolution of visual cortex: Where is V2? *Trends Neurosci*, 22, 242–8.

Rose, J.E. & Woolsey, C.N. (1948) The orbitofrontal cortex and its connections with the mediodorsal nucleus in rabbit, sheep and cat. *Res Publ Assoc Res Nerv Ment Dis*, 27 (vol. 1), 210–32.

Rose, K.D. (2006) *The Beginnings of the Age of Mammals*. Johns Hopkins, Baltimore.

Rudebeck, P.H., Putnam, P.T., Daniels, T.E., Yang, T., Mitz, A.R., Rhodes, S.E., & Murray, E.A. (2014) A role for primate subgenual cingulate cortex in sustaining autonomic arousal. *Proc Natl Acad Sci USA*, 111, 5391–6.

Schaeffer, D.J., Hori, Y., Gilbert, K.M., Gati, J.S., Menon, R.S., & Everling, S. (2020) Divergence of rodent and primate medial frontal cortex functional connectivity. *Proc Natl Acad Sci USA*, 117, 21681-89

Schall, J.D., Zinke, W., Closman, J.D., Schall, M.S., Pare, M., & Pouget, P. (2017) On the evolution of the frontal eye field: Comparisons of monkeys, apes and humans. In Kaas, J. (ed.) *Evolution of Nervous Systems*. Elsevier, New York.

Schmidt, D. (2010) Primate locomotor evolution: Biomechanical studies of primate locomotion and their implication for understanding primate neuroethology. In Plattt, M.L., Gharanfar, A.A. (eds) *Primate Neuroethology*. Oxford University Press, New York, pp. 31-63.

Schoenbaum, G., Roesch, M.R., Stalnaker, T.A., & Takahashi, Y.K. (2009) A new perspective on the role of the orbitofrontal cortex in adaptive behaviour. *Nat Rev Neurosci*, 10, 885-92.

Seamans, J.K., Lapish, C.C., & Durstewitz, D. (2008) Comparing the prefrontal cortex of rats and primates: Insights from electrophysiology. *Neurotox Res*, 14, 249-62.

Sears, K.E., Finarelli, J.A., Flynn, J.J., & Wyss, A. (2008) Estimating body mass in New World 'monkeys' (*Platyrhini, Primates*) with a consideration of the Miocene platyrhine, Chleceburs carrascoensis. *Amer Mus Novit*, 3617, 1-29.

Seiffert, E.R. (2006) Revised age estimates for the later Paleogene mammal faunas of Egypt and Oman. *Proc Natl Acad Sci USA*, 103, 5000-5.

Selemon, L.D. & Goldman-Rakic, P.S. (1985) Longitudinal topography and interdigitation of corticostriatal projections in the rhesus monkey. *J Neurosci*, 5, 776-94.

Silcox, M.T., Bloch, J.I., Boyer, D.M., Chester, S.G.B., & Lopez-Torres, S. (2017) The evolutionary radiation of plesiadapiforms. *Evol Anthropol*, 26, 74-94.

Silcox, M.T., Sargis, E.J., Bloch, J.I., & Boyer, D.M. (2015) Primate origins and supraordinal relationships: Morphological evidence. In Henke, K., Tattersall, I. (eds) *Handbook of Paleoanthropology*. Springer-Verlag, Berlin.

Simons, E.L., Seiffert, E.R., Ryan, T.M. & Attia, Y. (2007) A remarkable female cranium of the early Oligocene anthropoid Aegyptopithecus zeuxis (*Catarrhini, Propliopithecidae*). *Proc Natl Acad Sci USA*, 104, 8731-6.

Stepniewska, I., Pouget, P., & Kaas, J.H. (2018) Frontal eye field in prosimian galagos: Intracortical microstimulation and tracing studies. *J Comp Neurol*, 526, 626-52.

Struhsaker, T.T. (1980) Comparison of the behaviour and ecology of red colobus and retail monkeys in the Kibale Forest, Uganda. *Af J Ecol*, 18, 33-51.

Suga, N., Yan, J., & Zhang, Y. (1997) Cortical maps for hearing and egocentric selection for self-organization. *Trends Cogn Sci*, 1, 13-20.

Surridge, A.K. & Mundy, N.I. (2002) Trans-specific evolution of opsin alleles and the maintenance of trichromatic colour vision in Callitrichine primates. *Mol Ecol*, 11, 2157-69.

Sussman, R.W. (1991) Primate origins and the evolution of angiosperms. *Amer J Primatol*, 23, 209-23.

Uylings, H.B.M., Groenewegen, H.J., & Kolb, B. (2003) Do rats have a prefrontal cortex? *Behav.Brain Res*, 146, 3-17.

Wallis, J.D. & Kennerley, S.W. (2011) Contrasting reward signals in the orbitofrontal cortex and anterior cingulate cortex. *Ann NY Acad Sci*, 1239, 33-42.

Williams, B.A., Kay, R.F., & Kirk, E.C. (2010) New perspectives on anthropoid origins. *Proc Natl Acad Sci USA*, 107, 4797-804.

Wise, S.P. (2008) Forward frontal fields: Phylogeny and fundamental function. *Trends Neurosci*, 31, 599-608.

Wise, S.P. (2017) The evolution of the prefrontal cortex in early primates and anthropoids. In Krubitzer, L., Kaas, J. (eds) *The Evolution of Nervous Systems*. Elsevier, New York.

Wong, P. & Kaas, J.H. (2008) Architectonic subdivisions of neocortex in the gray squirrel (*Sciurus carolinensis*). *Anat Rec* (Hoboken), 291, 1301–33.

Wong, P. & Kaas, J.H. (2009) Architectonic subdivisions of neocortex in the tree shrew (*Tupaia belangeri*). *Anat Rec* (Hoboken), 292, 994–1027.

Wong, P. & Kaas, J.H. (2010) Architectonic subdivisions of neocortex in the Galago (*Otolemur garnetti*). *Anat Rec* (Hoboken), 293, 1033–69.

Wright, A., Scadeng, M., Stec, D., Dubowitz, R., Ridgway, S., & Leger, J.S. (2017) Neuroanatomy of the killer whale (*Orcinus orca*): A magnetic resonance imaging investigation of structure with insights on function and evolution. *Brain Struct Funct*, 222, 417–36.

Wu, C.W., Bichot, N.P., & Kaas, J.H. (2000) Converging evidence from microstimulation, architecture, and connections for multiple motor areas in the frontal and cingulate cortex of prosimian primates. *J Comp Neurol*, 423, 140–77.

Zuberbuhler, K. & Janmaat, K.R.L. (2010) Foraging cognition in nonhuman primates. In Platt, M.L., Gzazanfar, A.A. (eds), *Primate Neuroethology*. Oxford University Press, Oxford, pp. 64–83.

PART II
SUBAREAS OF THE PREFRONTAL CORTEX

3
Medial Prefrontal Cortex
Self-Generated Actions

Introduction

Like other animals, non-human primates must learn where to find the foods and other resources that they need. They can do so because remembering where they have found them in the past is usually a good guide as to where they will find them in the future. The ability to find their way depends on knowledge of the landscape as well as of landmarks such as branches, bushes, and trees.

However, primates differ from other mammals in having hands rather than paws or hooves. This means that they forage by reaching for and retrieving fruit and leaves with their hands. In monkeys and apes, the motor cortex supports the ability to move the fingers independently (Kuypers, 1981), to flex and extend the wrist (Hoffman & Strick, 1995), and to supinate the arm (Passingham et al., 1983). But the motor cortex needs to receive instructions as to the appropriate action. This chapter argues that the ability to generate actions on the basis of the memory of past searches depends on the medial prefrontal cortex and the medial premotor areas with which it is connected.

These actions are restricted to actions that are self-generated. This means that either there are no external cues to specify the action that is appropriate, or that the external cues that are present are not sufficient to specify the appropriate choice. In other words, it is up to the animal what it should do or what it should choose. The actions can be described as being 'voluntary' (Passingham, 1993).

This does not mean that the actions or choices occur in isolation. They depend on the ability of the animal to recall past actions and choices and their outcomes. The role of the medial prefrontal cortex (PF) is to support retrieval or recall in the absence of cues that tell the animal what action to perform and when to do so. As Chapter 1 describes, there are cells in the supplementary motor cortex that code for sequences of actions that are performed on the basis of memory (Mushiake et al., 1991).

There is no mystery concerning self-generated or 'voluntary' actions. They are not random. Dog owners know that their dog comes when they call, but on other occasions the dog approaches spontaneously. The reason is that there is a desired outcome, for example being petted, and the animal has been petted when it approached in the past. Or to take another example, chacma baboons (Papio ursinus) have been seen to move spontaneously from their resting place to their feeding sites, even though the journey is several kilometres, and the feeding sits are not visible when they start out (Noser & Byrne, 2007).

Area

Figure 3.1 shows the extent of the medial PF cortex in macaque monkeys and humans.

The primate medial PF cortex includes agranular areas that have homologues in other mammals (Devinsky et al., 1995; Vogt & Paxinos, 2014). These are the

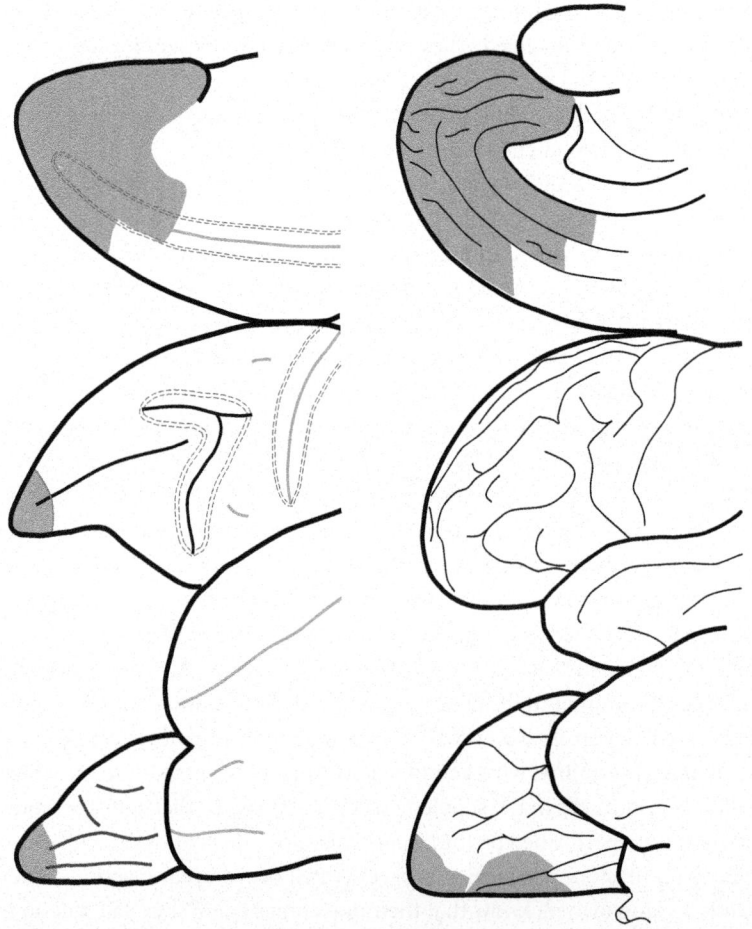

Figure 3.1 Medial PF cortex in macaque monkeys (left) and humans (right), indicated by shading

Reproduced from Passingham, R.E. & Wise, S.P. *Neurobiol Prefront Cortex*, p. 66, Figure 3.1 © 2012, Oxford University Press.

anterior cingulate cortex (ACC), area 24), the prelimbic cortex (area 32), and the infralimbic cortex (area 25). However, the medial PF cortex also includes the granular areas 9 and 10 that are unique to anthropoid primates. These are depicted in Figure 2.1 in light blue.

In this chapter, the sulcal anterior cingulate cortex (sACC) is contrasted with the gyral anterior cingulate cortex (gACC). The latter refers to the convexity cortex of the anterior cingulate (ACC) that lies ventral to the cingulate sulcus.

Recently, Laubach et al. (2018) have used volumetric graphical modelling to establish the homologies between the ACC, prelimbic, and infralimbic areas in rats, macaque monkeys, and humans. The medial aspect of primate brains differs from that of rats in that the corpus callosum has a more concave shape.

Laubach et al. therefore developed a topologically constrained graphical model of brain expansion in which the medial PF areas as seen in rodents maintain their relationship with the corpus callosum as the brain expands to human size and shape. Figure 3.2 (next page) illustrates their analysis.

This analysis supports the conclusions presented in Chapter 2. Rodent, macaque, and human brains, among other mammals, share homologues of the ACC, as well as the prelimbic and infralimbic areas.

The ACC also has subdivisions. In macaque monkeys, the rostral and caudal parts of area 24 differ in both cytoarchitecture and immunocytochemistry (Vogt et al., 2005), and the patterns of receptor binding reveal a similar distinction in humans (Palomero-Gallagher et al., 2009). The term mid-cingulate cortex is sometimes used to identify the caudal part of area 24, excluding the cingulate motor areas.

In many chimpanzee and human brains there is also a paracingulate sulcus which curves round the anterior cingulate sulcus (Amiez et al., 2019). This contains the dorsal and ventral prelimbic area 32. In hemispheres with no paracingulate gyrus, this tissue lies in the dorsal bank of the sACC. The existence of the paracingulate sulcus in chimpanzees and humans is evidence of the expansion of the prelimbic cortex.

Neubert et al. (2015) identified the functional subdivisions of the ACC by plotting the connectional fingerprints of different areas in thirty-five human and twenty-five monkey brains, using diffusion weighted imaging (DWI). As Chapter 1 describes, the assumption of this approach is that if areas can be found in the human and macaque monkey brain that have a similar pattern of overall connections, they can be treated as being homologous (Mars et al., 2016). Neubert et al. (2015) found that the area equivalent to the dorsal paracingulate (prelimbic) cortex in the human brain lies rostrally in the sACC in the macaque brain. The ventral paracingulate (prelimbic) cortex lies on the ventromedial surface in the macaque brain.

(A) Curvature

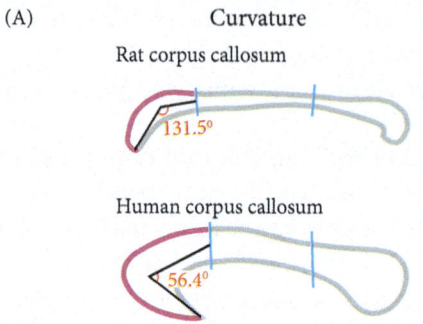

(B) Rodent areas in midsagittal section

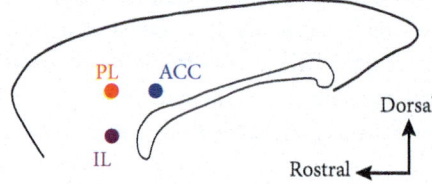

(C) Rotation of rodent areas by human callosal angle

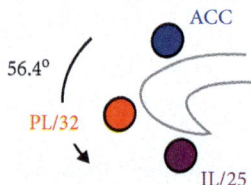

Figure 3.2 Homologies among agranular parts of the medial PF cortex in rodents and anthropoids

(A) Quantification of the curvature of rostral corpus callosum. The point of maximum curvature along the dorsal edge of the rostral third of the corpus callosum is 1.84 times greater in humans than in rats. The rat callosum has obtuse interior angles whereas the human callosum has acute interior angles. (B) Locations of three main parts of the rodent medial PF cortex shown on a midsagittal section. Abbreviations: ACC, anterior cingulate cortex; IL, infralimbic cortex; PL, prelimbic cortex (C) Rotation of the three points representing each cortical region by the measured curvature of the human callosum shifts the rodent areas into the approximate locations observed in human and other anthropoid brains. Abbreviations: pACC, pregenual cortex (area 32), sACC, subgenual cortex (area 25)

Reproduced from Laubach, M., Amarante, L.M., Swanson, K., & White, S.R. What, if anything, is rodent prefrontal cortex?. *eNeuro*, 5 (5), Doi: https://doi.org/10.1523/ENEURO.0315-18.2018 ©The Authors, 2018. Licensed under CC BY 4.0.

Connections

Figure 3.3 illustrates the main connections of the granular medial PF cortex in macaque monkeys. This plot, and the analogous ones for other areas in Chapters 4–7 aim to convey the most important points that emerge from the neuroanatomy literature. The illustrations emphasize the salient aspects of the connectional fingerprints and the connectome analyses of Markov et al. (2011; 2014).

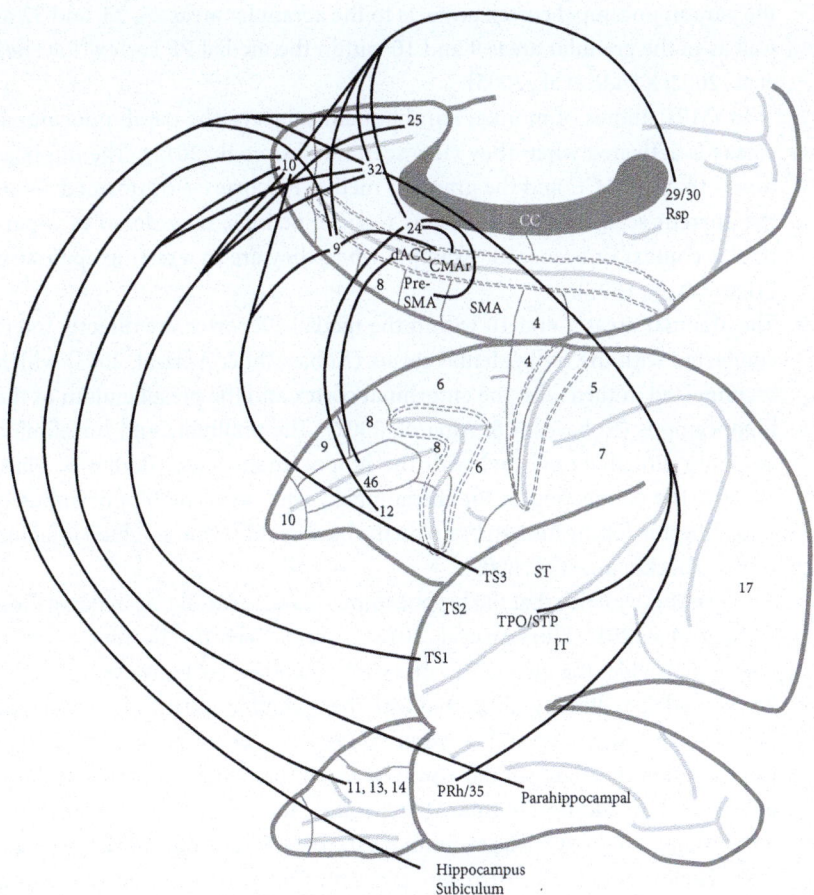

Figure 3.3 Selected connections of the medial PF cortex in macaque monkeys
The lines connect some of the areas that have direct axonal connections, assumed to be reciprocal.
Adapted from Passingham, R.E. & Wise, S.P. *Neurobiol Prefront Cortex*, p. 68, Figure 3.2 © 2012, Oxford University Press.

1. Unlike the orbital (Chapter 4) and ventral PF cortex (Chapter 7), the ACC does not receive direct visual information about the objects in foveal vision from the inferotemporal cortex (Morecraft et al., 2012) or perirhinal cortex (Kondo et al., 2005).

    The reason is that it supports the choice between actions rather than the choice between objects. Because these actions are performed from memory and without visual prompts, they can be said to be self-generated (Passingham et al., 2010).

2. Rather than receiving inputs from foveal vision, the visual inputs to the medial PF cortex reflect peripheral vision. Area V4 sends an input for the peripheral field to the parahippocampal cortex (Baizer et al., 1991); and in turn the parahippocampal cortex projects to the agranular areas 24, 25, and 32 as well as to the granular areas 9 and 10 within the medial PF cortex (Lavenex et al., 2002; Kondo et al., 2005).

    In fMRI studies of monkeys and human subjects, the parahippocampal cortex is activated when they view scenes (Nasr et al., 2011). The implication is that the ACC and the granular medial PF cortex are informed about the current scene. When monkeys retrieve actions, they do so in a particular context or scene, for example when they are in a testing apparatus (Thaler et al., 1995).

3. The granular areas 9 and 10 within the medial PF cortex are directly interconnected with the retrosplenial cortex (Kobayashi & Amaral, 2003) which is connected in turn with the entorhinal cortex and the presubiculum of the hippocampus (Kobayashi & Amaral, 2007). The prelimbic and infralimbic areas are directly connected with the hippocampus itself (Barbas & Blatt, 1995b). The connection of the infralimbic cortex with the hippocampus is also indicated for the human brain on the basis of diffusion weighted imaging (DWI) (Beckmann et al., 2009).

    Given the evidence that the hippocampus plays a critical role in navigation (Epstein et al., 2017; Hinman et al., 2018), it seems likely that the hippocampal complex provides the medial PF cortex with memories of journeys.

4. The amygdala is directly connected with the prelimbic and infralimbic cortex and with the ACC (Amaral & Insausti, 1992; Morecraft et al., 2007). DWI indicates that there are similar connections in the human brain (Beckmann et al., 2009). In addition to these direct routes, the infralimbic and prelimbic areas project to the nucleus accumbens in the ventral striatum (Ferry et al., 2000; Heilbronner et al., 2016), and the ventral striatum is connected in turn with the amygdala (Choi et al., 2017). Finally, there are connections between the medial PF and the orbital PF cortex (Cavada et al., 2000); and again, DWI indicates the same connectivity in the human brain (Beckmann et al., 2009). The amygdala is also strongly interconnected with the orbital PF (Amaral & Price, 1984).

As Chapter 4 describes, the orbital PF cortex encodes information about rewards or outcomes, such as the number of juice drops (Wallis & Miller, 2003). Thus, the ACC can support the generation of actions that are performed for reward or are based on internal motivation (Thaler et al., 1995).
5. In addition, the infralimbic and prelimbic areas have direct projections to the hypothalamus (Rempel-Clower & Barbas, 1998), and again this has been confirmed for the human brain using DWI (Beckmann et al., 2009).

The lateral hypothalamus plays a role in feeding as well as arousal (Qualls-Creekmore & Munzberg, 2018). In macaque monkeys, lesions of the infralimbic cortex have been shown to block sustained autonomic responses to stimuli that predict reward (Rudebeck et al., 2014).
6. The dorsal bank of the sACC receives a robust input from the dorsal bank of the principal sulcus (areas 46 and 9/46) (Saleem et al., 2014). And the areas 9 and 10 within the medial PF cortex also have dense connections with the dorsal PF cortex and the ACC (Petrides & Pandya, 1999; 2007). Cells in areas 46 and 9/46 encode locations (Yamagata et al., 2012), distances (Genovesio et al., 2011), durations (Genovesio et al., 2006), and order (Genovesio et al., 2012). These metrics are relevant for planning journeys as well as for planning a series of actions with the hands and arms.
7. The frontal polar cortex (area 10) (Petrides & Pandya, 2007) and the granular area 9 within the medial PF cortex (Barbas et al., 1999) project to the ACC which connects in turn with the rostral cingulate motor area (CMAr), the presupplementary (preSMA), and supplementary motor cortex (SMA) (Morecraft et al., 2012). In turn the SMA is interconnected with the caudal cingulate motor area (CMAc) (Luppino et al., 1993). Finally, the SMA and CMAc project to the primary motor cortex as well as to the spinal cord (He et al., 1995; Dum & Strick, 2002).

If the preSMA or SMA are inactivated with muscimol, the monkeys make errors when they are required to perform three movements in a sequence that they have memorized (Shima & Tanji, 1998). Yet, the animals can still perform the sequence accurately if the sequence is visually cued. Thus, the animals are impaired when they must choose between actions from memory but not when visual cues instruct the actions.
8. Finally, the superior part of the temporal pole (area (TS1)) projects to the ACC including the prelimbic cortex (Barbas et al., 1999; Saleem et al., 2008). The prelimbic cortex also receives an input from the dorsal bank of the superior temporal sulcus (STS) (Morecraft et al., 2012). There are cells in the dorsal bank of the STS that respond to biological motion, as when a macaque monkey views the actions of others (Jellema & Perrett, 2003).

The implication is that the prelimbic cortex has access to information about the actions of other animals. The dorsal bank of the STS also includes the superior temporal polysensory area (STP) (Scott et al., 2017) in which cells respond to both visual and auditory stimuli, and there are cells in the superior temporal cortex in which the responses to vocalizations are modulated by the sight of the face (Perrodin et al., 2014). Thus, the medial PF cortex also has access to information about the calls and facial expressions that other animals produce.

Summary

The connectional fingerprint of medial PF cortex (Figure 3.2) suggests the following conclusions:

- The medial PF cortex receives visual inputs about the scene or context rather than objects. It plays a role in retrieving actions from memory, based on internal motivation.
- Inputs to the medial PF cortex from the amygdala, the ventral striatum via the pallidum and thalamus, and the orbital PF cortex could provide these internal motivational signals.
- The direct and indirect connections of the medial PF cortex with the hippocampus suggest that it has access to memories about past journeys and sequences of events.
- The prelimbic cortex has access to information about the vocalizations, faces, and movements of conspecifics and predators. These connections probably evolved in anthropoids as they developed complex social systems and were faced an increased risk of predation (Chapter 2).

Choices

It is commonplace to say that the PF cortex is involved in making decisions. However, for the purposes of clarity, it is important to define the use of this and other terms. The text therefore adopts the distinctions suggested by Schall (2001).

He proposed that the term 'decision' be restricted to perceptual decisions. These apply to animals or humans when they discriminate among sensory inputs, for example deciding on the direction in which dots move coherently in a visual display. The rationale behind this definition is that decisions relate to what happens in the external sensory world

The term 'choice' is therefore used to refer to the choice between actions, which have a spatial goal, or the choice between objects, which have an object goal. The term 'goal' is used for the target of the choice. It is never used in the text as a synonym for the reward. Instead, the term 'outcome' is used to refer to the feedback that follows from a choice.

The terms 'decision' and 'choice' imply that there are alternatives. A common way of accounting for these is to use accumulator–racetrack models. These models were originally devised to account for perceptual decisions. For example, Roitman and Shadlen (2002) presented displays of moving dots, varying the degree to which the dots moved coherently. The monkeys had to decide on the overall direction in which the dots were moving.

Recordings form the intraparietal area LIP suggested that there were cell networks for the competing alternatives, some for dots moving up and others for dots moving down. Over time, the evidence for one accumulated until the neural network reached the threshold for producing an output. These networks act like a leaky integrator, accumulating evidence in favour of the output that they represent.

To find out which areas were critical for the decision, Hanks et al. (2006) applied microstimulation to area LIP while monkey had to decide on the direction of motion, and to do so in a speeded manner. Micro-stimulation of a cluster of cells with overlapping receptive fields influenced the choice towards the direction of the receptive fields of these cells. It also influenced the reaction time with which the decision was made.

Hanks and Summerfield (2017) have reviewed computational advances on the dynamics of neural integration. Their review also discusses the microcircuits that are involved in perceptual decision making.

The notion of accumulating alternatives can also be applied to choices between actions. For example, Zhang et al. (2012) studied fMRI activations as human subjects selected actions. In one condition, visual cues instructed the subjects which of three fingers to move; the other condition lacked such cues. Zhang et al. then contrasted activations for self-generated and externally cued movements.

In their analysis, Zhang et al. used a computational model to estimate the accumulated metabolic activity on a trial-by-trial basis. They assumed that on each trial three accumulators, one for each finger, increased in activation until one of them reached its threshold. They found voxels that matched this prediction in the sACC and SMA.

Choice of When to Act

Both animals and human subjects choose not only what action to perform but when to perform it. In other words, they also choose between alternative possible times to act.

Thaler et al. (1995) devised a laboratory task to test this ability by training monkeys to raise their arm whenever they wanted a peanut. The animals were tested in the dark so that there could be no possible external cues. When the monkey's arm broke an infra-red beam, a peanut was delivered to a foodwell below.

Thaler et al. then removed the preSMA and SMA in some animals, and the cortex in the ACC in others. Both groups of monkeys failed to raise their arm during the period of post-operative testing. This was not because they no longer wanted peanuts: they were quite willing to reach for peanuts that they could see.

However, the pre-SMA and SMA are premotor areas, and the CMAr lies in the sACC. So as to identify the PF area that influences these premotor areas, Khalighinejad et al. (2020a) used fMRI to scan monkeys on a task in which they had to choose when to press a button. The monkeys viewed a flow of dots and the more the dots descended in the display the greater the probability that pressing the button would be rewarded. The colour of the dots was also varied, and this told the animals the magnitude of the reward that was potentially available.

Two activations varied with the time to act. One was in the sACC, anterior to the CMAr. The other was in the basal forebrain.

Khalighinejad et al. suggested that the activation in the rostral sACC might be related to the influence of the reward magnitude on the time to act. The activation in the basal forebrain differed in that it integrated information about the present and past context, so determining the time to act. In this experiment, the present context concerned the current average probability of reward, whereas the past context was related to the response times on previous trials.

To demonstrate that these areas had a causal influence over the time to act, Khalighinejad et al. then applied deep ultrasound to temporarily inactivate these areas. Inactivation of the sACC led to increased response times when the rewards on offer were medium as opposed to small. Inactivation of the basal forebrain influenced the difference between the trial to trial variation in response times as predicted and as observed.

Khalighinejad et al. (2020b) went on to study the anatomy in more detail by scanning human subjects at 7 tesla while they performed a comparable task. The subjects watched as bubbles of different colours and size emerged from a water tank as it drained. The colour indicated the magnitude of monetary reward, and the size the probability of the reward. As in the experiment with monkeys, the human subjects pressed a button when they chose to do so.

In addition to the ACC and basal forebrain, there were other areas in which the activation varied with the time to act. These were the striatum, substantia nigra, habenula and pedunculopontine nucleus. As in the experiment with macaques, the basal forebrain integrated information about the current and past contexts.

However, the substantia nigra was activated early in the decision phase before movement, and the nucleus accumbens and putamen later in the decision phase. Khalighinejad et al. then looked for a psychophysiological interaction between the activations in the basal forebrain and substantia nigra. This analysis method looks for an influence of variation in the activation in one area on variation in activation in the other. The results were consistent with the proposal that the basal forebrain exerted its influence on the time of movement via the dopaminergic system of the substantia nigra.

This experiment was performed with human subjects, and it is not clear whether it would be ethical to use deep ultrasound to interfere with the activations in healthy human subjects. However, the substantia nigra projects to the putamen and nucleus accumbens (Hedreen & DeLong, 1991) and these are part of the reward circuitry, including the cortex of ACC (Haber & Knutson, 2010). Nixon and Passingham (1998) tested monkeys with lesions of the putamen and ventral striatum on the task described earlier in which the animals received food by raising their arm. The lesion caused a marked reduction in the number of times at which the animals performed this spontaneous action. In other words, the animals were severely impaired at initiating the action.

Summary

When animals or human subjects choose when to act or what action to perform there are alternatives. The evidence for each of these alternatives accumulates until one reaches a criterion threshold for action. The cortex in the sACC is involved in choosing which act to perform and when to perform it. This applies when the choices are self-generated or voluntary. The timing of the action takes into account both the present context and the times of past actions. It is the substantia nigra that actually initiates the action, influencing the tissue in the sACC via the striato-pallidal projections to the cortex.

Flexible Choice

Chapter 2 describes the need for anthropoids to respond flexibly, given the volatility of their resources. One way of assessing flexibility in the laboratory is to test monkeys on action reversals. The illustrations on the left of Figure 4.7 (page 135) in the next chapter illustrate two such tasks.

Deterministic Action Reversals

On one version of the action reversal task (Figure 4.7A, left), the monkeys are first taught to lift a handle for several consecutive trials. For every successful lift, they receive a food reward. Then, with no external signal to indicate that anything has changed, lifting the handle is not rewarded. Instead, the animal has to move the handle to the right to be rewarded. Over the course of many trials, the monkeys perform these two different actions on the same object, the handle. This means that the handle itself provides no clue as to the appropriate action.

The evidence that a reversal has occurred is the sudden lack of reward. The animal has then to generate the alternative action, with no external signal to prompt it.

Kennerley et al. (2006) tested macaque monkeys on a series of action reversals. After the first reversal, a series of further reversals followed in blocks of consecutive trials. For each reversal, the monkey needed to change its choice of action in order to produce a reward. Kennerley et al. then made lesions in the sACC, with the lesion extending from the rostral limit of the sulcus to the rostro-caudal level of the precentral sulcus. The effect was that the monkeys switched between the two actions more slowly than intact control monkeys.

One might suppose that the lesioned monkeys performed poorly because they repeated the response after an error, for example immediately after a reversal. And the bars on the right in Figure 3.4 (next page) show that indeed they did. The open clear and hatched bars show the pre- and post-operative data for the control animals; the solid grey and grey hatched bars show the pre- and post-operative data for the lesioned animals. After an error, the performance of the lesioned animals deteriorated after surgery.

But Kennerley et al. also studied how the animals behaved after a correct response. The bars on the left in Figure 3.4 show that the performance of the lesioned animals was also worse after surgery on the trial after a correct response. It turned out that the lesioned animals were also poor at making use of the fact that an action has just been rewarded.

In an fMRI study Glascher et al. (2009) scanned human subjects on reversals between two actions. These were either pressing or sliding a computer mouse. The subjects were also tested on reversals between the choice of one of two pictures. There was an activation in the ventral prelimbic cortex in both cases. However, there was also an activation in the sACC at the level of the mid-cingulate cortex, and this was greater for action reversals than object reversals. It is important to note that this does not necessarily mean that there was no activation for object reversals. Nonetheless, the experiment indicates that the tissue in the sACC has an especial role in the flexible choice between actions that are not specified by an external cue. The reversals are signed by a prediction error, the failure to obtain the expected reward.

Probabilistic Action Reversals

The fixed action reversal task measures flexibility, but it is not a good model of the conditions while foraging. In the real world the outcomes are often probabilistic.

To better mimic this situation, Kennerley et al. (2006) tested monkeys under conditions in which the two actions were associated with reward on a probabilistic basis. An example of such a task is illustrated on the left of Figure 4.7B (Chapter 4).

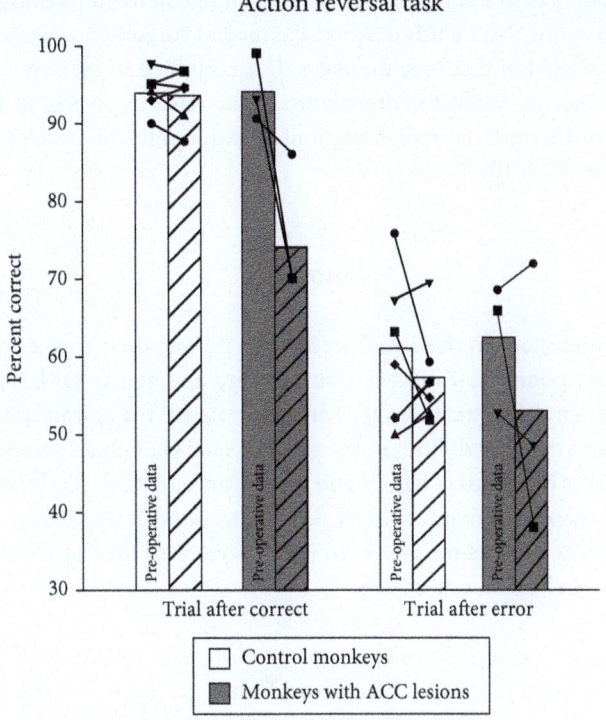

Figure 3.4 Reversal impairment for choices between two actions
Pre-operative (solid bars) and post-operative (hatched bars) data from macaque monkeys. Percent correct after a correct choice (left set of four bars) and after an incorrect choice (right set of four bars), for intact (control) monkeys (white) and lesioned monkeys (grey). Results from individual monkeys are shown by the symbols and are connected by lines.

Reproduced from Kennerley S.W., Walton M.E., Behrens T.E.J., Buckley M.J., & Rushworth M.F.S. Optimal decision making and the anterior cingulate cortex. *Nat Neurosci*, 9 (7), 940–7, Figure 2, Doi: 10.1038/nn1724 Copyright © 2006, Springer Nature.

There were four different ratios of reward for the two actions, 0.4:0.1, 0.5:0/2, 0.75:0/25, and 1:0. This meant that the monkeys had to sample both alternatives on a trial by trial basis so as to develop a sense of the utility of each action.

The animals with sACC lesions were slower to approach the optimum ratio of action choices. Furthermore, an analysis of their performance showed that the influence of the previous reward history was lower in the operated animals.

Behrens et al. (2007) followed up this study by scanning human subjects while the probability that choosing a red or green stimulus led to reward was varied over time; because the rewards were probabilistic, and there was no 'right' answer. There was an activation in the dorsal prelimbic cortex (area 32) that was related to the optimal estimate of the volatility of the rewards in a computational model.

It is not possible to use repetitive transcranial magnetic stimulation (rTMS) to disrupt activations that lie this deep on the medial surface. However, deep ultrasound can be used in macaque monkeys. If it is applied to the ventral prelimbic cortex on a task in which the probability of the rewards switch, it disturbs the translation of the internal representations of value into the behavioural choice (Fouragnan et al., 2019).

Summary

Monkeys with lesions in the sACC are slow to reverse their choices between actions. They are poor at learning both from reward and non-reward. If the rewards are given on a probabilistic basis, they are slow to learn the optimal ratio of action choices. When the probabilities of the rewards change, human subjects can estimate the degree of reward volatility, and there is an activation the dorsal prelimbic cortex that encodes this estimate. If deep ultrasound is applied to the ventral prelimbic cortex in macaque monkeys, it interferes with the translation of the internal value into behavioural choice.

Encoding Value

Action reversals depend on information about the probability of reward. So Kennerley et al. (2009) recorded the activity of cells in the sACC, while varying the probability of reward, the magnitude of reward, and the number of responses required to obtain the reward, a measure of effort. They then looked for correlations in the rates of activity with one or more of the three valuation variables. They found that cell populations in the sACC encoded all three choice variables.

Other neurophysiological studies have also found outcome-related signals in the sACC (Seo & Lee, 2007; Hayden et al., 2009; Hayden & Platt, 2010; Cai & Padoa-Schioppa, 2012). However, the rostral sACC makes a different contribution from the more caudal sACC. Both Kennerley et al. (2009) and Hayden et al. (2010) found value-encoding cells rostral to the genu of the corpus callosum in macaque monkeys. But Kennerley et al. (2009) found an abrupt fall-off in value coding in the sACC caudal to the genu. This drop-off in value coding corresponded to an increase in the representation of actions in the cingulate motor areas.

In addition to encoding the value of outcomes, cell populations in the sACC encode the difference between the expected and actual outcome. Prediction errors of this kind can serve as a 'common currency' for integrating multiple choice variables (Kennerley et al., 2011). When monkeys learn associations between actions and outcomes, cell activity in the sACC can encode either positive or negative prediction errors (Matsumoto et al., 2007). In the literature these are referred to as 'signed' prediction errors (Rescorla & Wagner, 1972).

In these experiments the animals were presented with a single choice, and they could act as soon as that choice was available. Hayden and his colleagues changed the conditions so that the monkeys were presented with two offers, one after the other. One offer appeared on the left and the other on the right; and the magnitude of the rewards and costs was varied.

Blanchard and Hayden (2014) analysed cell activity in the dorsal bank of the sACC and found that the activity reflected the choice made on the basis of both offers. Azab and Hayden (2017) showed that the activity after the first offer indicated that the animals had not yet committed to action; it only reflected commitment after the second offer had been presented.

Since the offers were presented on the left and right, the question arises whether the location of the offer is also reflected in the cell activity; and Strait et al. (2016) were able to demonstrate that it is. This is not surprising given that the dorsal PF cortex (areas 46 and 9/46) projects to the sACC (Petrides & Pandya, 1999; Saleem et al., 2014). Lesions of the dorsal PF cortex impair performance on tasks that require a choice between spatial locations (Goldman-Rakic, 1998).

Since the magnitude of the reward was varied, the final issue is whether the cell activity reflected that magnitude. Blanchard et al. (2015) analysed the cell activity while the monkeys waited for the expected reward and found cell activity that ramped up during this period. They suggested that this activity promoted a consistent commitment to a choice, and that it increased the motivation for making the same choice in the future.

However, though the studies reviewed here have shown that there is activity in the sACC that encodes value and commitment to action, they have not indicated in what way this differs from similar activity in the orbital PF cortex. To find out, Hunt et al. (2018) recorded simultaneously from cells in the orbital PF cortex (area 13), dorsal PF cortex (area 9/46) and the sulcal cortex of the ACC. As Chapter 1 mentions, the task involved attention-guided search and choice.

As would be expected, given that these areas are interconnected, similar cell activity could be found in all three areas. Therefore, to distinguish the pattern of activity for the populations a whole, Hunt et al. used representational similarity analysis (RSA). They found that the activity in the sACC differed in reflecting the emerging selection of which action to choose.

Summary

There are cells in the sACC that represent variables associated both with the probability and magnitude of reward and with choice. Value encoding is more common in the rostral sACC whereas the representation of actions is more common more caudally in the cingulate motor areas. Cell activity in the sACC ramps up as monkeys wait for an expected reward.

Exploitation and Exploration

The tasks that are described in the previous sections were devised without considering the choices that animals have to make in their natural environment. By contrast, behavioural ecologists have studied the foraging strategies that animals adopt in their natural habitats (Davies & Krebs, 2012), and they have also developed theories of optimal foraging that relate to economic theory (Monteiro et al., 2013; Vasconcelos et al., 2013).

In their natural habitat, animals gain an advantage by continuing to exploit known food resources when the environment is stable, but in a volatile environment it can pay to explore alternatives (Berger-Tal et al., 2014). The distinction between patchy and uniform resources plays an important role in foraging theory. Many primates forage for foods that occur in patches. The value of a patch diminishes if animals exhaust its resources, at which point it makes sense to search for another patch (Stephens & Krebs, 1986). The optimal foraging strategy also depends on other factors, such as the seasonality of resources. Foraging strategies can change as environmental conditions change.

Laboratory Model of Exploiting or Exploring

In a pioneering neurophysiological study, Hayden et al. (2011) designed a laboratory task to mimic the choice between exploiting a resource and exploring for new resources. Macaque monkeys had the option of choosing a blue rectangle, which meant 'stay', and they obtained a juice reward for this choice. The choice to stay was taken to correspond to the exploitation of a stable resource.

However, the amount of juice diminished when the monkeys made this choice repeatedly, reflecting the exhaustion of the resource in that patch. The animals could then opt to 'leave' by choosing a grey rectangle. This choice did not lead to an immediate reward; instead, it generated a long delay. A visual cue informed the monkeys about the duration of that delay. The choice to leave corresponded to the exploration of alternative resources, and the delay emulated the cost in terms of travel time.

Hayden et al. recorded in the sACC while monkeys performed this task. Whenever monkeys had to choose between exploiting a current but declining resource and switching to a new one, cells in the dorsal bank of the sACC increased their activity. For each delay interval, activity climbed toward a fixed threshold that correlated with the choice to switch to the new resource. This led Hayden et al. to conclude that the cells in the dorsal bank of the sACC encode the likelihood that monkeys will use information about resources to choose between exploitation and exploration.

A study by Quilodran et al. (2008) provides support for this conclusion. They showed that cells in the sACC encoded the feedback needed to switch between periods of exploration and exploitation. These cells ceased to encode rewards during periods of exploitation. At the beginning of new exploratory periods, the activity resumed.

The ACC (area 24) and the posterior cingulate cortex (area 23) have strong reciprocal connections (Vogt & Pandya, 1987) and the same kinds of representations can be found in the posterior cingulate cortex. Barack et al. (2017) recorded there while monkeys performed the task described earlier. Monkeys chose a rectangle of one colour to 'stay' or one of another colour to 'leave'. As expected, the monkeys chose to stay more frequently, the longer the delay that was imposed after a choice to leave. Activity in the posterior cingulate cortex predicted when the monkey chose to explore, and this neural signal developed several seconds before a choice to leave.

One can also study exploration with human subjects. In an fMRI experiment by Kolling et al. (2012) the subjects could choose between two stimuli on offer or decide to explore the consequences of choosing between alternative ones. Activations in the sACC encoded the choice variables related to exploitation or exploration. These results are as one would expect from the neurophysiological data on animals.

Nonetheless, Shenhav et al. (2014) objected that the activations simply reflected task difficulty. Yet, as Kolling et al. (2016) pointed out, in their original experiment task difficulty accounted for just 2% of the overall variance.

The reason why Shenhav et al. (2014) raised their objection is that they believed that the ACC monitored conflict (Botvinick et al., 2004). But the fact that there is more activation when decisions are made difficult due to conflicting evidence does not prove that the critical factor is conflict. The tissue in the sACC is involved in decisions, and as with any area, the more the attentional demands, the more the activation. Mansouri et al. (2017) have reviewed the studies that show an increase in cell activity under these conditions.

The decision whether to exploit or explore depends on detecting changes in the rates of reward. Wittmann et al. (2016) found that human subjects can perform this calculation implicitly. By studying fMRI activations, they found that opposing signals in the sACC represent current and previous reward rates as well as reward-prediction errors. Furthermore, a computational model could account for choices based on these representations.

Costs

There are costs involved in exploration. It takes effort to travel to new resources and it takes time to find them. In the laboratory task devised by Hayden et al. these were modelled by imposing a delay before the next choice became available. But in the natural environment the costs involve not only the time before the next choice (delay) but also the energy that is involved in travelling to the new resources (effort).

Hosokawa et al. (2013) recorded in the sACC and compared effort costs and delay costs. They found that activity in the majority of the cells in the sACC reflected effort costs. However, there was also some activity in the sACC that reflected delay costs.

This may have been derived from the orbital PF cortex. An earlier section mentions the study by Hunt et al. (2015) in which they recorded single-cell activity and local field potentials in the orbital PF cortex, dorsal PF cortex, and sACC. In their analysis, they devised an index to capture the trial-by-trial dynamics of the monkeys' choices. This index showed that when the choice involved effort costs, the sulcal ACC influenced activity in the dorsal PF cortex. But when the choice involved delay costs, it was the orbital PF cortex that exerted this influence.

Studies of fMRI activations have also looked for a role for the human ACC in evaluating efforts costs. In one experiment, Croxson et al. (2009) provided the subjects with information about the degree of effort required for particular choices. They found that activations in the sACC reflected the interaction between the expected reward and the effort costs.

In a later study, Klein-Flugge et al. (2016) varied effort in a different way by changing the grip force required to obtain a reward. The magnitude of reward was varied independently. They found activations in the sACC that correlated with the difference between the chosen option and the alternative in either effort cost, reward magnitude, or both. Furthermore, they could account for these results with a computational model that involved a comparator for effort costs under the different options.

Summary

Cell activity in the dorsal bank of the sACC is involved in the choice of whether to exploit the current resources or to explore so as to find new ones. There is also activity that reflects the energetic costs of exploration.

Anterior Cingulate Cortex and Social Behaviour

The previous sections have considered the role of the sACC in supporting choices between resources while foraging. These choices are self-generated in that there is no external cue to specify the choice that is appropriate. The role of the sACC in self-generated action is consistent with the relative lack of inputs to the ACC from the foveal representation in the ventral visual system.

However, the section on connections at the start of this chapter mentions that there are projections from the superior temporal cortex to the ACC, including the prelimbic cortex (Saleem et al., 2008). This input probably provides information about the vocalizations and faces of conspecifics (Perrodin et al., 2014). There is also a projection from the upper bank of the STS (Morecraft et al., 2012); this input probably provide information about the actions of others (Jellema & Perrett, 2003). Mars et al. (2013) suggested that these interconnected temporal and ACC areas comprise a neural network that supports social cognition.

They used resting-state covariance to show that in macaque monkeys the cortex in the middle part of the STS has connections with the prelimbic cortex. In the human brain, these connections come from the temporal–parietal junction (TPJ) which lies at the border between the temporal lobe and parietal lobe (Mars et al., 2013).

Social Interactions

There is evidence that the ACC encodes the value that a monkey attributes to conspecifics. Hadland et al. (2003) reported that monkeys with a lesion in the ACC approached other individuals in their social group less frequently. Rudebeck et al. (2007) also found that rats with excitotoxic lesions in the ACC showed a diminished interest in conspecifics.

However, an objective measure is needed of this lack of interest. Rudebeck et al. (2006) therefore designed an experiment in which the monkeys could obtain peanuts, but had to do so in the presence of a threatening male macaque. Under these conditions, intact control monkeys were slow to pick up the peanuts. But the monkeys with ACC lesions showed no such hesitation. This was taken as evidence of their lack of concern.

Given that this area plays a role in social valuations, Sallet et al. (2011) investigated whether the degree of its involvement depends on the number of animals with which a monkey interacts. They acquired structural MRIs of 23 macaque monkeys that lived in groups of varying size and found an increase in the grey matter in several areas that correlated positively with group size. These areas included the middle part of the STS, the superior part of the temporal pole, and the rostral part of the principal sulcus in the PF cortex (area 46). And critically, the larger the group size, the greater the degree of covariance between the STS and an area that included the rostral ACC and prelimbic cortex.

These changes presumably result from an increase in the frequency and complexity of the social interactions in groups of larger size. In macaque monkeys, these interactions include the establishment and maintenance of a dominance hierarchy. Therefore, Noonan et al. (2014) investigated whether the amount of grey matter in the STS and PF area 46 correlated positively with social dominance. They found that it did. Since area 46 projects to the prelimbic cortex (area 32) (Petrides & Pandya, 1999), presumably the prelimbic cortex also has access to information about social interactions.

That it does was proved by Sliwa and Freiwald (2017) who used fMRI to visualize the neural system in the macaque monkey brain that is involved when monkeys watched videos of other monkeys while they interacted socially. Figure 3.5 shows these areas included not only the dorsal PF area 46 but also the STS, the ACC and prelimbic cortex. The flat map in Figure 3.5B shows that this is same circuit in which the grey matter has been shown to vary as a function of the size of the social group (Sallet et al., 2011).

However, to interact successfully, monkeys need to interpret the actions of their conspecifics. It turns out that there are cells in the pregenual prelimbic cortex that show modulations in their activity while monkeys approach, make contact with, or groom conspecifics (Mao et al., 2017).

As they forage in groups, macaques both cooperate and compete for resources. To study cooperation, Chang et al. (2013b) taught monkeys to allocate rewards either to themselves or to another monkey. They found cells in the pregenual

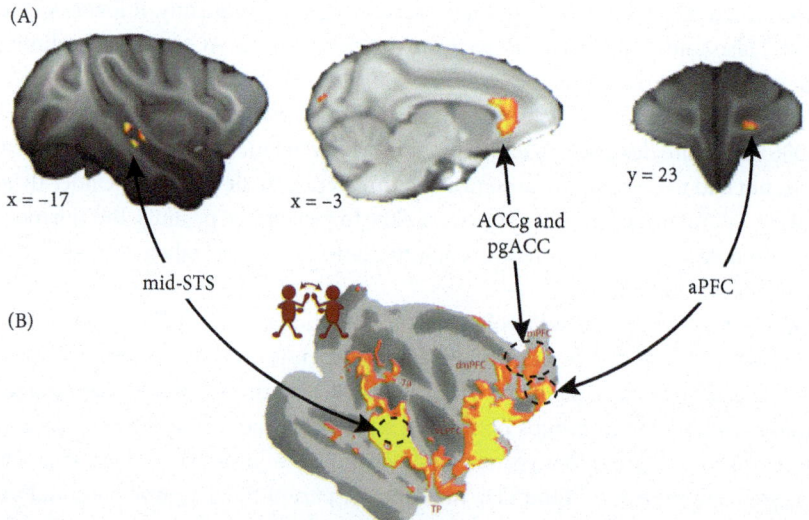

Figure 3.5 (A) Areas that were specifically activated when monkeys watched videos of other monkeys interacting. (B) Flat map of areas that showed a difference in grey matter depending on the size of the group
ACCg = anterior cingulate gyrus, prACC = pregenual ACC (prelimbic area 32), aPFC = anterior PF cortex.

Reproduced from Wittmann, M.K., Lockwood, P.L., & Rushworth, M.F.S. neural mechanisms of social cognition in primates. *Ann Rev Neurosci*, 41, 99–118, Figure 1, Doi: 10.1146/annurev-neuro-080317-061450 © 2018, Annual Reviews.

prelimbic cortex and the rostral sACC that encoded the decision to allocate rewards either to the monkey itself, to another monkey, or to both of them.

Macaques also need to be able to predict how others in their group will behave. Accordingly, Haroush and Williams (2015) taught monkeys to play a game based on the prisoner's dilemma. In this game, players can gain by cooperating, but they also obtain benefits from 'defecting' and pursuing their own interests. In the version of this game for monkeys, two macaques sat side by side and made choices, in turn, for juice rewards. They could not see the choice that the other monkey made until both of them had made a choice. The choice was self-generated; there were no cues to instruct the choice that would be best.

Haroush and Williams recorded cell activity in the ACC, although the exact location remains unclear from their report. As expected, the monkeys chose to cooperate more frequently after their partner monkey had just cooperated, and they chose to defect more frequently if their partner monkey had just defected. Haroush and Williams observed that 28% of the cells encoded a partner's choice and did so before the revelation of that choice. This compared with 11% of the cells that encoded the monkeys' own choice.

These results suggest that the cell activity supports the ability of macaque monkeys to predict the choices of a conspecific, and to choose whether to cooperate or compete. To find out if this was the case, Haroush and Williams used intracortical micro-stimulation to disrupt activity in the ACC. This manipulation reduced the likelihood that a monkey would cooperate. Basile et al. (2020) also found that a monkey with an anterior cingulate lesion could not learn to offer rewards to another monkey when the alternative was no reward for either monkey.

Behrens et al. (2008) set out to specifically compare the areas involved in social and non-social choices in the human brain. The subjects first saw two stimuli that differed in the probability of their association with reward. They then saw a 'suggestion' from a partner that supposedly conveyed the best choice, but this recommendation varied in its reliability. The subjects then made their choice and discovered the outcome.

Behrens et al. used a Bayesian model with parameters for reward-prediction errors, reward volatility, and outcome probability at a given time. Their model matched these parameters with activations for either social or non-social cues.

Correlations with the value of new information occurred in different parts of the ACC. When the subjects were learning on the basis of non-social cues, the correlation was in the sACC. This is as one would expect from the data from neurophysiology on monkeys. However, when the subjects were learning on the basis of social cues, the correlations occurred in the anterior cingulate gyrus (gACC).

These results suggest that non-social and social choices depend on similar cortical operations computed in nearby parts of the ACC. Chang et al. (2013a) have suggested that cortical mechanisms in the ACC that originally served non-social functions were co-opted to serve social ones. In a later review they suggested that this occurred both at the algorithmic and implementational level (Lockwood et al., 2020).

The assumption is that this functional augmentation occurred during anthropoid evolution as these species developed complex social systems. Co-option of this kind could occur simply by developing new inputs to a cortical area, such as those conveying representations of vocal calls, faces, the actions of conspecifics, and so on. The pre-existing input–output transform would simply operate on these new inputs as it does on the phylogenetically older ones

Summary

To survive, the anthropoid's species not only needed to search for food, but also to interact with other members of their social group. In anthropoids, interconnected cortical areas differ in size as a function of social interactions and dominance. The prelimbic cortex and the rostral part of the gACC play a critical role in this network.

These findings suggest that the cortical mechanisms of the ACC which originally evolved in support of foraging became co-opted for social function by the intrusion of new inputs into the ACC. These inputs drew on the ability of the ACC to represent actions, so enabling the animals to predict the actions of conspecifics.

The Medial Network

Along with the retrosplenial cortex and hippocampus, the granular areas 9 and 10 of the medial PF cortex are part of the so-called 'default network' (Raichle & Snyder, 2007; Fox et al., 2009). This network also includes the medial parietal cortex (7m) and the inferior parietal cortex.

The term 'default network' originally referred to a set of areas that showed a decrease in fMRI activation when human subjects were given tasks to perform. It has become common in the literature to contrast it with networks that are involved in 'task control' (Dosenbach et al., 2007; Vincent et al., 2008).

But unfortunately, the terms 'default' and 'task control' can give a misleading impression. They suggest that the medial network is activated when the subjects have *no* task to perform, as if the medial network was not involved in supporting task performance. But the proper distinction is not between a task and no task, but between a task that the experimenter devises and a task that is self-generated. The medial network is activated when human subjects think about the past, the present or the future, and do so spontaneously, of their own accord (Mason et al., 2007). When they do so, they are indeed carrying out a task.

Mars et al. (2012) used the BrainMap Database to visualize the medial cortical areas that showed a decrease in activation when macaque monkeys performed tasks that were instructed by external cues. They then compared this map with the map that showed areas with resting-state coactivation with the STS. They chose this region as a seed area because it shows an increase in grey matter with the size of the social group (Sallet et al., 2011). Mars et al. found an area of overlap between the two maps, and this is shown in blue in Figure 3.6.

In order to study the connections within the medial network, Li et al. (2013) used probabilistic DWI in macaque monkeys. They refer to the medial network as a 'connectional hub', a term that refers to cortical areas that have close connections with each other and that also communicate with several 'spoke' areas within an overall connectome (van den Heuvel & Sporns, 2013).

To validate their method, Li et al. compared the connections as delineated by DWI with the connections as identified by tracer methods. They found a reasonable correspondence. Figure 3.7 (top, on the next page) illustrates the medial hub as visualized by DWI. It includes the granular areas of the medial PF cortex (areas 9 and 10) as well as the parietal area 7m and the retrosplenial cortex.

Barks et al. (2015) visualized the medial network in chimpanzees by looking for areas that showed a decrease in activation when the chimpanzees performed a task. The task was the visual matching-to-sample task, in which the sample and choice items were short videos with social content. Barks et al. then anesthetized the animals and performed PET scans. As Chapter 1 explains, if 2-18F-fluoro-2-deoxyglucose is used as a positron emitting marker, changes in activation that occurred during task performance can be visualized during the subsequent PET scan.

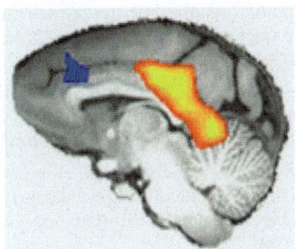

Figure 3.6 Overlap between medial (default) network and social network in macaques
Right: a midsagittal view with rostral to the left and dorsal up; left: a horizontal slice with rostral up. The areas in yellow, red, and orange contribute to the medial network, as defined by a decrease in fMRI activation during cue-guided tasks. The area in blue shows a region of overlap between the medial network and the part of the ACC that is increasingly recruited into the medial network when animals live in bigger groups.
Reproduced from Rogier B. Mars, Franz-Xaver Neubert, MaryAnn P. Noonan, Jerome Sallet, Ivan Toni, & Matthew F. S. Rushworth. On the relationship between the 'default mode network' and the 'social brain'. *Front Human Neurosci*, 6, 189, Figure 3, Doi: 10.3389/fnhum.2012.00189 © 2012 Mars, Neubert, Noonan, Sallet, Toni and Rushworth. Licensed under CC BY 4.0.

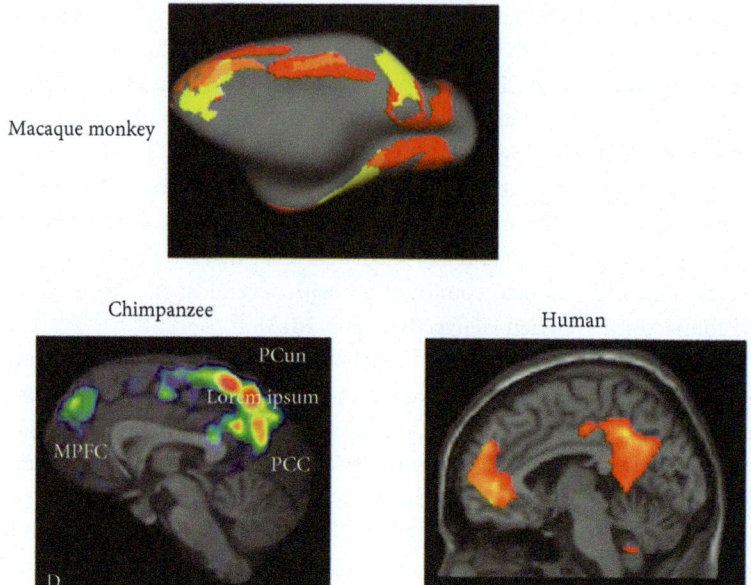

Figure 3.7 Medial network as visualized by fMRI
(Top) Network as shown by the use of DWI to show the interconnections in the macaque monkey brain. (Bottom left). The default network as visualized using PET in chimpanzees. MPFC = medial granular PF cortex; PCun = precuneus = medial parietal 7m; PCC = posterior cingulate cortex. (Bottom right) The medial network that is activated when human subjects recall events from their past life.

A: Reproduced from Longchuan Li, Xiaoping Hu, Todd M. Preuss, Matthew F. Glasser, Frederick W. Damen, Yuxuan Qiu, & James Rilling. Mapping putative hubs in human, chimpanzee and rhesus macaque connectomes via diffusion tractography, *Neuro Image*, 80, 462–74, Figure 3a, https://doi.org/10.1016/j.neuroimage.2013.04.024 Copyright © 2013, Elsevier Inc. All rights reserved.

B: Reproduced from Sarah K. Barks, Lisa A. Parr, & James K. Rilling, The default mode network in chimpanzees (pan troglodytes) is similar to that of humans. *Cereb Cortex*, 25 (2), 538–44, Figure 1a, Doi: https://doi.org/10.1093/cercor/bht253 Copyright © 2013, Oxford University Press.

C: Reproduced from Addis, D.R., Wong, A.T., & Schacter, D.L. Remembering the past and imagining the future: common and distinct neural substrates during event construction and elaboration. *Neuropsychologia*, 45 (7), 1363–77, Figure 2, Doi: 10.1016/j.neuropsychologia.2006.10.016 Copyright © 2006, Elsevier Ltd. All rights reserved.

As Figure 3.7 (bottom left) illustrates, Barks et al. found decreases in activation during performance of the task in the granular parts of the medial PF cortex (labelled MPFC), the medial praecuneus (PCun) or area 7m, the posterior cingulate cortex (PCC), and the retrosplenial cortex. The same network is also activated when chimpanzees are lying or sitting at rest (Rilling et al., 2007).

Figure 3.7 (bottom right) illustrates the medial network in humans. It shows the fMRI activations when human subjects retrieve events from their past life (Addis et al., 2007; Hassabis et al., 2007). Within this network the medial PF cortex includes the ventral prelimbic cortex (area 32) and the polar PF cortex (area 10) (Addis et al., 2007). DWI has been used to show that these areas are connected with the anterior and posterior cingulate cortex (Jackson et al., 2020).

Chapter 10 discusses these findings in detail, and in particular enquires what if anything is implied by the fact that the medial PF cortex is activated when chimpanzees are at rest. Here, it is sufficient simply to point out that the same medial network exists in macaque monkeys, chimpanzees, and humans.

The Hippocampal System and Navigation

Visual scenes play an important role in navigation, and Murray et al. (2018) have reviewed the role of hippocampus and its homologues in scene-guided navigation across vertebrate classes. However, most studies of the map-like representations that guide navigation come from rodents.

Visual cues that define a scene or a spatial context specify a foraging field. In the rodent hippocampus, the 'place cells' specify where the animal is within that scene. (Hartley et al., 2014). By contrast, in the entorhinal cortex there are 'grid cells', as well as 'head direction' and 'speed cells' (Rowland et al., 2016). Whereas the entorhinal cortex encodes movement through the whole environment, the place cells in the dentate encode a specific place within that environment.

By studying local field potentials, Foster and Wilson (2006) were able to demonstrate that sharp-wave ripples in the hippocampus replay the discovery of a goal. However, they do so in reversed order of the animal's movements towards that goal, as defined within a visual scene (Foster & Wilson, 2006).

So, this activity reflects recall of the path that had been taken. And when on the next occasion the rat returns, brief sequences of activity occur in hippocampal place cells that encode journeys towards that same spatial goal (Pfeiffer & Foster, 2013).

The primate hippocampus has similar properties. Sharp-wave ripples recorded from the hippocampus also show enhancement when macaque monkeys search for objects in remembered scenes (Leonard & Hoffman, 2017). And importantly, this activity leads to changes in the activation in the rest of the medial network (Kaplan et al., 2016).

There is also evidence for replay in the human hippocampus. Liu et al. (2019) taught classifiers to recognize individual objects in an experiment using magnetoencephalography (MEG). The subjects were first taught rules for ordering sequences of objects, and then required to apply them when shown sequences of new objects in a scrambled order. This was followed by a rest period of 5 minutes, and during this period it was possible to detect replay of the sequences. So, replay is also involved in remembering sequences that do not involve navigation.

There is even evidence for a grid-like organization of knowledge in the human brain. Constantinescu et al. (2016) scanned subjects with fMRI while they matched morphs of birds that differed in two continuous dimensions, the length of the neck and the length of the legs. Thus, the subjects learned about 'bird space' as opposed to navigational space. There was a signal in the entorhinal cortex and ventral prelimbic cortex that showed a grid-like or hexagonal symmetry as the subjects navigated through bird space. Furthermore, the subjects who showed greater hexagonal modulation in the ventral prelimbic cortex performed better on the task.

There is a relation between distance and the time taken to navigate. However, compared to the many experiments on place cells and grid cells, fewer investigations have examined cell activity that is related to timing even though it is no less important. Eichenbaum (2014; 2017) described hippocampus cells that had activity that encoded the temporal context and other temporal information such as the interval between events and the progress between landmarks. Cells that code for time have been described in rats and mice (Banquet et al., 2020. Optogenetic inactivation of the medial entorhinal cortex in rats also selectively impairs the temporal but not spatial representations in hippocampal cells (Robinson et al., 2017).

Furthermore, by using transgenic techniques to tag the active hippocampal cells in mice, Cai et al. (2016) showed that the active neuronal ensembles have more overlap for events that occurred within 1 day than for events that occurred 7 days apart. So, the representation of time could depend on leaky integrators that represent the current temporal factors, with the information for more remote events gradually becoming weaker (Howard et al., 2014).

Testing the Ability to Navigate

In rodents, the hippocampal system provides a mechanism for using past navigational sequences to repeat a previous journey to a goal. The Morris water maze provides a way to examining this ability. An array of external or extra-maze cues specifies the scene or context, and the rat must find a hidden platform from any location within that scene (Morris, 1984). The ability to return to a spatial goal has been taken to depend on NMDA-dependent long-term potentiation (LTP) (Vallejo et al., 2014).

However, it has been found that mutant rodents that lack hippocampal NMDA receptors can navigate to a single goal in the water maze (Bannerman et al., 2014). The impairment only showed up when they had to discriminate between platforms, of which one was stable and the other one sank in the water (Bannerman et al., 2014).

This indicates a role for the hippocampus in using cues from the external scene to discriminate between locations. Murray et al. (2018) reviewed evidence from macaque monkeys that rules out the idea that the hippocampus is necessary for spatial memory in a general sense. Complete hippocampus lesions can leave performance on several tests of spatial memory unchanged. Instead, the macaque hippocampus represents the kinds of spatial information that is useful for navigation, such as visual scenes.

Research using brain imaging has also confirmed a role in navigational representations for the human hippocampus (Epstein et al., 2017; Hinman et al., 2018). Hassabis et al. (2009) studied human subjects as they navigated in a virtual reality environment. It proved possible to decode the location of an individual within that environment from activations in the hippocampus.

In a pioneering fMRI study, Spiers and Maguire (2006) studied highly experienced taxi drivers while they 'drove' around a virtual reality representation of the streets of London. Remarkably, only one particular condition led to significant activations in the hippocampus. This was when the driver was instructed to go from one specific place to another. For example, there was an increase in hippocampal activation when the driver was told to drive from Trafalgar Square to the Bank of England. This activation presumably reflected both retrieving the memory of past routes and planning a new one.

To show that the hippocampus was critical for navigation, Maguire et al. (2006) studied a taxi driver who had suffered a stroke that caused bilateral damage to his hippocampus. He had driven through London for 40 years. Yet, when faced with the virtual-reality version of London, he became totally lost when asked to drive from St Paul's Cathedral to the Bank of England: yet this is only a short drive.

Summary

The hippocampal system plays a critical role in the ability to navigate. Cells in the hippocampus encode the place of the animal in the environment and cells in the entorhinal cortex the progression of the animal through the environment. There is cell activity in the hippocampus that replays sequences as well as activity that encodes a future path back to a spatial goal. There is a relation between distance and the time taken to travel, and there is also cell activity in the hippocampus that encodes time.

From Navigation to Episodic Retrieval

The previous section mentions that the medial network is activated not only when human subjects navigate but also when they retrieve memories of their past experiences. In the human literature this is referred to as retrieval from episodic memory. An individual might, for example, recollect buying a ticket at a cinema booth last Friday night. In fMRI studies of episodic retrieval, a typical control condition is remembering objects (Hassabis et al., 2007) or words (Addis et al., 2007). When subjects remember objects as opposed to events in their life, the activations are in the ventral visual system and its extension into the ventral PF cortex (Hassabis et al., 2007).

These results are as one would expect from the anatomical connections. The medial network receives inputs from peripheral vision via the parahippocampal and parietal cortex (Baizer et al., 1991), and these can be used to define a scene (Nasr et al., 2011). The ventral vision system receives inputs from foveal vision (Baizer et al., 1991), and these are involved when a subject views an object or word.

Episodic Memory

There is much discussion in the literature about whether animals have episodic memory. Clayton and Dickinson (1998) found that scrub jays (*Aphelocomoa californica*) can remember how much time has elapsed since they cached food; and they can also remember what the food was and where they cached it. Later studies have shown that there are other birds that have the same ability (Salwiczek et al., 2010).

The original finding led Clayton et al. (2003) to suggest that a necessary condition for episodic memory is the ability of an individual to remember not only what and where it did something, but also when it did so. The literature sometimes refers to these as what, where, when (WWW) memories. However, when people remember events from their past, they are not always able to place them in time or remember where the event occurred. In other words, not all episodic memories fulfil the WWW requirement.

The experiments on scrub jays exploited their natural behaviour, namely that they cache their food. So, the best way of seeing whether monkeys or apes can remember past events is to study foraging. Janson (2016) studied capuchin (*Cebus*) monkeys. He set up eight feeding sites and arranged that over time the amount of food increased at each site. The idea was to model the fruiting of trees. Over a test of 68 days, the monkeys used the time that had elapsed since their last visit to a site as a guide for when it would be best to revisit the site. This means that they could remember when they last visited, where the site was and what food was there.

This naturalistic experiment can be contrasted with the laboratory experiment by Hoffman et al. (2018). Using delayed matching, it took thousands of trials to teach macaque monkeys to remember what they had done, where they had done it, and how much time had elapsed since then. And this experiment only tested short-term, not long-term memory.

Sayers and Menzel (2012) carried out a semi-naturalistic experiment with chimpanzees. The animals were given the chance to see where experimenters were hiding ten, transparent bags of food at a variety of locations in a wooded area. The bags differed in the quantity and quality of the foods that they contained. Roughly 25 minutes later, the chimpanzees had the opportunity to guide a naïve person to different locations in the wooded area. They did so by using lexigrams and also pointing to the locations. The chimpanzees chose the most efficient paths for retrieving the foods that they preferred. These paths took into account the distance of the various foods as well as their value and so minimized the time and effort required to retrieve them.

The chimpanzees were therefore able to recall what foods had been hidden and the location from long-term memory. This test did not require them to distinguish when food had been hidden at a particular location. To demonstrate that the chimpanzees could also remember the time that elapsed it would be necessary to require them to indicate the order in which the bags were hidden.

This experiment only tested memory for an event that occurred less than an hour ago. However, Swartz et al. (2005) tested a western lowland gorilla (*Gorilla gorilla*) called King to see if he could remember what happened the day before. First, King had to learn the meanings of five cards that referred to different foods and three cards that referred to different people. By using these cards, King was able to indicate what food King had been given either 5 minutes or 24 hours ago and who gave him the food. This is a demonstration of 'source memory'.

As in human subjects, the retrieval of memories depends on the hippocampus in other primates. Hampton et al. (2004) tested macaque monkeys for their ability to remember where they saw someone hiding two foods in a spatial arena. The foods differed in how preferable they were. Since many trials were given, the animal has to remember not only where the foods had been hidden, but also where they had been hidden that day. Excitotoxic lesions of the hippocampus caused a very severe impairment on this task.

To study how information about events (what) and time (when) could be integrated, Naya and Suzuki (2011) presented monkeys with two objects in order. They recorded in the hippocampus, entorhinal, and perirhinal cortex, as well as in the inferior temporal cortex. A population of cells in the hippocampus encoded an estimate of the relative time since the presentation of the first cue as well as an estimate of the relative time for the presentation of the second cue. These cells provided an incremental timing signal.

At the same time, there were cells in the inferior temporal cortex that encoded one or other of the objects. This area projects to the perirhinal cortex (Webster et al., 1991), and it was only in the perirhinal cortex that cells integrated information about both time and item. In a later study, Naya et al. (2017) were able to demonstrate that information about the time and item was integrated earlier in the perirhinal cortex than in the PF cortex.

Given the fact that the hippocampus represents time as well as space, it is not surprising that in the human syndrome of retrograde amnesia, information is lost not only about both items and events but also the time at which they occurred.

However, though it was reported long ago that a bilateral hippocampal lesion caused a severe retrograde amnesia in patient HM (Scoville & Milner, 1957), it turned out that the lesion was not selective. As well as the hippocampus, it included the entorhinal cortex, and other tissue in the medial temporal lobe (Corkin et al., 1997).

There is, however, a more recent patient, Jon, who has a selective lesion of the hippocampus caused by a stroke during birth. As expected, he has a severe impairment in the retrieving events from his past life (Vargha-Khadem et al., 1997). Unlike healthy subjects he lacks the subjective imagery associated with past events.

In an fMRI experiment with healthy human subjects, MVPA has been used to demonstrate that, when they remember a film they have seen, it is possible to decode which specific film they are imagining at the time of retrieval (Bird et al., 2015).

The same method could be used to see if memories are reinstated in monkeys or chimpanzees. Of course, even if reinstatement were demonstrated, this would not prove that it was accompanied by any subjective experience. Indeed, the traditional view has been to deny that animals have such experiences (Suddendorf & Busby, 2003). But this is a hangover from the philosophical position taken by Descartes (1644), and Chapter 10 argues that at least for perception it is unjustified.

Navigation and Retrieval

The previous section has shown that the hippocampus is involved both in navigation and the retrieval of the memories for events. This raises an obvious question. How has a system that evolved for navigation been co-opted to support memories of events that do not involve navigation? The general answer is that once a particular kind of representation evolves, a new species will use it for a wide variety of functions unrelated to the selective factors that led to its emergence.

A more specific answer is that in monkeys and apes, new PF areas evolved, and these provided new inputs to the hippocampal system. Like other primates, these animals forage with their hands and arms. So, the actions that they perform are not confined to navigating from one place to another. It is true that rodents and carnivores use their paws to assist them in eating; but they lack the ability that primates have to use the fingers to manipulate food and other objects.

The actions that primates perform are defined by the spatial or object goal, and it is actions rather than the movements themselves that are remembered. Chapter 6 reviews the evidence that the dorsal PF cortex (area 46) specifies the goals for actions that are performed with the hand. This area is interconnected with the granular medial PF cortex (area 9) (Petrides & Pandya, 1999), as well as the preSMA (Wang et al., 2005), and CMAr (Morecraft et al., 2012); and the medial PF cortex is also connected with these areas (Wang et al., 2005; Morecraft et al., 2012).

It is critical, therefore, to note that the preSMA and CMAr represent the movements of the hand, but not the leg (He et al., 1995). The dorsal and medial PF cortices therefore provide new inputs to the retrosplenial cortex (Kobayashi & Amaral, 2003) and presubiculum of the hippocampus (Barbas & Blatt, 1995a). These inputs are linked with the new repertoire of movements that are available to primates.

Like navigation, these actions are also performed in a scene, and that scene provides the context for the memory. For example, when a human subject recalls buying a ticket at a cinema booth (Hassabis et al., 2007), the cinema acts as the scene in which the actions with the arm and hands occur.

The actions also have an outcome; for example, the ticket is handed out. In the case of non-human primates, the outcomes are usually food. Chapter 4 argues that the orbital PF cortex represents the value of foods, and the medial sector of the orbital PF cortex is connected with the retrosplenial, hippocampal, and parahippocampal cortex (Cavada et al., 2000).

Thus, the anatomical connections suggest that in monkeys the new granular medial PF areas could play a role in the spontaneous retrieval of actions that the animals have performed and the outcomes, and not simply the journeys that they have taken. Thus, a system that evolved for remembering the paths that lead to resources has been co-opted to remember events. These involve the scene, the actions, and the outcome.

Summary

Monkeys and apes are able to retrieve memories of actions that they have performed in the past and the outcome that followed. The system that evolved for navigation has been co-opted to the retrieval of events. This has been achieved by the evolution in monkeys and apes of the granular medial PF cortex which provides novel inputs to the medial network. The dorsal and medial PF cortices have outputs to the preSMA that represent movements of the hand and arms, but not the leg.

Medial Granular Prefrontal Cortex

There is a dearth of information from neurophysiology and neuropsychology on the functions of the granular medial PF cortex (areas 9 and 10) in macaque monkeys. Yumoto et al. (2011) trained macaque monkeys to indicate when a specific time interval had elapsed by pressing a button. The action was self-generated since there was no external cue to tell them when that time had elapsed. The monkey had to retrieve the interval of their own accord.

Yumoto et al. recorded in the medial PF cortex (area 9) when the monkeys generated the time interval. The study found cells that coded for specific intervals.

Furthermore, when the medial PF cortex was inactivated using muscimol, the performance of the monkeys was inaccurate. The intervals in this study were between 2 and 7 seconds. A study is therefore needed to find out whether the cells can code for longer intervals, and not simply for short-term memory.

There are also cells in the granular medial PF cortex (area 9) that code for order. Carpenter et al. (2018) taught monkeys a task on which visual stimuli cued three, four, or five locations in a specific series. The animals did not have to remember the locations because once any particular location appeared, it remained visible until the test period. During the recall or retrieval test, one of the locations turned blue, and the monkey had to move a cursor to the location that had followed this location in the series. So, the monkeys had to retrieve or recall the order of events from memory.

Carpenter et al. recorded cell activity in the lateral part of the medial PF cortex and the dorsal premotor cortex. During the presentation of the series, roughly 40% of the task-related cells in both areas encoded the conjunction of the location of the stimulus and its order in the sequence. The activity in the two areas differed in that the cells in the medial PF cortex were most active during the presentation phase, whereas the cells in the dorsal premotor cortex continued to encode the location and serial position during the test phase of the trial.

Thus, the experiments of Yumoto et al. (2011) and Carpenter et al. (2018) are consistent in suggesting that the granular medial PF cortex codes for events. The events were sequences that the animals observed, and which evolved in order and over time. In both cases memory for the events was assessed by recall.

Tsujimoto et al. (2012) recorded instead from cells in the frontal polar cortex. Macaque monkeys had to choose between two spatial responses. On the next trial, a visual cue told them whether to shift from or stay with that choice. But the cue did not specify the choice itself; the monkeys had to remember the choice that they had made on the previous trial. In other words, they had to generate their own choice from the memory of the previous event. The monkeys only found out whether they had made the correct choice after a fixed delay.

Figure 3.8 (next page) illustrates the key findings. In the polar PF cortex, cells only encoded the chosen action around the time of the reward or non-reward. When Tsujimoto et al. lengthened the interval between the action and the outcome, the time course of action coding shifted to this new outcome time. Unlike cells in orbital PF cortex, cells in the polar PF cortex had no sensory response to the visual cue that instructed them to stay or shift (Figure 3.8). Tsujimoto et al. took their finding to suggest that the frontal polar PF cortex plays a role in monitoring the outcome of actions that are self-generated.

To find out if the polar PF cortex is critical for learning an associative task, Boshchin et al. (2015) made lesions of the polar PF cortex after their monkeys had learned the 'object-in-scenes' task. On this task the monkeys saw a succession of twenty pairs of objects on repeated runs or trials, and they had to choose between

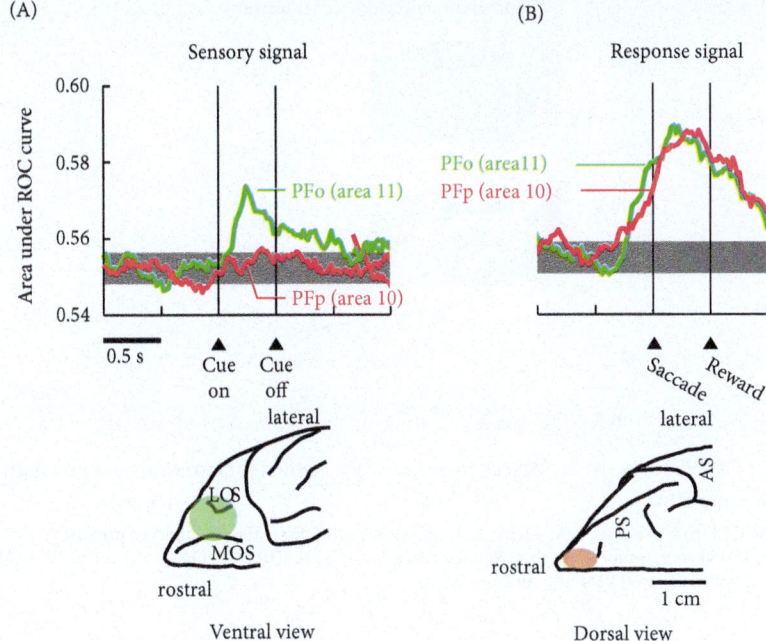

Figure 3.8 Cell activity encoding choices at feedback time for populations of cells in the polar PF cortex (area 10) (orange) and the orbital PF cortex (area 11) (green) The grey bar shows the level of population activity that fails to significantly encode the choice at each timepoint. An ROC value above the grey bar indicates significant encoding or either the strategy instructed by a visual cue (A) or the response (B). PFp, polar PF cortex; PFo, orbital PF cortex.

Data From Tsujimoto, S., Genovesio, A., & Wise, S.P. Neuronal activity during a cued strategy task: Comparison of dorsolateral, orbital, and polar prefrontal cortex. *J Neurosci*, 32 (32), 11017–31, Doi: https://doi.org/10.1523/JNEUROSCI.1230-12.2012, 2012.

the two 'objects' of each pair. To make the task easy, each pair of objects appeared against a unique and distinctly coloured background, and this served as a scene.

Figure 3.9 (next page) shows a series of example stimuli for two trials of this task. Gaffan (1992) proposed that monkeys solve this task by recalling the presentation of the objects in a unique scene. He argued that the task tests episodic memory.

Figure 3.10A (next page) shows the results of the study by Boschin et al. (2015). After the first presentation of the list of 20 scenes, the monkeys with polar PF lesions had a very severe impairment on the second presentation; this is shown by the bars labelled '1v2'. This finding suggests that they had forgotten which objects they had chosen in that single prior event, and this was taken to reflect an impairment in episodic memory.

Boschin et al. also tested monkeys on a task in which they touched two object-like images in succession to discover which choice led to a reward. This task differed

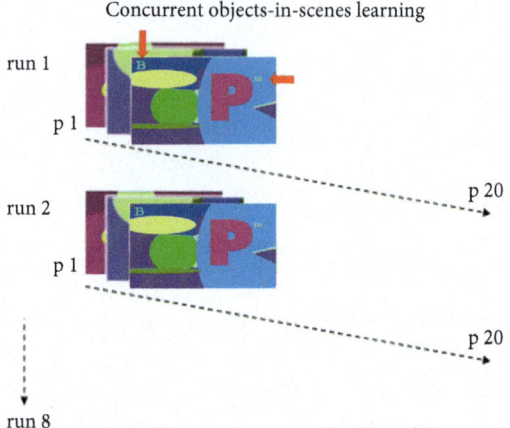

Figure 3.9 Pictures (p) for 'object-in-scenes' task, shown for two trials or runs, with 20 pictures per run.

Reproduced from Boschin, E.A., Piekema, C., & Buckley, M.J. Essential functions of primate frontopolar cortex in cognition. *Proc Nat Acad Sci USA*, 112 (9), E1020–7, Figure 2a, https://doi.org/10.1073/pnas.1419649112 © The Authors, 2015.

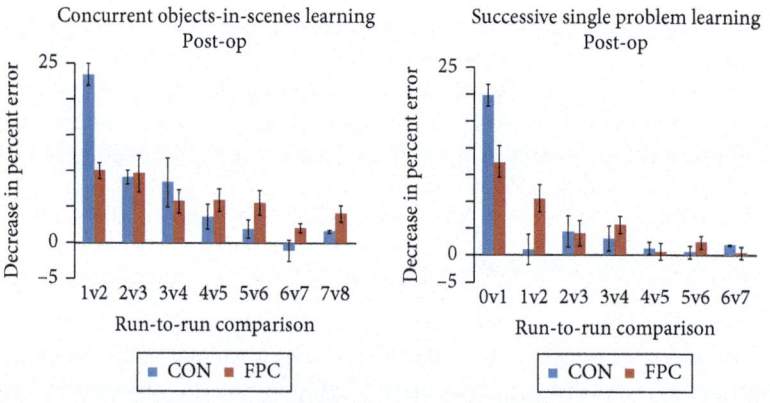

Figure 3.10 The effect of polar PF lesions in macaque monkeys
(A) Performance of the lesioned (red) and intact, control (blue) monkeys on a task in which they learn which of two 'objects' against a background scene is associated with reward. Each bar shows the increase percent-correct choices from one trial to the next. (B) As in part A, but for a task without background scenes. This task requires learning which of two picture-like stimuli is associated with reward from a single prior reward event.

Reproduced from Boschin, E.A., Piekema, C., & Buckley, M.J. Essential functions of primate frontopolar cortex in cognition. *Proc Nat Acad Sci USA*, 112 (9), E1020–7, Figures 2b and d, https://doi.org/10.1073/pnas.1419649112 © The Authors, 2015.

from the object-in-scenes task in two ways: it did not involve a background scene, and 10 choice screens, not 20, appeared in succession. However, like the object-in-scenes task, the monkeys had to make their choice based on the memory of a single event; in this case it was the discovery of which image in a pair had produced a reward when touched. The bars labelled '0v1' in Figure 3.10B (page 104) show that after polar PF lesions the monkeys also had a severe impairment on this task.

Yet, Figure 3.10 also shows that, after the first trial, monkeys with polar PF lesions could learn as rapidly as intact control monkeys when tested on successive trials. This finding indicates that there is a critical distinction between retrieving a single event and being able to learn associations via cumulative reinforcement. It is a defining feature of episodic memory that it involves retrieving the memory of a single event.

Given the connections of the granular medial PF cortex (areas 9 and 10) with the retrosplenial cortex (Kobayashi & Amaral, 2003; Petrides & Pandya, 2007), and thus indirectly with the hippocampus (Kobayashi & Amaral, 2007), one would expect hippocampal lesions to have a similar effect. Froudist-Walsh et al. (2018) also tested monkeys on the object-in-scenes task and made excitotoxic lesions of the hippocampus. After the lesion, the monkeys could learn new material at a normal rate, but they had a severe retrieval impairment; they could not remember choices that they had made 2 weeks previously. This finding reinforces the suggestion that the hippocampus is critical for recall of past events.

Summary

In macaque monkeys, the granular medial PF cortex comprises the medial part of areas 9 and 10. However, there is a paucity of neurophysiological and neuropsychology studies on this area. The limited evidence that we have indicates that the medial PF cortex (area 9) encodes the elapse of time and order, but it is not clear whether the cells can encode time over a longer period. Removal of the polar PF cortex impairs the ability of macaque monkeys to retrieve or recall single events.

Conclusions

Because animals are tested on laboratory tasks, it is standard to present them with cues to tell them what to do. But in their natural surroundings much of their behaviour is spontaneous, being controlled by the memory of past actions and by internal motivation. This chapter shows that the agranular and granular medial PF cortex are involved in the generation of actions and choices when there are no external cues to specify the appropriate response.

The tasks described are retrieval tasks, not recognition tasks. Consider action reversal as an example. The handle does not remind the animals of what action will currently achieve the desired outcome. The animal has to retrieve this from long-term memory. Lesions of the ACC cause a severe impairment on this task (Kennerley et al., 2006); and there are cells in the sACC that encode the action (Hunt et al., 2018) and the outcome (Kennerley et al., 2009).

The prelimbic and infralimbic cortex are directly connected with the hippocampus (Barbas & Blatt, 1995a). Studies on rodents show that there are cells in the hippocampus that encode the route that the animal took to find food (Foster & Wilson, 2006), and also cells that support the retrieval of the sequence of paths that will take the animal back there (Pfeiffer & Foster, 2013). In rodents there are sharp-wave ripples in the potentials that replay the route that an animal has taken, and these have also been found in the medial network in primates.

This system has been co-opted in monkeys and apes to support the retrieval of the sequence of actions when foraging with the arms and hands. This has occurred by the development of the granular medial PF cortex and dorsal PF cortex which have provided new inputs to the medial network, specifying actions performed with the hands and arms. Because, as Chapter 2 describes, the forelimb of primates dominates, these inputs support the actions that are involved in foraging, rather than simply in navigating from one place to another.

In experiments with human subjects, it can be shown that episodic retrieval or recall involves re-generating the representation of the previous action or event. Animals such as macaque monkeys can also generate an action that they have performed in the past, and they can learn from a single event (Boschin et al., 2015). But it is not clear how we could find out whether the retrieval is accompanied by any subjective experience. Chapter 10 takes up the issue.

Selective Advantage

Chapter 2 mentioned that there was a dramatic change in the weather patterns during the Oligocene. The result was that the resources on which the primates depended became much less reliable. This meant that it paid the anthropoids to be flexible depending on the outcomes. Kennerley et al. (2006) found that cell activity in the sACC in monkeys is involved in responding both to positive and negative outcomes.

When the animal finds food that is ripe and edible (positive feedback), it pays to continue to exploit it. But when the food is of poor quality or inedible (negative feedback) it pays to explore so as to try to find food that is more desirable. Given the volatility of their resources, there have been selection pressures for anthropoid primates to be able to respond rapidly to both positive and negative outcomes. The limiting case involves learning on the basis of recall of a single event.

Connectivity

It is the connectivity of the granular medial PF cortex that explains how it does what it does. Among its many connections, it has five that, collectively, make it unique. It has: 1) interconnections with the retrosplenial cortex and thus indirectly with the hippocampus; 2) inputs from the upper bank of the STS; 3) inputs from the amygdala; 4) interconnections with the orbital PF cortex; and 5) outputs to the medial premotor areas for the hand and arm.

Neural Operation

The following section in italics is a proposal concerning the transformation that is performed by the medial PF in anthropoid primates from inputs to outputs.

> *The medial PF cortex can receive information about the scene and time via several routes: from the posterior parietal cortex, parahippocampal and perirhinal cortex, as well as from the hippocampus itself. This information acts as the context for the retrieval and reinstatement of memories of actions and their outcomes.*
>
> *The sequences of actions are specified by the dorsal PF cortex, and this provides an input to the granular medial PF cortex which is indirectly connected in turn with the hippocampus. Inputs to the medial PF cortex from the orbital PF cortex and amygdala provide information about value, as well as about the outcomes of actions.*
>
> *The agranular medial PF cortex transforms the value into the action that is currently appropriate, and together with the substantia nigra and ventral striatum generates the timing of that action. The performance of sequences of self-generated actions is then achieved via outputs to the medial premotor areas and thence to the motor cortex.*
>
> *This system for the retrieval or recall of actions was co-opted in anthropoids so as to support social behaviour. This occurred via new inputs to the prelimbic cortex that bring information about the vocalizations, appearance, and movements of conspecifics and from the cortex of the STS. The prelimbic cortex uses these to retrieve memories of how other animals have behaved in the past and to generate predictions about their behaviour in the future.*

References

Addis, D.R., Wong, A.T., & Schacter, D.L. (2007) Remembering the past and imagining the future: Common and distinct neural substrates during event construction and elaboration. *Neuropsychol*, 45, 1363–77.

Amaral, D.G. & Insausti, R. (1992) Retrograde transport of D-[3H]-aspartate injected into the monkey amygdaloid complex. *Exp Brain Res*, 88, 375–88.

Amaral, D.G. & Price, J.L. (1984) Amygdalo-cortical projections in the monkey (*IMacaca fascicularis*). *J Comp Neurol*, 230, 465–96.

Amiez, C., Sallet, J., Hopkins, W.D., Meguerditchian, A., Hadj-Bouziane, F., Ben Hamed, S., Wilson, C.R.E., Procyk, E., & Petrides, M. (2019) Sulcal organization in the medial frontal cortex provides insights into primate brain evolution. *Nat Commun*, 10, 3437.

Azab, H. & Hayden, B.Y. (2017) Correlates of decisional dynamics in the dorsal anterior cingulate cortex. *PLoS Biol*, 15, e2003091.

Baizer, J.S., Ungerleider, L.G., & Desimone, R. (1991) Organization of visual inputs to the inferior temporal and posterior parietal cortex in macaques. *J Neurosci*, 11, 168–90.

Bannerman, D.M., Sprengel, R., Sanderson, D.J., McHugh, S.B., Rawlins, J.N., Monyer, H., & Seeburg, P.H. (2014) Hippocampal synaptic plasticity, spatial memory and anxiety. *Nat Rev Neurosci*, 15, 181–92.

Banquet, J-P., Gaussier, P., Cuperlier, N., Hok, V., Save, E., Poucet, B., Quoy, M., Wiener, S.I. (2020) Time as the fourth dimension in the hippocampus *Progr Neurobiol*, 11, doi: 10.1016/j.pneurobio.2020.101920.

Barack, D.L., Chang, S.W.C., & Platt, M.L. (2017) Posterior cingulate neurons dynamically signal decisions to disengage during foraging. *Neuron*, 96, 339–47e335.

Barbas, H. & Blatt, G.J. (1995a) Topographically specific hippocampal projections target functionally distinct prefrontal areas in the rhesus monkey. *Hippocampus*, 5, 511–33.

Barbas, H., Ghashghaei, H., Dombrowski, S.M., & Rempel-Clower, N.L. (1999) Medial prefrontal cortices are unified by common connections with superior temporal cortices and distinguished by input from memory-related areas in the rhesus monkey. *J Comp Neurol*, 410, 343–67.

Barks, S.K., Parr, L.A., & Rilling, J.K. (2015) The default mode network in chimpanzees (*Pan troglodytes*) is similar to that of humans. *Cereb Cortex*, 25, 538–44.

Basile, B.M., Schafroth, J.L., Karaskiewicz, C.L., Chang, S.W.C., & Murray, E.A. (2020) The anterior cingulate cortex is necessary for forming prosocial preferences from vicarious reinforcement in monkeys. *PLoS Biol*, 18, e3000677.

Beckmann, M., Johansen-Berg, H., & Rushworth, M.F. (2009) Connectivity-based parcellation of human cingulate cortex and its relation to functional specialization. *J Neurosci*, 29, 1175–90.

Behrens, T.E., Hunt, L.T., Woolrich, M.W., & Rushworth, M.F. (2008) Associative learning of social value. *Nature*, 456, 245–9.

Behrens, T.E., Woolrich, M.W., Walton, M.E., & Rushworth, M.F. (2007) Learning the value of information in an uncertain world. *Nat Neurosci*, 10, 1214–21.

Berger-Tal, O., Nathan, J., Meron, E., & Saltz, D. (2014) The exploration-exploitation dilemma: A multidisciplinary framework. *PLoS One*, 9, e95693.

Bird, C.M., Keidel, J.L., Ing, L.P., Horner, A.J., & Burgess, N. (2015) Consolidation of complex events via reinstatement in posterior cingulate cortex. *J Neurosci*, 35, 14426–34.

Blanchard, T.C. & Hayden, B.Y. (2014) Neurons in dorsal anterior cingulate cortex signal postdecisional variables in a foraging task. *J Neurosci*, 34, 646–55.

Blanchard, T.C., Strait, C.E., & Hayden, B.Y. (2015) Ramping ensemble activity in dorsal anterior cingulate neurons during persistent commitment to a decision. *J Neurophysiol*, 114, 2439–49.

Boschin, E.A., Piekema, C., & Buckley, M.J. (2015) Essential functions of primate frontopolar cortex in cognition. *Proc Natl Acad Sci USA*, 112, E1020–7.

Botvinick, M.M., Cohen, J.D., & Carter, C.S. (2004) Conflict monitoring and anterior cingulate cortex: An update. *Trends Cogn Sci*, 8, 539–46.

Cai, D.J., Aharoni, D., Shuman, T., Shobe, J., Biane, J., Song, W., Wei, B., Veshkini, M., La-Vu, M., Lou, J., Flores, S.E., Kim, I., Sano, Y., Zhou, M., Baumgaertel, K., Lavi, A., Kamata, M., Tuszynski, M., Mayford, M., Golshani, P., & Silva, A.J. (2016) A shared neural ensemble links distinct contextual memories encoded close in time. *Nature*, 534, 115–18.

Cai, X. & Padoa-Schioppa, C. (2012) Neuronal encoding of subjective value in dorsal and ventral anterior cingulate cortex. *J Neurosci*, 32, 3791–808.

Carpenter, A.F., Baud-Bovy, G., Georgopoulos, A.P., & Pellizzer, G. (2018) Encoding of serial order in working memory: Neuronal activity in motor, premotor, and prefrontal cortex during a memory scanning task. *J Neurosci*, 38, 4912–33.

Cavada, C., Company, T., Tejedor, J., Cruz-Rizzolo, R.J., & Reinoso-Suarez, F. (2000) The anatomical connections of the macaque monkey orbitofrontal cortex. A review. *Cer Cort*, 10, 243–51.

Chang, S.W., Brent, L.J., Adams, G.K., Klein, J.T., Pearson, J.M., Watson, K.K., & Platt, M.L. (2013a) Neuroethology of primate social behavior. *Proc Natl Acad Sci USA*, 110 (Suppl 2), 10387–94.

Chang, S.W., Gariepy, J.F., & Platt, M.L. (2013b) Neuronal reference frames for social decisions in primate frontal cortex. *Nat Neurosci*, 16, 243–50.

Choi, E.Y., Ding, S.L. & Haber, S.N. (2017) Combinatorial inputs to the ventral striatum from the temporal cortex, frontal cortex, and amygdala: Implications for segmenting the striatum. *eNeuro*, 4.

Clayton, N.S., Bussey, T.J., & Dickinson, A. (2003) Can animals recall the past and plan for the future? *Nat Rev Neurosci*, 4, 685–91.

Clayton, N.S. & Dickinson, A. (1998) Episodic-like memory during cache recovery by scrub jays. *Nature*, 395, 272–74.

Constantinescu, A.O., O'Reilly, J.X., & Behrens, T.E.J. (2016) Organizing conceptual knowledge in humans with a gridlike code. *Science*, 352, 1464–8.

Corkin, S., Amaral, D.G., Gonzalez, R.G., Johnson, K.A., & Hyman, B.T. (1997) H. M.'s medial temporal lobe lesion: Findings from magnetic resonance imaging. *J Neurosci*, 17, 3964–79.

Croxson, P.L., Walton, M.E., O'Reilly, J.X., Behrens, T.E., & Rushworth, M.F. (2009) Effort-based cost-benefit valuation and the human brain. *J Neurosci*, 29, 4531–41.

Davies, N.B. & Krebs, J.R. (2012) *An Introduction to Behavioural Ecology*. Wiley, New York.

Descartes, R. (1644) *Principia Philosophiae*. Elsevier. Amsterdam.

Devinsky, O., Morrell, M.J., & Vogt, B.A. (1995) Contributions of anterior cingulate cortex to behaviour. *Brain*, 118, 279–306.

Dosenbach, N.U., Fair, D.A., Miezin, F.M., Cohen, A.L., Wenger, K.K., Dosenbach, R.A., Fox, M.D., Snyder, A.Z., Vincent, J.L., Raichle, M.E., Schlaggar, B.L., & Petersen, S.E. (2007) Distinct brain networks for adaptive and stable task control in humans. *Proc Natl Acad Sci USA*, 104, 11073–8.

Dum, R.P. & Strick, P.L. (2002) Motor areas in the frontal lobe of the primate. *Physiol Behav*, 77, 677–82.

Eichenbaum, H. (2014) Time cells in the hippocampus: A new dimension for mapping memories. *Nat Rev Neurosci*, 15, 732–44.

Eichenbaum, H. (2017) Time (and space) in the hippocampus. *Curr Opin Behav Sci*, 17, 65–70.

Epstein, R.A., Patai, E.Z., Julian, J.B., & Spiers, H.J. (2017) The cognitive map in humans: Spatial navigation and beyond. *Nat Neurosci*, 20, 1504–13.

Ferry, A.T., Ongur, D., An, X., & Price, J.L. (2000) Prefrontal cortical projections to the striatum in macaque monkeys: Evidence for an organization related to prefrontal networks. *J Comp Neurol*, 425, 447–70.

Foster, D.J. & Wilson, M.A. (2006) Reverse replay of behavioural sequences in hippocampal place cells during the awake state. *Nature*, 440, 680–3.

Fouragnan, E.F., Chau, B.K.H., Folloni, D., Kolling, N., Verhagen, L., Klein-Flugge, M., Tankelevitch, L., Papageorgiou, G.K., Aubry, J.F., Sallet, J., & Rushworth, M.F.S. (2019) The macaque anterior cingulate cortex translates counterfactual choice value into actual behavioral change. *Nat Neurosci*, 22, 797–808.

Fox, M.D., Zhang, D., Snyder, A.Z., & Raichle, M.E. (2009) The global signal and observed anticorrelated resting state brain networks. *J Neurophysiol*, 101, 3270–83.

Froudist-Walsh, S., Browning, P.G.F., Croxson, P.L., Murphy, K.L., Shamy, J.L., Veuthey, T.L., Wilson, C.R.E., & Baxter, M.G. (2018) The rhesus monkey hippocampus critically contributes to scene memory retrieval, but not new learning. *J Neurosci*, 38, 7800–8.

Gaffan, D. (1992) Amnesia for complex naturalistic scenes and for objects following fornix transection in the rhesus monkey. *Eur J Neurosci*, 4, 381–8.

Genovesio, A., Tsujimoto, S., & Wise, S.P. (2006) Neuronal activity related to elapsed time in prefrontal cortex. *J Neurophysiol*, 95, 3281–5.

Genovesio, A., Tsujimoto, S., & Wise, S.P. (2011) Prefrontal cortex activity during the discrimination of relative distance. *J Neurosci*, 31, 3968–80.

Genovesio, A., Tsujimoto, S., & Wise, S.P. (2012) Encoding goals but not abstract magnitude in the primate prefrontal cortex. *Neuron*, 74, 656–62.

Glascher, J., Hampton, A.N., & O'Doherty, J.P. (2009) Determining a role for ventromedial prefrontal cortex in encoding action-based value signals during reward-related decision making. *Cereb Cortex*, 19, 483–95.

Goldman-Rakic, P.S. (1998) The prefrontal landscape: Implications of functional architecture for understanding human mentation and the central executive. In Roberts, A.C., Robbins, T.W., Weiskrantz, L. (eds) *The Prefrontal Cortex*. Oxford University Press, Oxford, pp. 117–30.

Haber, S.N. & Knutson, B. (2010) The reward circuit: Linking primate anatomy and human imaging. *Neuropsychopharmacology*, 35, 4–26.

Hadland, K.A., Rushworth, M.F., Gaffan, D., & Passingham, R.E. (2003) The effect of cingulate lesions on social behaviour and emotion. *Neuropsychol*, 41, 919–31.

Hampton, R.R., Hampstead, B.M., & Murray, E.A. (2004) Selective hippocampal damage in rhesus monkeys impairs spatial memory in an open-field test. *Hippocampus*, 14, 808–18.

Hanks, T.D., Ditterich, J., & Shadlen, M.N. (2006) Microstimulation of macaque area LIP affects decision-making in a motion discrimination task. *Nat Neurosci*, 9, 682–9.

Hanks, T.D. & Summerfield, C. (2017) Perceptual decision making in rodents, monkeys, and humans. *Neuron*, 93, 15–31.

Haroush, K. & Williams, Z.M. (2015) Neuronal prediction of opponent's behavior during cooperative social interchange in primates. *Cell*, 160, 1233–45.

Hartley, T., Lever, C., Burgess, N. & O'Keefe, J. (2014) Space in the brain: how the hippocampal formation supports spatial cognition. *Phil Trans R Soc London B*, 369, doi.org/10.C.1098/rstb.2012.0510.

Hassabis, D., Chu., Rees, G., Weiskopf, J., Molyneux, P.D. & Maguire, E.A. (2009) Decoding neuronal ensembles in the human hippocampus. *Curr Biol*, 19, 546–54.

Hassabis, D., Kumaran, D., & Maguire, E.A. (2007) Using imagination to understand the neural basis of episodic memory. *J Neurosci*, 27, 14365–74.

Hayden, B.Y., Pearson, J.M., & Platt, M.L. (2009) Fictive reward signals in the anterior cingulate cortex. *Science*, 324, 948–50.

Hayden, B.Y., Pearson, J.M., & Platt, M.L. (2011) Neuronal basis of sequential foraging decisions in a patchy environment. *Nat Neurosci*, 14, 933–9.

Hayden, B.Y. & Platt, M.L. (2010) Neurons in anterior cingulate cortex multiplex information about reward and action. *J Neurosci*, 30, 3339–46.

He, S.Q., Dum, R.P., & Strick, P.L. (1995) Topographic organization of corticospinal projections from the frontal lobe: Motor areas on the medial surface of the hemisphere. *J Neurosci*, 15, 3284–306.

Hedreen, J.C. & DeLong, M.R. (1991) Organization of striatopallidal, striatonigral and nigrostriatal projections in the macaque. *J Comp Neurol*, 304, 569–95.

Heilbronner, S.R., Rodriguez-Romaguera, J., Quirk, G.J., Groenewegen, H.J., & Haber, S.N. (2016) Circuit-based corticostriatal homologies between rat and primate. *Biol Psychiatry*, 80, 509–21.

Hinman, J.R., Dannenberg, H., Alexander, A.S., & Hasselmo, M.E. (2018) Neural mechanisms of navigation involving interactions of cortical and subcortical structures. *J Neurophysiol*, 119, 2007–29.

Hoffman, D.S. & Strick, P.L. (1995) Effects of a primary motor cortex lesion on step-tracking movements of the wrist. *J Neurophysiol*, 73, 891–5.

Hoffman, M.L., Beran, M.J., & Washburn, D.A. (2018) Rhesus monkeys (Macaca mulatta) remember agency information from past events and integrate this knowledge with spatial and temporal features in working memory. *Anim Cogn*, 21, 137–53.

Hosokawa, T., Kennerley, S.W., Sloan, J., & Wallis, J.D. (2013) Single-neuron mechanisms underlying cost-benefit analysis in frontal cortex. *J Neurosci*, 33, 17385–97.

Howard, M.W., MacDonald, C.J., Tiganj, Z., Shankar, K.H., Du, Q., Hasselmo, M.E., & Eichenbaum, H. (2014) A unified mathematical framework for coding time, space, and sequences in the hippocampal region. *J Neurosci*, 34, 4692–707.

Hunt, L.T., Behrens, T.E., Hosokawa, T., Wallis, J.D. & Kennerley, S.W. (2015) Capturing the temporal evolution of choice across prefrontal cortex. *Elife*, 4, doi.org/10.7554/eLife.11945.001.

Hunt, L.T., Malalasekera, W.M.N., de Berker, A.O., Miranda, B., Farmer, S.F., Behrens, T.E.J., & Kennerley, S.W. (2018) Triple dissociation of attention and decision computations across prefrontal cortex. *Nat Neurosci*, 21, 1471–81.

Jackson, R.L., Bajada, C.J., Lambon Ralph, M.A., & Cloutman, L.L. (2020) The graded change in connectivity across the ventromedial prefrontal cortex reveals distinct subregions. *Cereb Cortex*, 30, 165–80.

Janson, C.H. (2016) Capuchins, space, time and memory: An experimental test of what-where-shen memory in wild monkeys. *Proc Roy Soc B*, 283, 20161432. http://dx.doi.org/20161410-20161098/rspb.20162016.20161432.

Jellema, T. & Perrett, D.I. (2003) Cells in monkey STS responsive to articulated body motions and consequent static posture: A case of implied motion? *Neuropsychol*, 41, 1728–37.

Kaplan, R., Adhikari, M.H., Hindriks, R., Mantini, D., Murayama, Y., Logothetis, N.K., & Deco, G. (2016) Hippocampal sharp-wave ripples influence selective activation of the default mode network. *Curr Biol*, 26, 686–91.

Kennerley, S.W., Behrens, T.E., & Wallis, J.D. (2011) Double dissociation of value computations in orbitofrontal and anterior cingulate neurons. *Nat Neurosci*, 14, 1581–9.

Kennerley, S.W., Dahmubed, A.F., Lara, A.H., & Wallis, J.D. (2009) Neurons in the frontal lobe encode the value of multiple decision variables. *J Cogn Neurosci*, 21, 1162–78.

Kennerley, S.W. & Wallis, J.D. (2009) Evaluating choices by single neurons in the frontal lobe: Outcome value encoded across multiple decision variables. *Eur J Neurosci*, 29, 2061–73.

Kennerley, S.W., Walton, M.E., Behrens, T.E., Buckley, M.J., & Rushworth, M.F. (2006) Optimal decision making and the anterior cingulate cortex. *Nat Neurosci*, 9, 940–7.

Khalighinejad, N., Bongioanni, A., Verhagen, L., Folloni, D., Attali, D., Aubry, J.F., Sallet, J., & Rushworth, M.F.S. (2020a) A Basal Forebrain-Cingulate Circuit in Macaques Decides It Is Time to Act. *Neuron*, 105, 370–84.

Khalighinejad, N., Priestley, L., Jbabdi, S., Rushworth, M.F.S. Human decisions about when to act originate within a basal forebrain-nigral circuit. *Proc Nat Acad Sci*, 117, 11799–810.

Klein-Flugge, M.C., Kennerley, S.W., Friston, K., & Bestmann, S. (2016) Neural signatures of value comparison in human cingulate cortex during decisions requiring an effort-reward trade-off. *J Neurosci*, 36, 10002–15.

Kobayashi, Y. & Amaral, D.G. (2003) Macaque monkey retrosplenial cortex: II. Cortical afferents. *J Comp Neurol*, 466, 48–79.

Kobayashi, Y. & Amaral, D.G. (2007) Macaque monkey retrosplenial cortex: III. Cortical efferents. *J Comp Neurol*, 502, 810–33.

Kolling, N., Behrens, T., Wittmann, M.K., & Rushworth, M. (2016) Multiple signals in anterior cingulate cortex. *Curr Opin Neurobiol*, 37, 36–43.

Kolling, N., Behrens, T.E., Mars, R.B., & Rushworth, M.F. (2012) Neural mechanisms of foraging. *Science*, 336, 95–8.

Kondo, H., Saleem, K.S., & Price, J.L. (2005) Differential connections of the perirhinal and parahippocampal cortex with the orbital and medial prefrontal networks in macaque monkeys. *J Comp Neurol*, 493, 479–509.

Kuypers, H.G.J.M. (1981) Anatomy of the descending pathways. In Brooks, V. (ed.) *Handbook of Physiology*. American Physiological Society, Bethesda, pp. 597–666.

Laubach, M., Amarante, L.M., Swanson, K., & White, S.R. (2018) What, if anything, is rodent prefrontal cortex? *eNeuro*, 5.

Lavenex, P., Suzuki, W.A., & Amaral, D.G. (2002) Perirhinal and parahippocampal cortices of the macaque monkey: Projections to the neocortex. *J Comp Neurol*, 447, 394–420.

Leonard, T.K. & Hoffman, K.L. (2017) Sharp-wave ripples in primates are enhanced near remembered visual objects. *Curr Biol*, 27, 257–62.

Li, L., Hu, X., Preuss, T.M., Glasser, M.F., Damen, F.W., Qiu, Y., & Rilling, J. (2013) Mapping putative hubs in human, chimpanzee and rhesus macaque connectomes via diffusion tractography. *Neuroimage*, 80, 462–74.

Liu, Y., Dolan, R.J., Kurth-Nelson, Z., & Behrens, T.E.J. (2019) Human Replay Spontaneously Reorganizes Experience. *Cell*, 178, 640–52 e614.

Lockwood, P.L., Apps, M.A.J., & Chang, S.W.C. (2020) Is there a 'Social' brain? Implementations and algorithms. *Trends Cogn Sci*, 24, 802–13.

Luppino, G., Matelli, M., Camarda, R., & Rizzolatti, G. (1993) Corticocortical connections of area F3 (SMA-proper) and area F6 (Pre-SMA) in the macaque monkey. *J Comp Neurol*, 338, 114–40.

Maguire, E.A., Nannery, R., & Spiers, H.J. (2006) Navigation around London by a taxi driver with bilateral hippocampal lesions. *Brain*, 129, 2894–907.

Mansouri, F.A., Egner, T., & Buckley, M.J. (2017) Monitoring demands for executive control: Shared functions between human and nonhuman primates. *Trends Neurosci*, 40, 15–27.

Mao, C.V., Araujo, M.F., Nishimaru, H., Matsumoto, J., Tran, A.H., Hori, E., Ono, T., & Nishijo, H. (2017) Pregenual anterior cingulate gyrus involvement in spontaneous social interactions in primates-evidence from behavioral, pharmacological, neuropsychiatric, and neurophysiological findings. *Front Neurosci*, 11, 34.

Markov, N.T., Ercsey-Ravasz, M.M., Ribeiro Gomes, A.R., Lamy, C., Magrou, L., Vezoli, J., Misery, P., Falchier, A., Quilodran, R., Gariel, M.A., Sallet, J., Gamanut, R., Huissoud, C., Clavagnier, S., Giroud, P., Sappey-Mariner, D., Barone, P., Dehay, C., Toroczkai, Z., Knoblauch, K., Van Essen, D.C., & Kennedy, H. (2014) A weighted and directed interareal connectivity matrix for macaque cerebral cortex. *Cereb Cortex*, 24, 17–36.

Markov, N.T., Misery, P., Falchier, A., Lamy, C., Vezoli, J., Quilodran, R., Gariel, M.A., Giroud, P., Ercsey-Ravasz, M., Pilaz, L.J., Huissoud, C., Barone, P., Dehay, C., Toroczkai, Z., Van Essen, D.C., Kennedy, H. & Knoblauch, K. (2011) Weight consistency specifies regularities of macaque cortical networks. *Cereb Cortex*, 21, 1254–72.

Mars, R.B., Neubert, F.X., Noonan, M.P., Sallet, J., Toni, I., & Rushworth, M.F. (2012) On the relationship between the 'default mode network' and the 'social brain'. *Front Hum Neurosci*, 6, 189.

Mars, R.B., Sallet, J., Neubert, F.X., & Rushworth, M.F. (2013) Connectivity profiles reveal the relationship between brain areas for social cognition in human and monkey temporoparietal cortex. *Proc Natl Acad Sci USA*, 110, 10806–11.

Mars, R.B., Verhagen, L., Gladwin, T.E., Neubert, F.X., Sallet, J., & Rushworth, M.F. (2016) Comparing brains by matching connectivity profiles. *Neurosci Biobehav Rev*, 60, 90–7.

Mason, M.F., Norton, M.I., Van Horn, J.D., Wegner, D.M., Grafton, S.T., & Macrae, C.N. (2007) Wandering minds: The default network and stimulus-independent thought. *Science*, 315, 393–95.

Matsumoto, M., Matsumoto, K., Abe, H., & Tanaka, K. (2007) Medial prefrontal cell activity signaling prediction errors of action values. *Nat Neurosci*, 10, 647–56.

Monteiro, T., Vasconcelos, M., & Kacelnik, A. (2013) Starlings uphold principles of economic rationality for delay and probability of reward. *Proc Biol Sci*, 280, 20122386.

Morecraft, R.J., McNeal, D.W., Stilwell-Morecraft, K.S., Gedney, M., Ge, J., Schroeder, C.M., & van Hoesen, G.W. (2007) Amygdala interconnections with the cingulate motor cortex in the rhesus monkey. *J Comp Neurol*, 500, 134–65.

Morecraft, R.J., Stilwell-Morecraft, K.S., Cipolloni, P.B., Ge, J., McNeal, D.W., & Pandya, D.N. (2012) Cytoarchitecture and cortical connections of the anterior cingulate and adjacent somatomotor fields in the rhesus monkey. *Brain Res Bull*, 87, 457–97.

Morris, R. (1984) Developments of a water-maze procedure for studying spatial learning in the rat. *J Neurosci Methods*, 11, 47–60.

Murray, E.A., Wise, S.P., & Graham, K.S. (2018) Representational specializations of the hippocampus in phylogenetic perspective. *Neurosci Lett*, 680, 4–12.

Mushiake, H., Inase, M., & Tanji, J. (1991) Neuronal activity in the primate premotor, supplementary, and precentral motor cortex during visually guided and internally determined sequential movements. *J Neurophysiol*, 66, 705–18.

Nasr, S., Liu, N., Devaney, K.J., Yue, X., Rajimehr, R., Ungerleider, L.G., & Tootell, R.B. (2011) Scene-selective cortical regions in human and nonhuman primates. *J Neurosci*, 31, 13771–85.

Naya, Y., Chen, H., Yang, C., & Suzuki, W.A. (2017) Contributions of primate prefrontal cortex and medial temporal lobe to temporal-order memory. *Proc Natl Acad Sci USA*, 114, 13555–60.

Naya, Y. & Suzuki, W.A. (2011) Integrating what and when across the primate medial temporal lobe. *Science*, 333, 773–6.

Neubert, F.X., Mars, R.B., Sallet, J., & Rushworth, M.F. (2015) Connectivity reveals relationship of brain areas for reward-guided learning and decision making in human and monkey frontal cortex. *Proc Natl Acad Sci USA*, 112, E2695–704.

Nixon, P.D. & Passingham, R.E. (1998) The striatum and self-paced movements. *Behav Neurosci*, 112, 719–24.

Noonan, M.P., Sallet, J., Mars, R.B., Neubert, F.X., O'Reilly, J.X., Andersson, J.L., Mitchell, A.S., Bell, A.H., Miller, K.L., & Rushworth, M.F. (2014) A neural circuit covarying with social hierarchy in macaques. *PLoS Biol*, 12, e1001940.

Noser, R. & Byrne, R.W. (2007) Mental maps in chacma baboons (*Papio ursinus*): using inter-group encounters as a natural experiment. *Anim Cogn*, 10, 331–40.

Palomero-Gallagher, N., Vogt, B.A., Schleicher, A., Mayberg, H.S., & Zilles, K. (2009) Receptor architecture of human cingulate cortex: Evaluation of the four-region neurobiological model. *Hum Brain Mapp*, 30, 2336–55.

Passingham, R.E. (1993) *The Frontal Lobes and Voluntary Action*. Oxford University Press, Oxford.

Passingham, R.E., Bengtsson, S.L., & Lau, H.C. (2010) Medial frontal cortex: From self-generated action to reflection on one's own performance. *Trends Cogn Sci*, 14, 16–21.

Passingham, R.E., Perry, H., & Wilkinson, F. (1983) The long-term effects of removal of sensorimotor cortex in infant and adult rhesus monkeys. *Brain*, 106, 675–705.

Passingham, R.E. & Wise, S.P. (2012) *The Neurobiology of Prefrontal Cortex*. Oxford University Press, Oxford.

Perrodin, C., Kayser, C., Logothetis, N.K., & Petkov, C.I. (2014) Auditory and visual modulation of temporal lobe neurons in voice-sensitive and association cortices. *J Neurosci*, 34, 2524–37.

Petrides, M. & Pandya, D.N. (1999) Dorsolateral prefrontal cortex: Comparative cytoarchitectonic analysis in the human and the macaque brain and corticocortical connection patterns. *Eur J Neurosci*, 11, 1011–36.

Petrides, M. & Pandya, D.N. (2007) Efferent association pathways from the rostral prefrontal cortex in the macaque monkey. *J Neurosci*, 27, 11573–86.

Pfeiffer, B.E. & Foster, D.J. (2013) Hippocampal place-cell sequences depict future paths to remembered goals. *Nature*, 497, 74–9.

Qualls-Creekmore, E. & Munzberg, H. (2018) Modulation of feeding and associated behaviors by lateral hypothalamic circuits. *Endocrinology*, 159, 3631–42.

Quilodran, R., Rothe, M., & Procyk, E. (2008) Behavioral shifts and action valuation in the anterior cingulate cortex. *Neuron*, 57, 314–25.

Raichle, M.E. & Snyder, A.Z. (2007) A default mode of brain function: A brief history of an evolving idea. *Neuroimage*, 37, 1083–90.

Rempel-Clower, N.L. & Barbas, H. (1998) Topographic organization of connections between the hypothalamus and prefrontal cortex in the rhesus monkey. *J Comp Neurol*, 398, 393–419.

Rescorla, R.A. & Wagner, A.D. (1972) A theory of Pavlovian conditioning: Variation in the effectiveness of reinforcement and non-reinforcement. In Black, A.H., Procaksy, W.F. (eds) *Classical Conditioning in Current Research and Theory*. Appleton-Century-Crofts, New York.

Rilling, J.K., Barks, S.K., Parr, L.A., Preuss, T.M., Faber, T.L., Pagnoni, G., Bremner, J.D., & Votaw, J.R. (2007) A comparison of resting-state brain activity in humans and chimpanzees. *Proc Natl Acad Sci USA*, 104, 17146–51.

Robinson, N.T.M., Priestley, J.B., Rueckemann, J.W., Garcia, A.D., Smeglin, V.A., Marino, F.A., & Eichenbaum, H. (2017) Medial entorhinal cortex selectively supports temporal coding by hippocampal neurons. *Neuron*, 94, 677–88 e676.

Roitman, J.D. & Shadlen, M.N. (2002) Response of neurons in the lateral intraparietal area during a combined visual discrimination reaction time task. *J Neurosci*, 22, 9475–89.

Rowland, D.C., Roudi, Y., Moser, M.B., & Moser, E.I. (2016) Ten years of grid cells. *Annu Rev Neurosci*, 39, 19–40.

Rudebeck, P.H., Buckley, M.J., Walton, M.E., & Rushworth, M.F. (2006) A role for the macaque anterior cingulate gyrus in social valuation. *Science*, 313, 1310–12.

Rudebeck, P.H., Putnam, P.T., Daniels, T.E., Yang, T., Mitz, A.R., Rhodes, S.E., & Murray, E.A. (2014) A role for primate subgenual cingulate cortex in sustaining autonomic arousal. *Proc Natl Acad Sci USA*, 111, 5391–6.

Rudebeck, P.H., Walton, M.E., Millette, B.H., Shirley, E., Rushworth, M.F., & Bannerman, D.M. (2007) Distinct contributions of frontal areas to emotion and social behaviour in the rat. *Eur J Neurosci*, 26, 2315–26.

Saleem, K.S., Kondo, H., & Price, J.L. (2008) Complementary circuits connecting the orbital and medial prefrontal networks with the temporal, insular, and opercular cortex in the macaque monkey. *J Comp Neurol*, 506, 659–93.

Saleem, K.S., Miller, B., & Price, J.L. (2014) Subdivisions and connectional networks of the lateral prefrontal cortex in the macaque monkey. *J Comp Neurol*, 522, 1641–90.

Sallet, J., Mars, R.B., Andersson, J., O'Reilly, J.X., Jhabdi, S., Croxson, P.L., Miller, K.L., Jenkinson, M., & Rushworth, M.F. (2011) Social network size affects neural circuits in macaques. *Science*, 334, 697–700.

Salwiczek, L.H., Watanabe, A., & Clayton, N.S. (2010) Ten years of research into avian models of episodic-like memory and its implications for developmental and comparative cognition. *Behav Brain Res*, 215, 221–34.

Sayers, K. & Menzel, C.R. (2012) Memory and foraging theory: Chimpanzee utilization of optimality heuristics in the rank-order recovery of hidden foods. *Anim Behav*, 84, 795–803.

Schall, J.D. (2001) Neural basis of deciding, choosing and acting. 2, 33–42.

Scott, B.H., Leccese, P.A., Saleem, K.S., Kikuchi, Y., Mullarkey, M.P., Fukushima, M., Mishkin, M., & Saunders, R.C. (2017) Intrinsic connections of the core auditory cortical regions and rostral supratemporal plane in the macaque monkey. *Cereb Cortex*, 27, 809–40.

Scoville, W.B. & Milner, B. (1957) Loss of recent memory after bilateral hippocampal lesions. *J Neurol Neurosurg Psychiatr*, 20, 11–21.

Seo, H. & Lee, D. (2007) Temporal filtering of reward signals in the dorsal anterior cingulate cortex during a mixed-strategy game. *J Neurosci*, 27, 8366–77.

Shenhav, A., Straccia, M.A., Cohen, J.D., & Botvinick, M.M. (2014) Anterior cingulate engagement in a foraging context reflects choice difficulty, not foraging value. *Nat Neurosci*, 17, 1249–54.

Shima, K. & Tanji, J. (1998) Both supplementary and presupplementary motor areas are crucial for the temporal organization of multiple movements. *J Neurophysiol*, 80, 3247–60.

Sliwa, J. & Freiwald, W.A. (2017) A dedicated network for social interaction processing in the primate brain. *Science*, 356, 745–9.

Spiers, H.J. & Maguire, E.A. (2006) Thoughts, behaviour, and brain dynamics during navigation in the real world. *Neuroimage*, 31, 1826–40.

Stephens, D.W. & Krebs, J.R. (1986) *Foraging Theory*. Princeton University Press, Princeton.

Strait, C.E., Sleezer, B.J., Blanchard, T.C., Azab, H., Castagno, M.D., & Hayden, B.Y. (2016) Neuronal selectivity for spatial positions of offers and choices in five reward regions. *J Neurophysiol*, 115, 1098–111.

Suddendorf, T. & Busby, J. (2003) Mental time travel in animals? *Trends Cogn Sci*, 7, 391–6.

Swartz, B.L., Hoffman, M.L., & Evans, S. (2005) Episodic-like memory in a gorilla: A review and new findings. *Learn Motiv*, 36, 226–44.

Thaler, D., Chen, Y.-C., Nixon, P.D., Stern, C., & Passingham, R.E. (1995) The functions of the medial premotor cortex (SMA). I. Simple learned movements. *Exper Brain Res*, 102, 445–60.

Tsujimoto, S., Genovesio, A., & Wise, S.P. (2012) Neuronal activity during a cued strategy task: Comparison of dorsolateral, orbital, and polar prefrontal cortex. *J Neurosci*, 32, 11017–31.

Vallejo, M., Loyola, S., Contreras, D., Ugarte, G., Cifuente, D., Ortega, G., Cabrera, J.L., Zeise, M., Tonn, C., Carreno, M., Delgado, R., Morales, B., & Agnese, M. (2014) A new semisynthetic derivative of sauroine induces LTP in hippocampal slices and improves learning performance in the Morris Water Maze. *J Neurochem*, 129, 864–76.

van den Heuvel, M.P. & Sporns, O. (2013) Network hubs in the human brain. *Trends Cogn Sci*, 17, 683–96.

Vargha-Khadem, F., Gadian, D.G., Watkins, K.E., Connelly, A., van Paesschen, W., & Mishkin, M. (1997) Differential effects of early hippocampal pathology on episodic and semantic memory. *Science*, 277, 330–1 (comment); 376–30.

Vasconcelos, M., Monteiro, T., & Kacelnik, A. (2013) Context-dependent preferences in starlings: Linking ecology, foraging and choice. *PLoS One*, 8, e64934.

Vincent, J.L., Kahn, I., Snyder, A.Z., Raichle, M.E., & Buckner, R.L. (2008) Evidence for a frontoparietal control system revealed by intrinsic functional connectivity. *J Neurophysiol*, 100, 3328–42.

Vogt, B.A. & Pandya, D.N. (1987) Cingulate cortex of the rhesus monkey: II. Cortical afferents. *J Comp Neurol*, 262, 271–89.

Vogt, B.A. & Paxinos, G. (2014) Cytoarchitecture of mouse and rat cingulate cortex with human homologies. *Brain Struct Funct*, 219, 185–92.

Vogt, B.A., Vogt, L.J., Farber, N.B., & Bush, G. (2005) Architecture and neurocytology of monkey cingulate gyrus. *J Comp Neurol*, 485, 218–39.

Wallis, J.D. & Miller, E.K. (2003) Neuronal activity in primate dorsolateral and orbital prefrontal cortex during performance of a reward preference task. *Eur J Neurosci*, 18, 2069–81.

Wang, Y., Isoda, M., Matsuzaka, Y., Shima, K., & Tanji, J. (2005) Prefrontal cortical cells projecting to the supplementary eye field and presupplementary motor area in the monkey. *Neurosci Res*, 53, 1–7.

Webster, M.J., Ungerleider, L.G., & Bachevalier, J. (1991) Connections of inferior temporal areas TE and TEO with medial temporal-lobe structures in infant and adult monkeys. *J Neurosci*, 11, 1095–1116.

Wittmann, M.K., Kolling, N., Akaishi, R. Chau, B.K., Brown, J.W., Nelissen, N., & Rushworth, M.F.S. (2016) Predictive decision making driven by multiple time-linked reward representation in the anterior cingulate cortex. *Nature Communications*, 7, 1237.

Wittmann, M.K., Lockwood, P.L., & Rushworth, M.F.S. (2018) Neural mechanisms of social cognition in primates. *Annu Rev Neurosci*, 41, 99–118.

Yamagata, T., Nakayama, Y., Tanji, J., & Hoshi, E. (2012) Distinct information representation and processing for goal-directed behavior in the dorsolateral and ventrolateral prefrontal cortex and the dorsal premotor cortex. *J Neurosci*, 32, 12934–49.

Yumoto, N., Lu, X., Henry, T.R., Miyachi, S., Nambu, A., Fukai, T., & Takada, M. (2011) A neural correlate of the processing of multi-second time intervals in primate prefrontal cortex. *PLoS One*, 6, e19168.

Zhang, J., Hughes, L.E., & Rowe, J.B. (2012) Selection and inhibition mechanisms for human voluntary action decisions. *Neuroimage*, 63, 392–402.

4
Orbital Prefrontal Cortex

Evaluating Resources

Introduction

Chapter 3 discussed the choice between actions. But as primates forage, they also choose between foods such as fruit and leaves. To do this they have to be able to compare the value of the foods, and so choose the ones that will best satisfy their needs. In other words, they have to evaluate the resources that are available.

This evaluation requires a way of comparing the different resources in terms of current needs, whether they be fruit, leaves, bark, or insects. But the animals also have to be able to predict where and when they can find the foods they need. This means using signs such as twigs, branches, bushes, and trees to tell them where these are and to what extent they will satisfy their needs. However, since it is the type of food that they are looking for, the animals also need to be able to classify them together into categories, such as particular fruits.

The present chapter argues that it is the orbital prefrontal cortex that supports judgement of value and the use of signs to predict the value of foods. Later in the book, Chapter 7 describes the mechanisms that support the ability to classify them.

Area

Figure 4.1 (next page) illustrates the orbital frontal cortex in monkeys and humans. It has three main subdivisions (Figure 4.2, page 120):

- Medial orbital prefrontal cortex (PF) cortex (granular parts of area 14).
- Central orbital PF cortex (area 11 and granular parts of area 13).
- Lateral orbital PF cortex (orbital parts of area 47/12).

Traditionally, the orbital PF cortex was taken to include areas 11, 13, and 14 (Walker, 1940). However, based on the connections of what they called 'the orbital network', Carmichael and Price (1994) included anterior agranular insular areas (Ia) (many of the green areas in Figure 2.1A). The caudal parts of areas 13 and 14 also have an agranular cytoarchitecture. Together, the agranular Ia and the caudal parts of areas 13 and 14 compose the agranular orbital PF cortex.

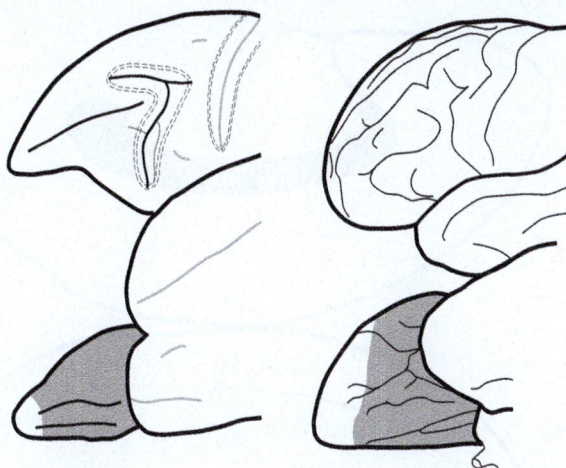

Figure 4.1 The orbital PF cortex in monkeys (left) and humans (right) indicated by shading
Adapted from Passingham, R.E. & Wise, S.P. *The Neurobiology of Prefrontal Cortex*, p. 98, Figure 4.1 © 2012, Oxford University Press.

The lateral sector of the orbital PF cortex was designated by Brodmann (1913) as area 47 in the human brain and as area 12 in the brain of monkeys. Petrides and Pandya (2002) therefore adopted the common designation 47/12. Taken as a whole, this area forms the inferior convexity of the PF cortex, including both the ventral surface and the lateral orbital PF cortex.

Connections

Figure 4.2 (next page) illustrates selected connections of the orbital PF cortex. The drawing emphasizes the connections that most illuminate the unique operations that it performs.

The main connections of the orbital PF cortex can be summarized as follows:

1. The agranular insular cortex (Ia), a major part of the agranular orbital PF cortex, receives relatively direct inputs from the gustatory cortex and the piriform (olfactory) cortex (Carmichael & Price, 1995b). It also receives visceral signals relayed from the brainstem and thalamus (Ray & Price, 1992). These inputs convey sensations reflecting an individual's metabolic state, such as hypoxia or hypoglycaemia, along with sensations arising from the lungs, heart, baroreceptors, and digestive tract (Craig, 2002).

   These findings indicate that the anterior insula contributes to analysing the smell, taste, and flavour of foods along with the consequences of

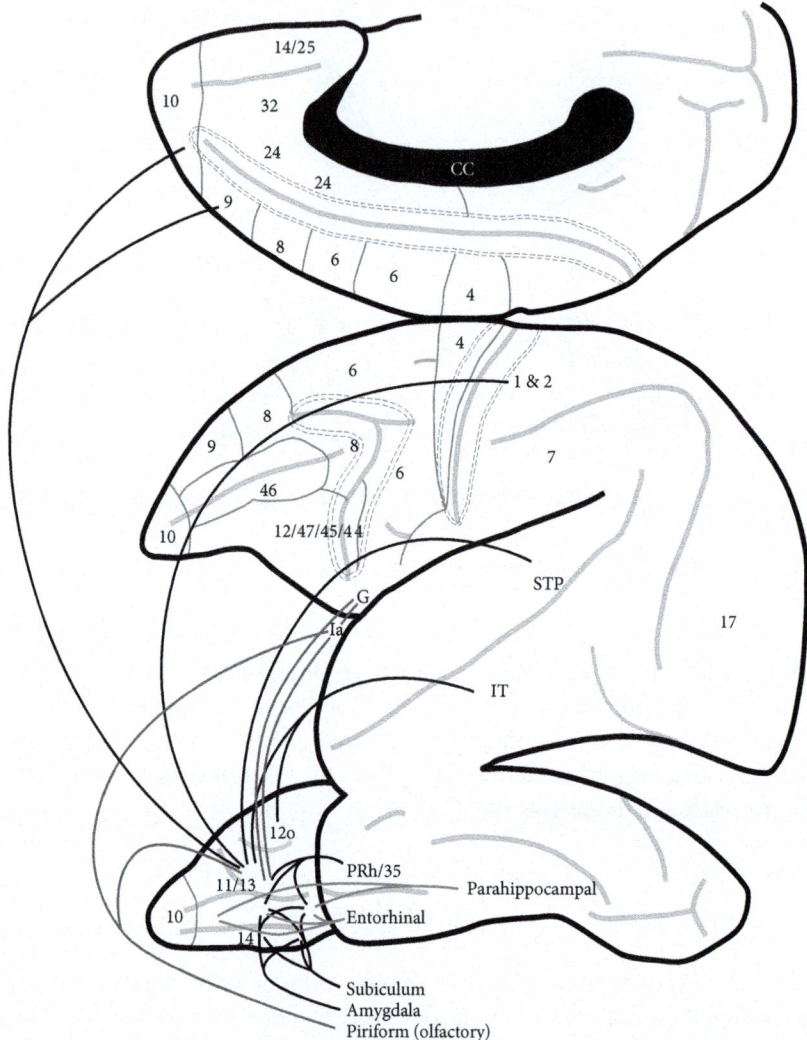

Figure 4.2 Selected connections of the orbital PF cortex
The lines connect some of the areas that have direct axonal connections with the orbital PF cortex, assumed to be reciprocal.

Adapted from Passingham, R.E. & Wise, S.P. *The Neurobiology of Prefrontal Cortex*, p. 99, Figure 4.2
© 2012, Oxford University Press.

consuming them. For example, there are activations in the anterior insula when human subjects experience disgust as the result of unpleasant tastes or smells (Craig, 2002).

2. Via the agranular orbital PF cortex, olfactory, gustatory, and visceral information can also reach the granular orbital PF cortex (Carmichael & Price,

1994). Somatosensory information from the primary somatosensory area (S1) also goes to the granular orbital PF cortex, including area 13. The lateral parts of S1, which represent the mouth, lips, and tongue provide much of this input, but additional inputs come from the hand representation of S1 (Carmichael & Price, 1994).

Taking points 1 and 2 together, these connections enable both the granular and agranular orbital PF cortex to process olfactory, gustatory, somatosensory, and visceral information about foods and fluids.

3. In contrast to the medial PF cortex, the granular orbital PF cortex also receives robust visual inputs concerning objects. These projections arise mainly from the perirhinal cortex (Kondo et al., 2005) and the inferior temporal area TE (Webster et al., 1994), two components of the ventral visual stream. There are also visual projections from these areas to the lateral sector of the orbital cortex, part of area 47/12 (Webster et al., 1991; Kondo et al., 2005).

    Projections from the perirhinal cortex provide the orbital PF cortex with visual information about objects, including food items (Murray et al., 2007). Projections from the inferior temporal cortex (Ungerleider et al., 2008) convey information about simple objects and other visual submodalities, such as colour (Huxlin & Merigan, 1998), shape (Wang et al., 1998), visual texture, glossiness, and translucence. In anthropoids, this information arises from foveal vision (Baizer et al., 1991), but strepsirrhines such as lemurs and bushbabies lack a fovea.

4. In contrast to robust visual inputs, the orbital PF cortex has a paucity of auditory inputs (Kondo et al., 2003).

    The reason is that different kinds of foods vary dramatically in appearance, but not in terms of sounds.

5. Both the granular and agranular orbital PF cortex as well as the agranular insula areas have dense interconnections with the basolateral nucleus of the amygdala. However, area 11 has fewer connections with the amygdala (Carmichael & Price, 1995a; Aggleton et al., 2015). There are also connections from the amygdala to the lateral sector of the orbital PF cortex (Amaral & Price, 1984).

    Connections with the amygdala provide access to information about the value of food items based on an individual's current state. Although much of the literature on the amygdala, especially in rodents, highlights its role in fear conditioning (LeDoux, 2003), studies of monkeys show that the amygdala contributes to learning about positive as well as negative outcomes (Baxter & Murray, 2002).

6. The lateral sector of the orbital PF cortex has outputs to the rostral part of the dorsal and ventral premotor cortex (Petrides & Pandya, 2002). However, the central sector of the orbital PF cortex has no such outputs (Morecraft et al., 1992).

The implication is that it is the lateral and not the central sector of the orbital PF cortex that is in a position to support the choice that the animals make.

The connectional fingerprint of the orbital PF cortex (Figure 4.2) leads to the following conclusions:

- Visual and olfactory inputs convey signals from the distant parts of the outside world.
- Tactile and oral inputs convey signals about the feel of the foods in the hand and mouth.
- Gustatory and visceral inputs provide information about foods as an individual consumes them.
- Connections with the amygdala allow primates to adjust to changes in the value of foods based on an individual's internal state.

The Desirability of Foods

Rolls (2004a; b) used single unit recording to show that the orbital PF cortex represents conjunctions of the visual, gustatory, and olfactory features of foods. As would be expected from the cortical connections, many cells in the caudal part of the orbital PF cortex in macaque monkeys respond to different tastes and smells (Rolls & Baylis, 1994; Rolls, 2006). There are cells that encode salty, bitter, sour, fatty, and astringent tastes (Rolls, 2004b) has also shown that the human orbital PF cortex is activated by the conjunction of smell and taste (de Araujo et al., 2003).

In macaque monkeys, cells in the granular orbital PF cortex encode the conjunction of vision and taste, or of vision and smell (Rolls & Baylis, 1994). The ability to associate the sight of foods with their tastes depends on combining visual inputs to the orbital PF cortex from the perirhinal cortex concerning objects (Bussey et al., 2005), with representations of their smell and taste in the agranular and granular orbital PF cortex (Carmichael & Price, 1995a).

Food Preferences

Information about the value of the foods is provided by the amygdala which is connected with the central sector of the orbital PF cortex (area 13). Machado and Bachevalier (2007a) gave monkeys the choice of foods and non-foods in a large enclosure. The monkeys were tested pre- and post-surgery. After 3 days of testing after surgery, the location of the boxes were moved within the enclosure; this was referred to as the 'shuffled phase'.

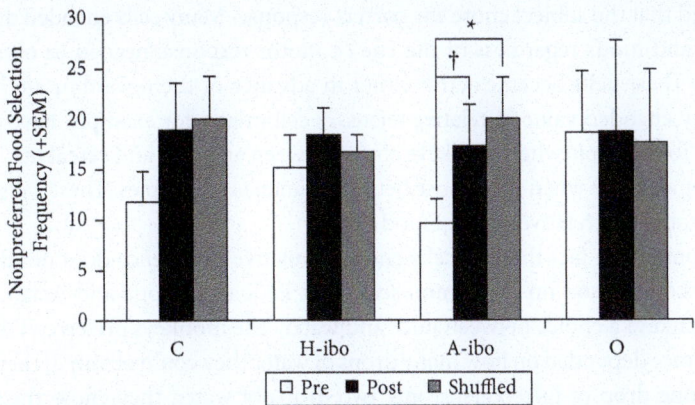

Figure 4.3 Total selection frequency for the category of nonpreferred foods (garlic, lemons, and monkey chow) for each experimental group across the pre-surgery, post-surgery, and shuffled testing phases
A maximum of 45 nonpreferred foods were available in each phase. Vertical bars indicate the standard errors of the mean. Pre, pre-surgery phase; Post, post-surgery phase; Shuffled = shuffled phase, H = hippocampus, A = amygdala, O = central sector of the orbital frontal cortex, Ibo = ibotenic acid, an excitotoxin.

Reproduced from Machado, C.J. & Bachevalier, J. Measuring reward assessment in a semi-naturalistic context: the effects of selective amygdala, orbital frontal, or hippocampal lesions. *Neuroscience*, 148 (3), 599–611, Figure 2, Doi: 10.1016/j.neuroscience.2007.06.035 Copyright © 2007 IBRO. Published by Elsevier Ltd. All rights reserved.

Figure 4.3 presents the results. Monkeys with lesions of the amygdala took and chewed the inedible items. However, monkeys with excitotoxic lesions of the central sector of the orbital PF cortex were not impaired.

This result might be correct. Alternatively, it might be necessary to remove the lateral as well as the central sector of the orbital PF cortex to find an impairment. The reason is that the amygdala also projects to the lateral as well as the central sector (Amaral & Price, 1984).

Common Currency

As they forage, primates have to compare the values of various resources: one food item versus another; fluids versus foods; and so forth. One way to compare food items is to compute an abstract value that is referred to as a 'common currency'. Neurophysiological evidence from macaque monkeys shows that the granular orbital PF cortex encodes valuations in this way.

For example, Tremblay and Schultz (1999) recorded from cells in areas 11 and 13 while a visual cue informed the monkeys about what food would become available, provided that the subject chose the correct response. Many cells encoded different liquids and foods regardless of the cue or motor response needed to obtain the reward. These cells became active either in advance of the reward or afterwards, and they encoded value in relative terms. A cell might, for example, have greater activity for an apple when monkeys chose between an apple and cereal but less activity for an apple when they chose between bananas and apples. These cells therefore encoded the relative value of food items.

Of course, the fact that the cells encoded relative preference does not demonstrate a comparison on a common scale. But Padoa-Schioppa and Assad (2006) gave monkeys a choice between juice and water. The monkeys preferred juice, but their choice depended on how many drops of water they could obtain. If they could obtain one drop of juice versus only two drops of water, they chose the much-preferred juice rather than the water. But if they could get six or ten drops of water, they chose the water rather than the juice.

During the experiment, Padoa-Schioppa and Assad used various juices, let us say juice A and juice B, and they fitted a sigmoid curve to the choices that a macaque monkey made. The x-axis plotted the ratio of juice A to juice B in terms of the number of drops; the y-axis plotted the monkey's choice of juice. This curve describes the relative value of the choices between all the different juices in terms of a common currency.

When the volumes of juice reach a certain ratio, monkeys chose equally between the two options. This 'indifference point' indicated that the monkey assigned equal value to the options. For example, the indifference point might reveal that, when described in terms of a common currency, two drops of apple juice have the same value as six drops of peppermint-flavoured water.

On each trial the monkeys saw yellow and blue symbols on a video screen. The colour of the symbols indicated the nature of the juice, and the number of symbols indicated how many drops of juice they would get. Padoa-Schioppa and Asaad recorded from cells in the granular part of area 13 while the monkeys viewed these arbitrary visual cues, which they called 'offers'. The monkeys then made their choice by executing a saccadic eye movement to fixate one set of symbols, for example three green squares. As expected, some cells encoded specific tastes. However, other cells encoded the value of the offer, and still others encoded the chosen value.

The activity of these cells did not encode the direction of the saccade or the spatial goal of that movement, at least in the period prior to the saccade. Instead they encoded the value of an offer cued by arbitrary combinations of colours and shapes. Because the activity matched the indifference-point curve, Padoa-Schioppa and Assad concluded that these cells encoded value in terms of a common currency.

In general, these cells did not encode value in absolute terms: they did so relative to the offers available. But this conclusion does not imply that all of the cells in area 13 have this property. Padoa-Schioppa and Assad (2008) used the metaphor of a 'menu' to explain this distinction. They contrasted menu-dependent (relative) valuations with menu-independent (absolute) valuations. They presented monkeys with the choice between a variety of pairs of juice. If the monkeys based their choice on value, then their choices should show transitivity. This means that if the monkey preferred juice A to juice B, and juice B to juice C, then it should also prefer to juice A to juice C. The monkeys made choices that conformed to these predictions.

However, transitivity alone does not demonstrate menu-independence. The critical test is whether the 'menu' affected neural responses to juice C, for example. In other words, for cells that encoded the value of juice C, did a 'menu' of juice A and juice C lead to the same activity as a 'menu' of juice B and juice C?

A population of cells in area 13 showed this invariance, which demonstrates menu-independent coding of value, rather than simply a relative preference (Padoa-Schioppa & Conen, 2017). Different populations of cells in the granular orbital PF cortex were found to have one or the other kind of value coding.

When choosing between two alternatives, individuals need to evaluate the value of each item in order to compute their relative value. Rich and Wallis (2016) recorded from areas 11 and 13 and used linear discriminant analysis of cell activity and local field potentials to decode values into four categories, from least to most desirable. The decoded values predicted how long the monkeys would take to decide between two alternatives. And as the monkeys made their choice, the local field potentials alternated between the states associated with each value. This finding suggests that the monkeys weighed each option in the expected manner.

Because recordings from cell populations only provides correlational data, Ballesta et al. (2020) disrupted activity in the orbital PF cortex by using electrical stimulation. This disrupted the ability of the animals to compare subjective values, as well as impairing the economic choices that they made.

Experiments on monkeys can tell us about the choices that they make, but they cannot tell us directly about their subjective states. But we can ask human subjects to rate things according to their subjective valuations. Grabenhorst and Passingham (2010) presented human subjects with different tastes or different degrees of temperature that were applied to their hand with a thermode. The subjects rated the tastes and temperatures on two scales, one for subjective pleasantness, the other for intensity.

Using this task in an fMRI study, Grabenhorst and Passingham examined activations related to subjective pleasantness; they looked for activations that were common to both taste and temperature. They chose three areas as regions of interest because they have been shown to be activated during the processing of rewards: these were the orbital PF cortex, the nucleus accumbens, and the prelimbic cortex (area 32). The amygdala projects to all of these structures (Aggleton et al., 2015; Choi et al., 2017).

In all three areas there were activations that correlated positively with the subjective pleasantness of a stimulus, regardless of whether it was a taste or temperature. This finding shows that these activations did not relate to the sensory features of stimuli but instead to their subjective pleasantness. In other words, they reflected an abstract valuation in a common currency. Because the activations showed no relation to the subjective intensity of the stimuli, we can rule out a relation between pleasantness and intensity as an account for these activations.

Summary

Monkeys with lesions of the central sector of the orbital PF cortex can distinguish between foods and non-foods. Three possible reasons are given to explain why their food preferences are unaffected. Choosing between different foods requires some way of comparing them in terms of value. There is evidence from neurophysiology that cells in the orbital PF cortex can encode the desirability of different foods in terms of a common currency.

Sensory Specific Satiety

If human subjects eat a given food to satiation, there is a decrease in their subjective ratings of 'pleasantness' at the sight of that food; but there is no change in the ratings for other foods (Rolls et al., 1983). For example, after eating several cheese sandwiches the subject is more likely to choose a sandwich with a different filling. The effect is known as 'sensory specific satiety'. The selective advantage is presumed to be that it promotes a varied and thus a balanced diet.

The orbital PF cortex is involved in this effect. Rolls and his colleagues (Critchley and Rolls, 1996; Rolls et al., 1986) studied changes in the cell activity in the orbital PF cortex, including the anterior insula. They fed monkeys one type of food to satiation and observed a gradual decrease in the responsiveness of the cells that encoded its taste or smell. Yet there was no corresponding decrease in the primary sensory areas (Rolls et al., 1989; Critchley & Rolls, 1996).

There are also activations in the lateral sector of the orbital PF cortex (area 47/12) that are related to both taste of a fluid and the subject's current motivational state in terms of hydration and dehydration (Kaskan et al., 2019). Batterham et al. (2007) measured the concentrations of peptide YY (PYY), a gut derived satiety signal. When the levels of PYY were high there was an activation in the caudolateral orbital PF cortex where it meets the anterior insula.

These results implicate the orbital PF cortex and anterior insula in representing the current value of the food based on the individual's current motivational state. If so, lesions of the orbital PF cortex should impair the ability of monkeys to represent the current value of foods. Machado and Bachevalier (2007b) examined the effects of satiation in a semi-naturalistic environment. An earlier section describes the set-up. Before testing, the animals were fed one of the two foods to satiation. The animals with lesions of the central sector of the granular orbital PF cortex continued to approach the box with the food that they had been fed, and unlike control animals they also ate the foods. By contrast, the control animals ate the foods in the other box. The implication is that in their natural environment the animals shift to another patch when they have fed to satiation on the resources in one place.

Rather than studying a shift in location, Izquierdo et al. (2004) gave monkeys the choice between two objects, one of which covered food 1 and the other food 2. So the objects served as signs for the foods that they covered. After the animals had been fed to satiety on their preferred food the control animals tended to choose the object that covered the other food. As shown by the black bar in the left half of Figure 4.4, they shifted their choice to the object associated with the food item that was more highly valued at the moment. The experiment was repeated with many object/food pairings, and because they saw each pair of objects only once, the monkeys could not learn the best choice during the testing session.

However, the blue bar in Figure 4.4 shows that the monkeys with lesions of the central sector of the orbital PF cortex behaved differently. They tended to choose objects associated with the food that they had originally preferred. For example, if a lesioned monkey originally preferred M&Ms˚, they tended to choose objects associated with M&Ms˚ even though they had just consumed these to satiety.

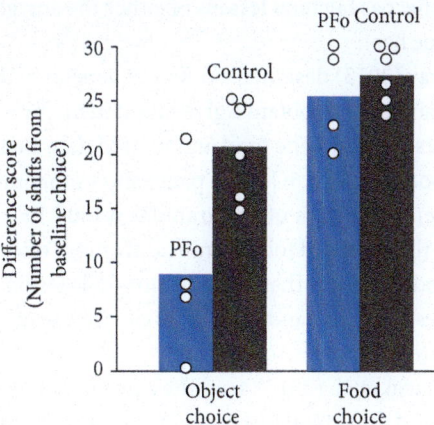

Figure 4.4 The effects of satiation on the choice between objects that were associated with particular foods
Black bar = data for control animals; blue bar = data for monkeys with lesions of the granular orbital PF cortex (PFo)

Reproduced from Izquierdo, A., Suda, R.K., & Murray, E.A. Bilateral orbital prefrontal cortex lesions in rhesus monkeys disrupt choices guided by both reward value and reward contingency. *J Neurosci*, 24 (34), 7540–8, Figure 4, Doi: https://doi.org/10.1523/JNEUROSCI.1921-04.2004 Copyright © 2004 by the Society for Neuroscience.

Yet the figure also shows that when the monkeys with orbital PF lesions were given the direct choice between the foods themselves, they tended to avoid the food that they had been fed to satiety. But in their experiment Machado and Bachevalier (2007a; b) found that once the lesioned monkeys had put their hand in the box with the food on which they had been fed, they were prepared to eat the food. The implication is that, unlike control animals, the lesioned animals are prepared to eat the food on which they been satiated, but will choose an alternative food instead if it is visible at the time.

Just as animals with amygdala lesions select inedible foods (Machado & Bachevalier, 2007a; b), so they also fail to shift between objects when the food under one of them is fed to satiation. Figure 4.5B (next page) shows that bilateral lesions of either the amygdala or the orbital PF cortex (Murray & Izquierdo, 2007) cause similar impairments. Compared to intact control monkeys (black bar), both groups of lesioned monkeys shifted their choices less frequently.

These results suggest that the devaluation effect depends on the interconnections between the granular orbital PF cortex and the amygdala. To test this idea directly, Baxter et al. (2000) made lesions in the amygdala in one hemisphere and the central sector of the orbital PF cortex in the other hemisphere. This is referred to in the figure as an 'amygdala x PFo disconnection'. To complete the disconnection, Baxter et al. cut the corpus callosum and anterior commissure as well. The third brain from the left depicts these commissurotomies with black rectangles. As the purple bar in Figure 4.4B (page 127) illustrates, the crossed-disconnection lesions had the same effect as bilateral lesions of either the amygdala (red bar) or the orbital PF cortex (blue bar).

Rhodes and Murray (2013) designed a different version of the devaluation task. Instead of choosing an object to obtain a given food item, the monkeys had to tap a video screen six times to obtain one food and to hold their hand still on the screen to obtain another food. Figure 4.5C (next page) shows that monkeys with lesions of the medial and central sectors of the granular orbital PF cortex chose the action associated with the devalued food significantly more often than intact control subjects. The coloured bars show that bilateral amygdala lesions (pink and red) and bilateral orbital PF lesions (light and dark blue) caused the monkeys to choose the two actions equally frequently.

In a later study, Fiuzat et al. (2017) used the same task and a disconnection between the amygdala and the orbital PF cortex. As Figure 4.5C (next page) illustrates (light and dark purple), monkeys with these disconnections continued to choose the action associated with a devalued food more often than control subjects did.

Gottfried et al. (2003) gave the devaluation task to human subjects. The subjects learned to associate one visual stimulus with one odour and to associate a different visual stimulus with another odour. They then had the opportunity to consume to satiety a food that had one of the two odours. On subsequent testing, there were

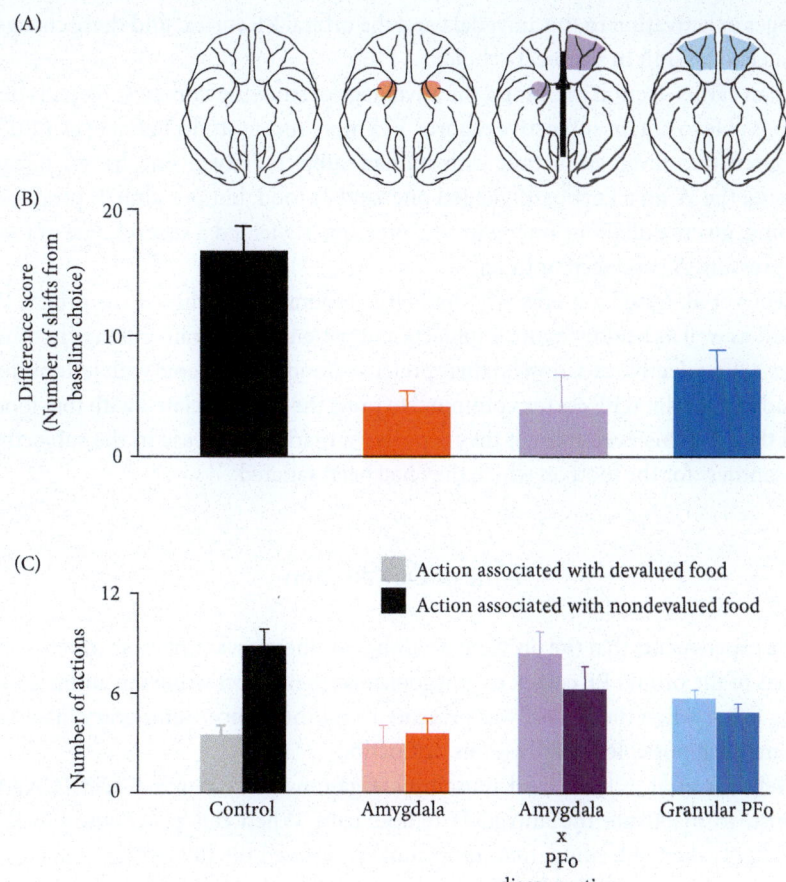

Figure 4.5 Effect of three lesions on choices in the devaluation task
(A) Lesions illustrated on a ventral view of a macaque monkey brain. The red shaded areas depict selective amygdala lesions (beneath the brain surface). (B) Results from the object version of the devaluation task. Shifts in object choice to the object associated with the non-devalued food. PFo, orbital PF cortex; error bars: SEM. (C) Results from the action version of the devaluation task. Monkeys chose between steady contact with a video touchscreen (hold) or repetitive brief contacts (tap). Choices of the action associated with the devalued food appears in lightly shaded bars; choices of the action associated with non-devalued food appears in the darkly filled bars.

Adapted from Murray E.A. & Izquierdo A. Orbitofrontal cortex and amygdala contributions to affect and action in primates. *Ann NY Acad Sci*, 1121 (1), 273–96, Figure 1, Doi: 10.1196/annals.1401.021 © 2007, John Wiley and Sons, with permission.

changes in activation in the amygdala and the orbital PF cortex, and these changes paralleled the shift in a subject's choice.

As an earlier section mentions, the advantage of studying human subjects is that it is possible to obtain subjective reports. In a more recent study, Reber et al. (2017) also used the devaluation task with human subjects. Given one fractal image, pressing the 'A' on a keyboard yielded one kind of food, but pressing 'L' produced nothing; given a different fractal image, pressing 'L' yielded a second type of food and pressing 'A' produced nothing.

Reber et al. tested patients who had large lesions that included the orbital PF cortex, as well as healthy control subjects and patients with brain damage in other areas. After selective satiation on their preferred food, the patients with lesions that included the orbital PF cortex continued to press the key associated with that food. And they did this even though they reported a marked decrease in the subjective pleasantness for the food on which they had been satiated.

Time of Devaluation

In the experiments that the previous section describes, it was not clear whether the lesions of the orbital PF cortex or amygdala had their effect while the animals was being fed or when the animal was presented with the choice. Temporary inactivation makes it possible to address this distinction.

Wellman et al. (2005) used muscimol, a gamma-aminobutyric acid (GABA) agonist, to inactivate the amygdala temporarily. When this procedure blocked cell activity during feeding, macaque monkeys showed the full devaluation effect, that is they failed to shift their choices normally. But when applied during the later choice test, muscimol had no effect. This finding shows that for a monkey to shift its choices normally, cells in the amygdala need to be capable of discharge activity as monkeys consume a food to satiety.

Murray et al. (2015) followed up this finding by using a different GABA agonist to temporarily inactivate different parts of the orbital PF cortex, either during the satiation procedure or during the choice test. Inactivation of the granular orbital area 13 had the same effect as inactivation of the amygdala, as previously reported by Wellman et al. (2005).

By contrast, inactivation of the more anterior orbital area 11 had no effect when applied during the devaluation procedure. Instead, it had a marked effect when applied during the subsequent choice test.

These findings are as expected from the connectivity. First, area 13 has robust connections with the amygdala, but area 11 does not (Price & Drevets, 2010). This distinction probably explains why temporary inactivation of area 13 and the amygdala had similar effects, but inactivation of area 11 had a different effect. Second, both area 11 and area 13 receive visual inputs from the perirhinal and inferior temporal cortex (Kondo et al., 2005). Accordingly, both areas have information about the visual features of objects and foods.

The Choice

It is unlikely that the central sector of the orbital PF cortex directly supports the choice of the object or the performance of the action. The reason is that it has no connections with the premotor cortex. It is the lateral sector and associated ventral PF cortex that are connected directly (Petrides & Pandya, 2002) and indirectly (Takahara et al., 2012) with the premotor areas. These connections are intact after a lesion of the central sector of the orbital PF cortex. The role of this sector is to support the ability to expect or predict food.

Summary

Monkeys with lesions of the orbital PF cortex are impaired at avoiding objects or actions that are associated with food when that food has become less desirable because they have been sated on it. However, the effects of lesions of area 13 and area 11 differ, depending on the time of satiation. Patients with lesions that include the orbital PF cortex continue to press a key to obtain food on which they have been satiated, even though when they report that the food itself is subjectively less desirable.

Signs of Resources

A number of studies have explored the neural encoding of signs that predict food in the orbital PF cortex. For example, Tremblay and Schultz (2000b) recorded there while macaque monkeys performed a 'go–no-go' task. A visual cue instructed the monkeys whether to press a lever to gain juice, which they received directly in their mouths. Some cells in the orbital PF cortex showed increased activity after the instruction cue, but other cells did so just before or just after the juice reward. This activity probably reflects a prediction that juice would become available.

However, very few cells encoded whether the monkey made or withheld a response. Likewise, Wallis and Miller (2003) found that cells in the orbital PF cortex encoded the number of juice drops, but not the action that led to the reward. This is as expected, given the lack of connections with the premotor cortex.

Tremblay and Schultz (2000a) also recorded from the orbital PF cortex as monkeys learned the cued go–no-go task, and they found that many cells increased their response to the instruction cues as learning progressed. In the same experiment,

Tremblay and Schultz included a condition in which reward sometimes failed to follow a given action. At first, many cells responded as if juice would become available on all trials. But, with learning, the responses of these cells became restricted to 'juice trials', that is the trials when cues indicated that a reward would occur.

Because the orbital PF cortex has strong connections with the amygdala (Aggleton et al., 2015), it is reasonable to suppose that the activity relating to the delivery of food arises from the amygdala. To test this idea, Rudebeck et al. (2013a) recorded from the orbital PF cortex in macaque monkeys before and after bilateral lesions of the amygdala. On each trial, the monkeys saw two object-like stimuli on a video screen in succession. By choosing between these two stimuli, the monkeys learned how many drops of water would follow the choice of each stimulus.

Figure 4.6 shows the population activity for cells in areas 11 and 13 of the central sector of the orbital PF cortex. As expected from previous work, many cells encoded the value of a stimulus as a function of a predicted quantity of reward, and they did this both after the stimulus appeared (Figure 4.6A) and in advance of the reward (Figure 4.6B).

If this activity in orbital PF cortex depends entirely on connections with the amygdala, then lesions of the amygdala should abolish it. And indeed, as Figure 4.6 shows, after the amygdala lesion (red curves), fewer cells encoded reward magnitude than in a comparable population of cells recorded before the lesion (blue curves). This was true both for responses to the object-like stimulus (Figure 4.6A)

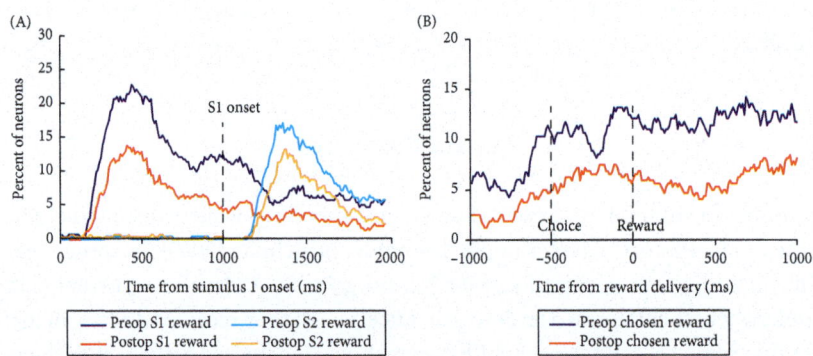

Figure 4.6 The effect of amygdala lesions on the encoding of reward in the orbital PF cortex
(A) The percentage of cells that encoded reward magnitude after the onset of the first of two object-like visual stimuli (S1). Similar data for the second stimulus (S2) are omitted for simplicity. Blue, encoding of reward magnitude before amygdala lesions; red, comparable data after the lesions. (B) The percentage of cells encoding reward magnitude relative to the time of reward delivery, in the format of A.

Adapted from Murray, E.A. & Rudebeck, P.H. Specializations for reward-guided decision-making in the primate ventral prefrontal cortex. *Nat Rev Neurosci*, 19 (7), 404–17, Figures 5a and 5b, Doi: https://doi.org/10.1038/s41583-018-0013-4 Copyright © 2018, Springer Nature.

and at the time of the reward (Figure 4.6B). However, surprisingly amygdala lesions did not totally abolish the coding of rewards in the orbital PF cortex.

To see if the same conclusion applies to new learning, Rudebeck et al. (2017a) recorded while the monkeys learned associations between novel stimuli and reward magnitude. Cells in the orbital PF cortex rapidly came to encode the magnitude of reward, which corresponds with their rapid learning of the value of each stimulus. Figure 4.6C illustrates both the pre-operative learning rate and the rate of learning after amygdala lesions. The lesions not only decreased the number of cells that encoded reward magnitude during learning (Figure 4.6A and B), but they also significantly slowed learning (Figure 4.6C).

These findings support the proposal that the amygdala makes a contribution to value coding in the orbital PF cortex; but nonetheless amygdala lesions did not eliminate these neural signals. The connections of the area probably explain why. The task involves arbitrary associations between object-like stimuli and rewards consisting of water, and this has sensory qualities. Because the cortico-cortical connections remained intact, the orbital PF cortex could still receive information about the visual stimuli from the ventral visual stream. It could also receive information about the feel of water in the mouth from the somatosensory cortex, and about the taste via connections with the gustatory cortex.

Negative Outcomes

For ethical reasons, most studies in the laboratory avoid causing harm to subjects. But the risk of harm played a crucial role in shaping the evolution of the primate brain. From early primates to ancestral anthropoids, the risk of predation and other deleterious events exerted a powerful selective force. Many primates, and especially many anthropoids, fail to make it to maturity. Primates of the past benefited from balancing the need for food and other resources with risks, as their descendants continue to do today.

To produce a laboratory model, Morrison and Salzman (2009) taught monkeys that different visual stimuli predicted either drops of juice or an aversive puff of air. Cells in the orbital PF cortex (area 13) showed increases in activity just before both the negative as well as the positive outcomes.

Instead of defining risk in terms of gains and losses, O'Neill and Schultz (2010) defined it in terms of the difference between gaining a high and low reward. They defined the degree of risk as the difference in magnitude between the two rewards, measured in terms of a volume of juice. In this case, a subject 'risks' losing the larger volume of juice; and the larger the difference the larger the 'risk'.

In their experiment, the monkeys chose between two cues. One cue consisted of a single bar that predicted a deterministic reward, with the height of the bar indicating its magnitude. The other cue had two bars, which indicated that both the high and low reward might be available, with each outcome being equally likely. The difference in the height of these two bars indicated the degree of 'risk'.

The monkeys tended to make the risky choice, and there were cells in the orbital PF cortex that encoded the degree of risk. However, a critic might object that these cells simply encoded value because the further apart the bars, the greater the potential gain. So O'Neill and Schultz examined this possibility by examining the activity of cells in the orbital PF cortex when the monkey chose the single bar. This indicated the magnitude of deterministic reward. Most of the risk-encoding cells had the same activity whether the single bar predicted a high or low reward. So these cells probably did not encode value per se. O'Neill and Schultz (2015) reviewed these and related results from subsequent papers, and concluded that indeed the orbital PF cortex represents risks as well as rewards.

Raghuraman and Padoa-Schioppa (2014) also presented macaque monkeys with choices in which cues specified the probabilities of different rewards. As in their previous work, they found activity in the orbital PF cortex that reflected the values of both the 'offer' and the monkey's choice, and this activity integrated information about the amount of juice and the probability that it would become available. But, as in the study by O'Neill and Schultz (2010), they also found cells that encoded the degree of risk undertaken by the monkey.

However, the definition of risk used in these experiments does not correspond to its common-sense definition. Normally, we think of risk as the probability of danger or some other serious cost. In these experiments, the monkey won something regardless of the choice they made.

So Rich and Wallis (2014) designed a task in which the monkeys could either win or lose. On each trial they saw a picture, and some of the pictures were designated as positive and others as negative. If the monkeys responded correctly to a positive picture, the length of a line increased. to represent the amount of juice available. If the monkeys responded incorrectly to a negative picture, the length of the line decreased. After blocks of six trials the monkeys received the amount of reward that they had earned. Rich and Wallis recorded from cells in all sectors of the orbital PF cortex and found activity that reflected both gains and losses throughout the area.

Summary

In macaque monkeys, cells in the orbital PF cortex respond to visual cues that predict a particular outcome, and they encode the expected value of outcomes. However, they do not code for the response. Removal of the amygdala decreases the proportion of cells that encode these valuations. However, it does not abolish this neural signal, and the reason is probably that cortico-cortical connections remain intact. There are also cells in the orbital PF cortex that encode negative outcomes.

Objects Reversals

In their natural surroundings, primates need to cope with resource volatility. There is variation in the availability of the foods at different times of the year or from year to year. Chapter 3 introduced an experimental analogue of resource volatility: this is the frequency with which a previously successful choice no longer produces a predicted and desired outcome. In the laboratory, reversal tasks provide one way of studying how an individual adapts to volatile outcomes.

Figure 4.7 illustrates an example of two reversal tasks. Action reversal tasks are shown to the left and object reversal tasks to the right. As Chapter 3 describes, on an action reversal task monkeys are taught to change their choice of actions. On an object reversal task the monkeys are taught to change their choice of objects.

The figure also illustrates two varieties of reversals, deterministic fixed (Figure 4.7A) and probabilistic (Figure 4.7B). On deterministic reversals, the action or

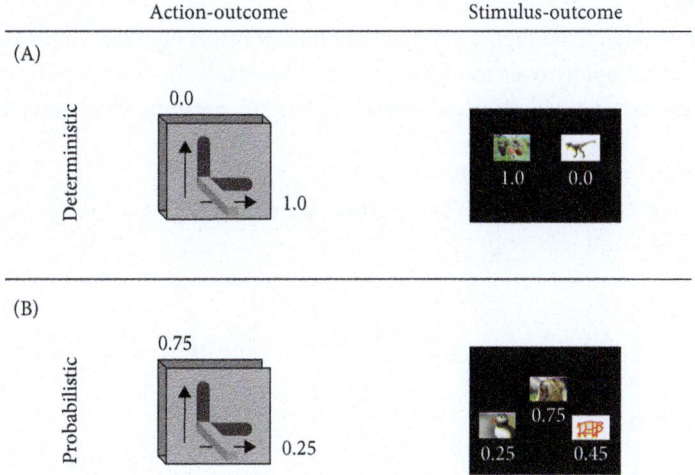

Figure 4.7 Action reversal and object reversal tasks
(A) Deterministic reward contingencies. During a series of consecutive trials, the outcome for each action or choice is certain: either a reward (1.0) or no reward (0.0). When a reversal occurs, these reward contingencies switch between the two actions or object choices (B) Probabilistic reward contingencies. Each action or choice is associated with a discrete probability of reward, and these probabilities change enough that the 'best choice' switches among alternatives. In the example, moving the handle up or choosing the lion produced a reward 75% of the time; moving the handle to the right or choose the puffin leads to reward 25% of the time.

Adapted from Rudebeck P.H., Behrens T.E., Kennerley S.W., Baxter M.J., Walton M.E., & Rushworth M.F. Frontal cortex subregions play distinct roles in choices between actions and stimuli. *J Neurosci*, 28 (51), 13775–85, Figure 1, Doi: 10.1523/JNEUROSCI.3541-08.2008 Copyright © 2008 by the Society for Neuroscience.

choice produces a reward for a series of consecutive trials, and the alternative never does. When a reversal occurs, the reward contingency switches between two stimuli without any additional cue to signal the change.

On probabilistic reversals, the relation between the action or object and the outcome changes in a graded way. One action or choice is the most likely to produce a reward at any given moment, but the best choice can change as reward probabilities change.

Deterministic Reversals

Chapter 3 mentions that Rudebeck et al. (2008) found that lesions of the anterior cingulate cortex (ACC) impaired the performance of macaque monkeys on action reversals. However, in that experiment lesions that removed the central and medial sectors of the orbital PF cortex did not have this effect. Instead, as Izquierdo et al. (2004) found, monkeys with a lesion of these areas have an impairment on object reversals.

It will be seen from Figure 4.8 that the intact controls monkeys improved their performance rapidly over the series of seven reversals. In other words, they acquired a 'learning set' (Harlow & Warren, 1952) meaning that over series of

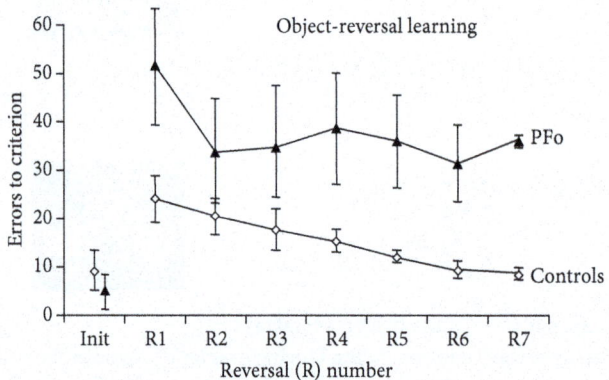

Figure 4.8 Impairment in object reversal set in monkeys after lesions of the central and medial sectors of the orbital PF cortex
Number of errors to criterion performance on a two-choice object discrimination task. Abbreviations: Init, learning of the original discrimination; R1, R2 ... R9, serial reversals, in which over time both objects become positive and neutral in alternating blocks of trials; PFo, orbital PF cortex, error bars: SEM.

Reproduced from Izquierdo, A., Suda, R.K., & Murray, E.A. Bilateral orbital prefrontal cortex lesions in rhesus monkeys disrupt choices guided by both reward value and reward contingency. *J Neurosci*, 24 (34), 7540–8, Figure 4, Doi: https://doi.org/10.1523/JNEUROSCI.1921-04.2004 Copyright © 2004 by the Society for Neuroscience.

problems they learned to solve new ones more and more quickly. In this case, the control animals acquired a 'reversal set'. But in the experiment by Izquierdo et al. (2004), monkeys with removal of the central and medial sectors of the orbital PF cortex showed little improvement across the series of reversals.

Figure 4.8 simply shows the number of errors per reversal. To find out how lesions of the orbital PF cortex impair performance, Rudebeck and Murray (2008) carried out a trial-by-trial analysis of the performance on object reversals, just as Kennerley et al. (2006) had done for the effect of ACC lesions on action reversals.

First, as Figure 4.9A (next page) illustrates, Rudebeck and Murray (2008) counted the number of errors made prior to the first correctly reversed choice. This measure reflects a tendency to perseverate, that is to continue with the previously correct choice. Monkeys with removal of the orbital PF cortex made more of these errors than did the control animals.

Then Rudebeck and Murray counted the errors after the first correct choice following an error. In their notation, '+ 1' designated the trial after a correct choice. When one correct (C) choice followed an error (E), they called the next trail 'EC + 1'. When two, three, or four correct choices following an error, they called the next trial EC (2) + 1, EC (3) + 1, or EC (4) +1.

Figure 4.9B (next page) illustrates the results of this analysis. Whereas the intact controls monkeys could use the feedback from correct choices efficiently, monkeys with surgical lesions of the orbital PF cortex were as likely to make an error after four correct responses as after one. In other words, they made errors both by failing to shift after an error and by failing to stay after a correct response.

Surgical versus Excitotoxic Lesions

The effects of surgical lesions of the orbital PF cortex can be misleading. The reason is that aspiration of the tissue in the central sector damages the fibres that run from the inferior temporal cortex and amygdala to the lateral sector (Folloni et al., 2019). The advantage of making excitotoxic lesions is that for the most part they avoid this problem. As Chapter 1 explains, excitotoxins cause cell death through excessive activity, but they spare the white matter because they have little or no effect on axons.

In a critical experiment, Rudebeck et al. (2013b) therefore made excitotoxic lesions of the central sector. The results were unexpected: the excitotoxic lesions failed to impair object reversals. However, this result can be explained by the connectivity. The lesions spared the projections to the lateral sector of the orbital PF cortex from the inferior temporal cortex (Folloni et al., 2019) and the amygdala (Amaral & Price, 1984).

To prove that these connections are critical, Rudebeck et al. made a 'strip lesion' by removing the tissue along the caudal limit of the orbital PF cortex. Because they used the aspiration method to make this lesion, this procedure interrupted

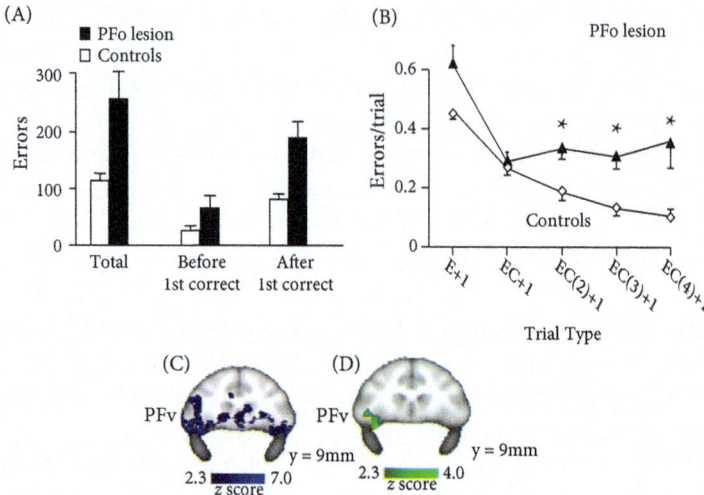

Figure 4.9 Errors made on the object reversal task, before and after the first correct choice
(A) Black bars show cumulative errors made by monkeys with lesions of the central and medial sectors of the orbital PF cortex. White bars contrast the scores for intact control animals. Most errors followed the first correct choice. (B) Trial-by-trial performance after an error followed by 1, 2, 3, or 4 correct choices. Asterisk: statistically significant difference. Error bars: SEM. (C) Regions of the lateral sector and associated inferior convexity (area 47/12) showing significantly increased fMRI activations related to outcome events in a macaque monkey (blue). (D) Regions of the lateral sector of the orbital PF cortex showing significantly increased activations related to win-stay and lose-shift strategies (green). PFv, ventral PF cortex.

A and B: Reproduced from Rudebeck P.H. & Murray E.A. Amygdala and orbitofrontal cortex lesions differentially influence choices during object reversal learning. *J Neurosci*, 28 (33), 8338–43, Figure 3, Doi: 10.1523/JNEUROSCI.2272-08.2008 Copyright © 2008 by the Society for Neuroscience.

C and D: Reproduced from Chau, B.K., Sallet, J., Papageorgiou, G.K., Noonan, M.P., Bell, A.H., Walton, M.E., & Rushworth, M.F. Contrasting roles for orbitofrontal cortex and amygdala in credit assignment and learning in macaques. *Neuron*, 87 (5), 1106–18, Figure 2a and b, Doi: 10.1016/j.neuron.2015.08.018 Copyright © 2015, The Authors. Licensed under CC BY 4.0.

the passage of the fibres to the lateral sector of the orbital PF cortex. This lesion left nearly all of the central and medial sectors of the orbital PF cortex intact; yet it caused an impairment on object reversal (Rudebeck et al., 2013b).

Further evidence that the lateral sector of the orbital PF cortex is involved comes from an fMRI study by Chau et al. (2015) in which the monkeys were tested on a series of object reversals. There were activations that were related to both the win-stay and lose-shift strategies in the lateral sector (Figure 4.9D). There were also activations that were related to the outcomes, distributed across both the lateral and central sectors (Figure 4.9C). Furthermore, successful performance of the task correlated positively with the degree of activation in the lateral sector of the orbital PF cortex.

The findings from the fMRI study by Chau et al. (2015) support the view that it is the lateral sector that is critical for reversal learning. However, they also suggest that the rest of the ventral PF cortex is involved.

Probabilistic Reversals and Credit Assignment

Walton et al. (2010) devised a novel probabilistic reversal task. On each trial, a macaque monkey chose among three pictures, each associated with a different probability of producing a reward if chosen. Walton et al. called this version of reversal learning the '3-arm bandit task', a reference to gambling machines known as one-arm bandits or slot machines.

The probabilities changed over time, in both a high-volatility and low-volatility mode. After about 145 trials, a stimulus that had not previously paid off at all began to do so. After a number of additional trials, its likelihood of paying off exceeded the other two possible choices.

Walton et al. argued that, given a series of choices, the monkey had to attribute each outcome to a particular choice, and they called this 'credit assignment'. A simple reinforcement learning model cannot assign credit to a particular choice because it works by cumulating the outcomes of past choices. This learning mechanism produces a broad average over several events, but it dispenses with information about any single event. The algorithm works well in environments with stable or slowly changing resources, but not in environments in which the resources are volatile.

Walton et al. carefully analysed the results for trials after the formerly low-value stimulus exceeded the others. At that point, it became the 'newly best' stimulus, in contrast to the 'previously best' stimulus. If, after its increase in value, the monkey selected this 'newly best' stimulus and received a reward, it should be more likely to choose the same stimulus again on the next trial.

Figure 4.10 (next page) shows what the monkeys did. Positive values on the y-axis show the degree to which the monkeys reverted to choosing the 'previously best' stimulus: called stimulus A in this plot. Because they had just received a reward for choosing stimulus B (denoted as B^+), the 'newly best' stimulus, a reversion to choosing stimulus A corresponded to a win-shift error. Negative values indicate how much they shifted their choice to the 'newly best' stimulus, stimulus B, the 'correct' choice. For intact control monkeys (black bars), receiving a reward for choosing the 'newly best' stimulus led a shift of their choices to that stimulus. For lesioned monkeys (blue bars), the opposite occurred. And the longer the history of choosing the 'previously best' (stimulus A), the more likely they were to choose A again after a trial on which B had been rewarded.

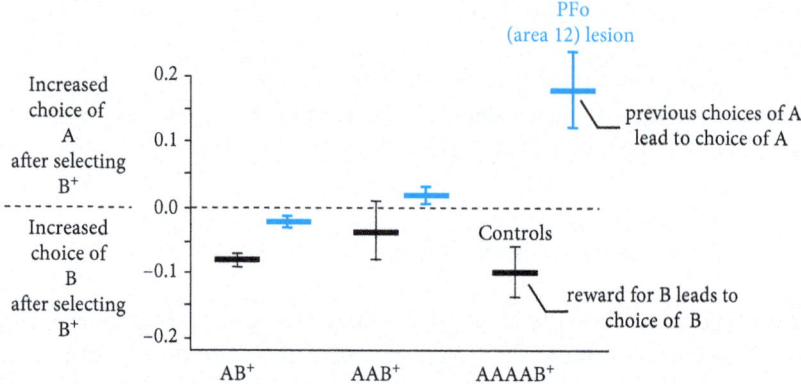

Figure 4.10 Performance on probabilistic reversals after lesions of the orbital PF cortex
The plot shows the mean (horizontal lines) and SEM for intact control monkeys (black) and macaque monkeys with lesions of the entire orbital PF cortex (blue). The *x*-axis divides trials according how many trials involving the choice of stimulus A preceded a trial on which choosing stimulus B led to a reward (+): once (AB⁺), twice (AAB⁺), or four times (AAAAB⁺). On the *y*-axis, positive values indicate the increased likelihood of choosing stimulus A after being rewarded for choosing stimulus B; negative values indicated the increased likelihood of choosing stimulus B in that circumstance (the correct choice). Preop, preoperative, PFo, orbital PF cortex.

Adapted from Walton M.E., Behrens T.E.J., Buckley M.J., Rudebeck P.H., & Rushworth M.F.S. Separable learning systems in the macaque brain and the role Preoof orbitofrontal cortex in contingent learning. *Neuron*, 65 (6), 927–39, Figure 6, Doi: 10.1016/j.neuron.2010.02.027 © 2010, The Authors. Licensed under CC BY 4.0.

Walton et al. concluded that the intact control monkeys recognized the causal link between a specific choice of the 'newly best' stimulus and the outcome. In the terminology used by Walton et al., the control monkeys could assign credit to a single choice. By contrast, monkeys with removal of the medial and central sectors of the orbital PF cortex could not remember these events and instead relied on cumulative averages over several events. They made their choices based on a longer history of having chosen the 'previously best' stimulus.

There are two reservations. The first is that the lesions were made by surgical aspiration. This means that, as explained earlier, they disconnected the lateral sector from the inferior temporal cortex and amygdala.

The second reservation concerns the use of the term 'credit assignment' to describe the findings. This term has traditionally been used to refer to a move that proves critical for the final outcome. A standard example is the kind of move that a chess master makes that determines the ultimate outcome many moves later. Walton et al. used the term in a narrower sense to refer to the ability to learn that a single trial determines the next choice.

Camille et al. (2011) studied credit assignment in this narrower sense in an fMRI study with human subjects. The tasks was a probabilistic reversal task. The subjects chose between a coloured deck of playing cards that either yielded the gain or loss of $50 of play money. A deck of one colour yielded a gain on 86% of trials, the other produced a gain on 14% of the trials. Camille et al. contrasted healthy control subjects with patients with lesions that included the orbital PF cortex.

On reversals between stimuli and outcomes, the patients with lesions that included the orbital PF cortex made more errors, failing to use a single positive trial to shift their choice. This impairment corresponds to the one that Figure 4.10 illustrates for macaque monkeys.

The Lateral Orbital Prefrontal Cortex

Experiments on macaque monkeys make it likely that the impairment in the study by Camille et al. (2011) was due to damage to the lateral sector of the orbital PF cortex. This is part of the ventral PF cortex (area 47/12).

To study the role of this area, Rudebeck et al. (2017b) used the probabilistic 3-arm bandit task In their experiment, Rudebeck et al. made excitotoxic lesions of either the central and medial sectors of the orbital PF cortex, or the ventral PF cortex including the lateral orbital sector.

They presented three object-like stimuli, and on each day the association between the choices or stimuli and probable outcomes changed several times in a graded manner, as in the task used by Walton et al. (2010). The coloured circles in Figure 4.11A (next page) show the pay-off likelihood of the 'currently best' stimulus. The change to red after trial 150 marks the trial when the choice of stimulus B became the best option.

As the bottom of Figure 4.11B (next page) shows, monkeys with selective excitotoxic lesions of the ventral PF cortex had a severe impairment. They reversed their choice from the 'previously best' stimulus to the 'newly best' stimulus much more slowly after the lesion than before it. By contrast, as the top of Figure 4.11B shows, the monkeys with excitotoxic lesions of medial and central sectors performed nearly identically with monkeys with no lesion.

As discussed earlier, in object reversal experiments the monkey receives two kinds of feedback indicating that the association between stimuli and outcomes has changed: an expected reward fails to occur, or an unexpected reward occurs. Both cases result in a 'reward-prediction error', and temporal-difference models provide one account for how these signals adjust future behaviour.

O'Doherty et al. (2003) used a temporal-difference model to analyse the activations in an fMRI study. Human subjects performed a task that used complex

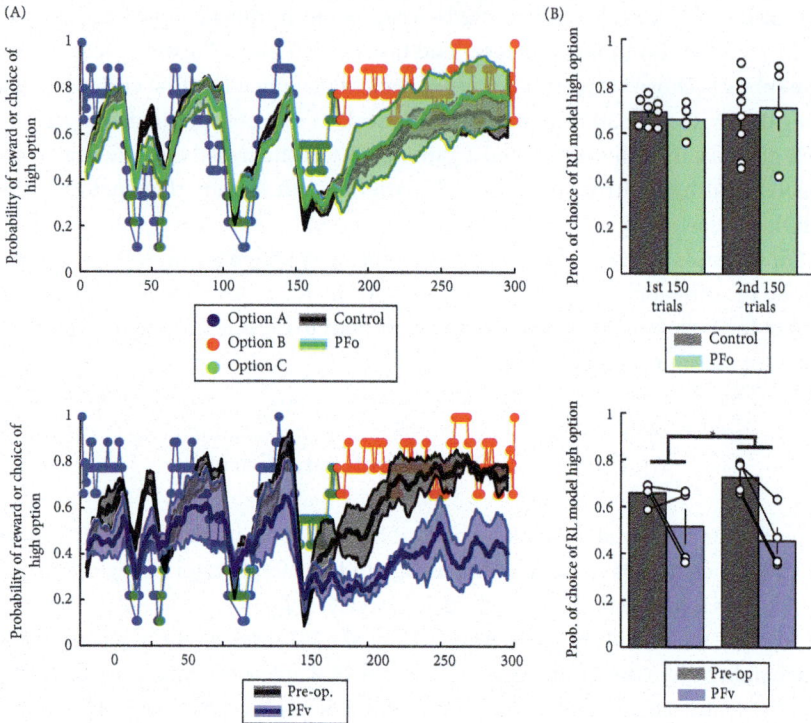

Figure 4.11 Performance on a probabilistic reversal learning task: the 3-arm bandit task

(A) Mean (± SEM) choice behavior of unoperated controls (gray, top), monkeys with orbital PF lesions (green, top), and monkeys before (gray, bottom) and after (blue, bottom) ventral and lateral orbital PF lesions. Note that in (A) (top), the gray curve and shading (control) are largely obscured by the overlying green curve and shading (orbital PF). Colored points represent the identity and probability of receiving a reward for selection of the high reward option. (B) Mean (± SEM) probability of choice of reinforcement learning estimated high reward option in the first and second sets of 150 trials for unoperated controls (n = 8) and monkeys with orbital PF lesions (top) and monkeys before and after ventral and lateral orbital PF lesions (bottom). Symbols show scores of individual subjects; SEM, standard error of the mean; PFo, animals with lesions of the central sector of the orbital PF cortex; PFv, animals with lesions of the ventral and lateral orbital PF cortex.

Reproduced from Rudebeck, P.H., Saunders, R.C., Lundgren, D.A., & Murray, E.A. Specialized representations of value in the orbital and ventrolateral prefrontal cortex: Desirability versus availability of outcomes. *Neuron*, 95 (5), Figures 2A and 2B, 1208–20, Doi: 10.1016/j.neuron.2017.07.042 © 2017 Elsevier Inc.

patterns of colours and shapes as stimuli. One of these stimuli signalled the delivery of a drop of sweet fluid, a second signalled no fluid, and a third signalled a neutral-tasting fluid. However, on some trials the predicted outcome failed to occur.

O'Doherty et al. looked for activations matching the pattern expected for reward-prediction error signals: activations that decreased when the sweet taste failed to occur as expected and increased when it occurred unexpectedly. They found activations in the orbital PF cortex and the ventral striatum that matched this pattern.

Bayesian Priors and Reversal Learning Set

Reversal learning involves two different kinds of predictions: the prediction of an outcome and the prediction of a reversal. Costa et al. (2015) suggested that the second type of prediction depends on Bayesian reasoning. They pointed out that after a long series of probabilistic reversals, macaque monkeys changed suddenly from one choice to another, in other words they shifted in one trial.

A Bayesian perspective can account for this finding, which indicates a strong reversal set. The analysis takes into account not only the information that results from receiving the outcome after a choice (called the posterior probability), but also the prior information that reversals will likely occur (called the prior probability). Costa et al. concluded that a naïve reinforcement model cannot account for the monkeys' behaviour; this family of models includes standard temporal-difference approaches. Specifically, the monkeys switched their choice in one trial. Costa et al. suggested that they could do this because of prior knowledge about the probability of a reversal. A representation of the prior probability of reversals corresponds to an estimate of the volatility of resources.

A Bayesian perspective also explains why macaques can learn to 'assign credit' to a single outcome on the 3-arm bandit task. Past experience tells them that reversals will happen. In the 3-arm bandit task this means that the probability of reward will change over time for a given stimulus. And this means that when a given stimulus becomes the 'newly best' one, individuals can quickly switch their choice to that stimulus rather than simply continuing to choose the 'previously best' stimulus.

Jang et al. (2015) re-analysed the data on reversals from monkeys with either surgical lesions of the medial and central sectors of the orbital PF cortex, or crossed-disconnection lesions of the amygdala and surgical removal of these sectors. Because the cortical lesions were surgical, they interrupted the projections to the ventral PF cortex from the inferior temporal cortex and amygdala.

Jang et al. combined the data for these animals with data from monkeys with lesions of the perirhinal and entorhinal cortex which provide visual inputs to the ventral and orbital PF cortex (Kondo et al., 2005). Figure 4.12 (next page) shows the data for this cluster in red (cluster 2).

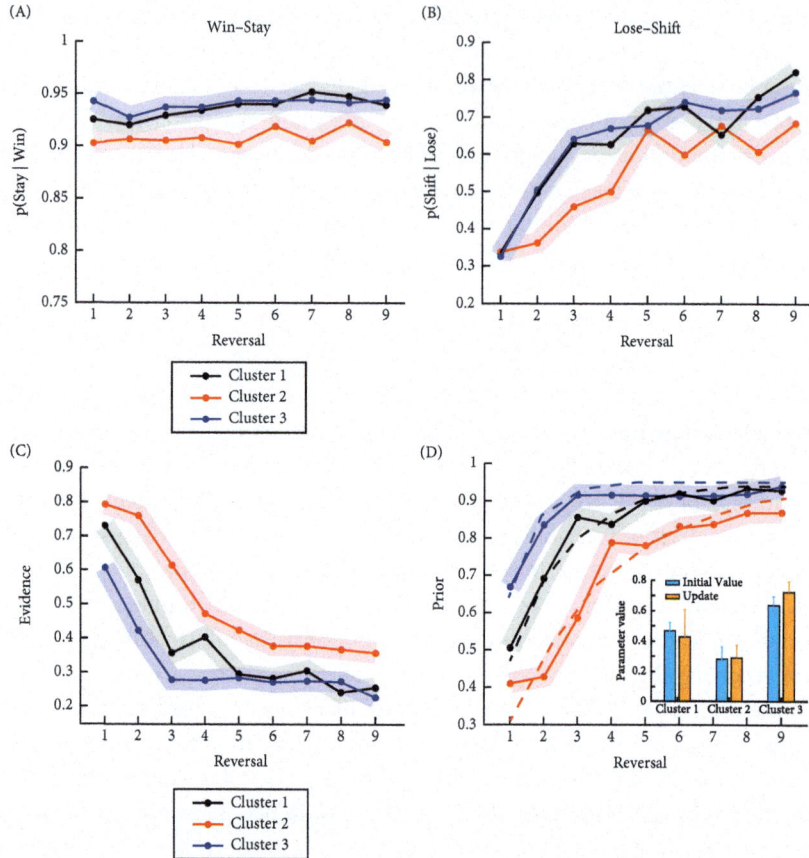

Figure 4.12 Data for object reversal learning for different groups of lesioned macaques
The colour code indicates the groups of macaques, as the text explains.
(A) Applications of a win-stay strategy. (B) Application of a lose-shift strategy. (C). Evidence that a reversal has occurred when monkeys reverses it choice. (D) Bayesian prior that a reversal will occur.

Adapted from Jang, A.I., Costa, V.D., Rudebeck, P.H., Chudasama, Y., Murray, E.A., & Averbeck, B.B. The role of frontal cortical and medial-temporal lobe brain areas in learning a Bayesian prior belief on reversals. *J Neurosci*, 35 (33), 11751–60, Figures 7a, 7b, 5, and 8b, Doi: 10.1523/JNEUROSCI.1594-15.2015 Copyright © 2015, The Authors. Licensed under CC BY 4.0.

Jang et al. (2015) also analysed the results for monkeys with excitotoxic lesions in the central sector of the orbital PF cortex. They grouped these monkeys with those having lesions of the medial PF cortex (areas 24, 25, and 32). Figure 4.12 displays the data for this cluster in blue (cluster 3).

Finally, Jang et al. (2015) analysed the data for intact control monkeys. They grouped these subjects with monkeys having lesions in the medial sector of the orbital PF cortex (area 14). In Figure 4.12 this cluster is black.

Figure 4.12A (page 144) depicts the data for the win-stay strategy, and Figure 4.12B does so for the lose-shift strategy. Monkeys with lesions that disrupt the function of the ventral PF cortex applied both strategies poorly. This is depicted by the red curves in Figure 4.12A and B.

Figure 4.12C shows the accumulated evidence that a reversal had occurred when the monkey reversed its choice, and Figure 4.12D estimates the prior probability that a reversal will occur. Macaques with dysfunction of the ventral PF cortex (red curves) had impairments on both of these measures. These findings demonstrate that lesioned monkeys in this cluster of subjects failed to learn that reversals were likely to occur. For this reason, they could not switch their choice based on a single trial in which the expected outcome, a reward, failed to occur.

A Bayesian perspective can also be applied to the results of a study by Noonan et al. (2017) who tested patients with lesions of the orbital PF cortex on a three-choice probabilistic reversal task. The lesions were centred on the lateral sector but also involved all of the ventral PF cortex either directly or by disconnection. Noonan et al. measured how strongly the combinations of choices and outcomes in the recent past influenced the subject's choices. In these patients, the choice and outcome on a given trial influenced the choice of the patients on the next trial less than in control subjects.

Noonan et al. interpreted this result as reflecting poor 'credit assignment'. But it could also result from a failure to learn the prior probability that reversals would occur. Given the expectation that reversals will occur, healthy subjects come to expect that an alternative choice or stimulus will come to supersede the 'currently best' choice or stimulus. Accordingly, they can alter their choices more rapidly than the patients.

These results indicate that the central sector of the orbital PF cortex and the ventral PF cortex, including the lateral sector, differ in their functions (Murray & Rudebeck, 2018). The central sector supports the ability to predict the desirability of the outcome, as measured for example by the devaluation task. This involves classical conditioning. The ventral PF cortex including the lateral sector supports the ability to change the choice between objects when the outcomes are volatile. This involves contingent or instrumental learning.

Summary

Surgical lesions of the central sector of the orbital PF cortex cause a severe impairment on object reversals, but the reason is that these lesions disconnect the inputs from the inferior temporal cortex and amygdala to the ventral PF cortex. Excitotoxic lesions of the central sector have no effect on object reversals. Instead, the central sector supports the ability to predict the desirability of the outcome. The ventral PF cortex, including the lateral sector supports the ability to modify behaviour when the outcomes change. Control animals can learn to reverse very rapidly, and the reason is that they can learn the prior probability that reversals will occur.

Conclusions

Both the central and lateral sectors of the orbital PF cortex receive inputs from the amygdala as well as from the perirhinal cortex and inferior temporal cortex. Nonetheless, they have very different functions.

The central sector supports predictions about resources and their desirability. There are three strands of evidence for this. The first is that if monkeys with lesions of the central sector are fed one food to satiety, they are poor at adjusting so as to look for the alternative food elsewhere. This suggests that when they see the location or object that is associated with the food, they fail to anticipate the food and its desirability.

The second strand of evidence comes from single unit recording. There are cells in the orbital PF cortex that change their activity when a stimulus appears that predicts food. There are also cells that change their activity just before the presentation of the food.

Finally, there are cells in the orbital PF cortex that encode the desirability of foods. These can compare different foods or fluids in a common currency.

However, though in these experiments the monkeys choose between locations or objects, cells in the central sector of the orbital PF cortex do not encode the responses themselves. In other words, this sector is involved in classical conditioning where the response of the animal is irrelevant.

There are no projections from the central sector of the orbital PF cortex to the premotor areas (Morecraft et al., 1992). Instead, it is the ventral PF cortex that projects both directly to the ventral premotor cortex (Petrides & Pandya, 2002) and indirectly to the dorsal premotor cortex via the presupplementary motor area (preSMA) (Takahara et al., 2012).

The ventral PF cortex includes the lateral sector of the orbital PF cortex. It supports the ability to choose objects or locations on the basis of the outcome. In other words, it is involved in contingent learning. It is particularly critical where there is a change in the outcome and the animal has to reverse its response. Given a series of reversals, intact control monkeys can learn to reverse in one trial. But monkeys with lesions of the ventral PF cortex are unable to do this.

Selective Advantage

It is important that primates obtain a varied diet. Silcox et al. (2015) concluded that many early primates and their immediate ancestors had an omnivorous diet. This is achieved via the mechanism of sensory specific satiety. Interactions between the amygdala and the central sector of the orbital PF cortex support the ability to learn to predict where alternative foods can be found when the animal has fed to satiety on one particular one.

Given the heightened resource volatility during a regime of global cooling faced by the ancestral anthropoids, they needed to shift their foraging choices rapidly. The need to do so would have acted as a selection pressure for the emergence of the lateral sector of the orbital PF cortex during the evolution of the anthropoids. Any reduction in the number of foraging excursions that proved fruitless (errors) would have been of selective advantage.

Connectivity

Among its many connections, the granular orbital PF cortex has four that, collectively, make it unique. It has: (1) gustatory, olfactory, and visceral inputs from the anterior insula; (2) inputs from the inferior temporal cortex; (3) inputs from the amygdala; and (4) both direct and indirect outputs to premotor areas from the lateral, but not central, sector of the orbital PF cortex.

Neural Operation

This connectional fingerprint suggests the following proposal concerning the operation that the orbital PF cortex performs, that is its input–outcome transform.

The amygdala represents the current desirability of a specific food item, given the internal state. It interacts with the central sector of the orbital PF cortex which supports the prediction of that value. Via its visual and gustatory inputs the orbital PF cortex associates this value with the appearance and taste of the food.

The inputs from the temporal lobe and amygdala enable the central sector to use visual signs or cues such as twigs or branches to predict the presence and value of food. These predictions depend on cell activity in the orbital PF cortex that responds to visual cues and also becomes active just before the appearance of the food.

The lateral sector of the orbital PF cortex emerged during the evolution of anthropoid brains. Together with the rest of the ventral PF cortex, this transforms inputs from the ventral visual stream into estimates of the likelihood that a response will lead to a particular resource. This area also encodes Bayesian priors concerning the volatility of the resources, based on the likelihood that outcomes will change.

The lateral sector generates the goal, and this is achieved via direct and indirect outputs to the premotor areas.

References

Aggleton, J.P., Wright, N.F., Rosene, D.L., & Saunders, R.C. (2015) Complementary patterns of direct amygdala and hippocampal projections to the macaque prefrontal cortex. *Cereb Cortex*, 25, 4351–73.

Amaral, D.G. & Price, J.L. (1984) Amygdalo-cortical projections in the monkey (IMacaca fascicularis). *J Comp Neurol*, 230, 465–96.

Baizer, J.S., Ungerleider, L.G., & Desimone, R. (1991) Organization of visual inputs to the inferior temporal and posterior parietal cortex in macaques. *J Neurosci*, 11, 168–90.

Ballesta, S., Shi, W., Conen, K.E., & Padoa-Schioppa, C. (2020) Valunes encoded in orbitofrontal cortex are causally related to economic choices. *Nature*, doi: 10.1038/s41586-020-2880-x.

Battercham, R.L., ffytche, D.H., Rosenthal, J.M., Zelaya, F.O., Barker, G.J., Withers, D.J., & Williams, S.C. (2007) PYY modulation of cortical and hypothalamic brain areas Predicts feeding behaviour in humans. *Nature*, 450, 106–9.

Baxter, M.G. & Murray, E.A. (2002) The amygdala and reward. *Nat Rev Neurosci*, 3, 563–73.

Baxter, M.G., Parker, A., Lindner, C.C.C., Izquierdo, A.D., & Murray, E.A. (2000) Control of response selection by reinforcer value requires interaction of amygdala and prefrontal cortex. *J Neurosci*, 20, 4311–19.

Brodmann, K. (1913) Neue Forchungsergebnisse der Grosshirnrindeanatomische mit besonderer Berucksichtung anthropologischer Fragen. *Gesselch. Deuts. Naturf. Artze*, 85, 200–40.

Bussey, T.J., Saksida, L.M., & Murray, E.A. (2005) The perceptual-mnemonic/feature conjunction model of perirhinal cortex function. *Q J Exp Psychol B*, 58, 269–82.

Camille, N., Tsuchida, A., & Fellows, L.K. (2011) Double dissociation of stimulus-value and action-value learning in humans with orbitofrontal or anterior cingulate cortex damage. *J Neurosci*, 31, 15048–52.

Carmichael, S.T. & Price, J.L. (1994) Architeconic subdivision of the orbital and medial prefrontal cortex in the macaque monkey. *J Comp Neurol*, 346, 366–402.

Carmichael, S.T. & Price, J.L. (1995a) Limbic connections of the orbital and medial prefrontal cortex in macaque monkeys. *J Comp Neurol*, 363, 615–41.

Carmichael, S.T. & Price, J.L. (1995b) Sensory and premotor connections of the orbital and medial prefrontal cortex of macaque monkeys. *J Comp Neurol*, 363, 642–64.

Chau, B.K., Sallet, J., Papageorgiou, G.K., Noonan, M.P., Bell, A.H., Walton, M.E., & Rushworth, M.F. (2015) Contrasting roles for orbitofrontal cortex and amygdala in credit assignment and learning in macaques. *Neuron*, 87, 1106–18.

Choi, E.Y., Ding, S.L., & Haber, S.N. (2017) Combinatorial inputs to the ventral striatum from the temporal cortex, frontal cortex, and amygdala: Implications for segmenting the striatum. *eNeuro*, 4.

Costa, V.D., Tran, V.L., Turchi, J. & Averbeck, B.B. (2015) Reversal learning and dopamine: A bayesian perspective. *J Neurosci*, 35, 2407–16.

Craig, A.D. (2002) How do you feel? Interoception: The sense of the physiological condition of the body. *Nat Rev Neurosci*, 3, 655–66.

Critchley, H.D. & Rolls, E.T. (1996) Hunger and satiety modify the responses of olfactory and visual neurons in the primate orbitofrontal cortex. *J Neurophysiol*, 75, 1673–86.

de Araujo, I.E., Rolls, E.T., Kringelbach, M.L., McGlone, F., & Phillips, N. (2003) Taste-olfactory convergence, and the representation of the pleasantness of flavour, in the human brain. *Eur J Neurosci*, 18, 2059–68.

Fiuzat, E.C., Rhodes, S.E., & Murray, E.A. (2017) The role of orbitofrontal-amygdala interactions in updating action-outcome valuations in macaques. *J Neurosci*, 37, 2463–70.

Folloni, D., Sallet, J., Khrapitchev, A.A., Sibson, N. Verhagen, L., & Mars, R.B. (2019) Dichotomous organization of amygdala/temporal-prefrontal bundles in both humans and monkeys. *Elife*, 8, doi.10.7554/eLife.47175

Grabenhorst, F., D'Souza, A., Parrris, B.A., Rolls, E., & Passingham, R.E. (2010) A common neural scale for the subjective value of different primary rewards. *Neuroimage*, 51,1265-74.

Harlow, H.F. & Warren, J.M. (1952) Formation and transfer of discrimination learning sets. *J Comp Physiol Psychol*, 45, 482–9.

Huxlin, K.R. & Merigan, W.H. (1998) Deficits in complex visual perception following unilateral temporal lobectomy. *J Cogn Neurosci*, 10, 395–407.

Izquierdo, A., Suda, R.K., & Murray, E.A. (2004) Bilateral orbital prefrontal cortex lesions in rhesus monkeys disrupt choices guided by both reward value and reward contingency. *J Neurosci*, 24, 7540–8.

Jang, A.I., Costa, V.D., Rudebeck, P.H., Chudasama, Y., Murray, E.A., & Averbeck, B.B. (2015) The role of frontal cortical and medial-temporal lobe brain areas in learning a Bayesian prior belief on reversals. *J Neurosci*, 35, 11751–60.

Kaskan, P.M., Dean, A.M., Nicholas, M.A., Mitz, A.R., & Murray, E.A. (2019) Gustatory responses in macaque monkeys revealed with fMRI: Comments on taste, taste preference, and internal state. *Neuroimage*, 184, 932–42.

Kennerley, S.W., Walton, M.E., Behrens, T.E., Buckley, M.J., & Rushworth, M.F. (2006) Optimal decision making and the anterior cingulate cortex. *Nat Neurosci*, 9, 940–7.

Kondo, H., Saleem, K.S., & Price, J.L. (2003) Differential connections of the temporal pole with the orbital and medial prefrontal networks in macaque monkeys. *J Comp Neurol*, 465, 499–523.

Kondo, H., Saleem, K.S., & Price, J.L. (2005) Differential connections of the perirhinal and parahippocampal cortex with the orbital and medial prefrontal networks in macaque monkeys. *J Comp Neurol*, 493, 479–509.

LeDoux, J. (2003) The emotional brain, fear, and the amygdala. *Cell Mol Neurobiol*, 23, 727–38.

Machado, C.J. & Bachevalier, J. (2007a) The effects of selective amygdala, orbital frontal cortex or hippocampal formation lesions on reward assessment in nonhuman primates. *Eur J Neurosci*, 25, 2885–904.

Machado, C.J. & Bachevalier, J. (2007b) Measuring reward assessment in a semi-naturalistic context: the effects of selective amygdala, orbital frontal or hippocampal lesions. *Neurosci.*, 148, 599–611.

Morecraft, R.J., Geula, C., & Mesulam, M.M. (1992) Cytoarchitecture and neural afferents of orbitofrontal cortex in the brain of the monkey. *J Comp Neurol*, 323, 341–358.

Morrison, S.E. & Salzman, C.D. (2009) The convergence of information about rewarding and aversive stimuli in single neurons. *J Neurosci*, 29, 11471–83.

Murray, E.A., Bussey, T.J., & Saksida, L.M. (2007) Visual perception and memory: a new view of medial temporal lobe function in primates and rodents. *Annu Rev Neurosci*, 30, 99–122.

Murray, E.A. & Izquierdo, A. (2007) Orbitofrontal cortex and amygdala contributions to affect and action in primates. *Ann NY Acad Sci*, 1121, 273–96.

Murray, E.A., Moylan, E.J., Saleem, K.S., Basile, B.M., & Turchi, J. (2015) Specialized areas for value updating and goal selection in the primate orbitofrontal cortex. *Elife*, 4.

Murray, E.A. & Rudebeck, P.H. (2018) Specializations for reward-guided decision-making in the primate ventral prefrontal cortex. *Nat Rev Neurosci*, 19, 404–17.

Noonan, M.P., Chau, B.K.H., Rushworth, M.F.S., & Fellows, L.K. (2017) Contrasting effects of medial and lateral orbitofrontal cortex lesions on credit assignment and decision-making in humans. *J Neurosci*, 37, 7023–35.

O'Doherty, J.P., Dayan, P., Friston, K., Critchley, H., & Dolan, R.J. (2003) Temporal difference models and reward-related learning in the human brain. *Neuron*, 38, 329–37.

O'Neill, M. & Schultz, W. (2010) Coding of reward risk by orbitofrontal neurons is mostly distinct from coding of reward value. *Neuron*, 68, 789–800.

O'Neill, M. & Schultz, W. (2015) Economic risk coding by single neurons in the orbitofrontal cortex. *J Physiol Paris*, 109, 70–7.

Padoa-Schioppa, C. & Assad, J.A. (2006) Neurons in the orbitofrontal cortex encode economic value. *Nature*, 441, 223–6.

Padoa-Schioppa, C. & Assad, J.A. (2008) The representation of economic value in the orbitofrontal cortex is invariant for changes of menu. *Nat Neurosci*, 11, 95–102.

Padoa-Schioppa, C. & Conen, K.E. (2017) Orbitofrontal cortex: A neural circuit for economic decisions. *Neuron*, 96, 736–54.

Passingham, R.E. & Wise, S.P. (2012) *The Neurobiology of Prefrontal Cortex*. Oxford University Press, Oxford.

Petrides, M. & Pandya, D.N. (2002) Comparative cytoarchitectonic analysis of the human and macaque ventrolateral prefrontal cortex and corticocortical connection pattern in the monkey. *Eur J Neurosci*, 16, 291–310.

Price, J.L. & Drevets, W.C. (2010) Neurocircuitry of mood disorders. *Neuropsychopharmacology*, 35, 192–216.

Raghuraman, A.P. & Padoa-Schioppa, C. (2014) Integration of multiple determinants in the neuronal computation of economic values. *J Neurosci*, 34, 11583–603.

Ray, J.P. & Price, J.L. (1992) The organization of the thalamocortical connections of the mediodorsal thalamic nucleus in the rat, related to the ventral forebrain-prefrontal cortex topography. *J Comp Neurol*, 323, 167–97.

Reber, J., Feinstein, J.S., O'Doherty, J.P., Liljeholm, M., Adolphs, R., & Tranel, D. (2017) Selective impairment of goal-directed decision-making following lesions to the human ventromedial prefrontal cortex. *Brain*, 140, 1743–56.

Rhodes, S.E. & Murray, E.A. (2013) Differential effects of amygdala, orbital prefrontal cortex, and prelimbic cortex lesions on goal-directed behavior in rhesus macaques. *J Neurosci*, 33, 3380–9.

Rich, E.L. & Wallis, J.D. (2014) Medial-lateral organization of the orbitofrontal cortex. *J Cogn Neurosci*, 26, 1347–62.

Rich, E.L. & Wallis, J.D. (2016) Decoding subjective decisions from orbitofrontal cortex. *Nat Neurosci*, 19, 973–80.

Rolls, E.T. (2004a) Convergence of sensory systems in the orbitofrontal cortex in primates and brain design for emotion. *Anat Rec A Discov Mol Cell Evol Biol*, 281, 1212–25.

Rolls, E.T. (2004b) The functions of the orbitofrontal cortex. *Brain Cogn*, 55, 11–29.

Rolls, E.T. (2006) Brain mechanisms underlying flavour and appetite. *Philos Trans R Soc Lond B Biol Sci*, 361, 1123–36.

Rolls, E.T. & Baylis, L.L. (1994) Gustatory, olfactory, and visual convergence within the primate orbitofrontal cortex. *J Neurosci*, 14, 5437–52.

Rolls, E.T., Rolls, B.J., & Rowe, E.A. (1983) Sensory-specific and motivation-specific satiety for the sight and taste of food and water in man. *Physiol Behav*, 30, 185–92.

Rolls, E.T., Sienkiewicz, Z.J., & Yaxley, S. (1989) Hunger modulates the response to gustatory stimuli of single neurons in the caudolateral orbitofrontal cortex of the macaque monkey. *Europ J Neurosci*, 1, 53–60.

Rudebeck, P.H., Behrens, T.E., Kennerley, S.W., Baxter, M.G., Buckley, M.J., Walton, M.E., & Rushworth, M.F. (2008) Frontal cortex subregions play distinct roles in choices between actions and stimuli. *J Neurosci*, 28, 13775–85.

Rudebeck, P.H., Mitz, A.R., Chacko, R.V., & Murray, E.A. (2013a) Effects of amygdala lesions on reward-value coding in orbital and medial prefrontal cortex. *Neuron*, 80, 1519–31.

Rudebeck, P.H. & Murray, E.A. (2008) Amygdala and orbitofrontal cortex lesions differentially influence choices during object reversal learning. *J Neurosci*, 28, 8338–43.

Rudebeck, P.H., Ripple, J.A., Mitz, A.R., Averbeck, B.B., & Murray, E.A. (2017a) Amygdala contributions to stimulus-reward encoding in the macaque medial and orbital frontal cortex during learning. *J Neurosci*, 37, 2186–202.

Rudebeck, P.H., Saunders, R.C., Lundgren, D.A., & Murray, E.A. (2017b) Specialized representations of value in the orbital and ventrolateral prefrontal cortex: Desirability versus availability of outcomes. *Neuron*, 95, 1208–20, e1205.

Rudebeck, P.H., Saunders, R.C., Prescott, A.T., Chau, L.S., & Murray, E.A. (2013b) Prefrontal mechanisms of behavioral flexibility, emotion regulation and value updating. *Nat Neurosci*, 16, 1140–45.

Silcox, M.T., Sargis, E.J., Bloch, J.I., & Boyer, D.M. (2015) Primate origins and supraordinal relationships: Morphological evidence. In Henke, K., Tattersall, I. (eds) *Handbook of Paleoanthropology*. Springer-Verlag, Berlin, pp. 1053–81.

Takahara, D., Inoue, K., Hirata, Y., Miyachi, S., Nambu, A., Takada, M. & Hoshi, E. (2012) Multisynaptic projections from the ventrolateral prefrontal cortex to the dorsal premotor cortex in macaques—anatomical substrate for conditional visuomotor behavior. *Eur J Neurosci*, 36, 3365–75.

Tremblay, L. & Schultz, W. (1999) Relative reward preference in primate orbitofrontal cortex. *Nature*, 398, 704–8.

Tremblay, L. & Schultz, W. (2000a) Modifications of reward expectation-related neuronal activity during learning in primate orbitofrontal cortex. *J Neurophysiol*, 83, 1877–85.

Tremblay, L. & Schultz, W. (2000b) Reward-related neuronal activity during go-nogo task performance in primate orbitofrontal cortex. *J Neurophysiol*, 83, 1864–76.

Ungerleider, L.G., Galkin, T.W., Desimone, R., & Gattass, R. (2008) Cortical connections of area V4 in the macaque. *Cereb Cortex*, 18, 477–99.

Walker, A.E. (1940) A cytoarchitectural study of the prefrontal areas of the macaque monkey. *J Comp Neurol*, 73, 59–86.

Wallis, J.D. & Miller, E.K. (2003) Neuronal activity in primate dorsolateral and orbital prefrontal cortex during performance of a reward preference task. *Eur J Neurosci*, 18, 2069–81.

Walton, M.E., Behrens, T.E., Buckley, M.J., Rudebeck, P.H., & Rushworth, M.F. (2010) Separable learning systems in the macaque brain and the role of orbitofrontal cortex in contingent learning. *Neuron*, 65, 927–39.

Wang, G., Tanifuji, M., & Tanaka, K. (1998) Functional architecture in monkey inferotemporal cortex revealed by in vivo optical imaging. *Neurosci Res*, 32, 33–46.

Webster, M.J., Bachevalier, J., & Ungerleider, L.G. (1994) Connections of inferior temporal areas TEO and TE with parietal and frontal cortex in macaque monkeys. *Cer Cort*, 4, 471–83.

Webster, M.J., Ungerleider, L.G., & Bachevalier, J. (1991) Connections of inferior temporal areas TE and TEO with medial temporal-lobe structures in infant and adult monkeys. *J Neurosci*, 11, 1095–116.

Wellman, L.L., Gale, K., & Malkova, L. (2005) GABAA-mediated inhibition of basolateral amygdala blocks reward devaluation in macaques. *J Neurosci*, 25, 4577–86.

5
Caudal Prefrontal Cortex
Searching for Objects

Introduction

Chapter 4 argues that the orbital prefrontal (PF) cortex assigns a value to objects such as foods while foraging. This chapter proposes that it is the caudal PF cortex that searches for those objects. These can be detected and attended to in peripheral vision, and anthropoid primates are then able to move the eyes so as to fixate them with foveal vision. This distinction is sometimes marked by distinguishing between covert attention and overt attention (eye-movements).

Most of this chapter deals with vision and eye movements, which evolved very early in vertebrate history. Evidence for eyes and extraocular muscles occurs in the oldest vertebrate and pre-vertebrate fossils, some dating from more than 500 Ma (Shu et al., 2003). But some important innovations in vision and eye movements evolved in the primates. Chapter 2 argues that the frontal eye field (FEF) first appeared in early primates, and this means that it predates the later development of the fovea and trichromatic colour vision in anthropoid primates.

Areas

In macaque monkeys the caudal PF refers to the cortex that lies in the anterior bank of the arcuate sulcus, as well as the convexity cortex that lies within the bow of this arcuate. Area 8Ad lies dorsal and 8Av ventral to the caudal tip of the principal sulcus on the cortical convexity (Figure 1.9). Figure 5.1 (next page) sketches the location of the caudal PF cortex.

Saccadic eye movements can be evoked by micro-stimulation of the anterior bank of the arcuate sulcus in macaque monkeys (Bruce et al. 1985), and it is this property that defines the FEF. However, there is a difference between the dorsal and ventral limbs of this sulcus. Stimulation of the dorsal limb (FEFd) evokes large saccades, whereas stimulation of the ventral limb (FEFv) evokes small ones (Stanton et al., 1995). Intracortical micro-stimulation of the posterior granular area of bushbabies also evokes saccades at comparable intensities (Wu et al., 2000; Stepniewska et al., 2018). This finding supports the conclusion that the FEF is homologous in bushbabies and macaques.

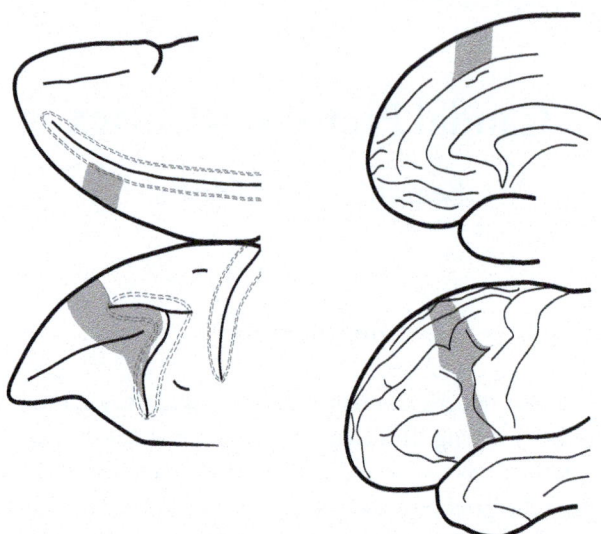

Figure 5.1 The caudal PF cortex in the macaque monkeys (left) and humans (right)
Adapted from Passingham, R.E. & Wise, S.P., *The Neurobiology of Prefrontal Cortex*, p. 133, Figure 5.1, Copyright © 2012, Oxford University Press.

In macaque monkeys, higher currents can evoke saccades from the convexity tissue within the bow of arcuate (Robinson & Fuchs 1969). However, it is assumed that stimulation in these areas evokes eye movements via the spread of the current to the FEF. It is generally accepted that it is the low-threshold region in the ventral limb of the arcuate sulcus that corresponds to the FEF (Bruce et al. 1985). Layer IV becomes less easy to discern in the depth of the anterior bank of the sulcus (Stanton et al. 1989).

It is important to note, however, that saccades can also be evoked in macaque monkeys by stimulation of the rostral part of the dorsal premotor cortex (Neromyliotis & Moschovakis, 2017). The reason this matters is that in the literature on functional brain imaging it has usually been assumed that the human FEF lies within the premotor cortex (area 6) rather than in the PF area 8.

Amiez et al. (2006) used fMRI to locate the peak of activation with respect to the cortical anatomy of individual human subjects. They compared activation during saccades with activation during fixation. They report that the FEF, as so defined, was located consistently in the ventral branch of the superior precentral sulcus and the dorsal branch of the inferior precentral sulcus. However, since there is both cell activity (Izawa et al., 2009) and activation (Henderson & Choi, 2015) in area 8 that are related to fixation, it could be that activation in area 8 is subtracted out when saccades are compared with fixation.

Since the FEF is defined by the effects of electrical stimulation in macaques, it is more appropriate to use the same method as that used to locate the FEF in the brains of non-human primates. Amiez and Petrides (2009) have reviewed studies that have localized the FEF in the human brain by electrical stimulation. Saccades can be evoked with low-threshold stimulation from an area that lies just rostral to the superior precentral sulcus, as well as from stimulation over that sulcus. This is suggestive evidence that, as in macaque monkeys, it is possible to evoke eye movements from two areas, a PF area rostrally and a premotor area caudally.

The issue appears to have been settled by a study by Ioannides et al. (2010) in which they used magneto-encephalography (MEG) to localize activity that is related to saccades in particular directions in the human brain. A Fourier analysis was applied to the localization of activity just before and after each saccade. The analysis revealed two areas, one just ventral to the caudal part of the superior prefrontal sulcus and the other in the superior limb of the precentral sulcus. This suggests that, as in the brain of the macaque monkey, there are two areas that are involved in controlling saccades, a PF area (the FEF proper), and a premotor area that lies caudally to this.

However, in addition, micro-stimulation shows that there is another premotor area, the supplementary eye field (SEF) (Schlag & Schlag-Rey, 1987). As in the FEF, cells in the SEF increase activity before saccades (Hanes et al., 1995). In macaques, the SEF lies in the dorsomedial frontal cortex, within the premotor area 6 (Schlag & Schlag-Rey, 1987; Olson & Gettner, 1999). And it has a similar location in the human brain (Amiez & Petrides, 2009).

Connections

Figure 5.2 (next page) shows a summary of selected corticocortical connections of the caudal PF cortex, including the FEF, in macaque monkeys.

1. The dorsal FEF (FEFd) is interconnected with the lateral intraparietal area (LIP) and ventral intraparietal area (VIP), as well as with the inferior parietal area PG and the higher order motion-encoding area MST (Stanton et al., 1995).

    The peripheral visual stream projects to the cortex of the intraparietal sulcus (IPS) (Baizer et al., 1991). The implication is that the area of the FEF from which large saccades can be evoked represents peripheral vision. These saccades move the eyes to objects that are detected in the periphery. The FEFd also lies close to the area from which smooth pursuit eye movements can be evoked (Neromyliotis & Moschovakis, 2017).

2. The ventral FEF (FEFv) receives information from low-order visual areas V2, V3, V4, as well as from the inferior temporal areas TE and TEO (Stanton et al., 1995).

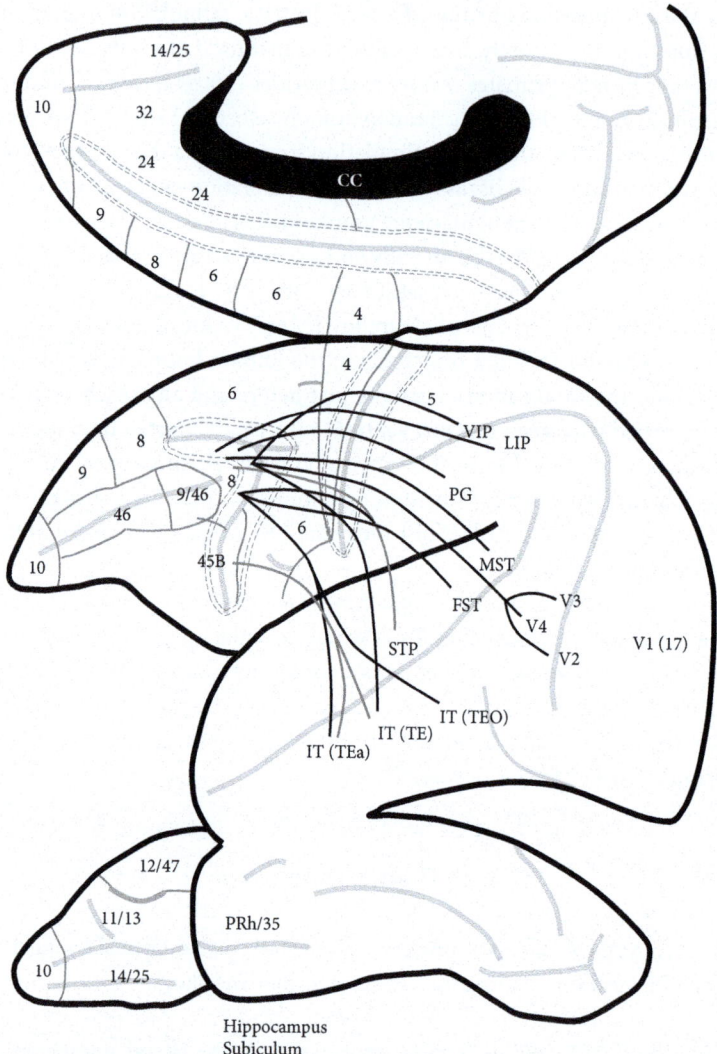

Figure 5.2 Selected connections of the caudal PF cortex
The lines show areas that have direct axonal connections with the caudal PF cortex, assumed to be reciprocal.

Reproduced from Passingham, R.E. & Wise, S.P. *The Neurobiology of Prefrontal Cortex*, p. 135, Figure 5.2, Copyright © 2012, Oxford University Press.

The foveal visual stream projects to the inferior temporal cortex (Baizer et al., 1991). The implication is that the area of the FEF from which small saccades can be evoked represents foveal vision. These saccades act to correct the position of the eyes so that the target can be accurately fixated.

3. Areas 8Ad, 8Av, and 8B are closely interconnected (Petrides & Pandya, 1999). They are alike in receiving an input from the inferior parietal area PG and the upper bank of the superior temporal sulcus (STS) (Petrides & Pandya, 1999). However, they differ in that 8Ad receives a widespread input from both banks of the intraparietal sulcus (IPS) whereas 8Av is connected with the inferior temporal areas TE and TEO as well as with area TEa in the ventral bank of the STS (Petrides & Pandya, 1999).

   These results suggest that the dorsal area 8Ad is like the FEFd in representing peripheral vision whereas the ventral area 8Av is like the FEFv in representing foveal vision.

4. Area 8Av is interconnected with the orbital PF cortex (areas 11 and 13) (Petrides & Pandya, 2002; Saleem et al., 2014).

   This means that it has access to information about the current needs of the animal and therefore the outcomes that are desirable.

5. Area 8Av is also interconnected with the ventral PF cortex (areas 47/12 and 45A) (Petrides & Pandya, 2002).

   As Chapter 7 discusses, the ventral PF cortex supports the ability to classify items, and thus to specify the type of food that it is currently desirable to search for.

6. The FEF (Kunzle et al., 1976; Huerta et al., 1986), areas 8Ad and 8Av (Fries, 1984) as well as area 45B (Gerbella et al., 2010) send direct projections to the intermediate layers of the superior colliculus. The superior colliculus orients the retina. So, these corticofugal connections point to a role of these areas in a network that is involved in controlling eye movements. The projections to the superior colliculus are excitatory; and they either terminate on fixation neurones or on inter-neurones that inhibit saccade-related neurones (Everling & Johnston, 2013).

7. The FEF also projects to the medial part of the caudate nucleus (Stanton et al., 1988); in turn this projects to the substantia nigral, pars reticulata (Hedreen & DeLong, 1991). The substantia nigra, pars reticulate, then projects to the superior colliculus where it exerts an inhibitory influence (Hikosaka & Wurtz, 1985).

8. Finally, the FEF sends direct projections to the oculomotor nuclei (Segraves, 1992). These connections come not only from areas from which saccades can be evoked but also from areas from which smooth pursuit movements can be evoked (Yan et al., 2001).

Summary

The connectional fingerprint of the caudal PF cortex in macaque monkeys suggests the following conclusions:

- The FEFd and area 8Ad receive inputs from the dorsal visual stream that represents peripheral vision.
- The FEFv and area 8Av receive inputs from the ventral visual stream that represents central or foveal vision.
- The FEF has direct outputs to the intermediate layers of the superior colliculus, influencing both saccadic eye movements and fixation,
- The FEF can also affect eye movements via projections from the substantia nigra, the pars reticulata, which has an inhibitory influence on the superior colliculus.
- The FEF has direct outputs to the oculomotor nuclei.

The FEF as Unique to Primates

The reader might object to the claim made in Chapter 2 that the FEF is unique to primates. After all, stimulation of the frontal cortex can evoke eye movements in cats (Schlag et al., 1974). But so can stimulation of the rostral part of the dorsal premotor cortex in macaque monkeys (Neromyliotis & Moschovakis, 2017) and human brain (Amiez & Petrides, 2009). So, the demonstration in cats does not prove that the stimulated area is a prefrontal as opposed to a premotor area.

The critical factor is not that primates have a fovea whereas cats do not. The evidence is that the FEF evolved in primates before the fovea (Shepherd & Platt, 2010). Anthropoid primates have a fovea but strepsirrhine primates do not. One possibility is therefore that it was the FEFd from which large saccades can be evoked that evolved first, and that the part from which small saccades can be evoked evolved later in anthropoid primates. It is the latter that receives an input from foveal vision. Only anthropoids have trichromatic colour vision. It is the ventral and not the dorsal limb of the arcuate sulcus that receives an input from the inferior temporal cortex, and the inferior temporal cortex is specialized for analyzing colour (Huxlin et al., 2000).

What, then, is special about the FEF? One answer comes from studying the way in which primates forage. They typically feed on fruit and leaves and these are found in a cluttered environment. This puts a premium on the ability to search for the food item that is currently edible. The situation is much like blackberry picking where the person has to search amongst an array of blackberries hidden amongst leaves, and the aim of the search is to find the blackberries that are ripe. It is true that hamadryas baboons (*Papio hamadryas*) and geladas (*Theropithecus gelada*) now feed on grasses and other plants in open ground (Napier & Napier, 1967), but this was not the ancestral way of life when the anthropoids evolved.

Evidence that the eyes alight on features of interest comes from studies on human subjects. If the eye movements of human subjects are measured while they view a display, the results show a complex pattern of saccades and micro-saccades (Martinez-Conde & Macknik, 2015). The eyes alight on and return to the salient features of the display more frequently than they do for other features. In anthropoid primates, this pattern of eye movements is the result of having a fovea. It reflects the movement of the fovea across the array, in other words the search path.

But carnivores do not seek for food in this way. A cat follows the movements of a mouse and lions chase until they can pick off a straggler. The eye movements that are needed are smooth pursuit movements. It is the small saccades that allow anthropoids to strategically sample the visual scene; and the generation of these is severely impaired by inactivation of the FEFv in macaque monkeys (Peel et al., 2016).

There is an indirect piece of evidence that the FEF is a prefrontal and not a premotor area in primates. As already mentioned, the FEFv in the ventral limb of the arcuate sulcus has inputs from the inferior temporal areas TE and TEO, as well as early visual areas (Stanton et al., 1995). However, the inferior temporal cortex is not directly interconnected with the premotor cortex. Instead, it projects to the ventral and orbital PF cortex (Webster et al., 1994; Kondo et al., 2005).

Summary

The FEF is a prefrontal area and it is unique to primates. In anthropoids such as macaque monkeys, the FEFv receives inputs from V4 and the inferior temporal cortex. These provide information about shape and colour, and it is these features that specify the target of search.

Visual Search

The previous sections review the evidence that, unlike the premotor cortex, the FEFv receives direct inputs from the lower-order visual areas V2, V3, V4, and MT (Stanton et al., 1995). These connections explain why some cells in the FEFv show increased activity after the presentation of sensory stimuli, and before movement (Schall, 1991). These can be classified as 'visual' (V) cells.

Of the cells that connect LIP with the FEFd, 78% have visual responses but no saccade-related activity (Ferraina et al., 2002). These are referred to as V cells. They provide information concerning peripheral visual space, independent of motor commands.

However, there are also cells in the FEF that have visuomotor activity (VM) or motor (M) activity (Schall, 1991). Sato and Schall (2003) taught monkeys to detect a visual pop-out stimulus in an array of distractors, and then to make a saccade either to that location (prosaccade trials) or to a saccade target in the opposite direction (antisaccade trials). The authors then compared the activity of cells in the FEF cells for the two tasks when the pop-out stimulus was located in the receptive field of the cell. Note that for the antisaccade task, the endpoint of the saccade was in the opposite location to the pop-out stimulus.

The activity of 57% of the task-related cells initially reflected the location of the stimulus. However, subsequently, 86% of these cells later coded for the endpoint of the saccade, even if it was antisaccade. So, these were VM cells.

There were also 27% of cells in the FEF that did not code for the location of the target. Instead, they coded for the endpoint of the antisaccade. So, these were M cells. However, the term 'motor cells' is strictly a misnomer because it suggests that the output was a motor command, whereas in fact it was the spatial goal.

However, the division of the population of cells in the FEF into V, VM, and M cells has turned out to be over-simplistic. Lowe and Schall (2018) used a clustering technique for unsupervised classification of the cells, and identified ten categories, rather than just three. The visual cells, for example, were subdivided according to the latency, magnitude, and sign of their visual response; the visuomotor cells were categorized as having sustained activity; and the motor cells were subdivided according to the dynamics, magnitude and direction of the saccade.

The role of the FEF is unlike that of the ventral premotor cortex which directs the movement of the hand to the mouth (Graziano et al., 2002). It is true that there are visual, visuomotor, and motor cells in the ventral premotor cortex (Murata et al., 2000). But whereas inactivation of the ventral premotor cortex abolishes the ability to shape the hand as it approaches an object (Fogassi et al., 2001), lesions of the FEF do not abolish saccadic eye movements (Schiller et al., 1979).

The reason why lesions of the FEF fail to abolish eye movements is probably that the SEF also projects to the superior colliculus (Shook et al., 1990). Saccades no longer occur if the superior colliculus is removed together with the FEF (Schiller et al., 1979). This combined lesion removes both the FEF and the terminations from the SEF.

There is a distinction between the SEF and the FEF in that the SEF is not involved in directing search. Though there are visual cells with receptive fields, only 2% discriminate between whether they contain a salient stimulus or a distractor in a pop-out search (Purcell et al., 2012).

The FEF differs in that there is cell activity that encodes visual salience so as to select the targets of saccades. The extent of this activity can be assessed by using fMRI to scan monkeys while they engage in a pop-out search. Wardak et al. (2010) found that the PF activation lay along the ventral extent of the arcuate sulcus, extending further forwards dorsally into the area 9/46 in the dorsal PF cortex, and into the ventral PF cortex (area 45A). There were also activations in parietal LIP and the inferior temporal area TE.

To demonstrate that the PF activation was causal, Bichot et al. (2015) inactivated the ventral pre-arcuate cortex. This impaired the ability of monkeys to find the target. And it also eliminated the response of the cells in the FEF to the relevant features.

Purcell et al. (2012) have put forward a model of how the search occurs. This assumes that the representation of stimulus salience drives a network of competing accumulators that are associated with each of the alternative stimuli. The model proposes that the conversion of salience to saccade targets involves tonic gated inhibition. Purcell et al. also presented data from single unit recording that were consistent with this model.

Nelson et al. (2016) confirmed that the discharge rate of cells in the FEF has to reach a constant threshold before a saccade is produced, as expected for an accumulator racetrack model. However, Heitz and Schall (2013) had to introduce modifications to the standard model so as to account for the trade-off between the speed of responding and accuracy. Accumulator models have typically explained this trade-off in terms of a strategic change in the response threshold.

Reppert et al. (2018) also used a task to assess the speed-accuracy tradeoff during search. In the run-up to accurate as opposed to fast trials, the cells in the FEF and superior colliculus had lower baseline firing rates. However, whereas the magnitude of the visual responses differed with the speed-accuracy trade-off, this was not found for cells in the superior colliculus. Instead, it varied with the peak velocity of the saccades.

Accumulator models are at the computational level as defined by Marr (1982). However, Heinzle et al. (2007) have proposed a microcircuit model of the FEF, and this is at the implementational level. It takes into account the cell types in columns from the superficial to the deep layers. Heinzle et al. specifically compared the pattern with the canonical microcircuit (Douglas et al., 1989) as established for the visual cortex, and found that it was necessary to introduce modifications. In the same way, the microcircuit of the SEF is also not identical to the canonical microcircuit (Godlove et al., 2014; Ninomiya et al., 2015).

The model proposed by Heinzle et al. (2007) distinguishes the different layers in terms of their visual input, fixation, target output, and their rule dependency. It also specifies the membrane dynamics with the synapses as decaying exponential conductances, as well as the connections within a layer.

The model was devised to account for the role of the FEF in visual search as well as in memory saccades and the rule governing the generation of prosaccades and antisaccades. The outlines of the model are illustrated in Figure 5.3 (next page). This figure omits the details of the recognition module.

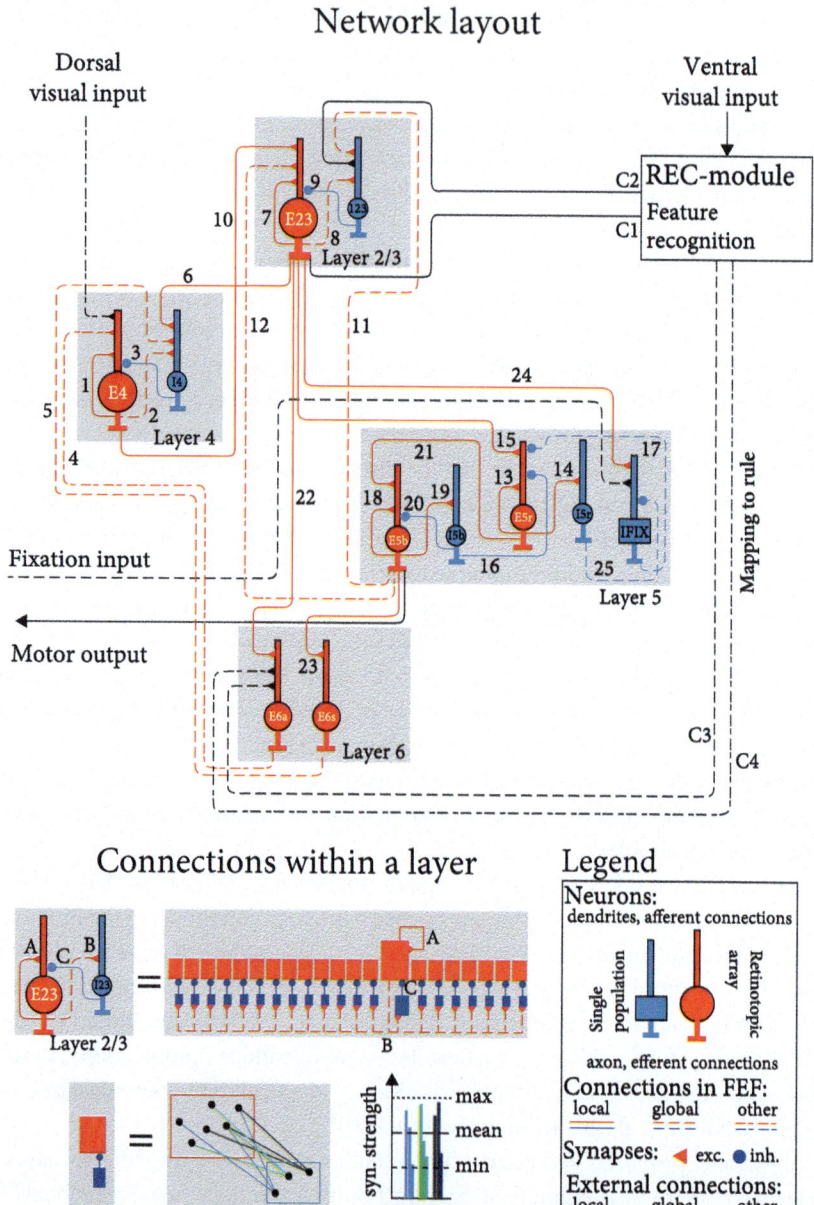

Figure 5.3 Layout of the FEF circuit
Top: Complete network architecture. Coloured circles are full retinotopic representations consisting of arrays of 21 populations of neurons. Coloured rectangles are single populations, e.g. fixation neurons (red, excitatory; blue, inhibitory). Layer 4 receives a visual input from the dorsal stream which is not feature specific. The fixation neurons receive a fixation input, and the motor output of the FEF is given by the bursting neurons in layer 5 of the FEF. The spatial pattern of the connections

(continued in footnote of next page)

Summary

The FEF plays a critical role in visual search. As primates forage on foods that are in a cluttered environment, they need to be able to identify the fruit and leaves that are edible as opposed to the distractor ones that are either unripe or poisonous. The FEFv has visual inputs from areas that analyze shape and colour, and so, as expected, many of the cells have visual properties. It also has visuomotor cells as well as motor cells that specify the endpoint of the saccade.

Covert Visual Attention

In the experiments on pop-out search described in the previous section, the monkeys were trained to fixate on a central spot, and then to make a saccade to the target item in the periphery. In this situation the animals could move their eyes as soon as they detected the peripheral target. It differs from the situation in which the subjects must focus attention on the periphery, but without shifting the gaze. This involves covert attention.

Figure 5.3 Continued

is summarized into three groups: local connections (solid lines) connected only to populations at the same retinotopic position, global connections (dashed lines) connected to all retinotopic positions, and the connections that could not be grouped into one of the two above, which are called other connections (dash-dotted line). External inputs and the connections to and from the recognition module are shown in black. The Recognition module receives a feature-specific visual input, which represents the ventral processing stream. Layers 2 and 3 are connected to the recognition module [C1] and in turn receive an input from it [C2]. Layer 6a of the FEF receives the rule input directly from antisaccade [C3] and no-go [C4] feature detectors in the recognition module. Bottom left: Retinotopic arrangement of the connections in layers 2 and 3.
(A) Local self-excitation (only shown for 1 retinotopic position).
(B) Global excitation of all inhibitory populations (only shown for 1 efferent excitatory population).
Ⓒ Local inhibitory connections.
Bottom: Illustration of the random selection of connections for three sample inhibitory neurons connecting randomly to 50% of the excitatory neurons. The distribution of the weights is indicated by the histogram on the right. The minimum, maximum, and mean of the uniform distribution are shown by the horizontal dashed lines.

Reproduced from Jakob Heinzle, Klaus Hepp, & Kevan A.C. Martin. A microcircuit model of the frontal eye fields, *J Neurosci*, 27 (35), 9341–53, Figure 2, Doi: https://doi.org/10.1523/JNEUROSCI.0974-07.2007 Copyright © 2007 by the Society for Neuroscience.

The advantage of retaining covert attention, even after the evolution of the fovea, is that it allows enhanced processing of a limited number of extrafoveal items, even as the most intensive processing is devoted to foveated items and places. The importance of covert attention and search lies in the possibility that all attended objects, and not just foveated ones, might become goals for future actions, so it provides a mechanism for supporting short-term memory of several items.

The classical way of testing covert attention in human subjects is to use the Posner task. This was devised to test patients with spatial neglect (Posner et al., 1984). The subjects are required to fixate a central spot and are told that a target will appear in the periphery to one side. If the subjects are instructed which side this will be, they are quicker to detect the target than if they are misled as to which side on which the target will appear. This difference in response times is strong evidence that the subjects can attend covertly, that is without moving their eyes. However, since the subjects may be tempted to peek, it is essential that in experiments of this sort high-resolution eye tracking is used to check that there are no small shifts in gaze (Kennett et al., 2007).

Monkeys can also be trained to maintain fixation while a stimulus is presented in the periphery. Bushnell et al. (1981) recorded from cells in the cortex of the IPS and found that the activity of the cells was enhanced when the monkeys attended covertly in this way, but only if the target was in the receptive field of the cell. Furthermore, the enhancement was the same as when the monkeys were allowed to move their eyes immediately to the target. In later studies, attentional enhancement was reported specifically for cells in the intraparietal area LIP (Colby et al., 1996; Colby & Goldberg, 1999).

In an early study of cell activity in the FEF, Goldberg and Bruce (1985) found the enhancement effect for eye movements, but not for covert attention. However in subsequent studies, cells in the FEF were shown to exhibit enhancement of their visual responsiveness on an attention task (Boch & Goldberg, 1989; Kodaka et al., 1997). Since the FEFd and area 8Ad are interconnected with the LIP (Stanton et al., 1995), it should not be surprising that cells in these areas are similar in showing enhanced sensory responses to both overtly and covertly attended locations.

In order to study covert attention, Hasegawa et al. (2000) presented a target in an array of distractors, but the monkeys could not respond until after a delay. Within 135ms of target presentation, there were cells in the caudal PF cortex (area 8Ad) that showed an enhancement in their response if the target appeared in the cell's receptive field. This enhancement did not occur if the target appeared outside the cell's receptive field, or if a distractor item appeared in the receptive field. This finding provides evidence that area 8Ad supports the search for targets in the periphery, and that it does so via attentional enhancement.

Jerde et al. (2012) used fMRI to visualize the areas in the human brain in which enhancement can be found during covert attention. They placed coils over the posterior parietal cortex and the precentral sulcus. They used a standard method for topographic mapping with a 'travelling wave' design. Engel (2012) showed that as a checker-board expands from central to peripheral vision, there is a wave of activation in the primary visual cortex (V1) that moves from the foveal to the peripheral representation of space.

In the experiment by Jerde et al. (2012), the subjects fixated a central spot, and travelling waves of activation were evoked by moving spots to eight equally spaced positions round the periphery. Within each spot there were coherently moving dots, and the task for the subjects was to discriminate the direction in which the dots were moving.

Jerde et al. found four topographically organized areas in the cortex of the IPS. There was also one anterior to the upper limb of the precentral sulcus and one anterior to the lower limb. Jerde et al. referred to these as 'priority maps'. Using MVPA, Jerde et al. were able to distinguish between the representation of the different locations.

Mackey et al. (2017) repeated the experiment with improved methods, and the results are shown in Figure 5.4. In the PF cortex the foveal representation of the FEFd lies at the intersection between the superior frontal sulcus and the upper

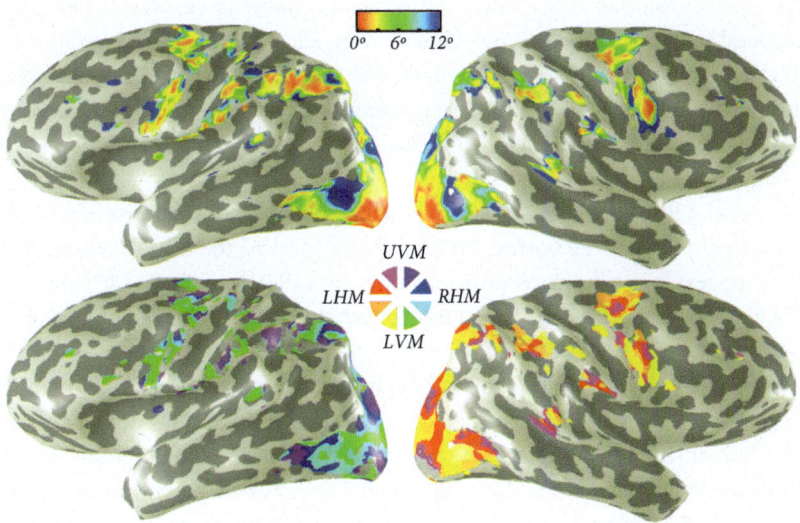

Figure 5.4 Topographic maps in the frontal eye fields and posterior parietal cortex in one subject
Top, eccentricity with colour scale for 0 to 12 degrees; Bottom, polar angle with colour wheel for 0 to 360 degrees.

Reproduced from Wayne E Mackey, Jonathan Winawer, & Clayton E Curtis. Visual field map clusters in human frontoparietal cortex. *eLife*, 6: e22974, Figure 5, Doi: 10.7554/eLife.22974 © 2017, Mackey et al. Licensed under CC BY 4.0.

limb of the precentral sulcus. The FEFd and FEFv in each hemisphere represent the contralateral hemifield.

Just as in the human brain, LIP is not a single area in macaque monkeys: it can be subdivided into LIPd and LIPv (Chen et al., 2016). The difference is that LIPv has stronger connections with FEF and the superior colliculus. While it was possible to record visual and saccadic responses in both areas, only inactivation of LIPv decreased the proportion of express saccades. Further work is needed to establish which of the four parietal areas that Jerde et al. (2012) identified in the human brain have homologues in the macaque monkey brain.

If the parietal cortex and FEF genuinely support covert attention, temporary inactivation should lead to an impairment when monkeys search for or attend to stimuli. So, Keller et al. (2008) inactivated the FEFv. The colour of a central cue told the animal which coloured target to search for in the periphery. When the inactivation was unilateral, the animals made errors, that is they searched in the wrong place. But, when the inactivation was bilateral, the number of these errors was greatly increased.

Wardak et al. (2011) compared the contribution of the FEF and LIP by teaching monkeys to detect a target in a field of distractors, while the animals maintained fixation on a central spot of light. Inactivation of either the FEF (Wardak et al., 2006) or LIP (Wardak et al., 2004) caused the monkeys to be slow to detect the peripheral target when no eye movements were allowed. However, Wardak et al. (2011) suggested that the two areas made different contributions. They proposed that LIP was mainly involved in the representation of salience and attentional selection, where the FEF was mainly involved in triggering attentional shifts and eye movements.

Gee et al. (2008) suggested that the enhancement in LIP was driven by a top-down signal from the PF cortex. But the main experimental data that were cited in this review came from recordings taken in LIP alone (Bisley & Goldberg, 2003; 2006). These are not sufficient to prove that that the enhancement occurred top-down.

One way of testing the direction of the effect is to record simultaneously in the FEF and posterior areas. Purcell et al. (2013) recorded the spiking activity of cells in the FEF during a pop-out search task and compared the timing of this activity with the generation of the macaque equivalent of the evoked potential (ERP) component linked to selective attention (N2pc) evoked potential in posterior cortex. They reported that the activity in the FEF preceded the macaque equivalent of the N2pc (m-N2pc).

Zhou and Desimone (2011) also compared the timing of activity, with the difference that they recorded cell activity in the FEFv and the ventral visual area V4. They used a conjunction search rather than a pop-out search. On a conjunction search the distractors share some, but not all, of the features of the target.

In the experiment by Zhou and Desimone the target was specified by a conjunction of colour and shape. The latencies for attentional enhancement were earlier in the FEF than V4. The difference was significant, 30ms during the early period, and 50ms during the later period of search.

Gregoriou et al. (2012) also measured enhancement in the FEFv and V4, with the difference that they distinguished between the V, VM, and M cells in the FEF. The distinction is important because the V cells are found in the supragranular and infragranular layers, and it is these layers that send a feedback signal to the posterior parietal and inferior temporal cortex (Thompson & Schall, 2000). Gregoriou et al. (2012) found that it was only the V cells that showed synchronization within the gamma range with the cell activity in V4.

However, though the differences in latency found by Zhou and Desimone (2011) can be taken to suggest causation, they do not prove it. To do this it is necessary to intervene. Armstrong and Moore (2007) applied intracortical micro-stimulation to the FEFv. This enhanced the response of cells in visual area V4 for a specific part of visual space. Thus, the effect of stimulating the FEF mimics what happens when monkeys attend covertly to an object that is presented in that location.

In human subjects, transcranial magnetic stimulation (TMS) can be used as an intervention. Kalla et al. (2008) applied double pulse TMS over the FEF and the posterior parietal cortex while healthy subjects were tested on a visual conjunction search task. TMS caused disruption over the FEF from 0–40ms but over the posterior parietal cortex from 120–160ms. This difference in timing is substantial.

The implication is that there is a top-down signal from the FEF. To prove it, Heinen et al. (2017) applied theta burst TMS to disrupt the activity in the FEF in human subjects. This method has the advantage that it can be applied before the task is performed because the effect lasts for some minutes (Di Lazzaro et al., 2008).

Heinen et al. recorded activations with fMRI. Inactivating the FEF led to a decrease in the enhancement in the IPS and inferior parietal cortex when the subjects shifted their attention. This was associated with a decrease in the functional connectivity between the FEF and posterior parietal cortex. Furthermore, this decrease in connectivity correlated with the behavioural impairment.

In a classical review, Duncan and Desimone (1995) suggested that the top-down signal acted to bias the competition between locations or items in posterior areas. So as to study this effect, Peelen and Kastner (2011) told human subjects to search for people in photos of cluttered scenes where there were cars as distractor items. There was also a condition on which trees were shown on their own, and these served as a neutral control.

The fMRI data were analyzed for the occipital and temporal cortex using a MVPA. It was easier to detect the representation of people than trees in the inferior temporal cortex, but more difficult to detect the representation of cars then trees. This suggests that the representations of target items were enhanced, but the representations of the distractor items were suppressed.

Dorsal and Ventral Attentional Systems

An earlier section mentions that there is a difference between the FEFd and area 8Ad on the one hand, and the FEFv and area 8Av on the other. Chapter 1 shows where these areas are in the brain of the macaque monkey (Figure 1.9).

These dorsal and ventral areas form part of two distinct attentional systems. The two systems can be visualized in the human brain by measuring resting state covariance (Fox et al., 2006). They are illustrated in Figure 5.5. The figure shows that the dorsal attentional system (blue) includes the FEFd as well as the cortex in the IPS, whereas the ventral attentional system (red) includes the FEFv as well as the cortex in the temporo-parietal junction (TPJ).

The areas in yellow indicate that the two systems overlap. The reason is that they are interconnected. This has been shown for the human brain in two ways. One is by studying resting state covariance (Sani et al., 2019); the other by demonstrating that TMS over the IPS influences activations in the ventral as well as the dorsal system (Leitao et al., 2015).

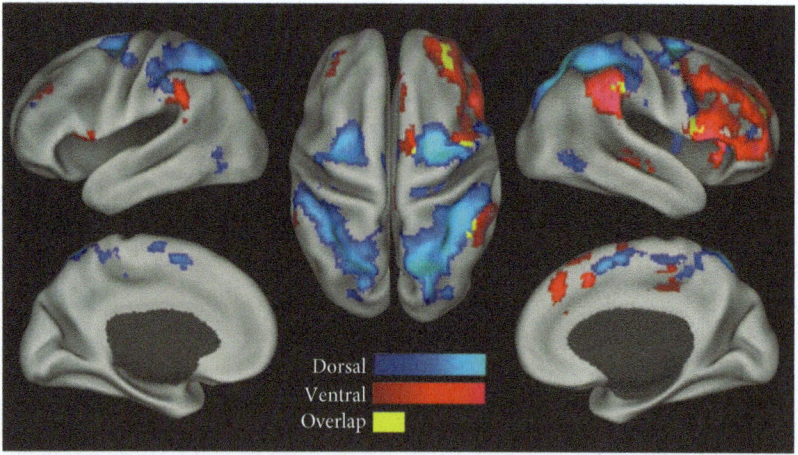

Figure 5.5 Intrinsically defined dorsal and ventral attention systems and the overlap between them
Voxels in the dorsal system (blue scale) are significantly correlated with both the IPS and FEF regions three resting state conditions (fixation, eyes open, and eyes closed). Voxels in the ventral system (red scale) are significantly correlated with both the temporo-parietal junction (TPJ) and ventral PF cortex in the same three resting-state conditions. Voxels significantly correlated with all four regions in all three conditions are shown in yellow.

Reproduced from Fox, M.D., Corbetta, M., Snyder, A.Z., Vincent, J.L., & Raichle, M.E. Spontaneous neuronal activity distinguishes human dorsal and ventral attention systems. *Proc Nat Acad Sci USA*, 103 (26), 10046–51, Figure 5, Doi: https://doi.org/10.1073/pnas.0604187103 Copyright © 2006, National Academy of Sciences.

Vossel et al. (2014) has proposed that the dorsal system acts top-down whereas the ventral system is stimulus driven when a stimulus is either salient or surprising. This proposal depends in part on the results of an fMRI study by Asplund et al. (2010). They reported enhanced activation in the FEFd and IPS during covert attention on the Posner task. By contrast, they reported enhanced activation in the cortex of the inferior frontal junction (IFJ) and the temporo-parietal junction (TPJ) when a surprising stimulus was presented in a stream of distractors.

The activation reported in the IFJ may actually have been in either the FEFv or area 45. The surprising stimulus was a face, and as Chapter 7 discusses there is cell activity in the ventral visual system (Taubert et al., 2017) and area 45 (Sugihara et al., 2006) when a monkey views a face.

However, the claim that the dorsal and ventral systems differ in terms of top-down versus bottom-up is not consistent with other results. It is true that cells in the FEFv respond during a pop-out search task in which the target is salient. But they also do so during a conjunction search task (Zhou & Desimone, 2011). Yet on a conjunction search task the target is not salient as on a pop-out task; the monkeys have to learn to recognize the target. Furthermore, there is a top-down signal from the FEFv to the ventral visual area V4 (Zhou & Desimone, 2011). In other words, there are top-down signals in both the dorsal and ventral attentional systems.

Summary

Covert attention refers to the ability to attend to a stimulus in the periphery while continuing to fixate a central point. During covert attention there is enhancement of activity or activation in the FEF as well as in the posterior parietal and the inferior temporal cortex and V4. The enhancement in posterior areas can be shown to be driven by a top-down signal from the PF cortex.

Neglect

Neglect is a failure to search for, attend to or notice stimuli that are presented contralateral to the lesion. In an early study, Crowne et al. (1981) removed area 8 together with the FEF from one hemisphere in macaque monkeys. The effect was that, when presented with an arc of lights arrayed in front of them, the monkeys were very slow to respond to the lights that were presented contralateral to the lesion. They recovered in a month or so, but the neglect was reinstated if the corpus callosum and anterior commissure were sectioned. The reason is that sectioning the commissures meant that if the monkeys fixated centrally, there was no way for visual information about the field contralateral to the lesion to reach the intact hemisphere.

We know from the single unit recordings taken by Funahashi et al. (1989) that the receptive fields of the cells in area 8 are organized such that the left hemisphere represents the right half of space, and the right hemisphere the left. So the recovery might have failed to occur had the monkeys been trained to fixate a central point as in the experiment by Wardak et al. (2006). In the experiment by Crowne et al. (1981) the animals were free to turn their head so that the stimuli contralateral to the lesion could reach the intact hemisphere.

Crowne et al. (1981) also assessed their animals on the extinction test. This task is frequently used to detect spatial neglect in patients. Two stimuli are presented simultaneously, one to each hemifield. The finding is that patients with left-sided neglect only report seeing the stimulus to their right.

Crowne et al. (1981) found that monkeys with a unilateral lesion of the FEF together with section of the commissures also showed a permanent impairment on the extinction test. When threatened with fear provoking objects, the monkeys only reacted to the one that was ipsilateral to the lesion. A similar impairment on the extinction test has been reported to follow a unilateral parietal lesion in monkeys (Lynch & McLaren, 1989). The lesion removed the inferior parietal cortex, including the ventral bank of the IPS.

The experiments by Crowne et al. (1981) studied the effect of unilateral lesions of the FEF. But Lawler and Cowey (1987) made bilateral lesions in area 8, including the FEF. The task for the monkey was to open a foodwell on the left if the central cue was black, and to open the foodwell on the right if the central cue was white. The lesioned monkeys failed to learn the task in 1000 trials. And Lawler and Cowey found the same result when they made bilateral lesions of the cortex in the IPS, including LIP. They suggested that the reason might be that the lesions caused a bilateral visual neglect.

In patients, neglect typically results from strokes that involve the TPJ (Mort et al., 2003) or the caudal part of the ventral PF cortex (Mannan et al., 2005). However, because strokes also damage the white matter, the arcuate fasciculus is also compromised (Machner et al., 2018). In macaque monkeys this runs from the temporo-parietal area Tpt to the FEF (Petrides & Pandya, 1988). The area Tpt in the macaque monkey brain may be homologous with the TPJ in the human brain (Galaburda et al., 1978).

However, even if this this is the case, there is a critical difference between the human and macaque monkey brain. It appears on the basis of a small sample of brains that the cortex of the TPJ is more extensive in the right than the left hemisphere in the human brain (Galaburda et al., 1978). Evidence for a functional asymmetry can also be seen in Figure 5.5. This figure shows that there is covariance between the TPJ and the FEF together with PF cortex more anteriorly (red in Figure 5.5) in the right, but not left hemisphere.

The TPJ in the right hemisphere is specialized for attention to both the contralateral and ipsilateral hemifields (Corbetta & Shulman, 2011). In an fMRI study by

Shulman et al. (2010), the subjects searched covertly for a complex object in the periphery that shared some features with the distractor objects. The subjects were cued as to which side to attend to by presenting either stay or shift cues. Activations were found in the right TPJ for shifts to the opposite visual field, whether right or left. By contrast, activation was only found in the left TPJ for shifts to the right visual field.

There were also widespread activations in the right hemisphere, including the cortex in the IPS, STS and area 8, when the target was detected in either visual field. Activations were found in the left hemisphere only for the detection of targets in the contralateral visual field.

A consequence of this difference is that, after lesions that involve the right hemisphere, targets in the contralateral visual field will no longer be represented in the attentional network. This is why it is a right hemisphere lesion that cause a hemineglect. Patients with hemineglect fail to move their eyes into the contralateral visual space (Mannan et al., 2005) (Figure 9.11). In other words, neglect reflects a failure to search.

Summary

Neglect can be produced in macaque monkeys by lesions of area 8 including the FEF or inactivation of area LIP. In the human brain resting-state covariance distinguishes between a dorsal and ventral attentional system. The dorsal system supports covert attention to the periphery whereas the ventral system supports attention to targets as defined by their colour and shape. However, in the human brain the cortex of the TPJ and area 8 support attention to either visual field, whereas the left hemisphere only supports the detect of targets in the right visual field. The result is that hemineglect is typically caused by strokes that involve the right TPJ or area 8.

Short Term Maintenance

In an experiment on covert attention, the subject is required to maintain fixation while attending to a peripheral target. During this period there is continuous cell activity (Colby et al., 1996). A simple change in the design of the experiment is to present the target briefly, but then to remove it for the duration of the delay so that there is a gap (Colby et al., 1996). The task for the animal is to make a saccade to the location in which that target had been presented. This task is referred to as an oculomotor delayed response task (ODR).

As illustrated in Figure 5.6 (next page), the task has three phases: cue, delay, and choice. In the first phase, a visuospatial cue appears briefly somewhere in the

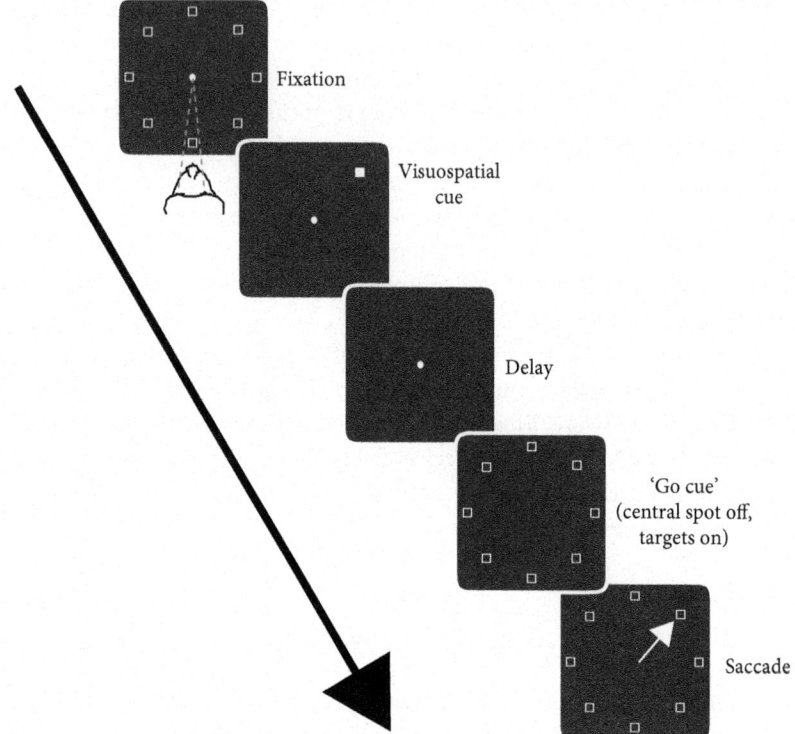

Figure 5.6 A common version of the oculomotor delayed response task
Each panel shows the screen at a different time during a trial, with unfilled white squares as potential spatial targets on other trials. The filled white square illustrates the cue for an example trial, filled white circle indicates the fixation point. Grey dashed lines and grey arrows show the monkey's fixation point. The inset at the lower left shows the spatial memory signal, recorded from a population of PF cortex cells by Lebedev et al. (2004), with the average activity for the preferred memory location in grey and that for the anti-preferred location in black.
Reproduced from Passingham, R.E. & Wise, S.P. *The Neurobiology of Prefrontal Cortex*, p. 143, Figure 5.5, Copyright © 2012, Oxford University Press.

subject's field of view. This stimulus indicates a location towards which the monkey must direct a saccade, but until the delay period ends the monkey must continue to fixate a central light spot. After a delay period, which lasts from a few 100ms to several seconds, a 'go' signal tells the monkey to make the saccade to the most recently cued location. Reward follows if the saccade the subject makes an accurate saccade.

There are two important things to note about this task. The first is that Figure 5.6 shows the locations as empty squares and indicates the cue with a filled square. However, as usually presented, the peripheral locations are not marked in any way, and this means that when the monkey responds to the 'go' signal, it makes a saccade into an unmarked location on the screen.

The second aspect of the design is that the screen is visible throughout. This means that, though the monkeys have to fixate a central spot, they can bridge the delay by attending covertly to the location at which the stimulus had just been presented. Thus, it is not surprising that Colby et al. (1996) found cell activity in LIP during the delay period on the ODR task, as well as enhanced activity while the monkeys attended covertly to a peripheral location.

Chafee and Goldman-Rakic (1998) reported cell activity during the delay in the caudal PF cortex (area 8A) as well as in the cortex of the IPS. They referred to these two areas as having matching patterns of activity. So as to visualize the areas in which cells were active, Inoue et al. (2004) used PET to compare the activations on the ODR task with the activations when the monkeys simply made a saccade to a target that was marked on the screen. There was differential activation in the FEF and dorsal PF cortex (area 46), as well as in the cortex of the IPS.

Chafee and Goldman-Rakic (1998) suggested that the delay-related activity reflected working memory. This is the term that Baddeley and Hitch (1974) introduced to refer to the use of short-term memory to solve tasks. However, since in the experiment by Chafee and Goldman-Rakic (1998) the screen was visible throughout, the task could also have been solved using covert attention alone.

To distinguish between these interpretations, Lebedev et al. (2004) introduced a novel experimental design. This is illustrated in Figure 5.7A (next page). The monkeys were taught to fixate a central point. A spot was presented in the periphery and the animals were required to remember its location. The spot then rotated to one of three other positions as shown in Figure 5.7A, and the monkeys had then to maintain covert attention to that location. The performance of the animals was then tested by either brightening the spot at the top or dimming it. If it was bright, the animals had to report the remembered location by making a saccade. If it was dim, the animals had to make a saccade to report the attended location.

Lebedev et al. recorded cell activity from the caudal PF cortex (area 8A) and the posterior part of the dorsal PF cortex (area 9/46). Of the task-related cells that encoded a spatial location, the majority (61%) coded selectively for the attended location rather than for the remembered one. This attentional signal is shown in Figure 5.7C (next page).

A minority (16%) of the cells coded selectively for the remembered location. However, there were also 23% of cells that were classified as hybrid, since they encoded both the remembered and attended locations. The percentage of the three types of cells are shown in histogram form in Figure 5.7B (next page). Whereas 61% coded for the attended location alone, a substantial number, that is 39% of the cells, coded either for the remembered location or both locations.

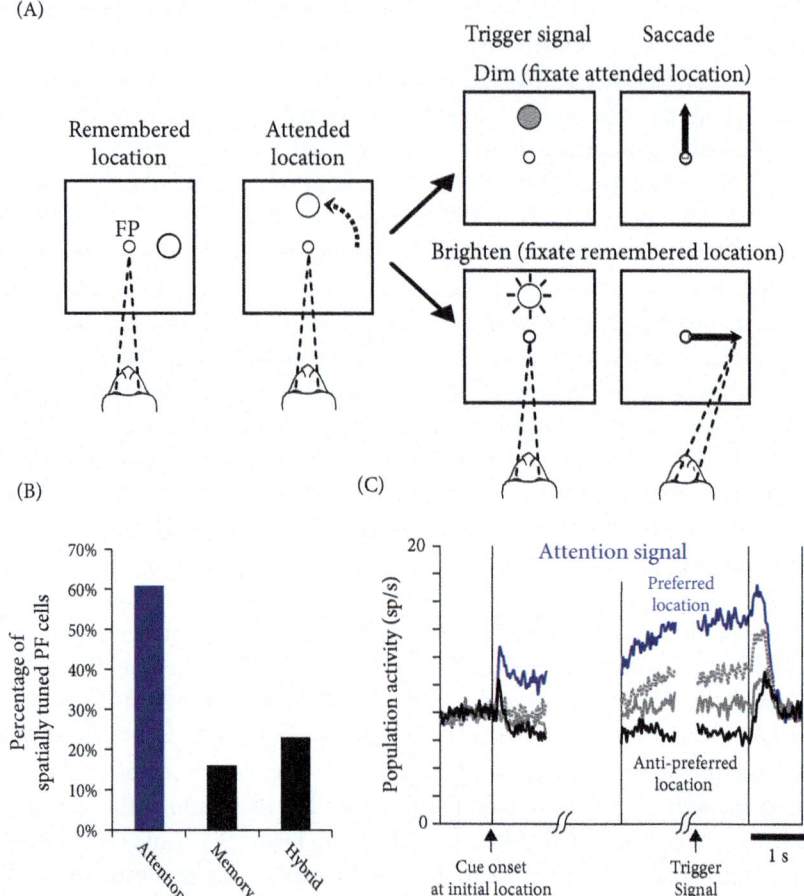

Figure 5.7 Attention versus memory coding in the PF cortex
(A) Task design. The arrow indicates the correct saccade in both conditions. Abbreviations: Rem, remembered-location; Att, attended-location. (B) Percentages of spatially tuned PF cortex neurons encoding the attended or remembered locations or both (hybrid). (C) Attentional signal in the format of the memory signal shown in Figure 5.6.

(A) Reproduced from and (C) modified from Lebedev M.A., Messinger A., Kralik J.D., & Wise S,P. Representation of attended versus remembered locations in prefrontal cortex. *PLoS Biol* 2 (11) e365. https://doi.org/10.1371/journal.pbio.0020365 © 2004, The Authors. Licensed under CC BY 4.0.

As an earlier section mentions, in an fMRI study Jerde et al. (2012) used MVPA to show that it was possible to train a classifier to distinguish between different locations in priority space during a covert attention task. They also found that the same classifier could predict the location on an ODR task. Since no screen was used, one explanation could be that the subjects could have maintained the location during the delay by using covert attention.

Disrupting Delay Activity

To find out if the caudal PF cortex and the cortex in the IPS were essential for supporting short-term memory, Chafee and Goldman-Rakic (2000) inactivated each area in turn by cooling the tissue with a cryoprobe. However, inaccurate saccades could reflect a low-level motor disability. To rule out this possibility, Chafee and Goldman-Rakic compared the effect of inactivation on the ODR task with the effect in a control condition. In the latter, there was no delay: the subject simply makes a saccade to a visible target.

The finding was that on the ODR task cooling the caudal PF cortex (area 8A) caused the saccades to be hypometric, that is to fail to extend out to the target location. Cooling the cortex of the IPS had little effect (Chafee & Goldman-Rakic, 2000).

However, it is critical to note that hypometric errors differ from frank errors. If an animal makes a hypometric error the saccade is towards the remembered location. This is very different from the situation in which the eyes move in the wrong direction. It is frank errors, not hypometric errors, that are evidence of a failure to distinguish between the locations in memory.

Rather than inactivating the cortex, Funahashi et al. (1993a) removed the tissue either in the caudal PF cortex (area 8A) or more anteriorly in the dorsal PF cortex (areas 46 and 9/46) in a series of monkeys. Figure 5.8 (next page) illustrates the results for one animal. In this monkey Funahashi et al. first removed areas 8A and area 9/46 from the left hemisphere and then removed area 46 in the right hemisphere.

This monkey had a greater impairment after the second lesion, but the animal rarely made frank errors, that is errors in the incorrect direction. Instead, most of the saccades simply ended short of the target location. At the longer delay of 6 seconds (not shown), the saccadic endpoints distributed more widely than with shorter delays, but the monkey still made mostly hypometric errors rather than frank errors.

The reason why the animals made few frank errors is probably that area 46 remained intact in one hemisphere. Chapter 6 reviews the evidence that this area is critical when monkeys are given spatial tasks to perform in which they have to remember a location across a delay. As that chapter describes, the spatial delayed response and alternation tasks differ from the ODR task in that during the delay the animals are unable to maintain covert attention on the display, because an opaque screen is lowered during the delay.

Even when the monkeys make frank errors on the ODR task, the reason may not be a failure to remember the target. Tsujimoto and Postle (2012) reported the effects of inactivating the caudal PF cortex (area 8A) with bicuculline while monkeys performed this task. If the animals made a frank error, they often chose the location that had been appropriate and rewarded on the previous trial. Furthermore, after the monkeys made a frank error, they made a corrective saccade to the correct location (Figure 5.9, page 177). This finding indicates that the monkeys had not forgotten that location.

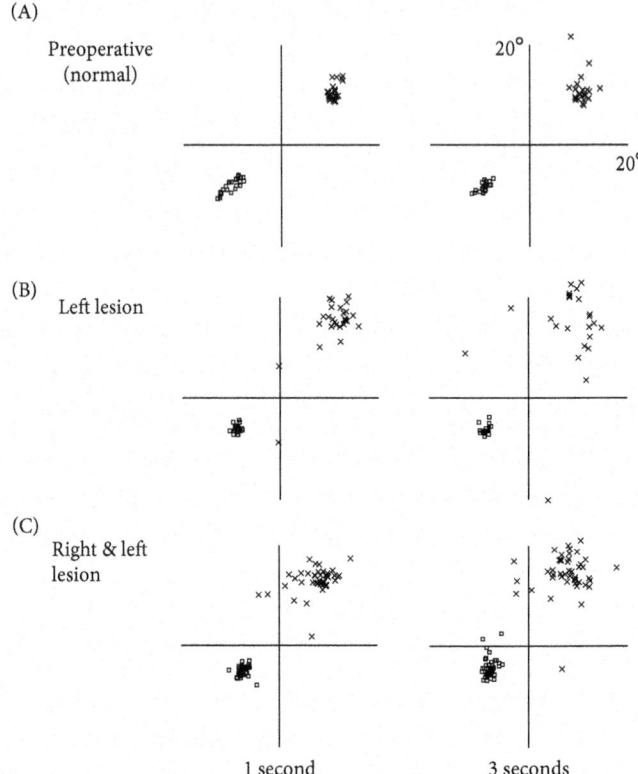

Figure 5.8 Performance on the oculomotor delayed response task for one lesioned monkey
(A) Preoperative (normal) performance. Squares indicate the saccade endpoints for targets at 225° (down and to the left); crosses indicate targets at 45° (up and to the right). (B) Performance after a unilateral lesion of area 9/46 the left hemisphere. (C) Performance after an additional lesion of the same area in the right hemisphere, in order to complete a bilateral lesion. Delay duration at bottom.

Reproduced from Funahashi S., Bruce C.J., & Goldman-Rakic P.S. Dorsolateral prefrontal lesions and oculomotor delayed-response performance: evidence for mnemonic 'scotomas'. *J Neurosci*, 13 (4), 1479–97, Figure 5, Doi: 10.1523/JNEUROSCI.13-04-01479.1993 © 1993, Society for Neuroscience.

Mackey et al. (2016b) found the same result when they compared the effects of lesions that included the caudal PF cortex, including the FEF, in patients. The primary saccade was inaccurate; but the patients then made corrective saccades that brought the endpoint much nearer to the target position.

Mackey et al. (2016a) also tested patients with parietal lesions on the same task. These patients also made frank errors, and even their corrective saccades were abnormal. The reason why the lesions caused an impairment whereas cooling in monkeys did not (Chafee & Goldman-Rakic, 2000) is probably that cooling over

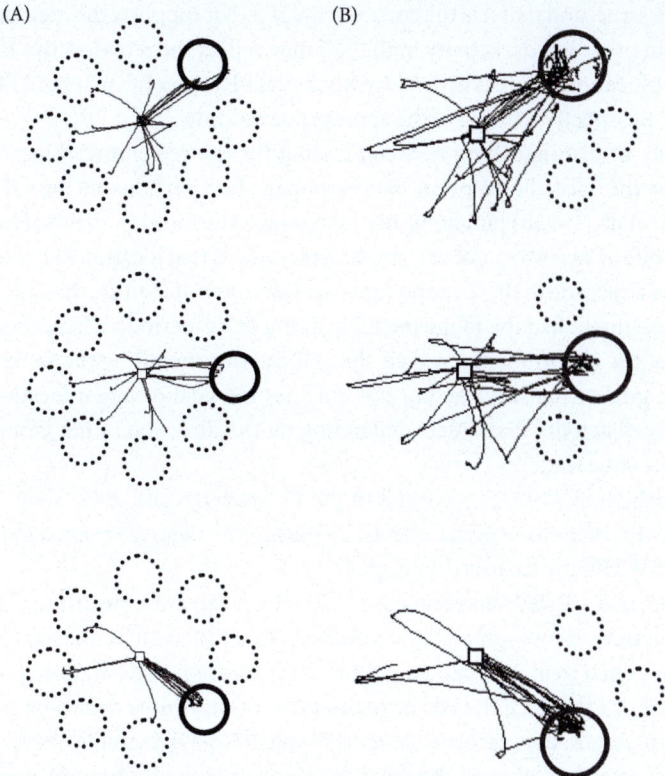

Figure 5.9 Corrective saccades after a frank error
(A) After inactivating the caudal PF cortex, the monkey made some saccades to incorrect goals. The correct goal for the top part of the figure was the upper-right location, for the middle part it was the right location, and for the bottom part it was the lower-right location. (B) The corrective saccades after the initial error.

Adapted from Tsujimoto S. & Postle B.R. The prefrontal cortex and oculomotor delayed response: A reconsideration of the mnemonic scotoma. *J Cog Neurosci*, 24 (3), 627–35, Figure 5.3a, Doi: 10.1162/jocn_a_00171.© 2012, Massachusetts Institute of Technology.

the IPS may not disrupt the activity of the tissue that lies deep in this sulcus. Yet strokes cause extensive damage to the cortex, as well as damage to the underlying white matter.

Retrospective versus Prospective Coding

It remains to explain how the monkeys and patients with lesions of area 8A were able to make corrective saccades that brought the eyes much nearer to the target location. One possibility is that the contribution of this area and the cortex of the IPS are not the same, and that it is the cortex of the IPS that supports the memory of the location. In other words, activity in the IPS may reflect the retrospective maintenance of the location whereas activity in the caudal PF cortex (area 8A and FEF) may reflect the prospective coding of the appropriate saccade.

One way of distinguishing between coding for the remembered location and coding for the saccadic response is to compare two versions of the ODR task. Funahashi et al. (1993b) taught monkeys to report the location that had been presented in one of two ways, either to make a saccade to that location (a prosaccade), or to make a saccade to the location that was 180° opposite (an antisaccade).

The experiment had the following logic. If the delay-period activity encodes the location of the cue in memory, then the cell activity should be the same regardless of the goal of the saccade. But if it encodes the goal or target location, there should be cell activity that differs depending on the direction of the saccade (saccade versus antisacade).

Funahashi et al. (1993b) recorded in the PF areas 8A and 9/46. They reported that of 51 cells with directional selectivity during the delay, 59% encoded the cue location and 25% the location of the goal.

In a later study, Takeda and Funahashi (2002) tried to draw the distinction with a different design. They compared the standard ODR task with a condition in which the monkeys had to make a saccade at 90° to the location that had been presented. They called this the R-ODR task, or rotated ODR task. This is shown on the left in Figure 5.10. Again, they recorded in area 8A and 9/46. Of a sample of 41 cells with directional selectivity during the delay, 86% encoded the location of the cue and 13% the direction of the saccade.

However, there was a difference between the delay-related activity on the two tasks. This could be shown by an analysis of the population activity. On both tasks there was a ramping of the activity towards the end of the delay; but this was more pronounced on the R-ODR task.

In order to visualize the change in the population activity on the R-ODR task, Takeda and Funahashi (2004) plotted the data in the form of population vectors. This type of analysis was introduced by Georgopoulos et al. (1986) to study the way in which the population of cells in motor cortex could specify particular directions, even though individually the cells have a broad tuning curve. By plotting population vectors, Takeda and Funahashi (2004) were able to show the evolution of the transformation from the specification of the cue location to the specification of the response location. Figure 5.10 (next page) shows the change in the vectors from the time that the cue was presented (bottom) to the time at which the monkey made the saccade at the end of the delay (top).

It is highly unlikely that the explanation of the progression of the population vectors reflected a change in the coding of individual cells. It is much more likely that it reflected an increase over time in the proportion of cells that coded for the final goal.

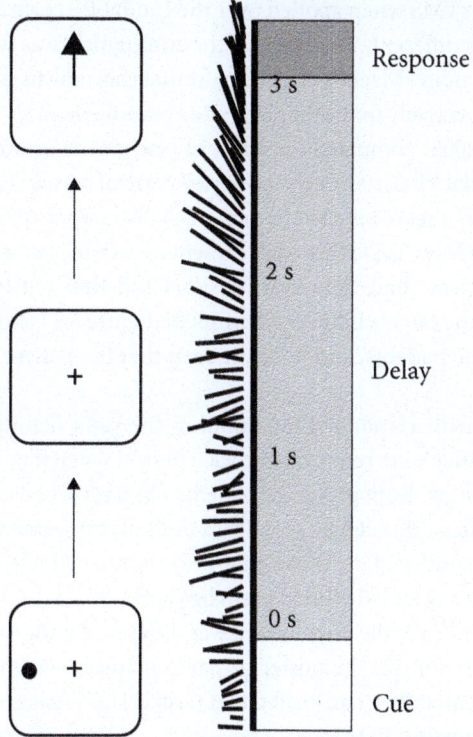

Figure 5.10 Change in population vectors on an oculomotor delayed response task (Left) Task on which the animal has to respond 90 degrees to the location that was presented. (Right) Change in the population vectors during the delay, plotted vertically

Reproduced from Takeda, K. & Funahashi, S. Population vector analysis of primate prefrontal activity during spatial working memory. *Cerebral Cortex*, 14 (12), 1328–39, Figure 6, Doi: https://doi.org/10.1093/cercor/bhh093 Copyright © 2004, Oxford University Press.

In these studies the analysis of cells that coded retrospectively and prospectively was based on a relatively small sample of cells. A more reliable way of distinguishing between retrospective and prospective coding is to study the effect of lesions or inactivation. Mackey et al. (2017) applied repetitive transcranial magnetic stimulation (rTMS) over the PF area 8A, including the FEF, or the IPS to see if the contribution of the two areas could be dissociated. The stimulation was applied during the delay-period and measures were taken of the initial saccade at the end of the delay and the end position after the corrective saccades that follow the initial saccade. When rTMS was applied to one of the parietal areas with a prioritized map (Jerde et al., 2012), the effect was that the end position of the correctional saccade was inaccurate. Mackey et al. (2017) suggest that this is consistent with a short-term memory impairment, that is a failure of retrospective memory.

However, when rTMS when applied over the caudal PF cortex (area 8A and the FEF) the effect was different. The effect of the stimulation was to cause the initial saccade to be inaccurate. Mackey et al. suggest that this may mean that are 8A and the FEF code prospectively for the appropriate eye movement.

Mannan et al. (2005) compared the effect of permanent lesions that were either centred on the caudal PF cortex or the posterior parietal cortex. The task for the patients was to identify a set of target letters amongst distractors. As Figure 5.11 shows, the targets were the letter T and they were presented with the letter L as distractors.

Because the patients had right-sided lesions and thus left hemineglect, they identified none of the targets in the left hemifield. Figure 5.11 shows the eye movements for one of the patients, and it can be seen that the patient failed to look towards the left at all.

Though the patients responded to stimuli in the right hemifield, they did not behave normally. They were required to click a button when they thought that they had found a new target. Both groups of patients often returned to targets that they had already identified. This can be seen from their eye movements, but it was also measured by the number of 're-clicks' that they performed, identifying targets as novel when in fact they had identified them before.

However, the pattern of the impairment was different for the two patient groups. The patients with lesions in the posterior parietal cortex were more likely to re-click the longer ago that they had visited that target. This is suggestive of a problem in short-term memory for locations.

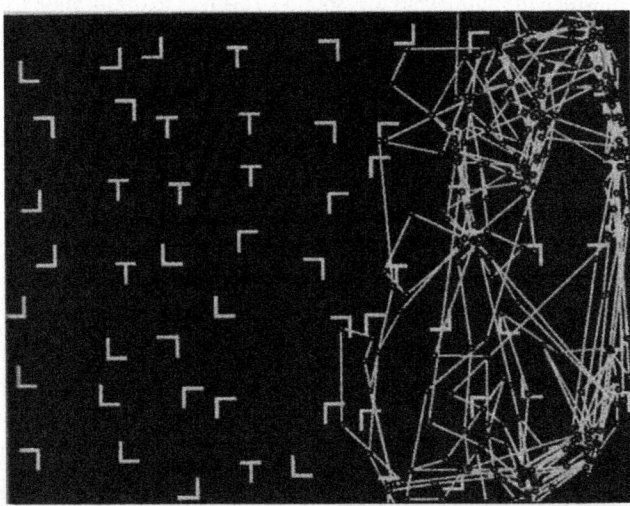

Figure 5.11 An example of a search array on which are superimposed the scan paths of one patient

Reproduced from Passingham, R.E. *Cognitive Neuroscience: A Very Short Introduction*, p. 31, Figure 8, Copyright © 2016, The Author.

The patients with lesions that included the caudal PF cortex also revisited targets. But there was no relation between the time that they last visited a target and the time that they returned to it. Mannan et al. suggested that their impairment might reflect a failure to inhibit saccades.

Collin et al. (1982) also reported return errors in monkeys with bilateral lesions of the caudal PF cortex (area 8A and the FEF). The animals faced a display of 25 opaque doors, with a peanut behind each door. However, once the monkey had opened the door and retrieved the peanut the door swung shut again. So as to retrieve the maximum number of peanuts, it therefore payed the animals not to return to doors that they had already opened.

Before surgery, the monkeys were reasonably efficient at retrieving the peanuts. However, after removal of the caudal PF cortex they kept returning to the same doors. An analysis of their search paths showed that the reason was that they had difficulty in shifting their attention to a new part of the display. They tended to search within a very delimited part of the display, with the result that they made return errors.

Taken together, these experiments support the proposal that it is the cortex in the IPS that is critical for maintaining spatial locations in memory. The role of the caudal PF cortex, including the FEF, is to shift eye movements or attention across the display.

Summary

On an ODR task there is delay-related activity in the cortex of the caudal PF cortex and IPS. Recordings from the caudal PF cortex suggest that some of the cells code for covert attention, some for memory and some for both. However, there appears to be a difference between the contribution of the caudal PF cortex and cortex in the IPS. The parietal cortex encodes both covert attention and retrospective activity concerning the location of the cue. The caudal PF cortex codes prospectively for the saccade or shift in attention.

Conclusions

The caudal PF areas 8A and the FEF support the ability to search for and maintain attention to items of interest. These could be interesting for two reasons. One is that they are salient or surprising; the other is that they are desirable, given current biological needs. The distinction is sometimes drawn as being between stimulus driven or goal directed attention (Corbetta & Shulman, 2002). It is assumed that the former is bottom-up and the latter top-down.

The FEF supports search and attention and does so in either case. On a pop-out search task, the target stands out because it is salient, and there is cell activity in the FEFv when monkeys search for the target (Sato & Schall, 2003). On a conjunction-search task, the target shares features with the distractors and again there is cell activity in the FEFv when monkeys attend to the target (Zhou & Desimone, 2011). This activity precedes attentional enhancement in V4 (Zhou & Desimone, 2011), and this is suggestive of a top-down signal.

In the laboratory, monkeys have to be taught what target to look for by rewarding them when they direct attention to the target or move their eyes to it. But the cortex of the IPS is not in a position to support the initial learning, the reason being that it has very sparse inputs from the amygdala (Amaral & Price, 1984) and no direct inputs from the orbital PF cortex. By contrast, there are heavy inputs to the central and lateral sections of the orbital PF cortex from the amygdala (Amaral & Price, 1984; Aggleton et al., 2015), in turn these areas are interconnected with the ventral PF cortex (areas 47/12 and 45A).

Though area 8Av is sparsely connected with the orbital PF cortex (Petrides & Pandya, 2002), this connection is not sufficient to specify the target items that are currently desirable. One reason is that in the natural environment monkeys are not looking for a singleton but for targets items that are defined in terms of their category or class. For example, a monkey might search for a particular type of fruit or leaf.

As Chapter 7 describes, it is the ventral PF cortex (areas 47/12 and 45A) that supports the ability to recognize items as being of a particular category. In experiments by Freedman and his colleagues, cells were found in the ventral PF cortex that encoded categories such as dogs, cats, or cars (Freedman et al., 2002; Cromer et al., 2010). And in the fMRI experiment with human subjects by Peelen and Kastner (2011), the target items for search were people, as opposed to cars or trees. The MVPA analysis revealed categorical coding of the targets in the inferior temporal cortex, an area which sends projections to the ventral PF cortex (Webster et al., 1994).

Selective Advantage

Primates forage for fruit and leaves and these are found in a cluttered environment. With a few exceptions, primates are unlike ungulates that forage on the grasslands. And though primates catch insects, they are unlike carnivores in that they do not chase their prey. Lions and cheetahs typically pick off the weakest that lag behind as the herd tries to escape. So the basic way in which primates forage differs in that they have to search for the edible fruits and leaves amongst distractors, the inedible ones.

Chapter 2 points out that the FEF evolved in strepsirrhine primates. The ancestral primates fed on fruit and gums (Silcox et al., 2015). The visual life of early primates might have resembled what we experience as peripheral vision (Shepherd & Platt, 2010). They foraged in dim light for scattered and hidden resources in the clutter of the fine-branch niche. The FEF provided advantages in visual foraging in these conditions.

But when the anthropoids evolved, they could use trichromatic colour vision and foveal vision to pick out the fruit and leaves that were edible. It is possible that it was at this time that the FEFv evolved.

Connectivity

Among its many connections, the caudal PF cortex has five that, collectively, make it unique. It has: (1) inputs from the posterior parietal cortex in the dorsal visual stream; (2) inputs from the inferior temporal cortex in the ventral visual stream; (3) interconnections from area 8Av with the orbital PF cortex; (4) interconnections with the ventral PF cortex; and (5) direct and indirect outputs to the superior colliculi as well as outputs to the oculomotor nuclei.

Neural Operation

This pattern of connections together with the response properties of the cells leads to the following proposal concerning the transformation that is performed by the caudal PF cortex.

The FEF receives visual inputs from both the dorsal and ventral streams, and as a result it contains cells that represent both peripheral and central vision. The visual cells enhance the representation of the item in posterior regions by a top-down signal.

Inputs to the ventral PF cortex and orbital PF cortex carry information about current biological needs, and inputs from the ventral PF cortex to the FEF carry information about the type or class of foods that will satisfy those needs.

Visuomotor cells and motor cells transform the visual representation of the target into the endpoint of a saccade, and the superior colliculus and the oculomotor nuclei then direct the eyes to that location.

References

Aggleton, J.P., Wright, N.F., Rosene, D.L., & Saunders, R.C. (2015) Complementary patterns of direct amygdala and hippocampal projections to the macaque prefrontal cortex. *Cereb Cortex*, 25, 4351–73.

Amaral, D.G. & Price, J.L. (1984) Amygdalo-cortical projections in the monkey (IMacaca fascicularis). *J Comp Neurol*, 230, 465–96.

Amiez, C., Kostopoulos, P., Champod, A.S., & Petrides, M. (2006) Local morphology predicts functional organization of the dorsal premotor region in the human brain. *J Neurosci*, 26, 2724–31.

Amiez, C. & Petrides, M. (2009) Anatomical organization of the eye fields in the human and non-human primate frontal cortex. *Prog Neurobiol*, 89, 220–30.

Armstrong, K.M. & Moore, T. (2007) Rapid enhancement of visual cortical response discriminability by microstimulation of the frontal eye field. *Proc Natl Acad Sci USA*, 104, 9499–504.

Asplund, C.L., Todd, J.J., Snyder, A.P., & Marois, R. (2010) A central role for the lateral prefrontal cortex in goal-directed and stimulus-driven attention. *Nat Neurosci*, 13, 507–12.

Baddeley, A.D. & Hitch, G. (1974) Working memory. In Bower, G.H. (ed.) *The Psychology of Learning and Motivation.* Academic Press, New York, pp. 47–90.

Baizer, J.S., Ungerleider, L.G., & Desimone, R. (1991) Organization of visual inputs to the inferior temporal and posterior parietal cortex in macaques. *J Neurosci*, 11, 168–90.

Bichot, N.P., Heard, M.T., DeGennaro, E.M., & Desimone, R. (2015) A source for feature-based attention in the prefrontal cortex. *Neuron*, 88, 832–44.

Bisley, J.W. & Goldberg, M.E. (2003) Neuronal activity in the lateral intraparietal area and spatial attention. *Science*, 299, 81–6.

Bisley, J.W. & Goldberg, M.E. (2006) Neural correlates of attention and distractibility in the lateral intraparietal area. *J Neurophysiol*, 95, 1696–717.

Boch, R.A. & Goldberg, M.E. (1989) Participation of prefrontal neurons in the preparation of visually guided eye Primate movements in the rhesus monkey. *J Neurophysiol*, 61, 1064–84.

Bruce, C.J., Goldberg, M.E., Bushnell, M.C., & Stanton, G.B. (1985) Primate frontal eye fields. II. Physiological and anatomical correlates of electrically evoked eye movements. *J Neurophysiol*, 54, 714–34.

Bushnell, M.C., Goldberg, M.E., & Robinson, D.L. (1981) Behavioral enhancement of visual responses in monkey cerebral cortex. I. Modulation in posterior parietal cortex related to selective visual attention. *J Neurophysiol*, 46, 755–72.

Chafee, M.V. & Goldman-Rakic, P.S. (1998) Matching patterns of activity in primate prefrontal area 8a and parietal area 7ip neurons during a spatial working memory task. *J Neurophysiol*, 79, 2919–40.

Chafee, M.V. & Goldman-Rakic, P.S. (2000) Inactivation of parietal and prefrontal cortex reveals interdependence of neural activity during memory-guided saccades. *J Neurophysiol*, 83, 1550–66.

Chen, M., Li, B., Guang, J., Wei, L., Wu, S., Liu, Y., & Zhang, M. (2016) Two subdivisions of macaque LIP process visual-oculomotor information differently. *Proc Natl Acad Sci USA*, 113, E6263–70.

Colby, C.L., Duhamel, J.R., & Goldberg, M.E. (1996) Visual, presaccadic, and cognitive activation of single neurons in monkey lateral intraparietal area. *J Neurophysiol*, 76, 2841–52.

Colby, C.L. & Goldberg, M.E. (1999) Space and attention in parietal cortex. *Annu Rev Neurosci*, 22, 319–349.

Collin, N.G., Cowey, A., Latto, R., & Marzi, C. (1982) The role of frontal eye-fields and superior colliculi in visual search and non-visual search in rhesus monkeys. *Behav Brain Res*, 4, 177–93.

Corbetta, M. & Shulman, G.L. (2002) Control of goal-directed and stimulus-driven attention in the brain. *Nature Neuroscience Reviews*, 3, 201–15.

Corbetta, M. & Shulman, G.L. (2011) Spatial neglect and attention networks. *Annu Rev Neurosci*, 34, 569–99.

Cromer, J.A., Roy, J.E., & Miller, E.K. (2010) Representation of multiple, independent categories in the primate prefrontal cortex. *Neuron*, 66, 796–807.

Crowne, D.P., Yeo, C.H., & Russell, I.S. (1981) The effects of unilateral frontal eye field lesions in the monkey: Visual-motor guidance and avoidance behaviour. *Behav Brain Res*, 2, 165–87.

Desimone, R. & Duncan, J. (1995) Neural mechanisms of selective visual attention. *Ann Rev Neurosci*, 18, 193–222.

Di Lazzaro, V., Pilato, F., Dileone, M., Profice, P., Oliviero, A., Mazzone, P., Insola, A., Ranieri, F., Tonali, P.A., & Rothwell, J.C. (2008) Low-frequency repetitive transcranial magnetic stimulation suppresses specific excitatory circuits in the human motor cortex. *J Physiol*, 586, 4481–87.

Douglas, R.J., Martin, K.A.C., & Whitteridge, D. (1989) A canonical microcircuit for neocortex. *Neural Comput*, 1, 480–8.

Engel, S.A. (2012) The development and use of phase-encoded functional MRI designs. *Neuroimage*, 62, 1195–200.

Everling, S. & Johnston, K. (2013) Control of the superior colliculus by the lateral prefrontal cortex. *Philos Trans R Soc Lond B Biol Sci*, 368, 20130068.

Ferraina, S., Pare, M., & Wurtz, R.H. (2002) Comparison of cortico-cortical and cortico-collicular signals for the generation of saccadic eye movements. *J Neurophysiol*, 87, 845–58.

Fogassi, L., Gallese, V., Buccino, G., Craighero, L., Fadiga, L., & Rizzolatti, G. (2001) Cortical mechanism for the visual guidance of hand grasping movements in the monkey: A reversible inactivation study. *Brain*, 124, 571–86.

Fox, M.D., Corbetta, M., Snyder, A.Z., Vincent, J.L., & Raichle, M.E. (2006) Spontaneous neuronal activity distinguishes human dorsal and ventral attention systems. *Proc Natl Acad Sci USA*, 103, 10046–51.

Freedman, D.J., Riesenhuber, M., Poggio, T., & Miller, E.K. (2002) Visual categorization and the primate prefrontal cortex: Neurophysiology and behavior. *J Neurophysiol*, 88, 929–41.

Fries, W. (1984) Cortical projections to the superior colliculus in the macaque monkey: A retrograde study using horseradish peroxidase. *J Comp Neurol*, 230, 55–76.

Funahashi, S., Bruce, C.J., & Goldman-Rakic, P.S. (1989) Mnemonic coding of visual space in monkey dorsolateral prefrontal cortex. *J Neurophysiol*, 61, 331–49.

Funahashi, S., Bruce, C.J., & Goldman-Rakic, P.S. (1993a) Dorsolateral prefrontal lesions and oculomotor delayed-response performance: Evidence for mnemonic 'scotomas'. *J Neurosci*, 13, 1479–97.

Funahashi, S., Chafee, M.V., & Goldman-Rakic, P.S. (1993b) Prefrontal neuronal activity in rhesus monkeys performing a delayed anti-saccade task. *Nature*, 365, 753–6.

Galaburda, A.M., Sanides, F., & Geschwind, N. (1978) Human brain. Cytoarchitectonic left-right asymmetries in the temporal speech region. *Arch Neurol*, 35, 812–17.

Gee, A.L., Ipata, A.E., Gottlieb, J., Bisley, J.W., & Goldberg, M.E. (2008) Neural enhancement and pre-emptive perception: The genesis of attention and the attentional maintenance of the cortical salience map. *Perception*, 37, 389–400.

Georgopoulos, A.P., Schwartz, A.B., & Kettner, R.E. (1986) Neuronal population coding of movement direction. *Science*, 233, 1416–19.

Gerbella, M., Belmalih, A., Borra, E., Rozzi, S., & Luppino, G. (2010) Cortical connections of the macaque caudal ventrolateral prefrontal areas 45A and 45B. *Cereb Cortex*, 20, 141–68.

Godlove, D.C., Maier, A., Woodman, G.F., & Schall, J.D. (2014) Microcircuitry of agranular frontal cortex: Testing the generality of the canonical cortical microcircuit. *J Neurosci*, 34, 5355–69.

Goldberg, M.E. & Bruce, C.J. (1985) Cerebral cortical activity associated with the orientation of visual attention in the rhesus monkey. *Vision Res*, 25, 471–81.

Graziano, M.S., Taylor, C.S., & Moore, T. (2002) Complex movements evoked by microstimulation of precentral cortex. *Neuron*, 34, 841–51.

Gregoriou, G.G., Gotts, S.J., & Desimone, R. (2012) Cell-type-specific synchronization of neural activity in FEF with V4 during attention. *Neuron*, 73, 581–94.

Hanes, D.P., Thompson, K.G., & Schall, J.D. (1995) Relationship of presaccadic activity in frontal eye field and supplementary eye field to saccade initiation in macaque: Poisson spike train analysis. *Exper Brain Res*, 103, 85–96.

Hasegawa, R.P., Matsumoto, M., & Mikami, A. (2000) Search target selection in monkey prefrontal cortex. *J Neurophysiol*, 84, 1692–6.

Hedreen, J.C. & DeLong, M.R. (1991) Organization of striatopallidal, striatonigral and nigrostriatal projections in the macaque. *J Com Neurol*, 304, 569–95.

Heinen, K., Feredoes, E., Ruff, C.C., & Driver, J. (2017) Functional connectivity between prefrontal and parietal cortex drives visuo-spatial attention shifts. *Neuropsychol*, 99, 81–91.

Heinzle, J., Hepp, K., & Martin, K.A. (2007) A microcircuit model of the frontal eye fields. *J Neurosci*, 27, 9341–353.

Heitz, R.P. & Schall, J.D. (2013) Neural chronometry and coherency across speed-accuracy demands reveal lack of homomorphism between computational and neural mechanisms of evidence accumulation. *Philos Trans R Soc Lond B Biol Sci*, 368, 20130071.

Henderson, J.M. & Choi, W. (2015) Neural correlates of fixation duration during real-world scene viewing: Evidence from Fixation-related (FIRE) fMRI. *J Cogn Neurosci*, 27, 1137–45.

Hikosaka, O. & Wurtz, R.H. (1985) Modification of saccadic eye movements by GABA-related substances. II. Effects of muscimol in monkey substantia nigra pars reticulata. *J Neurophysiol*, 53, 292–308.

Huerta, M.F., Krubitzer, L.A., & Kaas, J.H. (1986) Frontal eye field as defined by intracortical microstimulation in squirrel monkeys, owl monkeys, and macaque monkeys: I. Subcortical connections. *J Comp Neurol*, 253, 415–39.

Huxlin, K.R., Saunders, R.C., Marchionini, D., Pham, H.A., & Merigan, W.H. (2000) Perceptual deficits after lesions of inferotemporal cortex in macaques. *Cereb Cortex*, 10, 671–83.

Inoue, M., Mikami, A., Ando, I., & Tsukada, H. (2004) Functional brain mapping of the macaque related to spatial working memory as revealed by PET. *Cereb Cortex*, 14, 106–19.

Ioannides, A.A., Fenwick, P.B., Pitri, E., & Liu, L. (2010) A step towards non-invasive characterization of the human frontal eye fields of individual subjects. *Nonlinear Biomed Phys*, 4 (Suppl 1), S11.

Izawa, Y., Suzuki, H., & Shinoda, Y. (2009) Response properties of fixation neurons and their location in the frontal eye field in the monkey. *J Neurophysiol*, 102, 2410–22.

Jerde, T.A., Merriam, E.P., Riggall, A.C., Hedges, J.H., & Curtis, C.E. (2012) Prioritized maps of space in human frontoparietal cortex. *J Neurosci*, 32, 17382–90.

Kalla, R., Muggleton, N.G., Juan, C.H., Cowey, A., & Walsh, V. (2008) The timing of the involvement of the frontal eye fields and posterior parietal cortex in visual search. *Neurorep*, 19, 1067–71.

Keller, E.L., Lee, K.M., Park, S.W., & Hill, J.A. (2008) Effect of inactivation of the cortical frontal eye field on saccades generated in a choice response paradigm. *J Neurophysiol*, 100, 2726–37.

Kennett, S., van Velzen, J., Eimer, M., & Driver, J. (2007) Disentangling gaze shifts from preparatory ERP effects during spatial attention. *Psychophysiology*, 44, 69–78.

Kodaka, Y., Mikami, A. & Kubota, K. (1997) Neuronal activity in the frontal eye field of the monkey is modulated while attention is focused on to a stimulus in the peripheral visual field, irrespective of eye movement. *Neurosci Res*, 28, 291–8.

Kondo, H., Saleem, K.S., & Price, J.L. (2005) Differential connections of the perirhinal and parahippocampal cortex with the orbital and medial prefrontal networks in macaque monkeys. *J Comp Neurol*, 493, 479–509.

Kunzle, H., Akert, K., & Wurtz, R.H. (1976) Projection of area 8 (frontal eye field) to superior colliculus in the monkey. An autoradiographic study. *Brain Res*, 117, 487–92.

Lawler, K.A. & Cowey, A. (1987) On the role of posterior parietal and prefrontal cortex in visuo-spatial perception and attention. *Exper Brain Res*, 65, 695–8.

Lebedev, M.A., Messinger, A., Kralik, J.D., & Wise, S.P. (2004) Representation of attended versus remembered locations in prefrontal cortex. *PLoS Biol*, 2, e365.

Leitao, J., Thielscher, A., Tunnerhoff, J., & Noppeney, U. (2015) Concurrent TMS-fMRI reveals interactions between dorsal and ventral attentional systems. *J Neurosci*, 35, 11445–57.

Lowe, K.A. & Schall, J.D. (2018) Functional categories of visuomotor neurons in macaque frontal eye field. *eNeuro*, 5. doi: 10.1523/eneuro.0131.18.2018.

Lynch, J.C. & McLaren, J.W. (1989) Deficits of visual attention and saccadic eye movements after lesions of parietooccipital cortex in monkeys. *J Neurophysiol*, 61, 74–90.

Machner, B., Konemund, I., von der Gablentz, J., Bays, P.M., & Sprenger, A. (2018) The ipsilesional attention bias in right-hemisphere stroke patients as revealed by a realistic visual search task: Neuroanatomical correlates and functional relevance. *Neuropsychology*, 32, 850–65.

Mackey, W.E. & Curtis, C.E. (2017) Distinct contributions by frontal and parietal cortices support working memory. *Sci Rep*, 7, 6188.

Mackey, W.E., Devinsky, O., Doyle, W.K., Golfinos, J.G., & Curtis, C.E. (2016a) Human parietal cortex lesions impact the precision of spatial working memory. *J Neurophysiol*, 116, 1049–54.

Mackey, W.E., Devinsky, O., Doyle, W.K., Meager, M.R., & Curtis, C.E. (2016b) Human dorsolateral prefrontal cortex is not necessary for spatial working memory. *J Neurosci*, 36, 2847–56.

Mackey, W.E., Winawer, J., & Curtis, C.E. (2017) Visual field map clusters in human frontoparietal cortex. *Elife*, 6.

Mannan, S.K., Mort, D.J., Hodgson, T.L., Driver, J., Kennard, C., & Husain, M. (2005) Revisiting previously searched locations in visual neglect: Role of right parietal and frontal lesions in misjudging old locations as new. *J Cogn Neurosci*, 17, 340–54.

Marr, D. (1982) *Vision*. MIT Press, Cambridge.

Martinez-Conde, S. & Macknik, S.L. (2015) From exploration to fixation: An integrative view of Yarbus's vision. *Perception*, 44, 884–99.

Mort, D.J., Malhotra, P., Mannan, S.K., Rorden, C., Pambakian, A., Kennard, C., & Husain, M. (2003) The anatomy of visual neglect. *Brain*, 126, 1986–97.

Murata, A., Gallese, V., Luppino, G., Kaseda, M., & Sakata, H. (2000) Selectivity for the shape, size and orientation of objects for grasping in neurones of monkey parietal AIP. *J Neurophysiol*, 83, 2580–601.

Napier, J.R. & Napier, P.H. (1967) *A Handbook of Living Primates*. Academic Press, London.

Nelson, M.J., Murthy, A., & Schall, J.D. (2016) Neural control of visual search by frontal eye field: Chronometry of neural events and race model processes. *J Neurophysiol*, 115, 1954–69.

Neromyliotis, E. & Moschovakis, A.K. (2017) Saccades evoked in response to electrical stimulation of the posterior bank of the arcuate sulcus. *Exp Brain Res*, 235, 2797–809.

Ninomiya, T., Dougherty, K., Godlove, D.C., Schall, J.D., & Maier, A. (2015) Microcircuitry of agranular frontal cortex: Contrasting laminar connectivity between occipital and frontal areas. *J Neurophysiol*, 113, 3242–55.

Olson, C.R. & Gettner, S.N. (1999) Macaque SEF neurons encode object-centered directions of eye movements regardless of the visual attributes of instructional cues. *J Neurophysiol*, 81, 2340–6.

Peel, T.R., Hafed, Z.M., Dash, S., Lomber, S.G., & Corneil, B.D. (2016) A causal role for the cortical frontal eye fields in microsaccade deployment. *PLoS Biol*, 14, e1002531.

Peelen, M.V. & Kastner, S. (2011) A neural basis for real-world visual search in human occipitotemporal cortex. *Proc Natl Acad Sci USA*, 108, 12125–30.

Petrides, M. & Pandya, D.N. (1988) Associative fiber pathways to the frontal cortex from the superior temporal region in the rhesus monkey. *J Comp Neurol*, 273, 52–66.

Petrides, M. & Pandya, D.N. (1999) Dorsolateral prefrontal cortex: Comparative cytoarchitectonic analysis in the human and the macaque brain and corticocortical connection patterns. *Eur J Neurosci*, 11, 1011–36.

Petrides, M. & Pandya, D.N. (2002) Comparative cytoarchitectonic analysis of the human and macaque ventrolateral prefrontal cortex and corticocortical connection pattern in the monkey. *Eur J Neurosci*, 16, 291–310.

Posner, M.I., Walker, J.A., Friedrich, F.J., & Rafal, R.D. (1984) Effects of parietal injury on covert orienting of attention. *J Neurosci*, 4, 1863–74.

Purcell, B.A., Schall, J.D., Logan, G.D., & Palmeri, T.J. (2012) From salience to saccades: Multiple-alternative gated stochastic accumulator model of visual search. *J Neurosci*, 32, 3433–46.

Purcell, B.A., Schall, J.D., & Woodman, G.F. (2013) On the origin of event-related potentials indexing covert attentional selection during visual search: Timing of selection by macaque frontal eye field and event-related potentials during pop-out search. *J Neurophysiol*, 109, 557–69.

Reppert, T.R., Servant, M., Heitz, R.P., & Schall, J.D. (2018) Neural mechanisms of speed-accuracy tradeoff of visual search: Saccade vigor, the origin of targeting errors, and comparison of the superior colliculus and frontal eye field. *J Neurophysiol*, 120, 372–84.

Robinson, D.A., & Fuchs, A.F. (1969) Eye movements evoked by stimulation of the frontal eye fields. *J Neurophysiol*, 32, 637–48.

Saleem, K.S., Miller, B., & Price, J.L. (2014) Subdivisions and connectional networks of the lateral prefrontal cortex in the macaque monkey. *J Comp Neurol*, 522, 1641–90.

Sani, I., McPherson, B.C., Stemmann, H., Pestilli, F., & Freiwald, W.A. (2019) Functionally defined white matter of the macaque monkey brain reveals a dorso-ventral attention network. *Elife*, 8.

Sato, T.R. & Schall, J.D. (2003) Effects of stimulus-response compatibility on neural selection in frontal eye field. *Neuron*, 38, 637–48.

Schall, J.D. (1991) Neuronal activity related to visually guided saccades in the frontal eye fields of rhesus monkeys: Comparison with supplementary eye fields. *J Neurophysiol*, 66, 559–79.

Schiller, P.H., True, S.D., & Conway, J.L. (1979) Effects of frontal eye field and superior colliculus ablation on eye movements. *Science*, 206, 590–92.

Schlag, J., Petre-Quadens, O., De Lee, C., & Goffe, B. (1974) Eye movements and occipital electrocortical rhythms: Effects of stimulation of the frontal eye field in the cat. *J Physiol (Paris)*, 68, 343–50.

Schlag, J. & Schlag-Rey, M. (1987) Evidence for a supplementary eye field. *J Neurophysiol*, 57, 179–200.

Segraves, M.A. (1992) Activity of monkey frontal eye field neurons projecting to oculomotor regions of the pons. *J Neurophysiol*, 68, 1967–85.

Shepherd, S.V. & Platt, M.L. (2010) Neuroethology of attention in primates. In Platt, M.L., Ghazanfar, A.A. (eds) *Primate Neuroethology*. Oxford University Press, Oxford.

Shook, B.L., Schlag-Rey, M., & Schlag, J. (1990) Primate supplementary eye field: I. Comparative aspects of mesencephalic and pontine connections. *J Comp Neurol*, 301, 618–642.

Shu, D.G., Morris, S.C., Han, J., Zhang, Z.F., Yasui, K., Janvier, P., Chen, L., Zhang, X.L., Liu, J.N., Li, Y., & Liu, H.Q. (2003) Head and backbone of the Early Cambrian vertebrate Haikouichthys. *Nature*, 421, 526–9.

Shulman, G.L., Pope, D.L., Astafiev, S.V., McAvoy, M.P., Snyder, A.Z., & Corbetta, M. (2010) Right hemisphere dominance during spatial selective attention and target detection occurs outside the dorsal frontoparietal network. *J Neurosci*, 30, 3640–51.

Silcox, M.T., Sargis, E.J., Bloch, J.I., & Boyer, D.M. (2015) Primate origins and supraordinal relationships: Morphological evidence. In Henke, K., Tattersall, I. (eds) *Handbook of Paleoanthropology*. Springer-Verlag, Berlin.

Stanton, G.B., Bruce, C.J. & Goldberg, M.E. (1995) Topography of projections to posterior cortical areas from the macaque frontal eye fields. *J Comp Neurol*, 353, 291–305.

Stanton, G.B., Goldberg, M.E., & Bruce, C.J. (1988) Frontal eye field efferents in the macaque monkey: I. Subcortical pathways and topography of striatal and thalamic terminal fields. *J Comp Neurol*, 271, 473–92.

Stanton, G.B., Deng, S.Y., Goldberg, M.E., & McMullen, N.T. (1989) Cytoarchitectural characteristics of the frontal eye field in macaque monkey. *J Comp Neurol*, 282, 415–27.

Stepniewska, I., Pouget, P., & Kaas, J.H. (2018) Frontal eye field in prosimian galagos: Intracortical microstimulation and tracing studies. *J Comp Neurol*, 526, 626–52.

Sugihara, T., Diltz, M.D., Averbeck, B.B., & Romanski, L.M. (2006) Integration of auditory and visual communication information in the primate ventrolateral prefrontal cortex. *J Neurosci*, 26, 11138–47.

Takeda, K. & Funahashi, S. (2002) Prefrontal task-related activity representing visual cue location or saccade direction in spatial working memory tasks. *J Neurophysiol*, 87, 567–88.

Takeda, K. & Funahashi, S. (2004) Population vector analysis of primate prefrontal activity during spatial working memory. *Cereb Cortex*, 14, 1328–39.

Taubert, J., Wardle, S.G., Flessert, M., Leopold, D.A., & Ungerleider, L.G. (2017) Face pareidolia in the rhesus monkey. *Curr Biol*, 27, 2505–9 e2502.

Thompson, K.G. & Schall, J.D. (2000) Antecedents and correlates of visual detection and awareness in macaque prefrontal cortex. *Vision Res*, 40, 1523–38.

Tsujimoto, S. & Postle, B.R. (2012) The prefrontal cortex and oculomotor delayed response: A reconsideration of the 'mnemonic scotoma'. *J Cogn Neurosci*, 24, 627–35.

Vossel, S., Geng, J.J., & Fink, G.R. (2014) Dorsal and ventral attention systems: Distinct neural circuits but collaborative roles. *Neuroscientist*, 20, 150–9.

Wardak, C., Ibos, G., Duhamel, J.R., & Olivier, E. (2006) Contribution of the monkey frontal eye field to covert visual attention. *J Neurosci*, 26, 4228–35.

Wardak, C., Olivier, E., & Duhamel, J.R. (2004) A deficit in covert attention after parietal cortex inactivation in the monkey. *Neuron*, 42, 501–8.

Wardak, C., Olivier, E., & Duhamel, J.R. (2011) The relationship between spatial attention and saccades in the frontoparietal network of the monkey. *Eur J Neurosci*, 33, 1973–81.

Wardak, C., Vanduffel, W., & Orban, G.A. (2010) Searching for a salient target involves frontal regions. *Cereb Cortex*, 20, 2464–77.

Webster, M.J., Bachevalier, J., & Ungerleider, L.G. (1994) Connections of inferior temporal areas TEO and TE with parietal and frontal cortex in macaque monkeys. *Cer Cort*, 4, 471–83.

Wu, C.W., Bichot, N.P., & Kaas, J.H. (2000) Converging evidence from microstimulation, architecture, and connections for multiple motor areas in the frontal and cingulate cortex of prosimian primates. *J Comp Neurol*, 423, 140–77.

Yan, Y.J., Cui, D.M., & Lynch, J.C. (2001) Overlap of saccadic and pursuit eye movement systems in the brain stem reticular formation. *J Neurophysiol*, 86, 3056–60.

Zhou, H. & Desimone, R. (2011) Feature-based attention in the frontal eye field and area V4 during visual search. *Neuron*, 70, 1205–17.

6

Dorsal Prefrontal Cortex

Planning Sequences

Introduction

Chapter 2 explains that in the Oligocene the primate prefrontal (PF) cortex expanded in anthropoids as new areas appeared. Routine trichromacy developed in the stem catarrhines. Because they foraged by day it was easy to identify resources; but it also meant that they were at risk from predation. This placed a premium on foraging efficiently. And, as Chapter 5 shows, the development of the fovea enhanced the search for food items in a cluttered environment. When an object serves as a goal, the fact that it occupies a specific location means that it also has spatial coordinates. In anthropoids, these coordinates depend on a visual frame of reference with the fovea at its origin (Shadmehr & Wise, 2005).

However, this search poses two problems. The first is that it is necessary to remember previous locations that have been exploited most recently within a patch because once the fruit and leaves have been picked there, returning to the same location will not be productive. The second is that, given limited time, it is important to plan the search path that will be most efficient so as to exploit the richer patches before those that are less so. In other words, there is a premium on learning how to forage in an optimal way (Stephens & Krebs, 1986).

This chapter considers two areas that evolved in the dorsal PF cortex, areas 46 and 9/46. These support the generation of the sequence of goals during foraging.

Areas

Walker (1940) used the term area 46 for the cortex along the entire length of the principal sulcus. But more recently Petrides and Pandya (1999) claimed a distinction on cytoarchitectonic grounds between area 9/46 in the caudal third of this sulcus and area 46 in the middle and rostral part (Figure 1.9, page 24). In the human brain, these divisions lie between the superior and inferior frontal sulci (Figure 1.10, page 26). It is as if the expansion of the PF cortex in the human brain had flattened out the tissue that was homologous with the cortex of the principal sulcus in macaques. The effect is that the tissue that corresponds to the dorsal bank lies between the middle and superior frontal sulci in the human brain; whereas the

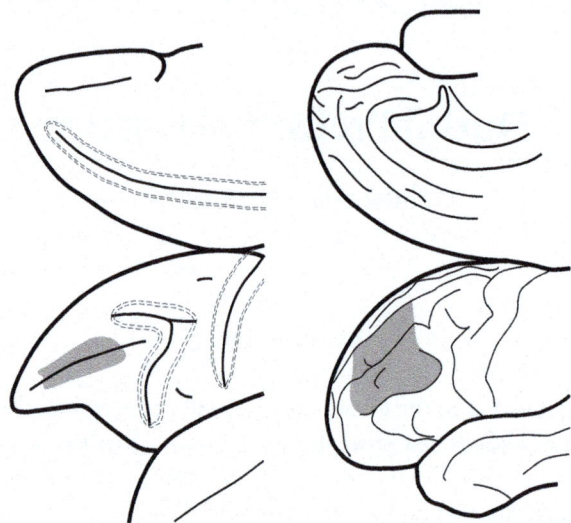

Figure 6.1 The dorsal PF cortex in macaque monkeys (left) and humans (right)
Adapted from Passingham, R.E. & Wise, S.P. *The Neurobiology of Prefrontal Cortex*, p. 158, Figure 6.1
© 2012, Oxford University Press.

tissue that corresponds to the ventral bank lies between the middle and inferior frontal sulci. This distinction corresponds to the difference between area 9/46d and area 9/46v (Figure 1.9).

The dorsal PF cortex is illustrated in Figure 6.1 for the macaque monkey and human brains. As used here, the term dorsal PF cortex excludes the tissue that lies dorsal to the principal sulcus in macaque monkeys. This is the lateral part of area 9, and this is included in the medial PF cortex (Chapter 1) because area 9 continues down the medial surface. In the human brain, the lateral part of area 9 lies above the superior frontal sulcus (Figure 1.10).

Thus, the dorsal PF cortex is not equivalent to the dorsolateral PF cortex as it has often been defined in the literature. This term originally referred to the whole lateral PF cortex in macaque monkeys (Pribram et al., 1952), but later came to mean just the cortex in and dorsal to the principal sulcus (Mishkin et al., 1969). In studies using fMRI, it has become common to refer to the dorsolateral PF cortex or DLPFC, but this term is often used very loosely, with little attention to anatomical landmarks. As a result, the term DLPFC should be avoided (Passingham & Rowe, 2016).

Connections

Figure 6.2 (next page) illustrates some of the cortico-cortical connections of the PF area 46. The extrinsic connections of two areas are different in important respects, and for this reason they are discussed separately in the following sections.

CONNECTIONS 193

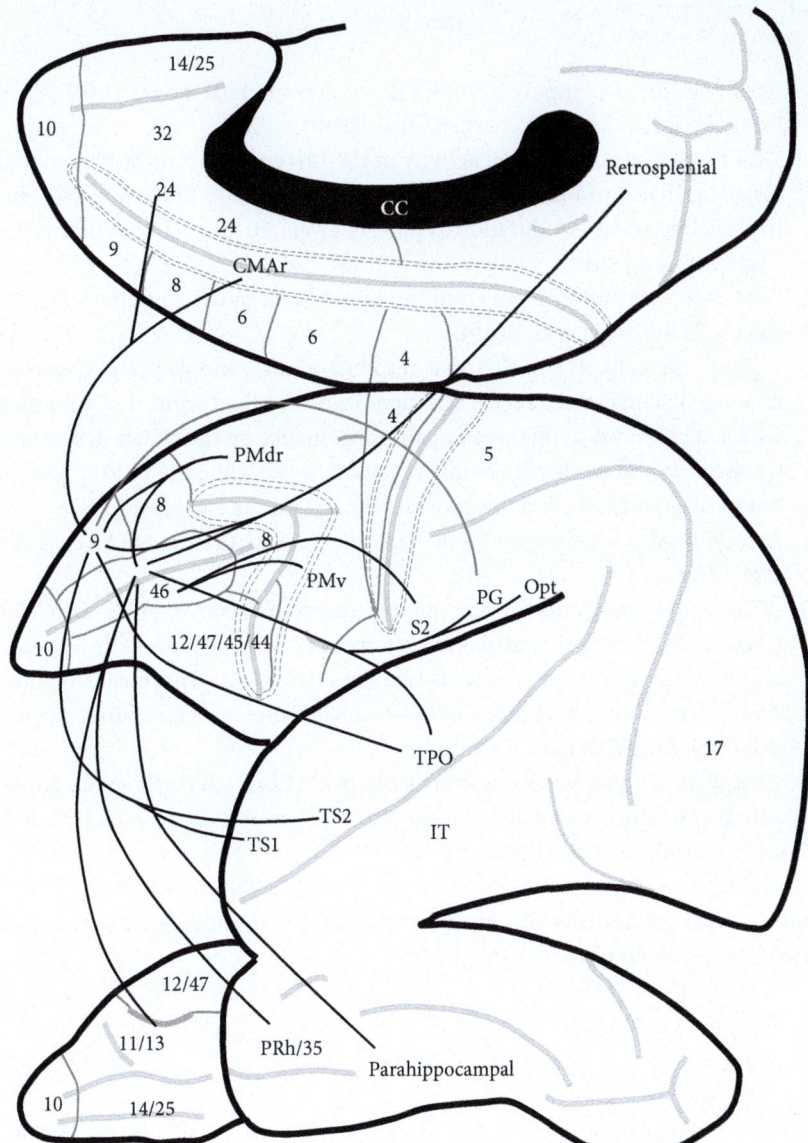

Figure 6.2 Selected connections of the dorsal PF area 46
The Lines connect some of the areas that have direct axonal connections with area 46, assumed to be reciprocal.

Adapted from Passingham, R.E. & Wise, S.P., *The Neurobiology of Prefrontal Cortex*, p. 160, Figure 6.2 © 2012, Oxford University Press.

Area 9/46

1. Area 9/46 receives inputs from the lateral intraparietal cortex (LIP) in the intraparietal sulcus (IPS) (Gerbella et al., 2013).

    In macaque monkeys, cell activity in the LIP encodes a priority map that assigns salience to location (Bisley & Goldberg, 2010). There are also cells that encode temporal duration (Finnerty et al., 2015) and temporal order (Aghdaee et al., 2014).

2. Area 9/46 also has an input from the upper bank of the superior temporal sulcus (STS) (Borra et al., 2019).

    There are cells in the STS that respond when a monkey sees a person moving (Jellema et al., 2000). The significance of the connection with area 9/46 is that lesions of this area impair performance on the classic delayed response task, and on this the monkeys watch the experimenter putting food in one or two foodwells; and this provides the cue for the correct response.

3. Area 9/46 also receives an input from the ventral PF area 45A (Gerbella et al., 2013).

    There are connections from the inferotemporal cortex to PF area 45A (Webster et al., 1994), and the inferotemporal cortex supports the analysis of the colour and shape of objects (Huxlin et al., 2000). This means that the ventral part of area 9/46 (area 9/46v) also has indirect access to information about colour and shape.

4. Area 9/46 has output to the frontal eye field (FEF) (Gerbella et al., 2013), and also to the intermediate layers of the superior colliculus (Selemon & Goldman-Rakic, 1988; Borra et al., 2014).

Just in areas 8A and the FEF, many cells in area 9/46 modulate their activity during eye movements (Tanila et al., 1993).

Area 46

1. Area 46 receives no direct inputs from the cortex in the IPS (Gerbella et al., 2013; Borra et al., 2019). Instead, it only receives visuospatial input from LIP indirectly via connections with area 8 (Gerbella et al., 2013) and area 9/46 (Gerbella et al., 2013; Borra et al., 2019). Unlike area 9/46, area 46 also lacks outputs to the superior colliculus (Fries, 1984).

    Thus, area 46 differs from area 9/46 in being much less closely associated with the oculomotor system.

2. The parietal connections to area 46 come from the inferior parietal cortex. The strongest inputs are with area PFG (7ab), though there are also lighter inputs from areas PF (7b) and the inferior parietal area PG (7a) (Gerbella

et al., 2013). Like the connection from the secondary somatosensory area (S2), the input is to the ventral bank of the principal sulcus. The intraparietal area (AIP) also projects to the ventral convexity cortex of area 46 (Gerbella et al., 2013).

Cells in PFG respond to hand and arm movements (Rozzi et al., 2008) and the cells in area 46 do the same (Tanila et al., 1993). S2 provides tactile information about the hand (Murray & Mishkin, 1984). Parietal area AIP also has visual and visuo-motor cells that fire when a monkey is shaping the hand in advance of picking up an object (Murata et al., 2000); and inactivation of this area prevents the hand from shaping in this way (Fogassi et al., 2001). The connections of area 46 therefore bring information about the hand and arm. However, there are also cells in area PFG that respond not only when a monkey reaches to a specific goal but also when it sees an experimenter doing so (Fogassi et al., 2005). In other words, these cells are 'mirror neurons'.

3. The ventral bank of the principal sulcus at the level of area 46 receives a projection from the orbital PF cortex whereas it is less clear whether area 9/46 does so (Gerbella et al., 2013; Saleem et al., 2014). The implication is that area 46 has access to information about the current needs of the animal.

   Cells have been found in area 46 hat code both for the goal of the action and the number of juice drops that are available (Wallis & Miller, 2003) and there are also cells that respond more vigorously the greater the expected reward (Leon & Shadlen, 1999).

4. Areas 46 is interconnected with the retrosplenial cortex (Kobayashi & Amaral, 2003; Saleem et al., 2014) and presubicular cortex (Morris et al., 1999), and thus indirectly with the hippocampus (Kobayashi & Amaral, 2007). The upper bank of areas 46 also projects to cingulate areas 24 and 23 (Borra et al., 2019). In turn, these cingulate areas are also interconnected with the retrosplenial cortex (Kobayashi & Amaral, 2003), and thus indirectly with the hippocampus (Kobayashi & Amaral, 2007).

   The indirect connection with the hippocampal system suggests that the PF area 46 may play some role either in memory or planning (see Chapter 3). There is also an incremental timing signal in the hippocampus (Naya et al., 2017) and lesions of the hippocampus impair the ability of monkeys to judge temporal order (Heuer & Bachevalier, 2013). Thus, the connections with the hippocampal system may also provide information about the temporal order of actions that are performed with the hand and arms.

5. The upper bank of the principal sulcus at the level of area 46 has an output to the dorsal premotor cortex and the presupplementary motor area (preSMA) (Borra et al., 2019), as well as to the rostral cingulate motor area (CMAr) (Morecraft et al., 2012). The ventral bank has an output to the ventral premotor cortex (Gerbella et al., 2013).

These connections indicate that area 46 plays a role in actions that are performed with the arm and hand. It is a critical observation that rostral part of the dorsal premotor cortex (Tachibana et al., 2004), ventral premotor cortex (He et al., 1995), preSMA (Luppino et al., 1991), and CMAr (He et al., 1995) all have representations of the hand and arm, but not of the leg.

The Division between Areas 46 and 9/46

Areas 46 and 9/46 are closely interconnected (Gerbella et al., 2013; Borra et al., 2019). However, as the previous review points out, they differ in that area 46 is connected with the manual system whereas area 9/46 is connection with the oculomotor system and covert attention. Given this distinction, it is important to know where in the human brain the border lies between these two areas.

Sallet et al. (2013) used resting state covariance to identify these two areas in both the macaque monkey and the human brain. As Chapter 1 describes, the rationale is that if the connectional fingerprint of an area is similar in the macaque and human brain, then we can take these areas to be homologous.

Figure 6.3 (next page) shows areas 46 and 9/46 in the human brain as identified by Sallet et al. (2013).

Cieslik et al. (2013) used resting state covariance to identify the border between these two areas in the human brain. Having identified two clusters on the basis of hierarchical cluster analysis, they then showed that the anterior area of the dorsal PF cortex (green in Figure 6.4A, page 198) co-activated with the anterior cingulate cortex. This was in contrast to the posterior area of the dorsal PF cortex (red in Figure 6.4B, page 198) that co-activated with the parietal cortex.

The coordinates for the centre of gravity in Montreal Neurological Institute (MNI) space of the two clusters were $x = 30$, $y = 43$, $z = 23$ for the anterior cluster and $x = 37$, $y = 33$, $z = 32$ for the posterior cluster. This chapter, therefore, uses these coordinates to judge whether, in fMRI studies of human subjects, the peaks of activation lie in areas 46 or 9/46.

It is clear from both Figures 6.3 and 6.4 that area 46 lies more anteriorly in the human brain than many have supposed. Petrides and Pandya (1999) had also previously maintained on the basis of cytoarchitecture that area 46 lies very anteriorly (Figure 1.10). This means that the identification of activations as lying in PF area 46 in the imaging literature is very frequently inaccurate because it has generally been assumed that area 46 lies more posteriorly than in fact it does. The error has been compounded by using the term DPLFC as if it was equivalent to area 46.

Some neuroscientists who use fMRI might argue that this kind of anatomical exactitude does not matter. But it does, because it is the exact identification of cortical areas that enables us to use data on single-cell activity from macaque monkeys when interpreting fMRI activations. The upshot is that, if

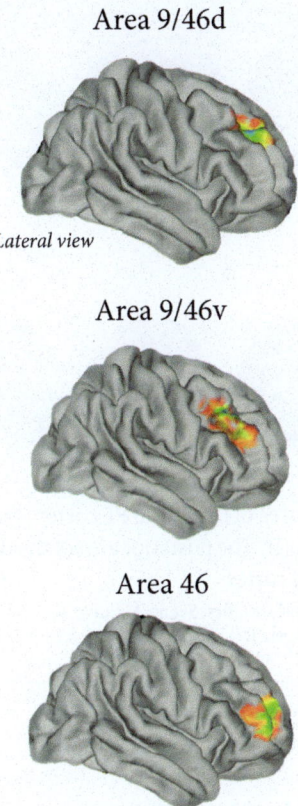

Figure 6.3 Areas 9/46 and 46 in the human brain as identified on the basis of the similarity of their connectional fingerprints to the same areas in a macaque monkey 9/46d, dorsal 9/46; 9/46v, ventral 9/46

Reproduced from Jérôme Sallet, Rogier B. Mars, MaryAnn P. Noonan, Franz-Xaver Neubert, Saad Jbabdi, Jill, X. O'Reilly, Nicola Filippini, Adam G. Thomas, & Matthew F. Rushworth. The organization of dorsal frontal cortex in humans and macaques. *J Neurosci*, 33 (30), 12255–74, Figures 9a and b, Doi: https://doi.org/10.1523/JNEUROSCI.5108-12.2013 Copyright © 2013 The Society for Neuroscience, with permission.

the localizations of the peaks in an fMRI study are wrong, the conclusions of the study are wrong. It would be as if neurophysiologists recorded from the medial intraparietal area (MIP) but wrongly attributed their findings to the LIP. They would then conclude that LIP encodes something about reaching movements rather than eye movements. This mistake would be easy to make because, like areas 9/46 and 46, MIP and LIP are both homotypical areas that are difficult to distinguish on the basis of cytoarchitecture.

This problem is by no means unique to fMRI studies. The thickness of the bone means that it is easier to mount an electrode assembly over area 9/46 than over area 46 in macaque monkeys. Accordingly, cell recording studies often attribute

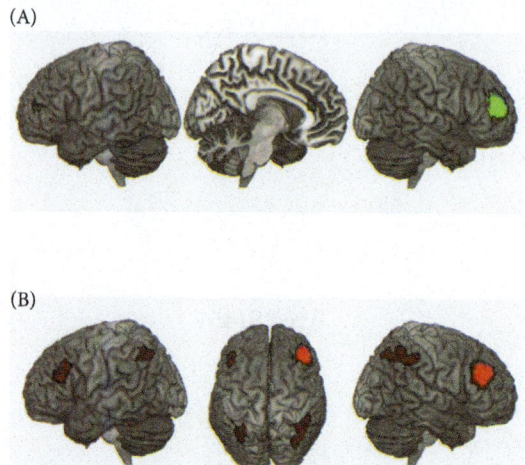

Figure 6.4 (A) The anterior area of the dorsal PF cortex (green) that co-activated with the anterior cingulate cortex. (B) The posterior area of the dorsal PF cortex (red) that co-activated with the parietal cortex

Reproduced from Edna C. Cieslik, Karl Zilles, Svenja Caspers, Christian Roski, Tanja S. Kellermann, Oliver Jakobs, Robert Langner, Angela R. Laird, Peter T. Fox, & Simon B. Eickhoff, Is there 'One' DLPFC in cognitive action control? Evidence for heterogeneity from co-activation-based parcellation. *Cereb Cortex*, 23 (11), 2667–89, Figures 3 and 4, Doi: https://doi.org/10.1093/cercor/bhs256, Copyright © 2012, Oxford University Press.

the properties of area 9/46 to area 46. Where the papers present figures of the recording sites, I have used my judgement on the basis of surface landmarks.

Summary

The connectional fingerprint of the dorsal PF cortex in macaque monkeys suggests the following conclusions:

- Areas 46 and 9/46 differ radically in their connections.
- Area 9/46 is interconnected with areas that are involved in overt and covert attention, whereas area 46 is interconnected with areas that are involved in reaching.
- The ventral bank of area 46 has inputs from the orbital PF cortex, and the dorsal bank has outputs to the dorsal and medial premotor areas for the arm and hand. Thus, this area is in a position to integrate information about actions and outcomes.
- Area 46 is indirectly connected with the hippocampal system and this suggests that it plays some role either in memory or planning.

Search

As Chapter 3 describes, Collin et al. (1982) devised a task to measure the efficiency of searching for food amongst an array of locations. There are twenty-five opaque doors and, on any one trial, there is a peanut behind each door. However, once the peanut has been retrieved from behind a door, the door swings shut. This means that it does not pay to return to that door on that trial. This task has the advantage that it serves as a laboratory model for search during foraging.

In an experiment using this task by Passingham (1985), the monkeys were given the opportunity to collect twenty of the twenty-five peanuts on each trial by opening the doors in any order. The search paths of the monkeys were plotted for many trials. And, so as to avoid any bias in the selection, the data were plotted for trials that were chosen randomly and before the search path had been visually inspected.

Before surgery the monkeys searched in a relatively systematic way. For example, they retrieved the peanuts by opening the doors round the outside of the display. This is an efficient strategy.

After the animals had been trained so that they knew the rules of the task, the dorsal PF cortex (areas 46 and 9/46) was removed in three animals. The search paths of these monkeys were compared with the performance of three unoperated control monkeys. The paths of the operated animals proved to be much less efficient. They were broken up because the animals made errors by revisiting locations that they had already visited. When the trials for all post-operative 40 trials were analysed, the three operated monkeys made roughly twice as many such errors as the control monkeys (Passingham, 1985).

Chapter 5 mentioned that animals with lesions of the FEF and area 8A are also impaired on this task. However, the reason was that they were poor at shifting their attention from a small area of the search space. The monkeys with lesions of the dorsal PF cortex did not show this same effect.

There could be two reasons why these animals were impaired. The first is that there was interference in memory. The operated animals did not make any more errors than the control animals on the first five moves. Instead, they made the majority of their errors on the last ten moves. As the task proceeds, the scope for interference in memory increases.

The second explanation is that the lesioned animals may have adopted a poor strategy. One advantage of searching in a regular way is that it decreases the demands on short-term memory. Taffe and Taffe (2011) developed a strategy score for macaque monkeys performing the spatial search task. The monkeys acquired the strategy of minimizing the distance between movements, and they made errors when they failed to use this strategy.

Harlow and Akert (1964) reported that monkeys with large lateral PF lesions were less systematic than control animals in the order in which they searched for a peanut hidden in one of four boxes, but the conclusions of the study are wrong. And, in two experiments on patients with frontal lobectomies, Owen et al. (1990; Owen et al., 1996) found that the patients were less likely than control subjects to start the search task regularly at the same location. These results suggest an impairment in planning.

Summary

While anthropoid primates forage, there is pressure to do so quickly and efficiently. Lesions of the dorsal PF cortex impair the ability to search for peanuts behind opaque doors without return to doors that they have already opened. The reason may be that either because of interference in memory or because of a failure to plan ahead.

Delayed Response and Alternation

The delayed response (DR) and delayed alternation (DA) tasks differ from the oculomotor delayed response task (ODR) (Chapter 5) in using an opaque screen to disrupt the ability of the monkeys to attend covertly to one side during the delay. Passingham (1971) monitored the movements of intact monkeys during the delay on the DR task, and found that, even in trials on which they responded correctly, they crossed to the other side on 40% of trials.

The classic version of the DR task is illustrated in Figure 6.5 (next page). The experimenter baits one of two food wells (+) as a monkey watches from its testing cage (Figure 6.5A). This action serves as the visual cue. The experimenter then lowers an opaque screen for the delay period which is typically 5 seconds (Figure 6.5B). After the experimenter lifts the screen, the monkey chooses between the two foodwells by displacing one of the covers. If the response is correct (+) it finds the peanut, but if it is incorrect (−) there is no reward (Figure 6.5C).

The DA task differs in that the monkey does not see the experimenter baiting the foodwells. Instead, the baiting is done while the opaque screen is lowered. The rule is that the monkey should reach to the foodwell on the left on one trial, the right on the next, the left on the next and so on, alternating between the two foodwells. Again, the delay period is typically 5 seconds.

It is a classic finding that bilateral lesions of the dorsal PF cortex cause a devastating impairment on these tasks (Goldman et al., 1971). For example, the animals are unable to relearn the DA task in 2,000 trials (Goldman et al., 1971). Nixon and Passingham (1999) also tested monkeys with removal of the dorsal PF cortex on a computerized version of the DA task. To disrupt covert attention, an opaque screen was interposed automatically during the delay. And again, the lesioned monkeys failed to relearn the task in 2,000 trials.

DELAYED RESPONSE AND ALTERNATION 201

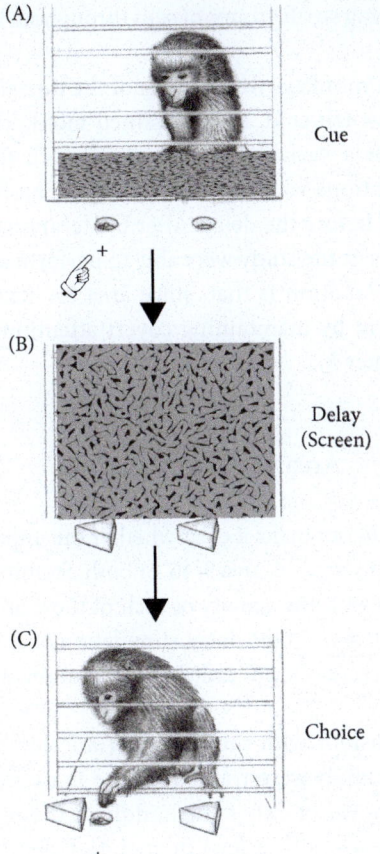

Figure 6.5 Testing procedure for the classic delayed response task in a Wisconsin general testing apparatus (WGTA)

Adapted from Murray E.A. Contributions of the amygdala complex to behavior in macaque monkeys. *Prog Brain Res* 87, 167–80, DOI: https://doi.org/10.1016/S0079-6123(08)63051-4 Copyright © 1991 Elsevier Science Publishers B.V. Published by Elsevier B.V. All rights reserved.

This is such a striking finding that it is likely that it holds the key to the function of the dorsal PF cortex. No theory is plausible unless it can explain the severity of this impairment. Animals with total PF lesions are totally unable to sequence their choices correctly.

Upright et al. (2018) tested monkeys on the DR task, but rather than making surgical lesions they inactivated cells in the dorsal PF cortex by using DREADDS, a designer receptor exclusively activated by designer drugs. This causes chemicogenetic inhibition. The result was that three of the five animals performed at chance level at one or other of the longer delays, 10, 15, or 20 seconds. And this was the case even though only 3% of the cells that were counted were found to be affected.

In other words, disrupting small groups of cells throughout the area was enough to lead to failure.

If no screen is used, monkeys with lesions of the dorsal PF cortex are not severely impaired. Ainsworth et al. (2018) trained monkeys to touch a cue on a computer screen. After a variable delay during which the monkeys fixated a central spot, three locations were presented, and the monkey had to touch the one that had appeared before the delay. After bilateral removal of the dorsal PF cortex, the two animals in the study were able to perform at a level of above 80% correct. The likely explanation is that, since area 8A was intact, the monkeys could solve the problem by maintaining covert attention to the cued location during the delay (Chapter 5).

Area 46 versus Area 9/46

The results cited so far do not make it clear whether the impairment on these tasks results from a lesion in the area 46, area 9/46, or both. Stamm (1969) used electrical stimulation to interfere with the activity of each of these areas in turn during the delay period. In the early part of the delay, stimulation of area 46 caused performance to fall to chance level on the DR task; but stimulation of area 9/46 had a lesser effect.

This finding is consistent with the results of a lesion study by Butters and Pandya (1969). The animals were trained on the DA task pre-operatively and then retested after removal of either the middle of the principal sulcus or the caudal third. The animals with lesions in area 46 failed to relearn the task in 1,000 trials, whereas the animals with lesions in area 9/46 were impaired but relearned in task in around 600 trials. The likely reason for the severity of the impairment with lesions of area 46 is that the responses on the classical DA task are manual, and it is this area that has outputs to the premotor areas for the arm and hand (Borra et al., 2019).

However, the animals are not impaired simply because they have to reach. It is critical that the reach is based on memory. Passingham (1985) devised a task in which both the cues and responses were spatial, but there was no delay. The cue was provided by a light on one of two panels in the centre of a display. The monkeys learned that if the upper panel was lit, it should touch a box on the left; whereas if the lower panel was lit, then it should touch the box on the right. Thus, the task involved a spatial cue and a spatial response just like the DR task; but unlike the DR task it did not include a delay period. The finding was that monkeys with a lesion of the dorsal PF cortex showed no impairment at all.

Summary

Monkeys with combined lesions of the PF areas 46 and 9/46 fail to learn or relearn the spatial DR and DA tasks, though the impairment is more severe after lesions of area 46. Yet they can perform a comparable task if there is no delay. This indicates that it is critical that the current goal be generated on the basis of an event in memory.

Short-term Memory

As on the search task, the impairment on the DR and DA tasks after removal of the dorsal PF cortex could be due to a failure of interference in short-term memory or planning. But there is another possibility, and this is that they cannot learn the task rules. As Chapter 1 mentions, the advantage of appealing to data on human subjects is that they do not have to learn these. They know the task instructions because these are provided before testing begins.

Retrospective Coding

If the impairment is due to an impairment in short-term memory, it should be possible to demonstrate delay-period activation in the dorsal PF cortex when human subjects are tested for their memory of a series of locations. In an fMRI study, Rowe and Passingham (2000) presented a display of three locations followed by a variable delay with a mean of 14 seconds. A line was presented at the end of the delay and the subjects had to move a cursor to the location that would have been bisected by the line. They were unable to plan their response during the delay because they could not know in advance what this would be.

Surprisingly, Rowe and Passingham found no significant delay-period activation in the dorsal PF areas 46 or 9/46. It could be argued that the experiment lacked sufficient power. But the power was enough to demonstrate sustained activations in the dorsal FEF and the cortex within the IPS, and the power was also enough to show activation in the dorsal PF cortex at the time of the manual response.

This was an early experiment and only six subjects were used. However, Blacker and Courtney (2016) carried out two fMRI experiments with a similar design, and they used thirty-two subjects. In one experiment the subjects viewed a display of two locations, and after an unfilled delay the subjects had to indicate whether a location would be on the line that could be drawn between those two locations. As in the experiment by Rowe and Passingham (2000), the activations during the delay period lay in the dorsal FEF and the cortex in the IPS. And the result was the same in a second experiment in which the memory load was increased to three items, and the subjects had to indicate whether a test location was one of the three locations in the display.

However, the lack of activation in the dorsal PF cortex in these experiments could be due to the fact that only two or three items were used. In a later study Silk et al. (2010) presented either two or five items simultaneously. The subjects had to say whether a probe location was one of the locations that had been presented. As in previous studies, there were no significant activations in the dorsal PF cortex.

Pochon et al. (2001) presented five locations in sequence rather than simultaneously, but the delay period was short, just 6 seconds. After the delay, the subjects were required to say whether a second sequence matched the first one or not. This matching task is shown in the top section of Figure 6.6. Because the condition involves matching, the subjects could not know in advance what response they are going to have to make.

As in the other experiments, there was sustained activation in the FEF, though Pochon et al. attributed this to the dorsal premotor cortex. There was also a sustained activation in the cortex of the IPS. Figure 6.7 (next page) shows that there was a slight rise in the activation of PF area 46 during the delay, but it was not statistically significant after whole-brain correction.

In order to find out if the PF cortex is necessary for correct performance on this matching task, Ferreira et al. (1998) tested patients with frontal lesions. There were eight patients of whom seven had lesions that included areas 46 and 9/46. None of them were impaired when tested for their recognition memory of locations. These

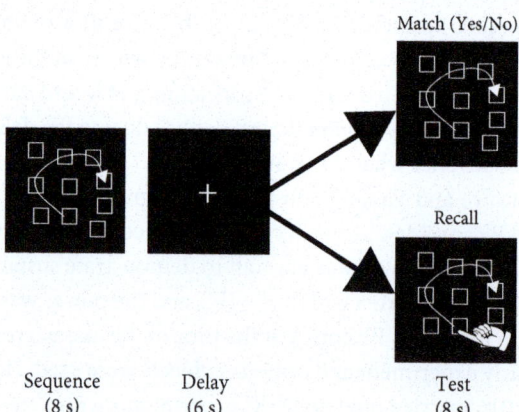

Figure 6.6 The top section shows the matching task and the bottom section the recall task that were used in an fMRI experiment

Reproduced from Pochon J.P., Levy R., Poline J.P., Crozier S., Lehéricy S., Pillon B., Deweer B., Le Bihan D., & Dubois B. The role of dorsolateral prefrontal cortex in the preparation of forthcoming actions: An fMRI study. *Cereb Cortex*, 11 (3), 260–6, Figure 1a, https://doi.org/10.1093/cercor/11.3.260 Copyright © 2001, Oxford University Press.

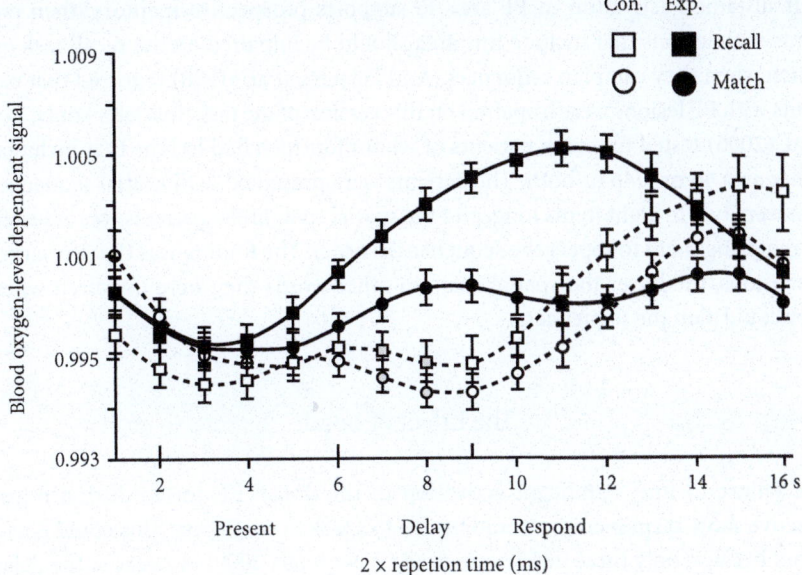

Figure 6.7 The course of the BOLD signal during the tasks that were illustrated in Figure 6.6
Filled squares: recall condition; unfilled squares: control for recall condition; filled circles: matching condition, unfilled circles: control for matching condition. Error bars: S.E.M. The time is shown in units of 2 x the TR or repetition time.

Reproduced from Pochon J.P., Levy R., Poline J.P., Crozier S., Lehéricy S., Pillon B., Deweer B., Le Bihan D., & Dubois B. The role of dorsolateral prefrontal cortex in the preparation of forthcoming actions: An fMRI study. *Cereb Cortex*, 11 (3), 260–6, Figure 5a, https://doi.org/10.1093/cercor/11.3.260 Copyright © 2001, Oxford University Press.

results are consistent with the claim that retrospective spatial short-term memory is supported by the posterior parietal cortex (Chapter 5).

Prospective Coding

Pochon et al. (2001) tested subjects on the recall task illustrated in the bottom section of Figure 6.6. The task for the subjects was to point to the locations at the end of the delay in the order in which they had been presented. Because the subjects knew what responses they needed to make, they could plan or prepare the sequence during the delay. This involves prospective coding.

Figure 6.7 shows the time course of the BOLD signal during the delay period for the recall task. This shows that there was highly significant delay-period activation in the PF area 46 when the subjects were able to plan the sequence.

If sustained activation in PF area 46 supports prospective-memory, then patients with lesions that include this area should be impaired on the recall task on which when they can plan sequences. And Ferreira et al. (1998) reported that patients with PF lesions were impaired on this version of the task. In a later study, Bor et al. (2006) tested nineteen patients of whom fourteen had lesions that included either area 46 or 9/46 or both. The patients were presented with spatial sequences that were two to eight items long, and the spatial span of the patients was assessed by requiring them to repeat the sequences in order. The finding was that the larger the lesions, the lower the spatial span, in other words they were impaired when they could plan the sequences.

The Effect of Load

The failure to find significant activation in the dorsal PF cortex during retrospective short-term memory could be due to a lack of sensitivity. This could occur either because only three items were used (Rowe et al., 2000) or because the delay was too short (Pochon et al., 2001), or an interaction between these two factors.

In a study by Leung et al. (2002), the subjects had to remember a series of either three or five locations, and recognition memory was tested 18 or 24 seconds later by presenting a probe location. Statistics were presented for 17 regions of interest, including the PF areas 46 and 9/46. Delay-period activation was reported in these two areas when the series was five items long but not when it was three items long.

However, the procedure of testing a large number of regions of interest is no longer an acceptable way of analysing data from an imaging study. Given the problems that imaging has faced with the reproducibility of results (Gorgolewski & Poldrack, 2016), the results need to be replicated using a large sample, as in the study by Blacker and Courtney (2016). And the analysis should be a whole brain analysis so as to ensure that the results are reliable.

Fortunately, Volle et al. (2005) ran a study using whole brain correction for the statistical results. They divided the delay into two periods of 8 seconds each. A sequence five long was presented, and on 50% of trials the subjects had to repeat this sequence at the end of the second delay. As expected, during the second delay there was activation in the PF area 46, reflecting prospective coding.

However, there was also an activation in area 46 during the first delay, but only when the sequence was five as opposed to three moves long. Yet, during this period, the subjects simply had to maintain the sequence. At this point they did not know whether they were going to have to repeat that sequence at the end of the second delay or not.

There is a possible explanation for the effect of load. A later section points out that there is a critical difference between remembering just three items in a sequence and remembering four or more. It is easy to remember the first and last items in a sequence of three, and thus also the item in the middle. But as the sequence lengthens, there is scope for interference in memory between the items in the middle of the sequence (Petrides, 1991; Amiez & Petrides, 2007; Osada et al., 2008).

Summary

When human subjects are tested, there is no delay-period activation in the dorsal PF cortex (areas 46 and 9/46) when retrospective memory is assessed for a set of three locations. The activations are restricted to FEF and the cortex in the IPS. However, the dorsal PF cortex is activated when there is scope for interference in memory because the sequence is longer. There is also significant delay-period activation in PF area 46 while subjects plan or prepare or plan the sequence during the delay.

Interference

On the DR and DA tasks there is high interference because only two locations are used throughout the experiment. This means that it is easy for the animal to confuse what happened on the current trial with what happened on the last trial or the trial before that.

To find out whether this was a factor, Diamond and Goldman-Rakic (1989) devised a modified version of the DR task. It was based on the 'A-not-B' task, originally invented by Piaget (1954) to study the ability of infants to understand object permanence. Diamond and Goldman-Rakic (1989) presented the food on one side (A) on two trials in a row, and then presented it on the other side (B). The monkeys performed at 50% correct, that is at the level of chance. Yet, if the location was the same on the third trial as the two previous trials, the monkeys performed at a level of 85% correct.

It is a critical observation that the lesioned monkeys did not perseverate, since perseveration would lead to below-chance performance. They failed to make their choice on the basis of a single event, the baiting of the foodwell on that trial. Instead, they based it on the fact that both possible choices been rewarded at roughly equal frequency, as averaged over recent trials.

The n-Back Task

There is also evidence for the role of interference from studies of humans. The n-back task was devised to study interference in memory (Cohen et al., 1994).

A series of items is presented continuously, and on an n-3 task, for example, the subjects have to press a button every time that an item appears that is the same as one that was presented three back in the series. Thus, if the series includes the letters L E R D E D, the subjects should press the key when the E repeats but not when the D repeats. The E was three letters back, whereas the D acts as a distractor item.

In a study using PET, Owen et al. (1999) presented an n-2 back task with locations. The subjects had to touch the locations as they occurred 2 back in the series. There was an activation in PF area 46 while the subjects did this task.

Pochon et al. (2002) used letters instead. They compared performance of n-1, n-2 and n-3 tasks with performance of a control task; they called this a 0-back task since the subjects simply had to press the key when an X appeared. The activation in PF area 9/46 was greater for the n-3 than the n-2 task, and the n-2 than the n-1 task; and area 46 was only activated during the n-3 task. These results support the claim that the dorsal PF cortex is activated more the greater the scope for interference in memory.

In order to find out which PF area was critical, Volle et al. (2008) used a lesion-mapping analysis. This is a statistical procedure that looks for an association between particular groups of voxels and the behavioural impairment. There was a significant association between damage to the PF area 9/46 and poor performance on the n-back task. And this was true whether locations or letters were used as the material.

Like objects, letters are defined by conjunctions of colour and shape, and there is no direct projection to area 9/46 from the inferior temporal cortex (Webster et al., 1994). However, there are heavy interconnections between the ventral and dorsal PF cortex (Petrides & Pandya, 2002; Gerbella et al., 2013). This means that the dorsal PF cortex has access to information about colour and shape when it is the order of the items that is critical.

The overall conclusion is that both the monkeys and patients with lesions of the dorsal PF cortex make errors as the result of interference in memory. They fail to distinguish between the temporal order of events.

Summary

Because only two locations are used throughout the DR and DA tasks, there is scope for interference in memory. When tested on a variant of the DR task, monkeys with removal of the dorsal PF cortex perform at chance when the same location is baited twice in a row and then the other side is baited. Patients with dorsal PF lesions make errors on the n-back task by responding to distractors. The dorsal PF cortex supports the ability to distinguish the temporal order of events.

Temporal Order

There is direct evidence that the PF cortex is involved in memory for temporal order. In an early study, Milner et al. (1991) presented a series of drawings or paintings, and the subjects were tested either for their ability simply to recognize the items or to judge which of two probe pictures had been presented more recently. Patients with frontal lobectomies had no problem in recognizing the items, but they were poor at judging which of two items was the more recent.

In order to find out which area was critical, Petrides (1991) removed the tissue in the PF area 46 together with the tissue that lay on the lateral surface above it. The monkeys were presented with sequences of up to five objects. They then had to choose which of two objects had been shown earlier in the sequence. This task cannot be solved by recognition memory alone.

When required to distinguish between the order of the second and third object in the sequence, the monkeys with the lesions that included area 46 were impaired. Whereas monkeys with lesions in area 8 and the FEF reached a criterion level of performance in a mean of 16 trials, the monkeys with lesions that included area 46 failed in 40 trials.

In this study, the monkeys made more mistakes when they had to distinguish between the order of objects in the middle rather than at the ends of the series. It was easier to confuse their order. In an fMRI study with monkeys, Osada et al. (2015) compared the activations at the time of recall for the two items in the middle of a sequence of objects compared with items at the beginning and end of the sequence. As expected, they found activations in the dorsal PF cortex (areas 46 and 9/46) for this contrast.

Amiez and Petrides (2007) used fMRI to study the memory for temporal order in human subjects. They measured activation during the presentation of a sequence of four objects. At the time of encoding there was more activation in the dorsal PF cortex (areas 46 and 9/46) during the presentation of the second and third objects compared with the first and last ones in the sequence. At the time of recall, these were the only areas that were activated.

In a more recent fMRI study, Roberts et al. (2018) tested the memory for temporal order by presenting four objects, in this case coloured kaleidoscopic images. The images were shown serially, but they were separated into two groups, marked by the labels 'Group 1' and 'Group 2'. A delay of 8 seconds followed before the test for recall. In one condition the subjects had to decide whether a probe image had been presented early (Group 1) or late (Group 2). In the other condition the subjects had to indicate whether a probe image had been presented first or second in its group. In the baseline condition, the subjects simply had to say whether a probe item was the same as one of the ones that had been presented.

As expected, there were extensive activations in the dorsal PF cortex both during the encoding period and during the delay; and this was the case for both conditions. However, there were also activations along the cortex of the IPS as well as in the perirhinal cortex and posterior hippocampus.

But there is a critical difference between this study and the one by Amiez and Petrides (2007). Amiez and Petrides (2007) tested the ability to distinguish between items in the middle of a series of four items, whereas Roberts et al. (2018) only presented a series of two items, either Group 1 versus Group 2, or the order of the two items within a group. In other words, the study by Roberts et al. (2018) did not measure activations for items in the middle of a sequence. The conclusion is that it is the dorsal PF cortex that supports the ability to distinguish serial order under conditions in which it is easy to confuse the relative order of the items.

Cell Activity Encoding Order

Neurophysiological studies support the claim that the dorsal PF cortex is involved both in the encoding and recall of the order in which items are presented. Funahashi et al. (1997) used spatial stimuli. Their monkeys had to choose between two locations according to the order in which they had appeared. There were three circles in a row on the screen, and two of them lit up in turn. After a delay, all three locations were lit, and the monkeys were trained to press keys to indicate the order in which the locations had been cued.

Funahashi et al. found two types of cells of interest in the dorsal PF cortex. They called the first 'position-dependent cells' because they coded for just one position in the order. The second type were called 'pair-dependent cells' because they coded for both positions in the order. Of the cells that showed changes during the delay after the sequence, 26% showed position-dependent activity and 54% pair-dependent activity. These cells were distributed across both area 46 and 9/46.

Cell activity can also be found in the dorsal PF cortex that encodes the order in which objects were presented. Ninokura et al. (2003) presented three coloured objects in sequence, and at the recall test the monkeys were required to touch them in the order in which they had been presented. In the PF areas 46 and 9/46, 43% of the cells with delay-period activity encoded the sequence in which the pictures had appeared.

In a follow-up study, Ninokura et al. (2004) found more sequence cells dorsally and more objects cells ventrally. However, since the diagram only shows the point of entrance of the electrode tracks, it is not clear how many of the object's cells lay in the ventral PF cortex and how many in the PF area 9/46v.

The tasks used in these experiments were recall tasks. So, Warden and Miller (2010) compared recall with recognition. They presented two objects in order. In the recognition version of the task, memory was tested after a delay by presenting a further sequence and requiring the animal to release a bar if it matched the sample sequence. In the recall version, three objects were presented after the delay and the animal had to make an eye movement to two out of three objects in order in which they had been presented. During the delay after presentation of the sequence, there was considerably more activity on the recall than the recognition task. This is consistent with the findings of the fMRI study by Pochon et al. (2001) (Figure 6.7).

Cell Activity Encoding the Elapse of Time

The ability to judge the temporal order of events is related to the ability to judge the elapse of time, since the time differs for items presented early rather than late in the order. Leon and Shadlen (2003) recorded cell activity in the parietal area LIP and found cell activity that discriminated between a short and long interval. In a later study, Jazayeri and Shadlen (2015) taught monkeys to reproduce a time interval. They found that the firing rate of cells in the LIP increased the longer the sample interval.

Consistent with the fMRI study by Robert and Ranganath (2018), cells have also been found in LIP that encode the temporal order of events. Aghdaee et al. (2014) presented two visual stimuli in rapid succession, one in the right field and one in the left. After a delay, one or other of the stimuli turned green and the monkeys had to release a lever if this was the earlier of the two stimuli. There were cells in LIP that responded differentially depending on whether the stimulus in a hemifield was reported as occurring first or second.

The parietal area LIP sends projections to the PF area 9/46 (Gerbella et al., 2013). Genovesio et al. (2006b) recorded in area 9/46 while the monkeys reported whether an interval had been short, intermediate, or long. As in the study by Leon and Shadlen (2003) the monkeys did this by making a saccade to an associated target. Of the total number of cells, about 10% showed activity levels that depended on the prior time interval, even though in this experiment the stimulus duration was irrelevant to the choice that the monkey made. Cells have also been reported in the PF area 9/46 that coded for whether the first or second stimulus had lasted longer (Genovesio et al., 2009).

The dorsal PF cortex (areas 46 and 9/46) connects with the medial area 9 in turn. As Chapter 3 mentions, Yumoto et al. (2011) found cells in area 9 that fired differentially when monkeys reproduced specific time intervals. And Carpenter et al. (2018) also reported cell activity in area 9 that encoded serial order on a recall task.

Areas 46 and 9 are also interconnected with the retrosplenial cortex (Kobayashi & Amaral, 2003) and presubiculum (Morris et al., 1999) and thus indirectly with the hippocampus (Kobayashi, 2007). When monkeys are tested for recall of a sequence of two objects, there is an incremental timing signal in the hippocampus as well as in the dorsal PF cortex (area 46), though this is more pronounced in the hippocampus (Naya et al., 2017).

Cruzado et al. (2020) compared the representation of time in the hippocampus and PF cortex. The monkeys were trained to associate two stimuli that were separated in time by around 1 second. Cells were found in both areas that represented both a specific stimulus and the time at which it had been presented.

Just as lesions of the dorsal PF cortex impair the ability of monkeys to judge the relative order of items (Petrides, 1991), so monkeys with lesions of the hippocampus are unable to judge the relative order of four items (Heuer & Bachevalier, 2013). And transection of the fornix also impairs the ability of monkeys to judge how recently they saw an item (Charles et al., 2004).

The conclusion is that there is cell activity encoding both temporal order and the elapse of time in a set of interconnected areas. These are the posterior parietal cortex, the dorsal PF cortex, the dorsomedial PF cortex, and the hippocampus. However, within this system it the dorsal prefrontal cortex and the hippocampus with which it is interconnected that are critical for encoding the order of items when that order can easily be confused in memory.

Summary

Monkeys with lesions that include the dorsal PF cortex (area 46) are impaired at distinguishing the temporal order in which objects were presented. While they can distinguish between the outliers, they are impaired at distinguishing the relative order of objects in the middle of the sequence. Cell activity has also been recorded in the posterior parietal and dorsal PF cortex that encodes the temporal order in which locations or objects are presented. There are also cells that encode the elapse of time. However, it is the dorsal PF cortex and the hippocampus with which it is interconnected that are critical for supporting the ability to encode the order of items when they can easily be confused in memory.

Mechanisms

Siegel et al. (2009) have suggested a mechanism via which the temporal order of objects could be encoded on a recall task. They presented two objects in sequence and, after a delay of 1 second, three objects appeared of which two had been in the original sequence. The task for the animal was to report the order in which they had been presented by making a saccade to each one in turn.

Siegel et al. measured both spiking activity and field potentials. They found that during the delay before recall, the field potentials were rhythmically synchronized at frequencies around 32 and 3 Hz. The maximal information carried by the spiking activity occurred earlier in the 32 Hz cycle for the first than the second object. This led Siegel et al. to suggest that encoding at different phases might play a role in distinguishing between the order of items when presented sequentially in short-term memory.

Maintaining Temporal Order

The literature on maintenance in memory has been dominated by the assumption that it is the sustained activity of cells that maintains that memory during the delay (Chafee & Goldman-Rakic, 1998; Constantinidis et al., 2018). However, the evidence comes from studying the ODR task on which only one location has to be remembered on each trial.

When a sequence of locations has to be remembered, the presentation of the later ones has the potential to disrupt the activity that is involved in maintaining the earlier ones. Qi et al. (2015) directly compared the effect of distraction on cell activity in the PF cortex and the posterior parietal cortex. In the frontal lobe they recorded cell activity in the area 9/46 and area 8A, whereas in the parietal lobe they recorded activity in the IPS and the inferior parietal area PG. Two locations were presented with a delay of 1.5 seconds between them. When the task was to remember the first location, a greater proportion of the cells in the PF cortex than the posterior parietal cortex continued to code for that location during presentation of the second one. This led Qi et al. to suggest that the PF cortex was better able to resist distraction.

However, Lundqvuist et al. (2018a) have pointed out that the delay-period activity is visualized by averaging the activity over many trials. They argued that this activity may not be representative of what happens on individual trials. In an earlier study, Lundqvist et al. (2016) presented two or three coloured locations in series. During the test phase the animals had to make a saccade to the location that had changed in colour.

The data were analysed for individual trials, and rather than finding continuous activity Lundqvist et al. found brief bursts of cell activity that differed in duration and timepoint during the delay. In the analysis they measured the percentage of explained variance, that is the proportion of the variance that could be explained by the location of the stimulus. All of the cells that were informative on this measure occurred at sites at which bursts of gamma activity could be recorded in the local field potentials. None of the informative cells were found at sites at which gamma bursts could not be recorded.

These results are shown in Figure 6.8 (next page). It can be seen that at sites at which spiking is informative, the gamma and beta bursts are mirror images of each other. At sites at which spiking is not informative, only the beta is modulated.

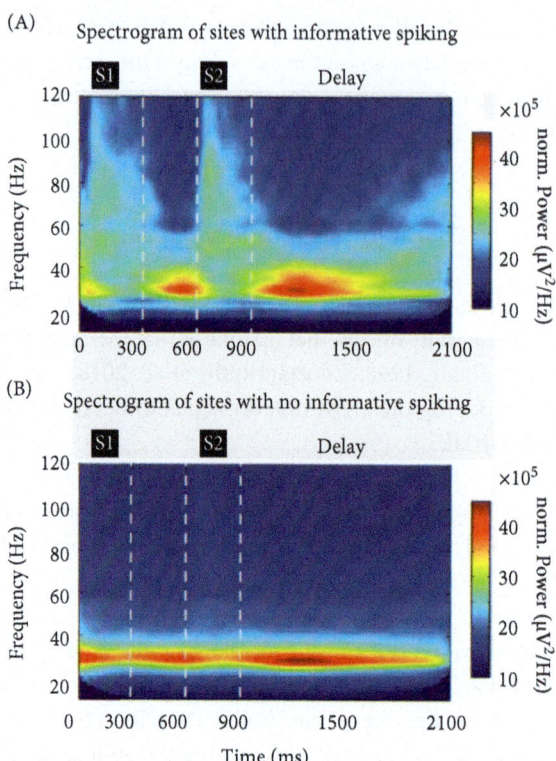

Figure 6.8 (A) The data for sites at which spiking activity was informative. (B) The data for sites at which the spiking activity was not informative
The numbers on the ordinate show the frequency in Hz. The colour scale shows the normalized power.
Reprinted from Earl K. Miller, Mikael Lundqvist, & André M. Bastos. *Neuron*, 100 (2), In *Working Memory* 2.0, pp. 463–75, Figure 1, https://doi.org/10.1016/j.neuron.2018.09.023 © 2018, Elsevier Inc.

In a related study, Bastos et al. (2018) recorded spike and local field potentials in the superficial and deep layers. Gamma activity was evident in the upper layers, and these are the layers that receive feedforward connections. Beta activity was evident in the deep layers, and these are the layers that send feedback connections. Sherfey et al. (2020) have proposed that oscillations in the superficial layers are associated with the maintenance of the order of the items, whereas oscillations in the deep layers act to gate the output so as to guide action.

The advantage of synchronous oscillations is that they provide a mechanism for the flexible formation and re-formation of neuronal ensembles by changing the pattern of synchrony (Buschman & Miller, 2020). It is a feature of short-term memory for order that there is a constant need for updating as well as remembering new sequences. For example, on the n-3 task the subject needs to keep a running and ever changing sequence of three or four items in mind.

To account for the memory for the order of multiple locations, Miller et al. (2018) have proposed a hybrid attractor dynamics and synaptic model. In this model, neuronal ensembles have inhibitory connections with other ensembles so that each item is expressed in brief bursts of spiking. However, the spiking induces brief changes in the synaptic weights. The suggestion is therefore that both spiking and short-term synaptic plasticity are involved in the short-term maintenance of multiple items.

Buschman and Miller (2020) have argued that the advantage of silent synaptic changes in the dorsal PF cortex is that they could provide a mechanism for combating interference between items in memory. This follows from this proposal that monkeys with lesions in the dorsal PF cortex would be particularly susceptible to interference in memory, and as an earlier section shows, this is indeed the case (Diamond & Goldman-Rakic, 1989).

Masse et al. (2019) have provided a computational model involving recurrent neural networks. The model suggests that synaptic plasticity may be able to maintain information over short time periods, but that persistent activity may be necessary when the information has to be manipulated in memory. However, Manohar et al. (2019) have proposed an alternative account. In their model persistent activity can hold one item in a focused or attended state, whereas the other items in a series can be stored in activity-silent synaptic traces.

Empirical evidence is needed to distinguish between these models. It is not enough for a model to account for the known data. It has to make testable predictions that other models do not make.

Recall and Re-activation

Lundqvist et al. (2018b) re-analysed the data from the study by Warden and Miller (2010) on memory for the order in which two objects had been presented. The aim was to see if the sites at which spike activity was informative were also informative at the time of the readout at the time of the test. They found gamma bursts that were associated with increased spiking and increased gamma activity at the time of recall.

Instead of presenting locations, Jacob and Nieder (2014) taught monkeys to remember the number of spots in a display. Their memory was later tested by presenting probe displays and requiring the monkeys to indicate whether or not they had the same number of spots. The size and location of the spots in the probe display was varied so as to require the animals to respond on the basis of their memory of numerosity alone. However, to make the task difficult, a distractor display of dots was presented in the interval before the test.

Jacob and Nieder recorded in the PF areas 8A and 9/46 as well as in the parietal area VIP which lies in the IPS. The cells in VIP coded for numerosity throughout the trial. However, the activity of the PF cells differed; at the time of recall and after the presentation of the distractor they re-activated information about the initial number again. Furthermore, the degree to which they did this correlated with the accuracy with which the monkeys performed.

Jacob et al. (2016) further investigated whether dopamine could be involved in the ability of the PF cells to re-activate the original number after distraction. They applied either a dopamine agonist or a dopamine antagonist to PF cells using micro-iontophoresis. The results differed depending on whether the agonist or antagonist were applied to single pyramidal neurons or to interneurons. When a dopamine D1 receptor antagonist was applied to pyramidal neurons, it enhanced the re-activation of the representation of the original number, whereas if it was applied to interneurons it inhibited this re-activation.

The implication is that the ability to maintain the representation of one item when it is followed by a second item or distractor depends on short-term synaptic changes. This means that at the time of recall, in the absence of persistent activity, the representation of any particular item can be re-activated.

Summary

Distinguishing the order of items in a sequence involves phase encoding. The storage of the sequence is best accounted for by a hybrid model which involves both spiking and short-term changes in synaptic weights. In the absence of persistent activity, the representation of any particular item can be re-activated from synaptic weights. Thus, persistent delay-period activity and temporary changes in synaptic weights combine to overcome the many distractions that occur during foraging. The recall or re-activation of the sequence is associated with gamma bursts at the sites that at which spiking activity was informative during the encoding of the sequence.

Distractor-Resistant Memory in Human Subjects

Human subjects can mentally rehearse a sequence of items in memory. If the items are locations, this rehearsal involves moving of the eyes to the locations in turn (Tremblay et al., 2006). If the items are letters or words it involves subvocal articulation (Baddeley & Hitch, 2019). The advantage of rehearsal is that it protects the memories against distraction.

Sakai et al. (2002a) showed this by using the task that is illustrated in Figure 6.9 (next page). The subjects watched while a sequence of five spatial locations lit up on a screen, as indicated by the small squares in the figure. A memory delay followed that varied between 8 and 16 seconds. On half the trials memory was then tested by showing an arrow that pointed from one location to another. The subjects had to say whether the arrow replicated the order in which these two locations had been presented.

However, unpredictably, on half the trials after the memory delay, a second sequence of five blue dots was presented. On these trials the subjects were first shown a probe blue asterisk and required to judge if this had been in the second sequence. Only after that were the subjects tested for their memory for the initial sequence.

Figure 6.9B shows that on distractor trials the degree of sustained activation in the PF area 46 during the memory delay was closely related to the degree of accuracy of performance. The same relation was not found for the activation in the PF area 8 or the cortex in the IPS. In those areas there was delay-period activation irrespective of whether the subjects performed correctly or not. These results indicate a clear distinction between the role of PF area 46 and the areas that support spatial working memory. It is delay-period activation in area 46 that acts to protect against distraction.

In a related experiment with a similar design, Sakai and Passingham (2004) compared two conditions, one with high distraction and the other with low distraction.

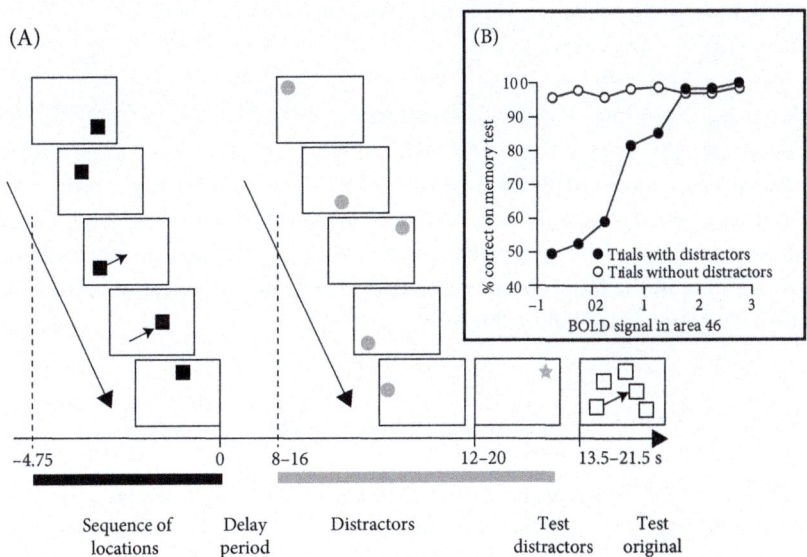

Figure 6.9 (A) The task used in an imaging experiment to test for the effect of distraction in memory. (B) The relation between sustained activation during the delay and the accuracy of performance on trials on which there were no distractors and trials on which there were distractors

Adapted from Sakai K, Rowe J.B., & Passingham R.E. Active maintenance in prefrontal area 46 creates distractor-resistant memory. *Nat Neurosci*, 5 (5), 479–84, Figures 1 and 5, Doi: 10.1038/nn846 Copyright © 2002, Springer Nature.

The subjects saw a sequence of four letters, and there was then a variable memory delay of 8 to 16 seconds. On the high distraction condition another sequence of four different letters followed, whereas on the low distraction condition it was a series of four numbers. Memory for the original sequence was then tested as in the previous experiment.

At the time of the memory test, there was significantly more activation in the PF area 46 in the high than low distraction condition. Sakai and Passingham interpreted this to suggest that this PF area plays a role in distinguishing between the items to be remembered and the distractors in memory.

However, there was also a signal in the hippocampus at the time of the memory test. In a related experiment, Sakai et al. (2002b) presented a sequence of five letter followed by a variable memory delay. After the delay, on some trials five numbers were presented and the subject had to add them before being tested for their memory of the letter sequence. In the control condition, five zeroes were presented, and this meant that the subjects could continue rehearsing the letters until the test for memory of the original sequence.

At the time of the test, Sakai et al. reported an activation that they located as being in the parahippocampal gyrus. However, it is clear from the recent atlas by Petrides (2018) that the peak was actually in the hippocampus itself. This activation reflected the re-activation of the letter sequence after distraction.

Summary

Human subjects can mentally rehearse sequences. This rehearsal acts to protect the items in memory against distraction. The PF area 46 acts to distinguish between the relevant items and distractors in memory. At the time of the memory test there is also a signal in the hippocampus, and this reflects the re-activation of the material that was presented before the distractor items.

Generating Sequences

In the experiments that have been described so far, the order of events was determined by the experimenter. But both monkeys and human subjects can also generate ordered sequences of their own accord.

Watanabe et al. (2006) devised a free choice task for monkeys. Four locations were lit, and after a delay the animal was given a free choice to make a saccade at the end of the delay to whichever location, they had chosen. In a comparison condition, one of the four light lit to specify the location to which the monkey should respond.

Many of the cells in the dorsal PF cortex showed directional selectivity and this meant that it was possible to study the evolution of that selectivity during the delay on the task on which the animals had a free choice. There were 58 cells that had directionally selective activity during the cue-period on the externally-specified task; but of these only five did so on the free-choice task. On both tasks, during the delay period, there was a gradual increase in the specificity for coding the direction of the response; but this ramping activity was more marked on the free-choice task.

Unfortunately, Watanabe and Funahashi (2006) failed to analyse their data to see to what extent the choice that the animal made on trial one depends on its choice on trial n−1. However, Barraclough et al. (2004) designed an experiment in which a computer programme determined whether a choice would be rewarded, and the rewards changed over time so as to encourage the monkey to switch to the alternative target. A significant proportion of the cells in the dorsal PF cortex encoded the choice on the previous trial. These cells continued to do so throughout the delay before the monkey made its next choice.

Genovesio et al. (2006a) used a different design to study free choice. They taught monkeys to choose among three targets, and as in the study by Barraclough et al. (2004) a reward might or might not follow a given choice. When the subjects saw a 'stay' cue, the experimenters rewarded a saccade to the same target as on the previous trial. But, when the subjects saw a 'change' cue, the subjects never received a reward for repeating their previous choice. Instead, they had a free choice between two remaining spatial goals. Many cells in the dorsal PF cortex encoded the location of the goal; some did so for the goal on the previous trial, and some for the goal on the current trial. Tsujimoto and Sawaguchi (2004) also found a subset of cells in the dorsal PF cortex that coded for the conjunction of the previous response and its outcome.

This finding made it possible for Tsujimoto et al. (2008) to study pairs of cells in the dorsal PF cortex, with each member of the pair encoding either the previous goal or the future goal. At the time that monkeys selected the current goal, pairs of cells encoding that goal exhibited transient synchrony and usually represented the same goal.

The more interesting pairs consisted of one cell that encoded the previous goal and the other the current goal. These cells also showed transient synchrony when monkeys chose the current goal, and these correlations became stronger when monkeys changed goals as opposed to repeating a previous one. Such pairs usually differed in the spatial goal they encoded. Tsujimoto et al. concluded that their correlated activity could enhance the transformation of a previous goal into a prospective code for the current goal.

Unlike macaque monkeys, human subjects can choose freely without the confounding factor of instrumental conditioning and therefore of rewards. In an fMRI study by Rowe et al. (2008), the subjects were given a free choice; in one experiment they chose which of two buttons to press, and in the other they chose between four buttons. In the control task, the choice was externally specified by a light. The analysis of the fMRI data contrasted these two conditions. As expected from the single unit data, there were activations in the dorsal PF cortex.

Thus, the results of cell recording in monkeys and fMRI with human subjects agree in suggesting that the dorsal PF cortex plays a role in generating goals. However, it is important to note that in all these experiments the subjects were required to generate a sequence of goals, and thus that the movements be varied from trial to trial. So, it is not clear whether the activity or activation reflects free choice or the need to produce a varying sequence.

To find out, Rowe et al. (2010) gave human subjects a free choice as to which finger to move, but they examined activation on the first trial alone. They did this for data that had been accumulated for fifty-seven subjects. The critical observation is that on the first trial there was no significant activation in either the PF areas 46 or 9/46. To meet the objection that this might be due to a lack of statistical power, Rowe et al. selected a single trial in the middle of the series; they were able to show that on this trial they could detect significant activation in the dorsal PF cortex. So, the lack of activation on the first trial was not due to a lack of statistical power; instead, it reflected the fact that as yet there was no sequence.

Rowe et al. (2010) calculated an 'equitability index' to measure the degree to which the sequences were random. This is a measure of the evenness of the distribution of the responses. If the distribution was random, the response on trial n should be independent of the response on trial n-1. The index is shown for an activation in the dorsal PF area 46 (Figure 6.10A). This activation was greater the smaller the value of this index (Figure 6.10B). This means that the PF cortex was engaged the more that the response on trial n was influenced by the response on

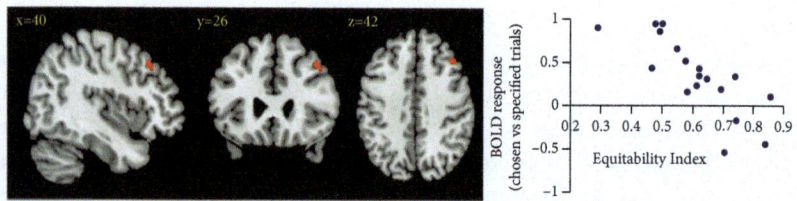

Figure 6.10 (A) The activation in the dorsal PF cortex on a task in which human subjects decide of their own accord which finger to move. (B) The plot of the degree of activation as a function of the equipotentiality index

Reproduced from Rowe, J.B., Hughes, L., & Nimmo-Smith, I. Action selection: A race model for se-lected and non-selected actions distinguishes the contribution of premotor and prefrontal areas. *Neuroimage*, 51 (2), 888–96, Figure 3, Doi: 10.1016/j.neuroimage.2010.02.045 Copyright © 2010 The Authors. Licensed under CC BY 40.

trial n-1. So, the subjects were trying to vary their responses rather than flipping a coin independently on each trial.

In a follow-up study using magnetoencephalography (MEG), Phillips et al. (2018) measured 'selection entropy'. The less the selection entropy, the more the subjects were monitoring their previous finger movements. Consistent with the results from the fMRI study, the activity in the dorsal PF cortex was greater the less the selection entropy.

Because MEG has a much better temporal resolution than fMRI, it was possible to analyse the time-course of the activity. It turned out that there was a negative correlation between choice entropy and activity both before and after the subjects made their response. So, the activity in the dorsal PF cortex reflected monitoring of the current choice once it had been made, as well as monitoring of previous choices.

Summary

When monkeys are given a free choice as to which target to select, the activity of cells in the dorsal PF cortex ramps up during the delay. There is cell activity that codes for the choice on the previous trial, the outcome of that choice, and the transformation from the previous to the current choice. When human subjects are scanned while they are given a free choice as to which of four fingers to move, there is activation in the dorsal PF cortex. However, there is no activation on the first trial, since at this point there is no sequence. Studies using fMRI and MEG show that the activity in these areas is greater the more the subjects monitor what they have done on previous trials.

Planning Sequences

The last section describes delay-period activity or activation in the dorsal PF cortex when subjects generate sequences of their own accord. These are influenced by the choice on the previous trial. However, when subjects generate sequences it is also possible to find activity in the dorsal PF cortex that reflects planning future moves.

Averbeck et al. (2006) taught monkeys sequences of three eye movements that they learned by trial and error. The monkeys were able to learn such short sequences within just a few trials. There was cell activity in the dorsal PF area 9/46 that encoded the specific sequence that the monkey planned, and that did so just before the animal performed the sequence. Furthermore, by using a simple decoding algorithm, it was possible to predict when the monkey would make errors and what that error would be. These results are evidence that the monkeys planned a sequence of movements in advance.

The sequences in the study by Averbeck et al. (2006) involved eye movements. Shima et al. (2007) also taught monkeys sequences, but these differed in that they involved actions performed with the hand. The monkeys manipulated a handle which they could turn, push, or pull. Each sequence was made up of four movements. The task is shown in Figure 6.11A. On each day, the monkey learned a sequence by following visual cues, and then repeated it from memory.

Over 40% of the task-related cells showed activity just before the monkeys generated the sequences from memory. Since Shima et al. trained the monkeys on eleven sequences in all, this meant that they could compare the activity that occurred prior to each sequence.

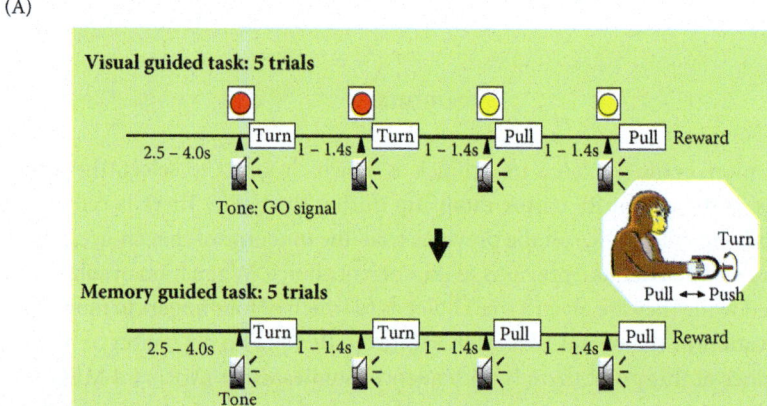

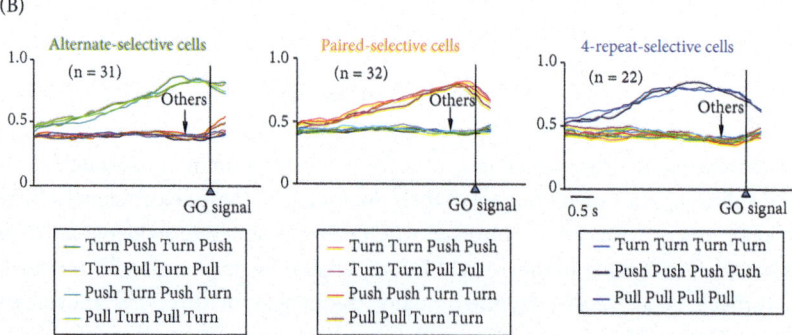

Figure 6.11 (A) Task in which the monkeys used a handle to make sequences of four movements. (B) Activity of cells that encoded sequences with a particular abstract structure

For description see text.

Reproduced from Tanji, J., Shima, K., & Mushiake, H. Concept-based behavioral planning and the lateral prefrontal cortex. *Trends Cog Sci*, 11 (12), 528–34, Figure 1, https://doi.org/10.1016/j.tics.2007.09.007 Copyright © 2007, Elsevier Ltd. All rights reserved.

Figure 6.11B shows that there were cells that coded for the sequences, but that they coded not for the specific movements but for abstract structure of the sequences. There were cells that coded for the abstract structure ABAB (left), AABB (middle), and AAAA (right). These cells did not code for sequences with a different structure. The sequences were abstract because, for example, the same cell increased its activity for 'turn, push, turn, push' and 'pull, turn, pull, turn'. Chapter 8 returns to the issue of abstract coding and argues that it is critical for understanding what is special about the prefrontal cortex when considered as a whole.

In these sequences, it was the experimenter who specified the order of elements in a sequence. But one can modify the task so that the monkey starts with a goal and must work out a sequence of moves that are needed to achieve it. For example, Mushiake et al. (2001) presented monkeys with a visual maze (Figure 6.12) and required them to use a handle to move a cursor through the maze so as to reach a final goal. It is important to note that there was no requirement for short-term memory since the locations and the cursor were all in full view throughout.

Once the monkeys had learned a sequence that reached the goal, the experimenters introduced a visual obstacle so that the monkeys had to plan a new route. This meant that the animals had to plan a series of subgoals of their own accord. Mushiake et al. (2006) recorded cell activity in the dorsal PF cortex, focusing their analysis on the delay period just prior to the movement of the cursor. During this period, the cells encoded the movement of the cursor. And, if achieving the final goal required a sequence of three subgoals, different subpopulations of cells coded for each subgoal (Figure 6.13A, next page).

But it could be that the cell activity coded for the movement of the handle rather than the cursor. To check, Mushiake et al. (2006) manipulated the spatial transform between the movements of the handle and cursor. For example, movement of

Figure 6.12 Visual maze task
The monkey needed to make hand movements that drove a cursor from its current position on a trial (grey square) to a final goal (black square).
Redrawn from Mushiake H., Saito N., Sakamoto K., Sato Y., & Tanji J. Visually based path-planning by Japanese monkeys. *Cogn Brain Res*, 11 (1), 165–9, Figure 1a, https://doi.org/10.1016/S0926-6410(00)00067-7 Copyright © 2001, Elsevier Science B.V. All rights reserved.

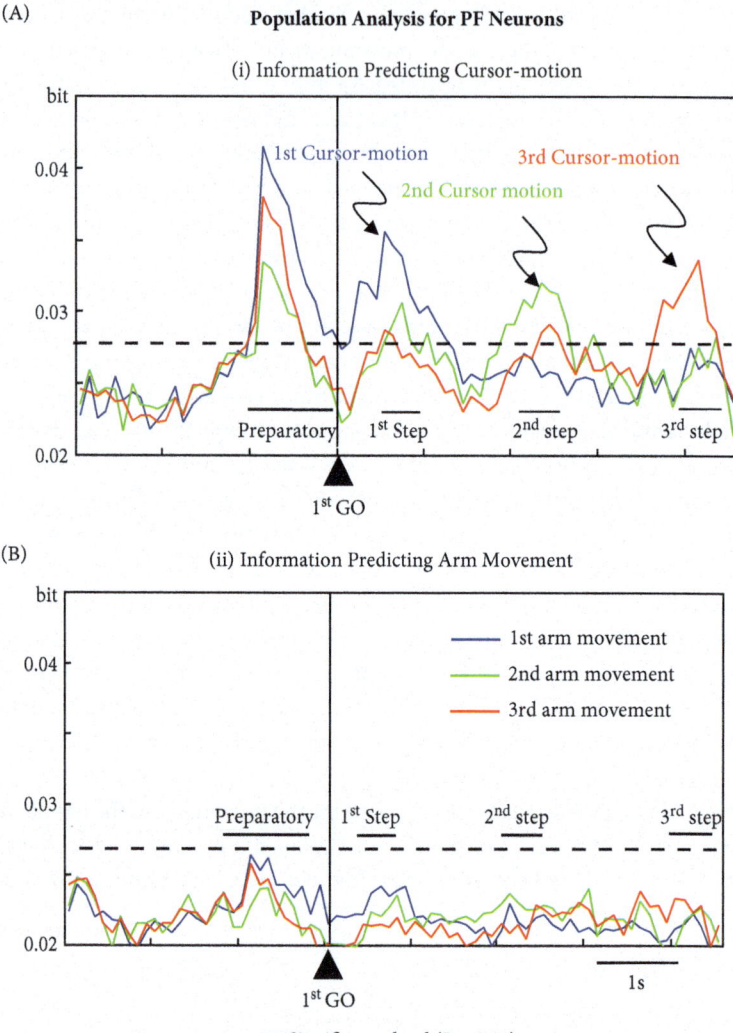

Figure 6.13 Population analysis of cells in the dorsal PF cortex during planning A = time course of information that predicted the first (plotted in blue), second (green), and third (red) movements of the cursor (goals). B = time course of information predicting the direct of the movement of the arm (kinematics).

Reproduced from Hajime Mushiake, Naohiro Saito, Kazuhiro Sakamoto, Yasuto Itoyama, & Jun Tanji. Activity in the lateral prefrontal cortex reflects multiple steps of future events in action plans. *Neuron*, 50 (4), 631–41, Figure 5a, https://doi.org/10.1016/j.neuron.2006.03.045 Copyright © 2006, Elsevier Inc. All rights reserved.

the handle to the left could cause the cursor to move either to the left or to the right, depending on the spatial transform in operation at the moment.

The results of this control are shown in Figure 6.13B (page 224). The population of cells did not code for the direction of the arm movement. In other words, the cells coded for the goals or targets, not for the movements that were required to reach them.

To solve the problem posed by the maze task, the monkey had to calculate both subgoals and final goals. It has been shown that there are cells in the dorsal PF cortex that code for one, the other, or for both (Saito et al., 2005). This made it possible for Sakamoto et al. (2008) to isolate pairs of cells, one coding for a subgoal and the other for the final goal. The synchrony in the discharge between the pairs of cells peaked when a transition occurred between one cell encoding a subgoal and the other encoding the final goal. This finding is reminiscent of the demonstration of cell activity coding for the transformations from a previous goal to the current one (Tsujimoto et al., 2008).

Summary

When monkeys learn sequences of eye movements, it is possible to predict to use the cell activity in the dorsal PF cortex to predict when they are going to make an error. This is evidence that the monkeys planned the sequences. A large proportion of the cells in the dorsal PF cortex also encode the abstract structure of sequences of hand movements and do so before any movements have been performed. Subpopulations of the cells code for specific subgoals during a sequence; these code for the subgoals, not for the hand movements themselves.

Conclusions

It has long been thought that the dorsal PF cortex supports working memory (Goldman-Rakic, 1995). However, the term 'working memory' meant no more than short-term retrospective memory. The assumption was that monkeys with lesions of the dorsal PF cortex fail DR and DA because they do not remember what they have just seen (DR) or done (DA). The critical finding was that monkeys with these lesions could succeed if there was no delay (Goldman & Rosvold, 1970; Passingham, 1985).

But though this shows that a delay is necessary, it does not show that it is sufficient. The evidence reviewed in this chapter indicates that short-term memory for spatial items depends on the caudal PF cortex and the parietal area LIP. It is these areas that support the 'visuo-spatial' scratchpad that forms part of the working memory model introduced by Baddeley and his colleagues (Baddeley & Hitch, 1974; Baddeley & Logie, 1999).

The dorsal PF cortex is only recruited when the memory involves the order or timing of the items in a sequence. Rather than continuing to use the term working memory (Miller et al., 2018), it would advance mental hygiene if the term 'order memory' was used.

It could be objected that the term working memory is still useful because the working memory model has a central executive, and this could account for the evidence that there is prospective activity in the dorsal PF cortex that is involved in preparation or planning. But, as Baddeley (1996) himself admitted, the central executive was introduced as an inner homunculus. Yet, neuroscience cannot tolerate a homunculus because it suggests that operations could just happen without cause.

The way to banish a homunculus is to show how areas *transform* inputs to outputs. On the DR task, there is a transformation from viewing an action to generating one. On the DA task, there is a transformation from performing one action to generating the alternative one.

Both the dorsal PF areas 46 and 9/46 are critical for performance of these tasks. For example, on the DR task the monkey watches while the experimenter baits one of the two foodwells. Area 9/46 can receive information about the action as viewed from the upper bank of the STS (Jellema & Perrett, 2003), whereas area 46 can receives this information from the inferior parietal area PFG (Fogassi et al., 2005).

However, it is important to recognize that 9/46 and 46 have complementary functions on the DR and DA tasks. Area 9/46 is equipped to generate the goal for saccadic eye movements, whereas area 46 is equipped to generate a goal for reaching. Since on the classical versions of these tasks the response is to move the cover off one of the two foodwells, it is area 46 that is essential for correct performance since it is this area that projects to the hand and arm regions of the various premotor areas.

Nonetheless, areas 46 and 9/46 are, of course, closely interconnected (Borra et al., 2019). It is for this reason that when cell recordings are taken in the intact brain, cells with similar properties can be found in both areas. As Chapter 1 explains, some of the activity in either area will be derived from the other one.

If fMRI is used to scan human subjects while they generate sequences, there is no activation on the first trial (Rowe et al., 2010). The reason is that as yet there is no sequence and thus no order. But when monkeys learn the DR and DA tasks, they are tested for many trials a day, and this means that the animals have to generate a sequence of choices. The consequence is that there is scope for interference in memory. The correct response depends on what the animals last saw (DR) or did (DA), and it is easy for the monkeys to be confused because only two locations are used throughout testing.

The ability to retrieve the order of events can be tested more directly by presenting either a series of locations or objects and requiring the animal to recall the sequence. As expected, there is cell activity dorsal PF cortex when monkeys retrieve the order, either by reaching to the objects in the order presented (Ninokura et al., 2003) or by making a saccade to the objects in the order presented (Warden & Miller, 2010). These studies are like the recall condition in the studies on human subjects Pochon et al. (2001; Volle et al., 2005). When memory is assessed by recall, the subjects know in advance what responses they are going to be required to make; and this means that they can code for them prospectively.

There is therefore a similarity between the dorsal and medial PF cortex. Both are interconnected indirectly with the hippocampus. And both support both the memory of the order of events in the past and planning a sequence of actions for the future. One difference is that the dorsal PF cortex does this for the short term whereas the medial PF cortex does so for the long term.

When subjects have to remember the order of events, the goal at time t depends on or *is conditional* on an event or action that occurred at time t-n. On the DR and DA task n is typically 5 seconds; on the n-back task, the target is also conditional on an event at t-n, but in this case, n is the number of items. These are conditional tasks.

However, there is a difference between the experiments with monkeys and human subjects; and this is that the monkeys are taught via rewarding them. This means that on the DR and DA tasks the current goal is conditional on the *conjunction* of what the animal saw or did and the outcome; and there are cells in the dorsal PF cortex that code both for the last goal and its outcome (Tsujimoto & Sawaguchi, 2004). The population of cells that encodes a previous goal differs from the population that encodes a current goal (Genovesio et al., 2006a). Separate neural encoding of current and previous goals is essential to order conditionality because the relevant transforms requires distinguishing input representations from output representations. However, there is a correlation between the activity of the population of cells that codes for the last goal and the population that codes for the current goal (Tsujimoto et al., 2011), and this correlation is greater when the monkeys switch between goals.

So, the fundamental role of the dorsal PF cortex is to perform a *transformation* from previous goals to current goals, and to do so across time. The reason why it supports sequences is that both arm and eye movements necessarily follow each other in time.

Selective Advantage

Primates search for the fruit and leaves that are currently edible, and as the previous chapter explains, these are found in a cluttered environment. Once a piece of fruit or a leaf is identified as desirable, primates retrieve it by using their hands. However, for many primates the heat stress means that it is essential that they search in an efficient way. The effect is that it pays to plan the search rather than picking randomly. Furthermore, as on the laboratory search task, it will not pay to return to locations from which they have already picked the fruit and leaves. Additionally, when foraging for distant resources, it is important to remember the order and recency with which they have exploited patches of resources because they take time to replenish. So, the current action depends on encoding the order of previous actions.

Furthermore, anthropoids live in groups and this means that there is competition for food within the group. This means that the animals need to watch where others are foraging and avoid the sites that the more dominant animals are exploiting. Thus, the current action also depends on events that the animals view.

Connectivity

Among its many connections, the dorsal PF cortex has five that, collectively, make it unique. It has (1) inputs from the superior bank of the STS and the inferior parietal area PFG; (2) inputs to area 9/46 from the cortex in the IPS; (3) indirect inputs from the inferior temporal cortex that run via the ventral PF cortex; (4) interconnections with the orbital PF cortex; (5) connections with the retrosplenial cortex and thus the hippocampus; and (6) outputs to the oculomotor system and system for manual reaching.

Neural Operation

Like previous chapters, this one offers a proposal concerning the transformation that the area performs.

Area 9/46 receives inputs from the STS that carry information about actions as viewed; and area 46 has access to similar information via inputs from area PFG. Indirect inputs from the inferior temporal cortex carry information about objects as viewed. Connections with the parietal area LIP and hippocampus specify the temporal order of events and actions. In combination, areas 46 and 9/46 encode the order of events and actions in short-term memory in a way that protects against confusion in memory. Area 46 is also connected with the orbital PF cortex, and this connection carries information about food and liquids.

In combination areas 46 and 9/46 can therefore code conjunctively for the goal and the outcome. Areas 46 and 9/46 transform the conjunction of the memory of the previous goal or sequence of goals into a prospective code, and this can be reactivated if there is a distracting event. There are cells that code for the previous goal and outcome, and their activity correlates with the activity of cells that encode the next goal.

Areas 46 and 9/46 combine in generating the next goal for reaching and viewing. These goals are achieved through the outputs from area 46 to the premotor areas and from area 9/46 to the FEF. The outputs to the retrosplenial cortex and thus to the hippocampal system support the interaction between the mechanisms for recall of the order of events from short-term and long-term memory.

References

Aghdaee, S.M., Battelli, L., & Assad, J.A. (2014) Relative timing: From behaviour to neurons. *Philos Trans R Soc Lond B Biol Sci*, 369, 20120472.

Ainsworth, M., Browncross, H., Mitchell, D.J., Mitchell, A.S., Passingham, R.E., Buckley, M.J., Duncan, J., & Bell, A.H. (2018) Functional reorganisation and recovery following cortical lesions: A preliminary study in macaque monkeys. *Neuropsychol*, 119, 382–91.

Amiez, C. & Petrides, M. (2007) Selective involvement of the mid-dorsolateral prefrontal cortex in the coding of the serial order of visual stimuli in working memory. *Proc Natl Acad Sci USA*, 104, 13786–91.

Averbeck, B.B., Sohn, J.W., & Lee, D. (2006) Activity in prefrontal cortex during dynamic selection of action sequences. *Nat Neurosci*, 9, 276–82.

Baddeley, A. (1996) Exploring the central executive. *Q J Exp Psychol*, 49, 5–28.

Baddeley, A. & Logie, R.H. (1999) Working-memory: The multiple-component model. In Miyake, A., Shah, P. (eds) *Models of Working Memory: Mechanisms of Active Maintenance and Executive Control*. Cambridge University Press, Cambridge, pp. 28–61.

Baddeley, A.D. & Hitch, G. (1974) Working memory. In Bower, G.H. (ed.) *The Psychology of Learning and Motivation*. Academic Press, New York, pp. 47–90.

Baddeley, A.D. & Hitch, G.J. (2019) The phonological loop as a buffer store: An update. *Cortex*, 112, 91–106.

Barraclough, D.J., Conroy, M.L., & Lee, D. (2004) Prefrontal cortex and decision making in a mixed-strategy game. *Nat Neurosci*, 7, 404–10.

Bastos, A.M., Loonis, R., Kornblith, S., Lundqvist, M., & Miller, E.K. (2018) Laminar recordings in frontal cortex suggest distinct layers for maintenance and control of working memory. *Proc Natl Acad Sci USA*, 115, 1117–22.

Bisley, J.W. & Goldberg, M.E. (2010) Attention, intention, and priority in the parietal lobe. *Annu Rev Neurosci*, 33, 1–21.

Blacker, K.J. & Courtney, S.M. (2016) Distinct neural substrates for maintaining locations and spatial relations in working memory. *Front Hum Neurosci*, 10, 594.

Bor, D., Duncan, J., Lee, A.C., Parr, A., & Owen, A.M. (2006) Frontal lobe involvement in spatial span: Converging studies of normal and impaired function. *Neuropsychol*, 44, 229–37.

Borra, E., Ferroni, C.G., Gerbella, M., Giorgetti, V., Mangiaracina, C., Rozzi, S., & Luppino, G. (2019) Rostro-caudal connectional heterogeneity of the dorsal part of the macaque prefrontal area 46. *Cereb Cortex*, 29, 485–504.

Borra, E., Gerbella, M., Rozzi, S., Tonelli, S., & Luppino, G. (2014) Projections to the superior colliculus from inferior parietal, ventral premotor, and ventrolateral prefrontal areas involved in controlling goal-directed hand actions in the macaque. *Cereb Cortex*, 24, 1054–65.

Buschman, T.J. & Miller, E.K. (2020) How working memory works. In Poeppel, D., Mangun, G.R., Gazzaniga, M. (eds) *The Cognitive Neurosciences*. MIT Press, Cambridge.

Butters, N. & Pandya, D. (1969) Retention of delayed-alternation: Effect of selective lesions of sulcus principalis. *Science*, 165, 1271–3.

Carpenter, A.F., Baud-Bovy, G., Georgopoulos, A.P., & Pellizzer, G. (2018) Encoding of serial order in working memory: Neuronal activity in motor, premotor, and prefrontal cortex during a memory scanning task. *J Neurosci*, 38, 4912–33.

Chafee, M.V. & Goldman-Rakic, P.S. (1998) Matching patterns of activity in primate prefrontal area 8a and parietal area 7ip neurons during a spatial working memory task. *J Neurophysiol*, 79, 2919–40.

Charles, D.P., Gaffan, D., & Buckley, M.J. (2004) Impaired recency judgments and intact novelty judgments after fornix transection in monkeys. *J Neurosci*, 24, 2037–44.

Cieslik, E.C., Zilles, K., Caspers, S., Roski, C., Kellermann, T.S., Jakobs, O., Langner, R., Laird, A.R., Fox, P.T., & Eickhoff, S.B. (2013) Is there 'one' DLPFC in cognitive action control? Evidence for heterogeneity from co-activation-based parcellation. *Cereb Cortex*, 23, 2677–89.

Cohen, J.D., Forman, S.D., Braver, T.S., Casey, B.J., Servan-Schreiber, D., & Noll, D.C. (1994) Activation of the prefrontal cortex in a nonspatial working memory task with functional MRI. *Hum Brain Map*, 1, 293–304.

Collin, N.G., Cowey, A., Latto, R., & Marzi, C. (1982) The role of frontal eye-fields and superior colliculi in visual search and non-visual search in rhesus monkeys. *Behav Brain Res*, 4, 177–93.

Constantinidis, C., Funahashi, S., Lee, D., Murray, J.D., Qi, X.L., Wang, M., & Arnsten, A.F.T. (2018) Persistent spiking activity underlies working memory. *J Neurosci*, 38, 7020–8.

Cruzado, N.A.Z., Tikanj, Z., Brincat, S.L., Miller, E.K., & Howard, M.W. (2020) Conjunctive representation of what and when in monkey hippocampus and lateral prefrontal cortex during an associative memory task. *Hippocampus*, Doi: 10.1002/hipo.23282.

Diamond, A. & Goldman-Rakic, P.S. (1989) Comparison of human infants and rhesus monkeys on Piaget's AB task: Evidence for dependence on dorsolateral prefrontal cortex. *Exper Brain Res*, 74, 24–40.

Ferreira, C.T., Verin, M., Pillon, B., Levy, R., Dubois, B., & Agid, Y. (1998) Spatio-temporal working memory and frontal lesions in man. *Cortex*, 34, 83–98.

Finnerty, G.T., Shadlen, M.N., Jazayeri, M., Nobre, A.C., & Buonomano, D.V. (2015) Time in cortical circuits. *J Neurosci*, 35, 13912–16.

Fogassi, L., Ferrari, P.F., Gesierich, B., Rozzi, S., Chersi, F., & Rizzolatti, G. (2005) Parietal lobe: From action organization to intention understanding. *Science*, 308, 662–7.

Fogassi, L., Gallese, V., Buccino, G., Craighero, L., Fadiga, L., & Rizzolatti, G. (2001) Cortical mechanism for the visual guidance of hand grasping movements in the monkey: A reversible inactivation study. *Brain*, 124, 571–86.

Fries, W. (1984) Cortical projections to the superior colliculus in the macaque monkey: A retrograde study using horseradish peroxidase. *J Comp Neurol*, 230, 55–76.

Funahashi, S., Inoue, M., & Kubota, K. (1997) Delay-period activity in the primate prefrontal cortex encoding multiple spatial positions and their order of presentation. *Behav Brain Res*, 84, 203–23.

Genovesio, A., Brasted, P.J., & Wise, S.P. (2006a) Representation of future and previous spatial goals by separate neural populations in prefrontal cortex. *J Neurosci*, 26, 7305–16.

Genovesio, A., Tsujimoto, S., & Wise, S.P. (2006b) Neuronal activity related to elapsed time in prefrontal cortex. *J Neurophysiol*, 95, 3281–5.

Genovesio, A., Tsujimoto, S., & Wise, S.P. (2009) Feature- and order-based timing representations in the frontal cortex. *Neuron*, 63, 254–66.

Gerbella, M., Borra, E., Tonelli, S., Rozzi, S., & Luppino, G. (2013) Connectional heterogeneity of the ventral part of the macaque area 46. *Cereb Cortex*, 23, 967–87.

Goldman, P.S. & Rosvold, H.E. (1970) Localization of function within the dorsolateral prefrontal cortex of the rhesus monkey. *Exper Neurol*, 27, 291–304.

Goldman, P.S., Rosvold, H.E., Vest, B., & Galkin, T.W. (1971) Analysis of the delayed-alternation deficit produced by dorsolateral prefrontal lesions in the rhesus monkey. *J Comp Physiol Psychol*, 77, 212–20.

Goldman-Rakic, P.S. (1995) Cellular basis of working memory. *Neuron*, 14, 477–85.

Gorgolewski, K.J. & Poldrack, R.A. (2016) A practical guide for improving transparency and reproducibility in neuroimaging research. *PLoS Biol*, 14, e1002506.

Harlow, H.F. & Akert, K. (eds) (1964) *The Effect of Bilateral Prefrontal Lesions on Learned Behavior of Neonatal, Infant and Preadolsecent Monkeys.* McGraw-Hill, New York.

He, S.Q., Dum, R.P., & Strick, P.L. (1995) Topographic organization of corticospinal projections from the frontal lobe: Motor areas on the medial surface of the hemisphere. *J Neurosci*, 15, 3284–306.

Heuer, E. & Bachevalier, J. (2013) Working memory for temporal order is impaired after selective neonatal hippocampal lesions in adult rhesus macaques. *Behav Brain Res*, 239, 55–62.

Huxlin, K.R., Saunders, R.C., Marchionini, D., Pham, H.A., & Merigan, W.H. (2000) Perceptual deficits after lesions of inferotemporal cortex in macaques. *Cereb Cortex*, 10, 671–83.

Jacob, S.N. & Nieder, A. (2014) Complementary roles for primate frontal and parietal cortex in guarding working memory from distractor stimuli. *Neuron*, 83, 226–37.

Jacob, S.N., Stalter, M., & Nieder, A. (2016) Cell-type-specific modulation of targets and distractors by dopamine D1 receptors in primate prefrontal cortex. *Nature Comm*, 7, 13218.

Jazayeri, M. & Shadlen, M.N. (2015) A Neural Mechanism for Sensing and Reproducing a Time Interval. *Curr Biol*, 25, 2599–609.

Jellema, T., Baker, C.I., Wicker, B., & Perrett, D.I. (2000) Neural representation for the perception of the intentionality of actions. *Brain Cogn*, 44, 280–302.

Jellema, T. & Perrett, D.I. (2003) Cells in monkey STS responsive to articulated body motions and consequent static posture: A case of implied motion? *Neuropsychol*, 41, 1728–37.

Kobayashi, Y. & Amaral, D.G. (2003) Macaque monkey retrosplenial cortex: II. Cortical afferents. *J Comp Neurol*, 466, 48–79.

Kobayashi, Y. & Amaral, D.G. (2007) Macaque monkey retrosplenial cortex: III. Cortical efferents. *J Comp Neurol*, 502, 810–33.

Leon, M.I. & Shadlen, M.N. (1999) Effect of expected reward magnitude on the response of neurons in the dorsolateral prefrontal cortex of the macaque. *Neuron*, 24, 415–25.

Leon, M.I. & Shadlen, M.N. (2003) Representation of time by neurons in the posterior parietal cortex of the macaque. *Neuron*, 38, 317–27.

Leung, H.C., Gore, J.C., & Goldman-Rakic, P.S. (2002) Sustained mnemonic response in the human middle frontal gyrus during on-line storage of spatial memoranda. *J Cogn Neurosci*, 14, 659–71.

Lundqvist, M., Herman, P., & Miller, E.K. (2018a) Working memory: Delay activity, yes! Persistent activity? Maybe Not. *J Neurosci*, 38, 7013–19.

Lundqvist, M., Herman, P., Warden, M.R., Brincat, S.L., & Miller, E.K. (2018b) Gamma and beta bursts during working memory readout suggest roles in its volitional control. *Nature Comm*, 9, 394.

Lundqvist, M., Rose, J., Herman, P., Brincat, S.L., Buschman, T.J., & Miller, E.K. (2016) Gamma and beta bursts underlie working memory. *Neuron*, 90, 152–64.

Luppino, G., Matelli, M., Camarda, R.M., Gallese, V., & Rizzolatti, G. (1991) Multiple representations of body movements in mesial area 6 and the adjacent cingulate cortex: An intracortical microstimulation study in the macaque monkey. *J Comp Neurol*, 311, 463–82.

Manohar, S.G., Zokaei, N., Fallon, S.J., Vogels, T.P., & Husain, M. (2019) Neural mechanisms of attending to items in working memory. *Neurosci Biobehav Rev*, 101, 1–12.

Masse, N.Y., Yang, G.R., Song, H.F., Wang, X.J., & Freedman, D.J. (2019) Circuit mechanisms for the maintenance and manipulation of information in working memory. *Nat Neurosci*, 22, 1159–67.

Miller, E.K., Lundqvist, M.O., & Bastos, A.M. (2018) Working memory 2.0. *Neuron*, 100, 463–75.

Milner, B., Corsi, P., & Leonard, G. (1991) Frontal-lobe contribution to recency judgements. *Neuropsychol*, 29, 601–18.

Mishkin, M., Vest, B., Waxler, M., & Rosvold, H.E. (1969) A re-examination of the effects of frontal lesions on object alternation. *Neuropsychol*, 7, 357–64.

Morecraft, R.J., Stilwell-Morecraft, K.S., Cipolloni, P.B., Ge, J., McNeal, D.W., & Pandya, D.N. (2012) Cytoarchitecture and cortical connections of the anterior cingulate and adjacent somatomotor fields in the rhesus monkey. *Brain Res Bull*, 87, 457–97.

Morris, R., Pandya, D.N., & Petrides, M. (1999) Fiber system linking the mid-dorsolateral frontal cortex with the retrosplenial/presubicular region in the rhesus monkey. *J Comp Neurol*, 407, 183–92.

Murata, A., Gallese, V., Luppino, G., Kaseda, M., & Sakata, H. (2000) Selectivity for the shape, size and orientation of objects for grasping in neurones of monkey parietal AIP. *J Neurophysiol*, 83, 2580–601.

Murray, E.A. & Mishkin, M. (1984) Relative contributions of SII and area 5 to tactile discrimination in monkeys. *Behav Brain Res*, 11, 6–83.

Mushiake, H., Saito, N., Sakamoto, K., Itoyama, Y., & Tanji, J. (2006) Activity in the lateral prefrontal cortex reflects multiple steps of future events in action plans. *Neuron*, 50, 631–41.

Mushiake, H., Saito, N., Sakamoto, K., Sato, Y., & Tanji, J. (2001) Visually based pathplanning by Japanese monkeys. *Brain Res Cogn Brain Res*, 11, 165–9.

Naya, Y., Chen, H., Yang, C., & Suzuki, W.A. (2017) Contributions of primate prefrontal cortex and medial temporal lobe to temporal-order memory. *Proc Natl Acad Sci USA*, 114, 13555–60.

Ninokura, Y., Mushiake, H., & Tanji, J. (2003) Representation of the temporal order of visual objects in the primate lateral prefrontal cortex. *J Neurophysiol*, 89, 2868–73.

Ninokura, Y., Mushiake, H., & Tanji, J. (2004) Integration of temporal order and object information in the monkey lateral prefrontal cortex. *J Neurophysiol*, 91, 555–60.

Nixon, P.D. & Passingham, R.E. (1999) The cerebellum and cognition: Cerebellar lesions do not impair spatial working memory or visual associative learning in monkeys. *Eur J Neurosci*, 11, 4070–80.

Osada, T., Adachi, Y., Kimura, H.M., & Miyashita, Y. (2008) Towards understanding of the cortical network underlying associative memory. *Philos Trans R Soc Lond B Biol Sci*, 363, 2187–99.

Osada, T., Adachi, Y., Miyamoto, K., Jimura, K., Setsuie, R., & Miyashita, Y. (2015) Dynamically allocated hub in task-evoked network predicts the vulnerable prefrontal locus for contextual memory retrieval in macaques. *PLoS Biol*, 13, e1002177.

Owen, A., Herrod, N.J., Menon, D.K., Clark, J.C., Downey, S.P.M.J., Carpenter, A., Minhas, P.S., Turkhemier, F.E., Williams, E.J., Robbins, T.W., Sahakian, B.J., Petrides, M., & Pichard, J.D. (1999) Redefining the functional organization of working memory processes within human lateral prefrontal cortex. *Europ J Neurosci*, 11, 567–74.

Owen, A.M., Downes, J.J., Sahakian, B.J., Polkey, C.E., & Robbins, T.W. (1990) Planning and spatial working memory following frontal lobe lesions in man. *Neuropsychol*, 28, 1021–34.

Owen, A.M., Morris, R.G., Sahakian, B.J., Polkey, C.E., & Robbins, T.W. (1996) Double dissociation of memory and executive functions in working memory tasks following frontal excisions, temporal lobe excisions or amygdala-hippocampectomy in man. *Brain*, 119, 1597–615.

Passingham, R.E. (1971) *Behavioural Changes after Lesions of Frontal Granular Cortex in Monkeys ({IMacaca mulatta})*, London, PhD thesis.

Passingham, R.E. (1985) Memory of monkeys ({IMacaca mulatta}) with lesions in prefrontal cortex. *Beh Neurosci*, 99, 3–21.

Passingham, R.E. & Rowe, J.B. (2016) *A Short Guide to Brain Imaging*. Oxford University Press, Oxford.

Passingham, R.E. & Wise, S.P. (2012) *The Neurobiology of Prefrontal Cortex*. Oxford University Press, Oxford.

Petrides, M. (1991) Functional specialization within the dorsolateral frontal cortex for serial order memory. *Proc Biol Sci*, 246, 299–306.

Petrides, M. (2018) *Atlas of the Morphology of the Human Cerebral Cortex on the Average MNI Brain*. Academic Press, New York.

Petrides, M. & Pandya, D.N. (1999) Dorsolateral prefrontal cortex: Comparative cytoarchitectonic analysis in the human and the macaque brain and corticocortical connection patterns. *Eur J Neurosci*, 11, 1011–36.

Petrides, M. & Pandya, D.N. (2002) Comparative cytoarchitectonic analysis of the human and macaque ventrolateral prefrontal cortex and corticocortical connection pattern in the monkey. *Eur J Neurosci*, 16, 291–310.

Phillips, H.N., Cope, T.E., Hughes, L.E., Zhang, J., & Rowe, J.B. (2018) Monitoring the past and choosing the future: The prefrontal cortical influences on voluntary action. *Sci Rep*, 8, 7247.

Piaget, J. (1954) *The Construction of Reality in the Child*. Basic Books, New York.

Pochon, J.-B., Levy, R., Fossati, P., Lehericy, S., Poline, J.-B., Pillon, B., Le Bihan, D., & Dubois, B. (2002) The neural system that bridges reward and cognition in humans: An fMRI study. *Proc Nat Acad Sci*, 16, 5669–74.

Pochon, J.-B., Levy, R., Poline, J.-B., Crozier, S., Lehericy, S., Pillon, B., Deweer, B., Bihan, D.L., & Dubois, B. (2001) The role of dorsolateral prefrontal cortex in the preparation of forthcoming actions: An fMRI study. *Cereb Cortex*, 11, 260–6.

Pribram, K.H., Mishkin, M., Rosvold, H.E., & Kaplan, S.J. (1952) Effects on delayed-response performance of lesions of dorsolateral and ventromedial frontal cortex of baboons. *J Comp Physiol Psychol*, 45, 565–75.

Qi, X.L., Elworthy, A.C., Lambert, B.C., & Constantinidis, C. (2015) Representation of remembered stimuli and task information in the monkey dorsolateral prefrontal and posterior parietal cortex. *J Neurophysiol*, 113, 44–57.

Roberts, B.M., Libby, L.A., Inhoff, M.C., & Ranganath, C. (2018) Brain activity related to working memory for temporal order and object information. *Behav Brain Res*, 354, 55–63.

Rowe, J., Hughes, L., Eckstein, D., & Owen, A.M. (2008) Rule-selection and action-selection have a shared neuroanatomical basis in the human prefrontal and parietal cortex. *Cereb Cortex*, 18, 2275–85.

Rowe, J., Toni, I., Josephs, O., Frackowiak, R.S.J., & Passingham, R.E. (2000) Prefrontal cortex: Response selection or maintenance within working memory. *Science*, 288, 1656–60.

Rowe, J.B., Hughes, L., & Nimmo-Smith, I. (2010) Action selection: A race model for selected and non-selected actions distinguishes the contribution of premotor and prefrontal areas. *Neuroimage*, 51, 888–96.

Rozzi, S., Ferrari, P.F., Bonini, L., Rizzolatti, G., & Fogassi, L. (2008) Functional organization of inferior parietal lobule convexity in the macaque monkey: electrophysiological characterization of motor, sensory responses and their correlation with cytoarchitectonic areas. *Eur J Neurosci*, 28, 1569–88.

Saito, N., Mushiake, H., Sakamoto, K., Itoyama, Y., & Tanji, J. (2005) Representation of immediate and final behavioral goals in the monkey prefrontal cortex during an instructed delay period. *Cereb Cortex*, 15, 1535–46.

Sakai, K. & Passingham, R.E. (2004) Prefrontal selection and medial temporal lobe reactivation in retrieval of short-term verbal information. *Cereb Cortex*, 14, 914–21.

Sakai, K., Rowe, J.B., & Passingham, R.E. (2002a) Active maintenance in prefrontal area 46 creates distractor-resistant memory. *Nat Neurosci*, 5, 479–84.

Sakai, K., Rowe, J.B., & Passingham, R.E. (2002b) Parahippocampal reactivation signal at retrieval after interruption of rehearsal. *J Neurosci*, 22, 6315–20.

Sakamoto, K., Mushiake, H., Saito, N., Aihara, K., Yano, M., & Tanji, J. (2008) Discharge synchrony during the transition of behavioral goal representations encoded by discharge rates of prefrontal neurons. *Cereb Cortex*, 18, 2036–45.

Saleem, K.S., Miller, B., & Price, J.L. (2014) Subdivisions and connectional networks of the lateral prefrontal cortex in the macaque monkey. *J Comp Neurol*, 522, 1641–90.

Sallet, J., Mars, R.B., Noonan, M.P., Neubert, F.X., Jbabdi, S., O'Reilly, J.X., Filippini, N., Thomas, A.G., & Rushworth, M.F. (2013) The organization of dorsal frontal cortex in humans and macaques. *J Neurosci*, 33, 12255–74.

Selemon, L.D. & Goldman-Rakic, P.S. (1988) Common cortical and subcortical targets of the dorsolateral prefrontal and parietal cortices in the rhesus monkey: Evidence for a distributed neural network subserving spatially guided behavior. *J Neurosci*, 8, 4049–68.

Shadmehr, R. & Wise, S. (2005) *The Computational Neurobiology of Reaching and Pointing*. MIT Press, Cambridge.

Sherfey, J., Ardid, S., Miller, E.K., Hasselmo, M.E., & Kopell, N.J. (2020) Prefrontal oscillations modulate the propogation of neuronal activity required for working memory. *Neurobiol Learn Mem*, 173, 107228.

Shima, K., Isoda, M., Mushiake, H., & Tanji, J. (2007) Categorization of behavioural sequences in the prefrontal cortex. *Nature*, 445, 315–18.

Siegel, M., Warden, M.R., & Miller, E.K. (2009) Phase-dependent neuronal coding of objects in short-term memory. *Proc Natl Acad Sci USA*, 106, 21341–6A.

Silk, T.J., Bellgrove, M.A., Wrafter, P., Mattingley, J.B., & Cunnington, R. (2010) Spatial working memory and spatial attention rely on common neural processes in the intraparietal sulcus. *Neuroimage*, 53, 718–24.

Stamm, J.S. (1969) Electrical stimulation of monkeys' prefrontal cortex during delayed-response performance. *J Comp Physiol Psychol*, 67, 535–46.

Stephens, D.W. & Krebs, J.R. (1986) *Foraging Theory*. Princeton University Press, Princeton.

Tachibana, Y., Nambu, A., Hatanaka, N., Miyachi, S., & Takada, M. (2004) Input-output organization of the rostral part of the dorsal premotor cortex, with special reference to its corticostriatal projection. *Neurosci Res*, 48, 45–57.

Taffe, M.A. & Taffe, W.J. (2011) Rhesus monkeys employ a procedural strategy to reduce working memory load in a self-ordered spatial search task. *Brain Res*, 1413, 43–50.

Tanila, H., Carlson, S., Linnankoski, I., & Kahila, H. (1993) Regional distribution of functions in dorsolateral prefrontal cortex of the monkey. *Behav Brain Res*, 53, 63–71.

Tanji, J., Shima, K., & Mushiake, H. (2007) Concept-based behavioral planning and the lateral prefrontal cortex. *Trends Cogn Sci*, 11, 528–34.

Tremblay, S., Saint-Aubin, J., & Jalbert, A. (2006) Rehearsal in serial memory for visual-spatial information: Evidence from eye movements. *Psychon Bull Rev*, 13, 452–57.

Tsujimoto, S., Genovesio, A., & Wise, S.P. (2008) Transient neuronal correlations underlying goal selection and maintenance in prefrontal cortex. *Cereb Cortex*, 18, 2748–61.

Tsujimoto, S., Genovesio, A., & Wise, S.P. (2011) Comparison of strategy signals in the dorsolateral and orbital prefrontal cortex. *J Neurosci*, 31, 4583–92.

Tsujimoto, S. & Sawaguchi, T. (2004) Neuronal representation of response-outcome in the primate prefrontal cortex. *Cereb Cortex*, 14, 47–55.

Upright, N.A., Brookshire, S.W., Schnebelen, W., Damatac, C.G., Hof, P.R., Browning, P.G.F., Croxson, P.L., Rudebeck, P.H., & Baxter, M.G. (2018) behavioral effect of chemogenetic inhibition is directly related to receptor transduction levels in rhesus monkeys. *J Neurosci*, 38, 7969–75.

Volle, E., Kinkingnehun, S., Pochon, J.B., Mondon, K., Thiebaut de Schotten, M., Seassau, M., Duffau, H., Samson, Y., Dubois, B., & Levy, R. (2008) The functional architecture of the left posterior and lateral prefrontal cortex in humans. *Cereb Cortex*, 18, 2460–9.

Volle, E., Pochon, J.B., Lehericy, S., Pillon, B., Dubois, B., & Levy, R. (2005) Specific cerebral networks for maintenance and response organization within working memory as evidenced by the 'double delay/double response' paradigm. *Cereb Cortex*, 15, 1064–74.

Wallis, J.D. & Miller, E.K. (2003) Neuronal activity in primate dorsolateral and orbital prefrontal cortex during performance of a reward preference task. *Eur J Neurosci*, 18, 2069–81.

Warden, M.R. & Miller, E.K. (2010) Task-dependent changes in short-term memory in the prefrontal cortex. *J Neurosci*, 30, 15801–10.

Watanabe, K., Igaki, S., & Funahashi, S. (2006) Contributions of prefrontal cue-, delay-, and response-period activity to the decision process of saccade direction in a free-choice ODR task. *Neural Netw*, 19, 1203–22.

Webster, M.J., Bachevalier, J., & Ungerleider, L.G. (1994) Connections of inferior temporal areas TEO and TE with parietal and frontal cortex in macaque monkeys. *Cer Cort*, 4, 471–83.

Yumoto, N., Lu, X., Henry, T.R., Miyachi, S., Nambu, A., Fukai, T., & Takada, M. (2011) A neural correlate of the processing of multi-second time intervals in primate prefrontal cortex. *PLoS One*, 6, e19168.

7
Ventral Prefrontal Cortex
Associating Objects

Introduction

As Chapter 2 points out, primates forage on resources that are not always dependable. There is therefore a premium on learning rapidly what types of fruit and leaves are edible, inedible, or dangerous. When they search for fruit and leaves in a patch, they need to be able to identify the same type of fruit or leaf and to learn which are alike in terms of their nutritional value.

Most anthropoid primates have trichromatic vision and the cones are especially dense in the fovea (Perry & Cowey, 1985). The inferior temporal area (TE) is specialized for colour and form vision (Huxlin et al., 2000) and it sends connections to the ventral prefrontal (PF) cortex (Petrides & Pandya, 2002). It is the ventral PF cortex, in combination with orbital PF cortex, that enables the animals to identify rapidly the types of fruit and leaves that are currently desirable and edible, since it is their colour, size, and shape that provide the best evidence.

Areas

In macaque monkeys, the ventral PF cortex lies on the inferior convexity, ventral to the principal sulcus and extending to the lateral orbital sulcus (Figure 7.1, next page). In humans, the homologous area lies in the inferior frontal gyrus.

The ventral PF cortex includes areas 47/12 and 45A (Figure 1.9, page 24). The term 47/12 reflects the view of Petrides and Pandya (2002) that area 12 in the macaque monkey brain is homologous with area 47 in the human brain. Area 47/12 lies rostrally and ventrally in both the macaque brain (Figure 1.10B) and human brain (Figure 1.10A, as shown on page 26).

It has been suggested that bushbabies also have an area 45 (Preuss & Goldman-Rakic, 1991) (Chapter 2). But it has not been shown whether this merely corresponds to the posterior area 45B in macaques or whether it also includes the more anterior area 45A (Figure 1.10B).

It is important to note that area 45A does not reach dorsally as far as the principal sulcus in macaque monkeys. Area 9/46v extends a few millimeters below the

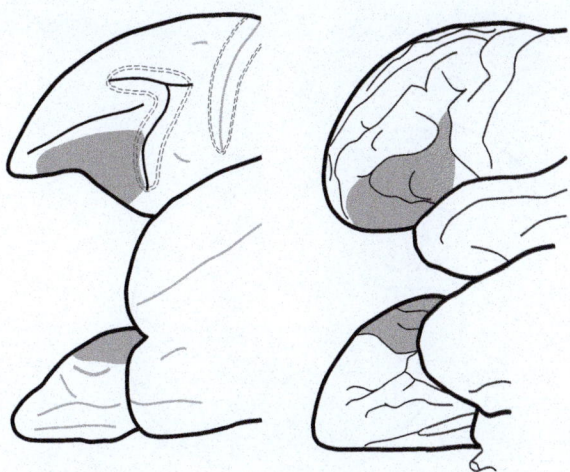

Figure 7.1 The ventral PF cortex in macaque monkeys (left) and humans (right).
Reproduced from Passingham, R.E. & Wise, S.P. *The Neurobiology of Prefrontal Cortex*, p. 196, Figure 7.1 © 2012, Oxford University Press.

principal sulcus. The dorsal extent of area 45A is roughly at the point of the inferior prefrontal dimple (Petrides & Pandya, 2002) (Figure 1.10B).

In the macaque monkey brain, area 45B lies caudal to 45A in the ventral limb of the arcuate sulcus. Area 44 lies in the fundus of this sulcus (Frey et al., 2014), anterior to the ventral premotor area 5a (Sharma et al., 2019) (Figure 7.2, next page).

In the human brain, area 45 B lies in the posterior section of pars triangularis, whereas area 45A forms the anterior section (Figure 1.10A). These two areas are divided by the triangular sulcus (Petrides, 2018). Area 44 lies just anterior to the descending limb of the precentral sulcus (Figure 1.10A). Areas 44 and 45B are included in the inferior caudal PF cortex (Table 1.1).

In the human brain areas 45A and 45B extend dorsally to the inferior frontal sulcus (Amunts et al., 1999). More anteriorly this sulcus forms the border between area 9/46v and the ventral PF cortex. The upper bank of the inferior frontal sulcus is part of area 9/46v whereas the lower bank is included in the ventral PF cortex. Since, in fMRI studies, the peaks of activations are frequently described in the tissue in the inferior frontal sulcus, it is usually unclear whether they are located in the ventral PF cortex or not.

It not always appreciated how far forwards area 45A can extend in the human brain. In terms of MNI coordinates it extends as far as $y = 40$ (Petrides, 2018). The most anterior extent of area 45B is around $y = 32$, and the most anterior extent of area 44 around $y = 20$ (Petrides, 2018).

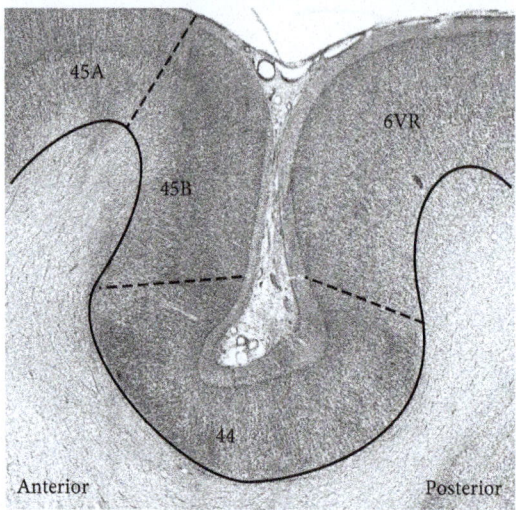

Figure 7.2 Cross section through the ventral limb of the arcuate sulcus on an MRI scan in a macaque monkey
Shows area 45A on the convexity of the ventral PF cortex, 45B in the anterior bank of the sulcus, 44 in the fundus and the ventral premotor area F5a in the posterior bank of the sulcus.

Reproduced from Frey, S., Mackey, S., & Petrides, M. Cortico-cortical connections of areas 44 and 45B in the macaque monkey. *Brain Lang*, 131, 36–55, Figure 1, Doi: 10.1016/j.bandl.2013.05.005 Copyright © 2013, Elsevier Inc. All rights reserved.

Area 47/12 lies anterior to this in the human brain, but also continues posteriorly on the ventrolateral surface into the frontal operculum to roughly the back of position of area 45A (Amunts et al., 2010). There is then an opercular region which Amunts et al. (2010) label op 9 that lies under area 45 at the level of the triangular sulcus. The opercular regions op 8 and op 7 continue under area 44 (Amunts et al., 2010). Activation peaks in this general region are often wrongly identified as being in the insula: the insula is the flattened cortex at the bottom of the sylvian fissure. At the level of area 45B, the frontal opercular surface of pars triangularis lies above the caudal part of area 47/12 (Petrides, 2018).

In macaque monkeys, there is also a granular opercular area (GrFO), as well as a precentral opercular area (PrCo) and dorsal opercular area (DO) (Gerbella et al., 2016). These may provide limbic inputs for the mouth and hand (Gerbella et al., 2016).

Kostopoulos and Petrides (2008) report activations in area 47/12 at a level of $z = -2$ and in area 45A at a level of $z = 10$. These coordinates are helpful in distinguishing peaks that lies in area 47/12 as opposed to area 45A.

However, there is variability between different brains in this region, even when the brains are co-registered to MNI space using the best registration algorithm

(Klein et al., 2009). This means that the only safe way of localizing the activations is by inspecting the details of the sulcal pattern of the individual brains (Sprung-Much & Petrides, 2018).

Neubert et al. (2014) used resting state covariance to compare the ventral PF cortex in the macaque monkey and human brain. They were able to identify corresponding areas in almost all regions. Like Amunts and Zilles (2012), they treat the inferior frontal junction (IFJ) as being area 44d. The IFJ lies at the border between the descending limb of the precentral sulcus and the posterior part of the inferior frontal sulcus (Brass et al., 2005).

Neubert et al. (2014) were able to identify areas 45A (Figure 7.3, orange) and 47/12 (Figure 7.3, light blue) that corresponded in the human and macaque brain. However, there was one area in the human brain for which Neubert et al. (2014) could not find a corresponding area in the macaque brain. This was the ventrolateral polar cortex, though they suggested that this had some similarity in its connections with area 46.

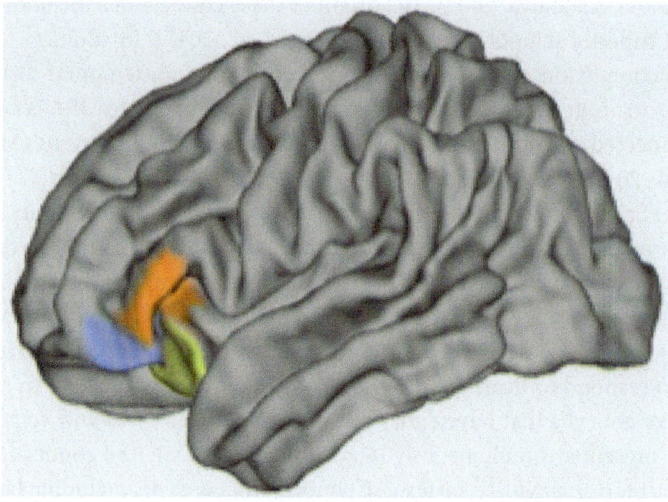

Figure 7.3 Areas in the human ventral PF cortex that correspond with areas in the macaque monkey, as demonstrated by the functional fingerprint based on resting state covariance
Area 45A orange; area 47/12 light blue; opercular cortex yellow.

Reproduced from Neubert, F.X., Mars, R.B., Thomas, A.G., Sallet, J., & Rushworth, M.F. Comparison of human ventral frontal cortex areas for cognitive control and language with areas in monkey frontal cortex. *Neuron*, 81 (3), 700–13, Figure 6, https://doi.org/10.1016/j.neuron.2013.11.012 Copyright © 2014, Elsevier Inc. All rights reserved.

Connections

Figure 7.4 (next page) shows the corticocortical connections of the ventral PF cortex (areas 45 and 47/12) in the macaque monkey. Since it is interconnected with the inferior caudal PF cortex (areas 45B and 44) (Frey et al., 2014), the following summary discusses the connections of these areas separately.

The Ventral Prefrontal Cortex (Areas 45A and 47/12)

1. Areas 47/12 and area 45A are closely interconnected (Petrides & Pandya, 2002).
2. Area 47/12 has auditory inputs from the superior temporal cortex as well as visual ones from the inferior temporal (Petrides & Pandya, 2002) and the perirhinal cortex (Saleem et al., 2008). Somatosensory inputs come from the secondary somatosensory area S2 (Petrides & Pandya, 2002). Thus, area 47/12 receives auditory, visual, and tactile information, including the representation of objects, whether viewed or palpated (Murray & Richmond, 2001).
3. Area 45A has strong inputs from the non-primary auditory cortex, including the dorsal bank of the superior temporal sulcus (STS) (Romanski et al., 1999; Petrides & Pandya, 2002). The origin of these connections includes the rostral superior temporal area (STp) (Scott et al., 2017). In addition, area 45A has connections with the anterior part of the inferior temporal cortex (area TE), including area TEa in the rostral part of the STS. Area 45A is also interconnected with the middle part of the anterior cingulate gyrus (Morecraft et al., 2012).

    O'Scalaidhe et al. (1997) described cells with response selectivity for faces in area 45A. As Chapter 3 mentions, there are also cells in area STp that show enhanced activity when faces were presented, accompanied by the sound of monkey calls. The connections to area 45A from the STS include areas that show such face-selective activation (Ku et al., 2011), and cells in area 45A also respond to hearing vocalizations (Romanski et al., 2005). In addition, there are cells that have enhanced responses when faces and vocalizations are presented simultaneously (Sugihara et al., 2006). The connections with the anterior cingulate cortex (ACC) mean that area 45, including both areas 45A and 45B, can interact with the cingulate vocalization area (Gavrilov et al., 2017).
4. Both area 47/12 and area 45A have connections with the orbital PF cortex (Petrides & Pandya, 2002). The lateral orbital sector of area 47/12 also has a direct input from the amygdala (Amaral & Price, 1984).

    Accordingly, the ventral PF cortex has access to information about specific outcomes predicted on the basis of visual information (Chapter 4), and

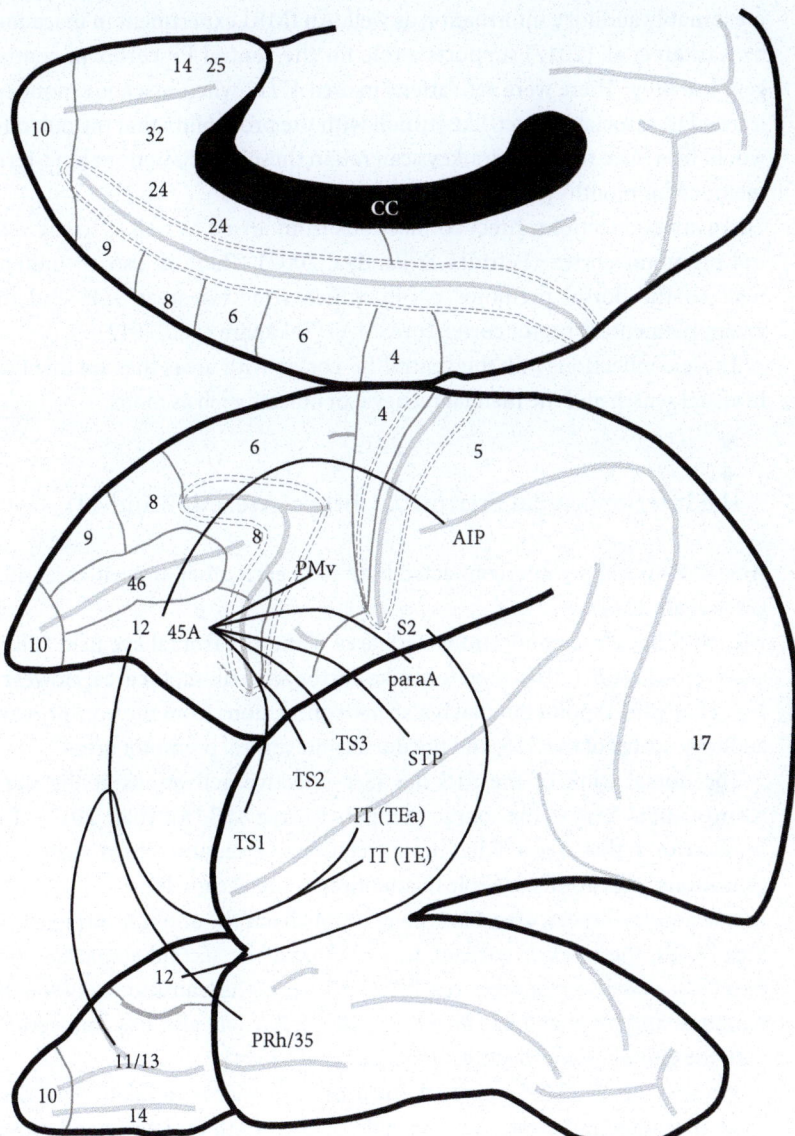

Figure 7.4 Selected connections of the ventral PF cortex
The lines connect some of the areas that have direct axonal connections with the ventral PF cortex, assumed to be reciprocal.

Adapted from Passingham, R.E. & Wise, S.P. *The Neurobiology of Prefrontal Cortex*, p. 197, Figure 7.2 © 2012, Oxford University Press.

presumably auditory information as well. An fMRI experiment in macaques by Kaskan et al. (2017) supports a role for the ventral PF cortex in contingent learning. There were activations in area 47/12 when macaque monkeys learned to associate object-like stimuli with the probability that the outcome would be a juice reward. Monkeys can retain these associations in long-term memory for months (Ghazizadeh et al., 2018).

5. There are connections direct connections from area 8 as well as to the ventral premotor cortex (Petrides & Pandya, 2002). There is also an indirect relay to the dorsal premotor cortex which runs via areas 9/46 and the presupplementary motor cortex (preSMA) (Takahara et al., 2012).

These connections link the ventral PF cortex with areas that are involved in visual search and the manual retrieval of objects such as foods.

The Inferior Caudal Prefrontal Cortex (Areas 45B and 44)

1. Area 45B and 44 are interconnected both with each other and with area 45A (Frey et al., 2014).
2. Area 45B has close connections with area 8 and the frontal eye field (FEF), and Gerbella et al. (2010) therefore take it to be an oculomotor area. However, Frey et al. (2014) point out that it also has connections from the non-primary auditory cortex of the STS, and the dorsal and ventral premotor areas.

   The dorsal bank of the STS has face-selective activations at the same rostro-caudal level as the source of inputs to area 45B (Ku et al., 2011). The implication is that area 45B has inputs relating to faces and the associated vocalizations. This points to a role in attention to social signals.
3. In macaque monkeys, area 44 receives inputs from the anterior intraparietal area (AIP), the inferior parietal area PFG, and the second somatosensory area (S2), as well as from the area Tpt which lies at the transition between the temporal and parietal lobe (Frey et al., 2014). Area 44 also has connections with the cingulate vocalization area (Frey et al., 2014).

   Petrides et al. (2005) applied intracortical micro-stimulation to area 44 in macaque monkeys, and were able to elicit both face and hand movements. Some of these effects could have resulted from current spread to the premotor areas. However, Sharma et al. (2019) found cell activity in area 44 and the ventral premotor area (F5) while monkeys either used their hands or ingested flavoured liquids. Nonetheless, there was more activity for hand movements in area 5F5p of the ventral premotor cortex, and more activity for mouth movements in area F5c. In the latter, Coudé et al. (2011) found cells in the ventral premotor area F5 that showed an increase in activity either before or during instrumentally conditioned vocalizations. At about a quarter of the sites from which this activity was recorded, intracortical micro-stimulation

evoked movements of the larynx as well as the lips, tongue, and mouth. It is clear that area 44 forms part of a widespread network that supports vocalization.
4. In macaque monkeys, the granular opercular area GrFO has connections with the other opercular areas, PrCO and DO, as well as with the convexity and orbital parts of area 47/12 (Gerbella et al., 2016). It also has connections with area 44 and the hand and mouth areas of the ventral premotor cortex (Gerbella et al., 2016). The agranular opercular areas differ in being mainly connected with a ventral premotor area, area F5c, which is specialized for the control of mouth movements.

Chapter 9 describes the connections of areas 44 and 45B in chimpanzees and humans.

Summary

- The ventral PF cortex (areas 45A and 47/12) receives visual information about objects and conjunctions of colours and shapes, as well as auditory and tactile information. Visual and acoustic inputs provide information at a long distance, as well as locally; tactile inputs come from interacting with objects and surfaces within reach.
- The ventral PF cortex also receives inputs from the perirhinal cortex about both the feel and appearance of complex objects.
- The inferior caudal PF area 45B and the ventral PF area 45A process information about vocalizations and the facial expressions that accompany them.
- Areas 47/12 and 45A also have connections with the orbital PF cortex (areas 11 and 13), which places the ventral PF cortex in a position to influence choices on the basis of their expected outcomes.
- There are direct outputs to the FEF and to the ventral premotor cortex from areas 45B and 45A.

Matching-to-Sample

Monkeys with complete removal of the ventral PF cortex are able to relearn the delayed alternation task at a normal rate (Passingham, 1975). This is strong evidence that the ventral and dorsal PF differ fundamentally in their specialization. Unlike the dorsal PF cortex, the ventral PF cortex does not base choices on the order of a sequence. Instead the choice is conditional on the association between one stimulus and another, irrespective of order and irrespective of whether there is a delay or not.

Simultaneous Matching

The role of the ventral PF cortex in associative learning can be illustrated by considering simultaneous matching. On a simultaneous matching task, the subject is presented with a sample object and two choice objects, and is required to choose the choice object that matches the sample. This is an associative task because the subject has to associate stimulus A in one place with stimulus A rather than B in another place.

An early study (Passingham, 1975) demonstrated that monkeys could relearn this task without trouble after removal of the dorsal PF cortex. The reason is that there was no delay between the presentation of the sample and the choice.

But, in the same study, monkeys with removal of the ventral PF cortex failed to relearn simultaneous matching in 1,000 trials (Passingham, 1975). It was later found that if the lesions were incomplete, the monkeys can relearn the task at the normal rate; but the animals failed if the lesions included all of the lateral and orbital sectors of area 47/12 as well as area 45A (Rushworth et al., 1997). The results are shown in Figure 7.5.

Given these results, it is not surprising that Bussey et al. (2001) also found that monkeys with combined lesions of the ventral and orbital PF cortex were also impaired at simultaneous matching. Bussey et al. (2002) also demonstrated that the critical input comes from the inferior temporal cortex. They made cross-lesions, removing the inferior temporal cortex in one hemisphere and the ventral and orbital PF cortex in the other. The animals were again impaired on matching.

In a PET study, human subjects were required to perform a simultaneous matching task involving faces (Haxby et al., 1994). The faces were presented in small square, located at different positions within a larger square. In the face matching condition, the subjects had to choose the face that matched the sample in identity. In the comparison condition, they had to choose the face that was located in the same position within the larger square as in the sample.

When the subjects matched the faces by identity as opposed to location, there were activations in area 45A and more anteriorly in the lateral orbital sulcus which forms the border between area 47/12 and area 11. The activation in the ventral PF area 45A agrees with the neurophysiological data: there are cells in area 45A that have activity that is selective to the presentation of faces (O.Scalaidhe et al., 1997).

Delayed Matching

There are also cells in the ventral PF area 45A that continue firing after the offset of faces (O.Scalaidhe et al., 1999). Furthermore, in a study in which monkeys were tested on a delayed matching task for objects, there was delay-period activity in area 45A (Rainer & Miller, 2002). Findings such as these have led Constantinidis

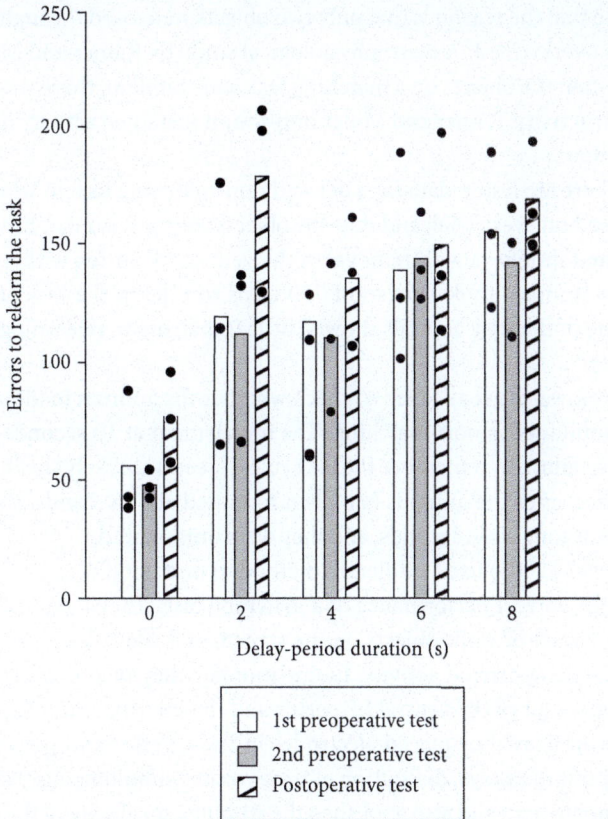

Figure 7.5 Effect of lesions of the ventral PF cortex on simultaneous matching Preoperative (preop) and postoperative (postop) performance. The results for preoperative performance show the trials to relearn the task after a break in training. Filled circles show the data for individual monkeys.

Reproduced from Rushworth M.F, Nixon P.D., Eacott M.J., & Passingham R.E. Ventral prefrontal cortex is not essential for working memory, *J Neurosci*, 17 (12), 4829–38, Figure 6, Doi: 10.1523/JNEUROSCI.17-12-04829.1997 Copyright © 1997, Society for Neuroscience.

and Qi (2018) to support the claim made by Wilson et al. (1993) that the function of the ventral PF cortex is to support working memory for objects, including faces.

However, this claim was tested in the lesion study by Rushworth et al. (1997). Though the monkeys with ventral PF lesions failed to relearn simultaneous matching in 1,000 trials, they succeeded with further testing. This meant that it was possible to find out whether the lesion caused an impairment when a delay period was interposed between the presentation of the sample and the choice. It turned out that the lesioned monkeys could perform normally when tested on delayed matching with delays of up to 8 seconds (Figure 7.5).

It could be that the reason is that only two objects were used throughout, so that the task was overlearned. A neurophysiological study by Rainer and Miller (2002) found that when the objects on a matching task were familiar, there was little if any delay-period activity. Rainer and Miller only found sustained activity if the objects were unfamiliar.

Luckily, there are lesion data for a delayed non-matching task in which new objects were used on every trial, and thus the objects were unfamiliar. Kowalska et al. (1991) retested monkeys with removal of the ventral PF cortex with a delay of 10 seconds. The animals took a mean of 790 trials to relearn the task at this delay. However, one of the four animals showed little impairment, relearning the task in just 160 trials.

However, Kowalska et al. failed to check whether the animals made errors when tested on simultaneous non-matching. The impairment at 10 seconds could have been due to a failure to remember the task rule. It seems unlikely that it was due to the delay since, after the animals had been retrained at 10 seconds, they were re-tested delays as long as 30 seconds, and they were unimpaired.

Sobotka et al. (2005) used a different method to find out if delay-period activity was critical for correct performance of a matching task. They tested two monkeys with a stimulus set of eight objects on the screen, and used electrical stimulation to disrupt the delay-period activity. In one monkey they placed electrodes in the lateral orbital sector of the ventral PF cortex and the inferior temporal cortex, and in the other monkey they placed electrodes in the anterior and posterior hippocampus. In both monkeys, disruption at the two sites simultaneously caused a behavioural impairment and also abolished the stimulus specificity of the cell activity during the delay period. However, unfortunately Sobotka et al. did not compare the effects of disrupting the ventral PF cortex alone or the inferotemporal cortex alone. Thus, the experiment fails to show that it was stimulation of the ventral PF cortex that was critical.

Fortunately, Plakke et al. (2015) inactivated area 45A on its own using cooling. This area was chosen because it is where cell activity has been found that continued after monkeys have been presented with faces (O.Scalaidhe et al., 1999). So Plakke et al. (2015) tested their monkeys on delayed matching for faces. The finding was that the animals were still able to perform the task successfully during the inactivation. This provides further evidence that the ventral PF cortex is not critical for working memory for objects, including faces.

Short-Term Memory in Humans

Yet, there are many fMRI experiments that have claimed to find delay-period activations in the ventral PF cortex when human subjects are tested on visual matching. However, little attention has been paid to exactly where the activations were located. For example, Sala et al. (2003) compared delayed matching for faces with delayed matching for buildings. They attributed the peak for delay-period activity to the ventral PF cortex, but in fact it lay in the inferior caudal PF cortex at the inferior frontal junction (IFJ or 44d).

Rama et al. (2004) also used a matching to sample design to test memory for either the identify of a voice or the location at which it was presented. They looked specifically for delay-period activation for this contrast. Again, the peak for the delay-period activation lay in the inferior caudal PF cortex (area 44), not in the more anterior ventral PF cortex.

In a follow up study, Rama and Courtney (2005) tested subjects on delayed matching for faces or voices. On all trials the subjects saw a sample face and heard words, but on some trials the subjects had to match the face, and on other trials the identity of the speaker. When the memory for faces was compared with the control condition, the peaks of activation lay in the inferior caudal PF cortex (areas 44 and 45B), as well as in the operculum at that level. When the memory for voices was compared with the control condition, the peak was in the operculum at the level of area 44.

Schon et al. (2008) tested subjects on delayed matching for objects rather than faces, with similar results. There were activations in the frontal operculum of the inferior caudal PF cortex. However, there was also peak in the lateral orbital sulcus; this forms the boundary between the orbital PF cortex (area 11) and the ventral PF cortex (area 47/12) (Petrides, 2018).

Taken together, these results show why lesions of the ventral PF cortex (areas 47/12 and 45A) in monkeys fail to impair delayed matching. The reason is that the inferior caudal PF cortex (areas 45B and 44) was deliberately not included in the lesions (Passingham, 1975; Kowalska et al., 1991; Rushworth et al., 1997) because the focus was on the ventral PF cortex (areas 47/12 and 45A).

The same account explains the results of an early study by Milner et al. (1991). As Chapter 6 describes, Milner et al. tested patients with large frontal lobectomies on recency judgements. But they also tested the patients on delayed matching for the pictures. The patients were unimpaired at recognizing the pictures that had been shown. However, inspection of the diagrams of the lobectomies (McAndrews & Milner, 1991) shows that the surgeon deliberately avoided the inferior caudal PF cortex (areas 44 and 45B) so as to leave speech unimpaired.

The conclusion is that, just as short-term memory for spatial locations depends in part on the caudal PF cortex (areas 8Ad and the FEF) (Chapters 5 and 6), so short-term memory for objects, faces and vocalizations depends in part on the inferior caudal PF cortex (areas 45B and 44).

Maintenance in Posterior Association Areas

Delay-period activity that encodes objects and faces occurs widely in the cortex and is by no means exclusive to the inferior caudal PF cortex. For example, sustained delay-period activity has been found in the inferior temporal cortex while monkeys were tested on a delayed matching task for coloured objects (Fuster & Jervey, 1982).

Miller and Desimone (1994) also reported delay-period activity in the perirhinal cortex. They presented a series of objects, the first of which served as the sample; and the monkeys had to identify the ones that matched the sample object. Miller and Desimone found suppression of cell responses when irrelevant (distractor) items repeated but enhancement when a relevant (sample) item was repeated. They took these results to suggest that short-term retention depends both on an automatic process (suppression for irrelevant repeats) and a controlled or top-down process (enhancement for relevant repeats).

Evidence for a top-down process was provided by an experiment with human subjects in a different laboratory. Miller et al. (2008) scanned the subjects using both fMRI and EEG. When the subjects retrieved memories of faces, there was a signal in the PF cortex that preceded that in fusiform face area.

One way of finding out what information is carried by delay-period activity is to use representational similarity analysis (RSA) (Kriegeskorte et al., 2008a) to decode the delay-period activations. Sreenivasan et al. (2014c) gave subjects short-term memory tasks in which they were presented with two faces and two scenes; they had to remember the faces, the scenes, or both the faces and the scenes. The assumption of RSA was that the patterns that encoded sensory representations of faces should be more similar to patterns that encoded faces and scenes than patterns that encoded scenes alone.

Sreenivasan et al. defined two a priori regions of interest. These involved extrastriate areas including the inferior temporal cortex and the ventral PF cortex as defined by the *Harvard-Oxford Probabilistic Atlas*. The analysis revealed that during the delay it was possible to decode the sensory representations in the extrastriate cortex, but not in the ventral PF cortex. Unfortunately, the study did not include a region of interest centered on the inferior caudal PF cortex (areas 45B and 44).

This is critical as can be seen from a further study by Sreenivasan et al. (2014b). They studied memory for faces alone. To make the task difficult, the targets involved morphs of the faces that varied in their similarity to the sample faces. On each trial, one of these morphs was paired with a novel face, and the subjects had to choose the morphed face. On a control task the subjects saw the same faces, but simply had to judge whether the choice faces were upright or not.

Sreenivasan et al. found that the activations in the fusiform face area and the inferior caudal PF cortex (IFJ and frontal operculum) increased the more similar the morphed targets to the sample faces. They argued that this suggests that these areas are involved in maintenance of the perceptual representations of the sample. They also performed an analysis to study Granger causality. A measure of the influence of the fusiform face area on the IFJ predicted the behavioural performance very closely, and Sreenivasan et al. took to indicate a bottom-up influence on the caudal PF cortex.

Other findings using multivoxel pattern analysis (MVPA) have also suggested that temporal and posterior parietal areas are involved in the maintenance of visual representations (Sreenivasan et al., 2014a; Nee & D'Esposito, 2018). However, Xu (2017) has objected that, since in everyday life there is a continuous stream of visual inputs, storing information online in sensory areas alone would be inefficient. They reasoned that 'distracting' stimuli would disrupt the stored information and thereby cause a memory loss.

In this context, a 'distracting' stimulus is defined as one that is irrelevant to the task in hand, and it is the PF cortex that represents what is relevant. Lorenc et al. (2015) applied TMS at theta frequency to the inferior caudal PF cortex (area 44). The subjects were then presented with faces and scenes, followed by a delay. In one condition they had to press keys to indicate whether they recognized a face, and in the other condition a scene. Lorenc et al. then used MVPA to decode activity relating to faces or scenes in posterior areas. The effect of the TMS was to decrease the ability to decode whether faces or scenes were relevant in the extrastriate cortex and posterior parietal cortex.

It is the ventral PF cortex that is in a position to learn the current task or rule. The reason is that monkeys learn the rule via reinforcement, and the ventral PF cortex has an input from the orbital PF cortex which represents outcomes. On a matching-to-sample task, the animal has to learn that the rule is to respond to the match, not the distractor item. On delayed matching, cell activity in the ventral PF cortex has been shown to be unaffected by distracting stimuli that are presented during the delay (Miller et al., 1996).

Summary

Monkeys with removal of the ventral PF cortex are severely impaired at learning visual simultaneous matching. This task requires them to learn to associate two identical stimuli. However, with further testing the monkeys can relearn the task rules. They are then no longer impaired when delays are introduced between the presentation of the sample and the choice objects. The reason is probably that the inferior caudal PF cortex (area 45B) and the inferior temporal cortex are intact.

Paired-Associate Learning

Paired-associate learning differs from matching in that none of the choice objects are presented as the sample. Instead, the task requires learning to associate stimulus A with stimulus C, stimulus B with stimulus D, and so on. Despite this difference, paired-associate learning and non-matching to sample are alike in that the choice is conditional upon the stimulus that appears at the beginning of a trial.

Visual-Visual Associations

Rainer et al. (1999) tested monkeys on paired associates and delayed matching. Each trial consisted of an initial picture and a series of probe pictures that appeared after various delay intervals. The subjects had to decide whether to respond to a probe picture, either because it was the correct associate of the initial picture (in the paired-associate task) or because it matched the initial picture (in the matching-to-sample task). This design meant that it was possible to distinguish what cell activity encoded during the delay.

Rainer et al. recorded cell activity mainly in the ventral PF area 45A. Early in the delay period, the cells tended to code for the initial picture in both tasks, which corresponds to retrospective sensory coding. However, as the delay period progressed, the majority of the cells coded prospectively for the associated picture. Only the paired-associate task could reveal this transition because in the matching-to-sample task the initial and goal pictures do not differ. Results for the matching- and paired-associate tasks differed in that when the sample picture reappeared as a choice stimulus, cell activity tended to decrease, whereas when the paired associate appeared, cell activity tended to increase. Rainer et al. suggested that the suppression resulted from a passive process related to stimulus repetition whereas the enhancement reflected an active process related to the retrieval of the associate from long-term memory.

Rainer et al. did not find any cells that responded similarly to the cue and its associate, like the 'pair coding neurons' that have been found in the rostral part of the inferior temporal cortex (Sakai & Miyashita, 1991). But the study by Rainer et al. (1999) only included three paired associates, and this made it difficult to detect pair coding. In a later study, Andreau and Funahashi (2011) taught monkeys twelve paired associates between pictures. Of 68 cells recorded in the PF cortex that showed a selective response to the cue, 32% responded similarly to the associate. Unfortunately, Andreau and Funahashi did not provide a map of exactly where they were recording from in the PF cortex, but it is likely that the sites included areas 9/46v and 45A.

To find out if the PF cortex was critical for learning visual-visual associations, Gutnikov et al. (1997) made lesions. They used coloured ACSCII characters as stimulus material, with each cue consisting of a distinctive conjunction of colours and shapes. Before surgery the monkeys were taught eight paired associates. Then the uncinate fascicle was severed which connects the rostral part of the inferior temporal cortex with the ventral PF cortex (Ungerleider et al., 1989; Folloni et al., 2019).

Unlike the experiments on cell recording (Sakai & Miyashita, 1991; Rainer et al., 1999; Funahashi & Andreau, 2013), Gutnikov et al. (1997) did not impose a delay period between presentation of the cues and their associates. As in the simultaneous matching task, two choice objects appeared at the same time as the cue. Despite the lack of an imposed memory period, cutting the uncinate fascicle caused a significant impairment on relearning the eight paired associates. And it also led to an impairment in learning eight novel pairs.

Severing the uncinate fascicle cuts off the input to the ventral PF cortex from the inferior temporal and perirhinal cortex, but it also blocks the reciprocal projection. Accordingly, this lesion interferes with top-down influences from ventral PF cortex as well as bottom-up information to the ventral PF cortex.

In order to study the top-down influence, Tomita et al. (1999) cut the caudal part of the corpus callosum, while leaving the commissural connections between the left and right PF cortex intact. In spite of the caudal section, presentation of a cue to the right inferior temporal cortex generated pair-coding activity in the left inferior temporal cortex. These results imply that information about the cue went from the right inferior temporal cortex to the right ventral PF cortex; and that it was then able to pass to the left ventral PF cortex via the rostral callosum and finally top-down to the left temporal lobe. In all, about 60% of the cells studied were influenced by this top-down signal.

Hasegawa et al. (1998) studied the behavioural performance of macaque monkeys with a transection of the caudal part of the corpus callosum. So long as the rostral part of the corpus callosum remained intact, the animals could succeed on a paired-associate task if the experimenters presented the cue to one hemisphere and the associate to another. However, after the whole of the corpus callosum and the anterior commissure was cut, their performance fell to chance level.

These experiments indicate that the ventral PF cortex sends a top-down signal to the temporal lobe that supports the retrieval of the associate from long-term memory. The fibres that convey these signals run to the perirhinal cortex. The evidence is that the perirhinal cortex then influences the inferior temporal cortex. Higuchi and Miyashita (1996) found that lesions placed in the perirhinal and entorhinal cortex abolished pair-coding cells in the inferior temporal cortex.

Like the cells in the ventral PF cortex (Rainer et al., 1999), there are also cells that code prospectively on a paired-associate task in the perirhinal cortex (Naya et al., 2001) and inferior temporal cortex (Naya et al., 1996). However, the activity in the two temporal areas differ in timing. Activity related to the cue occurs earlier in the inferotemporal cortex; activity related to the associate occurs earlier in the perirhinal cortex (Naya et al., 2001). These timing differences indicate that a top-down signal passes from the perirhinal cortex to the inferior temporal cortex, and subsequent work has revealed that this signal enters the inferior temporal cortex in the supragranular layers (Takeuchi et al., 2011).

The results of these experiments support the contention that the ventral PF cortex is critical for the cued retrieval of sensory representations on a paired-associate task, just as on a matching-to-sample task. It does so by sending a top-down signal encoding the associate to the perirhinal cortex, and this is then passed on to the inferior temporal cortex.

The neurophysiological experiments described so far recorded from cells when the associations had already been learned. But Brincat and Miller (2015) also recorded cell activity during learning. The recordings were taken from the ventral PF cortex as well as the PF areas 9/46 and 46 and 8, and also from the hippocampus. During each session the performance of the animals went from chance to a level of roughly 80% correct. There were responsive cells in all these areas.

In the period after presentation of the first object, the signal in the hippocampus led the signal in the PF cortex, and this was confirmed by a later multivariate cross-correlation analysis by Rodu and Miller (2018). However, Brincat and Miller (2015) found prospective activity for the associate in the PF cortex but not the hippocampus, and this activity increased as performance increased. At the same time there was synchrony between the local field potentials in the two areas. As learning progressed, there was a shift in this synchrony from the theta to the alpha and beta bands.

Further analysis showed that neural information about the externally presented stimuli decreased with learning (Brincat & Miller, 2016). There was a corresponding increase in beta power and synchrony. This was taken to represent internal or top-down processing.

In order to estimate the information that was carried by the neural population, Liu et al. (2020) used a measure of the 'normalized distance'. This is a dimensionless metric that is based on the geometry of the neural code. It was computed while the animals learned the associations, that is during the period of encoding. Compared with the hippocampus, the population of cells in the PF cortex exhibited more sustained encoding for the cue, the associate, and the decision. So, the PF cortex is involved both in encoding associations and in retrieving the associate from long-term memory after learning.

Visual-Auditory Associations

Rather than presenting visual-visual associations, Hwang and Romanski (2015) showed audiovisual clips of faces with accompanying vocalizations. After a delay period, a probe face was presented accompanied by a vocalization, and the monkeys had to indicate whether the combination of the face and sound had changed from the initial presentation. Changes could include either the face, the vocalization, or both.

Hwang and Romanski recorded in the ventral PF area 45A where cells had been found that responded specifically to faces (O.Scalaidhe et al., 1999). At the matching phase, a population of cells showed increased activity when one component had changed, thus reflecting a change in the association between a specific face and a specific vocalization. Furthermore, inactivating the ventral PF area 45A by cooling caused an impairment in performance for face–vocalization conjunctions, but not for faces alone. This indicates that the impairment related to specific face–sound associations rather than other aspects of the task.

Gaffan and Harrison (1991) also taught monkeys auditory-visual associations with the difference that the associations were arbitrary. The animals had to choose among six different visual stimuli, each of which was associated with a specific sound. Gaffan and Harrison then disconnected the PF cortex from the superior temporal cortex by making a PF cortex lesion in one hemisphere and a superior temporal cortex lesion in the other. They also cut the commissures between the two cerebral hemispheres to complete the disconnection. The result was that the monkeys could not perform the auditory-visual task above chance level.

Visual-Spatial Associations

The ventral PF cortex is also involved in learning associations between visual stimuli and movements. The relevant tasks are sometimes referred to as 'visuo-motor' tasks. However, the term 'visuo-spatial' associations captures the underlying computation better. The evidence comes from two studies. As Chapter 6 mentions, Mushiake et al. (2006) found that cells in the dorsal PF cortex coded for the spatial goal rather than the movement (Figure 6.14), and Shen and Alexander (1997) found the same for cells in the dorsal premotor cortex. Thus, when monkeys learn to move a lever in one direction or another given a visual cue, it seems likely that the ventral PF cortex encodes the spatial goal rather than the movements that are needed to achieve that goal.

Wang et al (2000) taught monkeys to move a handle to the left given one cue and to the right given another. The cue remained present as the animal moved the handle; thus, as in simultaneous matching there was no delay. When the monkeys had learned the task, Wang et al. then injected bicuculline into areas 45A and 47/12. This is an antagonist for the receptors of (GABA), and the effect is to cause hyperactivity of the cells in the injected area and so to disrupt its neural operations. During the inactivation, the monkeys were slow to learn the tasks with novel cues, but they were still able to perform the familiar one on which they had been trained beforehand.

In a similar study, Bussey et al. (2001) taught monkeys a series of visuo-spatial problems for either 48 or 52 trials each. On each problem, one of three or four coloured characters were presented on a video monitor and each was associated with a different spatial goal. Two monkeys moved a cursor to the goal by manipulating a joystick, and two others reached to it by touching the screen. Eight problems were presented a day.

After training the subjects on a series of these problems, Bussey et al. removed both the ventral and orbital PF cortex. After surgery, the monkeys failed to learn new problems within 48 trials. As Figure 7.6 illustrates, they remained at chance level throughout, although control subjects reduced their error rate to less than 10% with the same amount of training. By contrast with the results of Wang et al. (2000), these lesioned monkeys showed little, if any, retention of familiar associations they had overlearned before surgery. The reason may be that Bussey et al. (2001) removed the orbital PF cortex together with the ventral PF cortex, whereas Wang et al. only inactivated tissue within the ventral PF cortex.

The information about the cue comes from the inferior temporal cortex. So, Bussey et al (2002) removed the inferior temporal cortex in one hemisphere and the ventral and orbital PF cortex in the other hemisphere, thus disconnecting the two areas. These monkeys were severely impaired at learning new visual-spatial associations after surgery.

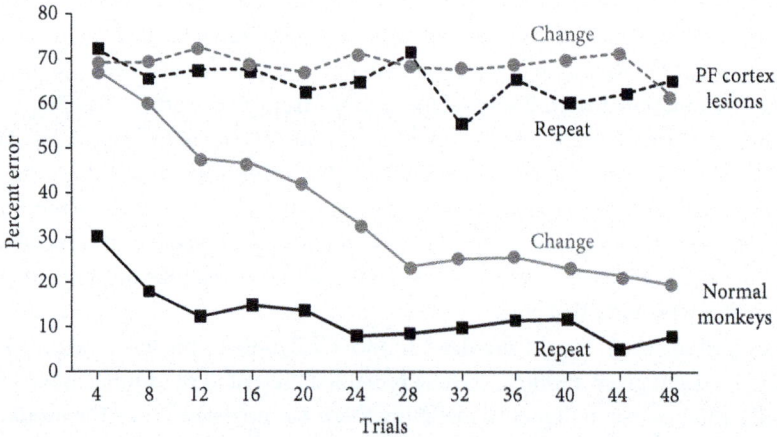

Figure 7.6 Performance across trials within problems on a series of visuo-spatial problems before and after ventral PF surgery
The data are shown separately for repeat trials in which the cue remained the same as that on the previous trial, and change trials in which it differed. 70% is the chance level, given the mean of three or four responses.

Adapted from Bussey T.J., Wise S.P., & Murray E.A. The role of ventral and orbital prefrontal cortex in conditional visuomotor learning and strategy use in rhesus monkeys (*Macaca mulatta*). *Behav Neurosci*, 115 (5), 971–82, Doi: 10.1037//0735-7044.115.5.971 Copyright © 2001, American Psychological Association.

Once the ventral PF cortex specifies a spatial goal, the animal can achieve it via indirect connections with the dorsal premotor cortex (Takahara et al., 2012) and thence to the motor cortex (Muakkassa & Strick, 1979). By injecting muscimol into the dorsal and ventral premotor cortex on different days, Kurata and Hoffman (1994) showed that it was inactivation of the dorsal premotor cortex that caused errors in the direction of the animal's movements.

Evidence from fMRI with human subjects supports the view that the ventral PF cortex is involved during learning. In a study by Toni et al. (2001), the subjects were taught to move different fingers in response to different visual cues. The subjects learned the arbitrary associations by trial and error during the scanning session, with feedback being given at the end of each trial. In a control condition, arrows pointed to the appropriate finger and the subjects merely need perform as instructed, without the need to learn any arbitrary associations. When learning was compared with the control condition, there were activations in the tissue in the inferior frontal sulcus and the inferior temporal cortex that increased as learning progressed.

In a later experiment, Boettiger and D'Esposito (2005) presented complex coloured visual cues which could be categorized into sets. Each set was associated with a different manual response. In one condition the subjects were scanned while they learned these associations, and in the other they were scanned while they responded for sets that they had learned on the previous days. When new learning was contrasted with retention, the peaks of activation were more extensive, lying in tissue in the inferior frontal sulcus as well as in the dorsal PF area 46. However, when familiar associations were contrasted with new learning, the activation was in the dorsal premotor cortex.

Leaning-related increases in activation could reflect something specific about associations between cues and the goals of action, but they could simply reflect an increase in attention either to the stimuli or to the outcomes alone. The issue can be resolved by recording from cells in monkeys. Asaad et al. (1998) taught monkeys to associate arbitrary cues with saccades to the left or right. There was a delay of 1 second between the presentation of the cues and the initiation of the saccade. Of the task-related cells in the ventral and dorsal PF cortex, 44% encoded the association between a specific cue and the goal of the associated saccade. Because the recordings were taken in intact brains, the cell activity in the dorsal PF cortex could have been derived from the ventral PF cortex.

As learning progressed, the activity that coded for the response occurred progressively earlier during the delay period. This demonstrates that the animals could retrieve the appropriate response more rapidly with training.

In this experiment, the monkeys learned new visual-spatial associations as the investigators repeatedly reversed the associations between two cues and two spatial goals. But the change in associations is not essential. Cromer et al. (2011a) taught several new associations between cues and saccades using novel stimuli. The monkeys learned rapidly, and learning-related activity occurred in the PF cortex in parallel with the change in performance. As in the earlier experiment by Asaad et al. (1998), the selectivity occurred earlier in the trial as learning progressed.

Puig et al. (2014) applied antagonists to dopamine D1 or D2 receptors while the animals learned similar associations. This disrupted the performance of the animals. Furthermore, the blockade decreased the neural information about the association between the cues and responses, and also exaggerated alpha and beta oscillations. However, the manipulation had no effect once the associations had been overlearned.

Summary

The ventral PF cortex plays a key role in learning new associations. These can be visual-visual, visual-auditory or visual-spatial. Even though the tasks are often taught with a delay between the presentation of the cue and its associate, the delay is not critical. Lesions of the ventral PF cortex severely disrupt the learning of visual-spatial associations even when there is no delay. There is a suggestion that the ventral PF cortex plays a greater role in new learning of associations than in retention of associations that have been overlearned.

Association via Categories

The previous section describes a study by Brincat and Miller (2015) in which they taught macaque monkeys paired associates. In this study animals had to associate two object-like stimuli (A and A') with a third one (B). The stimuli A and A' can be regarded as a category even though they had no physical similarity.

Learning Categories

In order to directly study the ability to learn to form categories, Freedman et al. (2001; Freedman et al., 2002) taught monkeys to distinguish drawings of cats and dogs. They produce a graded series of morphed drawings, and the monkeys were taught the boundary between the category 'cat' and the category 'dog'. The boundary between cats and dogs is labelled as '2-category boundary' in Figure 7.7A (next page), and the morphs of the cats and dogs are shown in Figure 7.7B.

Freedman et al. recorded in the ventral PF area 45A while the monkeys were given a delayed matching task in which the animals had to pick the morph of a cat if the sample was a cat, and the morph of a dog if the sample was a dog. There was a delay of 1 second between the sample and the match.

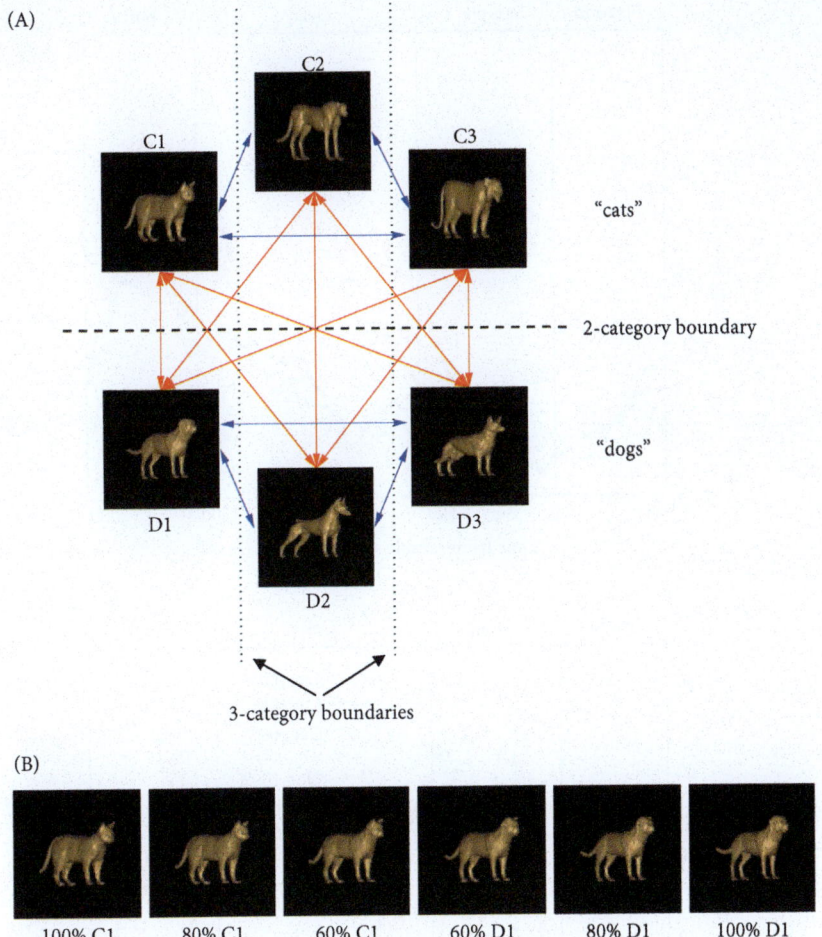

Figure 7.7 Stimuli used in a categorization task
A The heavy stippled horizontal line (2-class boundary) shows the boundary between cats and dogs. Alternatively, the lightly stippled vertical lines (3-class boundaries) show that the animals could be categorized arbitrarily in a different way, so that for example dog 2 (D2) and cat 2 (C2) are defined as being in the same category. B Morphs of cats and dogs showing the differences in the proportions.

Reproduced from Freedman D.J., Riesenhuber M., Poggio T., & Miller E.K. Visual categorization and the primate perceptual cortex: Neurophysiology and behavior. *J Neurophysiol* 88 (2), 929–41, Figure 1, Doi: 10.1152/jn.2002.88.2.929 Copyright © 2002, The American Physiological Society.

Freedman et al. found cells that encoded one or the other category, that is cats or dogs, and this was true both during the delay and test period. An example of a cell that coded for the category dog is shown in Figure 7.8A (next page). This shows that the activity was greater for dogs than cats, irrespective of whether the choice

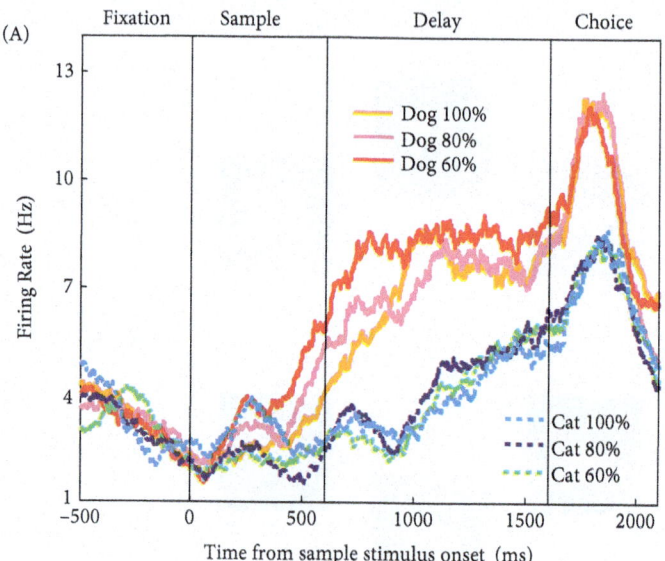

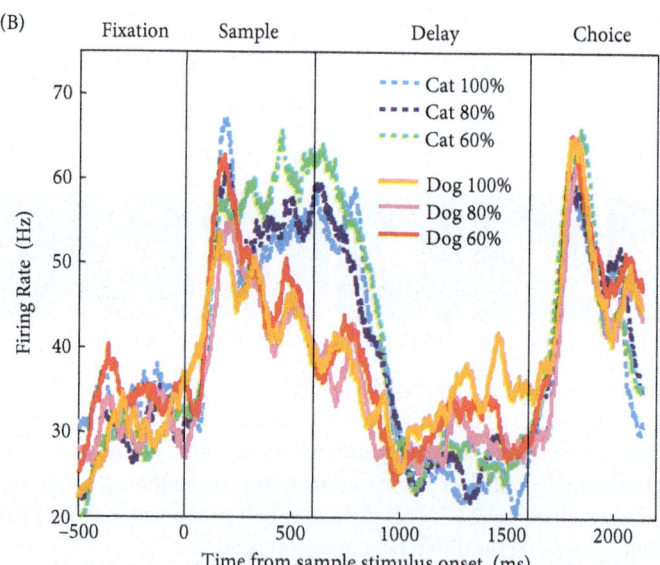

Figure 7.8 (A) A cell in the ventral PF cortex that encoded the category 'dog'. (B) A cell in the ventral PF cortex that encoded the category 'cat'
(A) Solid red, mauve and orange lines show the cell's activity for trials having a sample stimulus was an identical drawing (100%) or a morph (80% or 60%) of a dog. Dashed blue lines show activity for trials entirely or mainly consisting of cat drawings. (B) Dashed blue and green lines show the cell's activity for trials having a sample stimulus was an identical drawing (100%) or a morph (80% or 60%) of a cat. Solid red, mauve and orange lines show activity for trials entirely or mainly consisting of dog drawings.

Reproduced from Freedman D.J., Riesenhuber M., Poggio T., & Miller E.K. Visual categorization and the primate perceptual cortex: Neurophysiology and behavior. *J Neurophysiol*, 88 (2), 929–41, Figure 7a, Doi: 10.1152/jn.2002.88.2.929 Copyright © 2002, The American Physiological Society.

stimulus was a dog (100%) or a morph of a dog (80%, 60%). Figure 7.8B illustrates a cell that coded for cats rather than dogs.

Though there was cell activity that encoded categories in the ventral PF cortex, this does not show that that activity is essential for learning. One possibility is that the inferior temporal cortex could learn the categories and the ventral PF cortex could retrieve the information from there. But Freedman et al. (2003) found that cells in the inferior temporal cortex more often simply encoded the visual features of the stimuli.

Nonetheless, Minamimoto et al. (2010) have argued that the PF cortex is not necessary for learning new categories. They trained monkeys to perform a task in which stimulus A predicted one level of reward and stimulus B predicted another. With a little experience, the monkeys learn to generalize; that is, they learned to treat a stimulus with some of the features of stimulus A' as if it was stimulus A. Yet, combined lesions of the ventral and dorsal PF cortex did not block the ability to learn to generalize in this way.

But, unlike the task used by Freedman et al. (2002), the task used by Minamimoto et al. (2010) did not require the monkey to *compare* items by presenting a sample and then requiring a choice of the associated picture. The experiment simply involved stimulus generalization or 'feature generalization' as Buckley and Sigala (2010) call it. In stimulus generalization, if the animal has learned that a stimulus predicts some outcome, it will also predict the same outcome when presented with a stimulus that shares some of its features. Furthermore, the more the shared features, the stronger the prediction. This basic perceptual phenomenon exists in all vertebrates, but it does not involve learning to form associations between different stimuli.

This point is best made by showing that the ventral PF cortex is involved in learning to associate objects in the same category even when they are not perceptually similar, since generalization depends on perceptual similarity. After teaching monkeys the categories dogs and cats, Freedman et al. (2002) taught them new category boundaries, as depicted by the vertical light stippled lines in Figure 7.7 (page 257). Thus, the monkeys had to learn that dog C1 belonged to the same category as cat C1 (category A), dog C2 to the same category as cat C2 (category B), and dog D3 to the same category as cat C3 (category C). The effect was that Freedman et al. were now able to find cells in the ventral PF cortex that coded for one or other of the three new categories.

Furthermore, they no longer coded for the irrelevant categories dog and cat. Figure 7.9A (next page) shows an example of a cell that now coded for the category which included dog D3 and cat C3, while Figure 7.9B shows it no longer coded for the category dog versus cat.

Cromer et al. (2010) made the same point by teaching monkeys two independent categories. One category involved dogs and cats, with different degree of morphing; the other involved sports cars and limousines, again with different degrees of morphing. After the monkeys had learned both categories, Cromer et al. were able to find many cells that coded both categories. They provided an

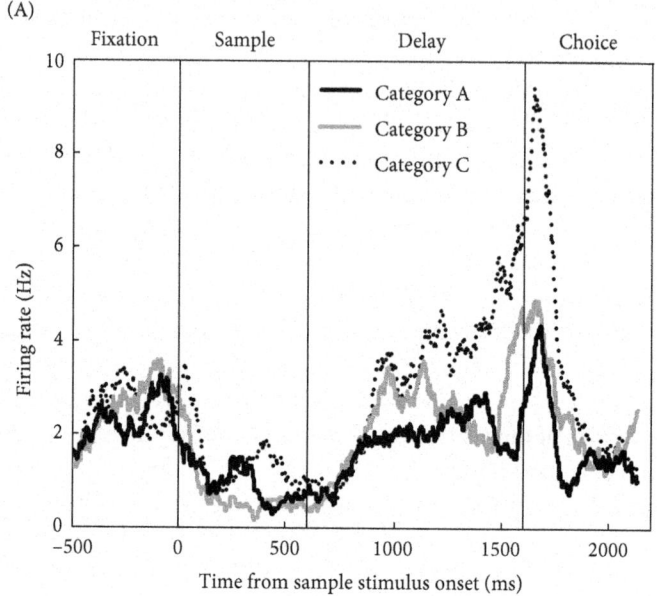

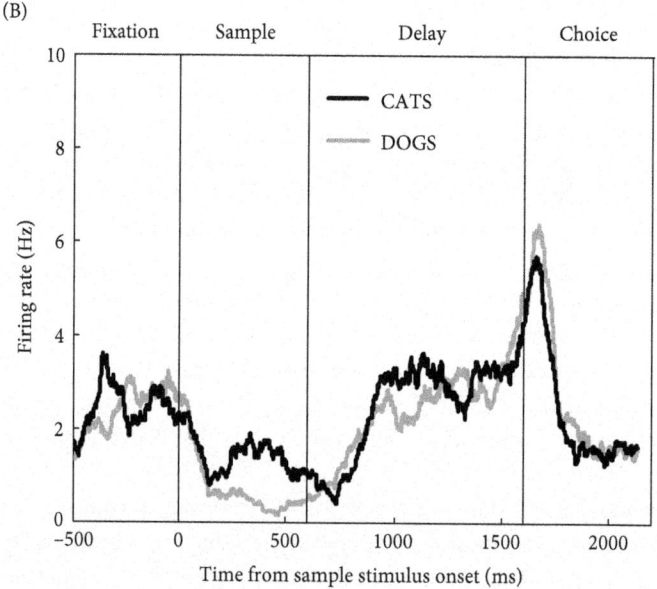

Figure 7.9 A cell in the ventral PF cortex that encoded arbitrary categories, demarcated by the lightly stippled vertical lines in Figure 7.7
(A) this cell preferred category C, that is the category of morphed dog D3 and morphed cat C3; (B) it no longer coded for the previous category dog versus cat.

Reproduced from Freedman D.J., Riesenhuber M., Poggio T., & Miller E.K. Visual categorization and the primate perceptual cortex: Neurophysiology and behavior. *J Neurophysiol*, 88 (2), 929–41, Figure 11a, Doi: 10.1152/jn.2002.88.2.929 Copyright © 2002, The American Physiological Society.

illustration of a cell that coded for cats and limousines. Yet, the association between these two categories was purely arbitrary.

As Hebb (1949) suggested in his classical book on the organization of behaviour, the coding of a stimulus depends not on single cells but on a network of cells. He called this a cell assembly and proposed that there was overlap between cell assemblies that code for different stimuli. In the experiment by Cromer et al. (2010), the population of cells that coded for cats overlapped with the populations that coded for limousines. However, Roy et al. (2010) reported that roughly 40% of the cells responded selectively to just one category.

Synchrony

Stanley et al. (2018) measured the synchrony between local field potentials recorded with different electrodes in the ventral and dorsal PF cortex while monkeys categorized cats or dogs. There was an increase in synchrony in the beta band when a particular category was relevant. When the animals had to make a difficult judgement near the category boundary, the synchrony increased. Stanley et al. suggested that this implied the recruitment of additional resources.

Wutz et al. (2018) varied the difficulty of the task in a different way. They used patterns of dots that formed two categories, and by introducing distortion it was possible to make the distinction more difficult. Categories that were cued by low-level distortion looked alike, whereas categories that were produced by high distortion looked less so.

During the sample period, there was an increase in gamma power in the ventral PF cortex when the level of distortion was low. On the other hand, during the delay period, there was an increase in beta power in the dorsal PF cortex. Wutz et al. suggested that the difference could be related to the degree of abstractness of the category. But it could also have been related to whether there was a delay or not. Delay is a critical factor for the dorsal PF cortex (Chapter 6), but not for the ventral PF cortex (this chapter).

The Representation of Categories in Posterior Areas

There is evidence that categories can be represented in posterior association areas. In an fMRI study, Kriegeskorte et al. (2008b) compared the representation of categories in the inferior temporal cortex in monkeys and human subjects. They presented ninety-two coloured photographs of natural and artificial inanimate objects, as well as the faces and bodies of humans and non-human animals. There was no overt task for the subjects to perform. Nonetheless, RSA could decode the distinction between animate and inanimate categories in the inferior temporal cortex, both in monkeys and humans.

However, the distinction between animate and inanimate objects is one with which the monkeys and human subjects would be familiar. Thus, it is possible that it was the ventral PF cortex that learned the categories in the first place, but that once they had been learned the information was transferred to the inferior temporal cortex.

In an fMRI study, Degutis and D'Esposito (2009) compared the learning of new categories with distinguishing categories that the subjects had already learned. They varied the position of the features in faces, and the subjects were told the categories to distinguish, for example higher eyebrows and lower mouths, versus lower eyebrows and higher mouths. The subjects were then given 250 trials of practice in drawing these distinctions, with feedback provided. Finally, they were overtrained for a further 4,250 trials. When overlearned performance was compared with new learning, there were activations in the middle and inferior frontal gyrus.

In these experiments the distinction between categories involved the overall shape (animals and objects) or internal pattern (faces). The parietal cortex also represents categories, but according to a different metric. Jackson and Woolgar (2018) presented human subjects with 'spiky objects' and 'smooth objects'. The spiky ones could be categorized in terms of their orientation or length, and the smooth ones in terms of their height or breadth. MVPA was used to decode the task-relevant from the task-irrelevant features. Jackson and Woolgar found that there were voxels in the cortex of the IPS, as well as inferior frontal sulcus, that represented the distinctions for both spiky and smooth objects.

However, they did not find any such voxels in the inferior temporal cortex. The reason is probably that the features that were task-relevant or task-irrelevant involved metrics such as orientation, height, breadth, and length. It is the posterior parietal cortex that is specialized for representing metrics (Genovesio et al., 2014), whereas the inferior temporal cortex is specialized for representing shape and colour.

Evidence that the posterior parietal cortex represents metrics comes from a neurophysiological experiment by Fitzgerald et al. (2012) who recorded in the parietal area LIP. There were two quite different categories. One involved the direction in which dots were moving, the two categories differing around an axis of 180 degrees (Figure 7.10B, next page). The other involved the arbitrary association between pairs of objects (Figure 7.10A). The plot shows the average activity for each category, with the categories being distinguished by colour.

There were cells that coded for both types of category. This can be seen from Figure 7.10C which plots the explained variance for the motion categories against the explained variance for the shape pairs or categories.

The coding for object-pairs could be derived from the inferior temporal area (TEO), since TEO and LIP are connected (Distler et al., 1993), and as an earlier section describes there are pair coding cells in the inferior temporal cortex (Miyashita et al., 1998). Alternatively, the coding for object-pairs could be derived

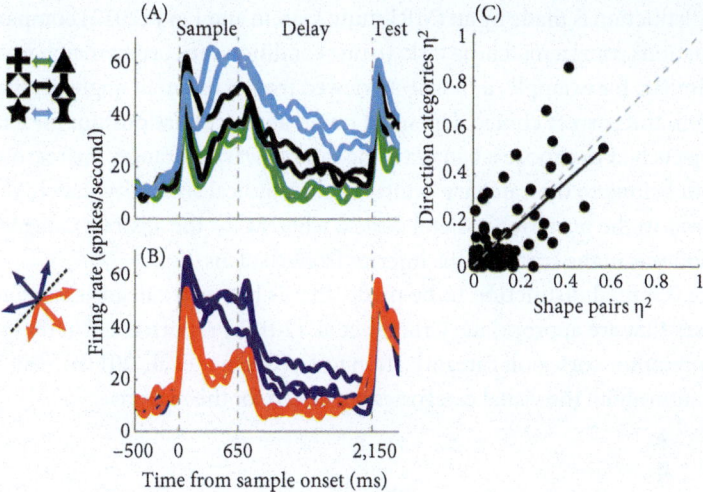

Figure 7.10 Single cells reflecting both shape-shape associations and motion direction categories
(A) The activity of a single LIP cell as a monkey associated six shapes into three pairs in a delayed-match-to-pair task. The average neuronal activity evoked by each sample shape is plotted, and same-colour traces correspond to associated pairs of shapes.
(B) The activity of the same cell while the animal performed a delayed-match-to-category task. Average activity evoked by each sample motion direction is shown, and same-colour traces correspond to directions in the same category. (C) The association or category strength, as measured by explained variance (η^2) for direction categories versus shape pairs, during the first half of the delay period for all neurons tested in both tasks. The solid line is a regression fit, and the dashed line has a slope of 1.
Reproduced from Fitzgerald, J.K., Swaminathan, S.K., & Freedman, D.J. Visual categorization and the parietal cortex. *Front Integ Neurosci*, 6, 18, Figure 1, Doi: 10.3389/fnint.2012.00018 © 2012, Fitzgerald, Swaminathan and Freedman. Licensed under CC BY 4.0.

directly from the PF cortex since the recordings were taken after the monkeys had been overtrained on the categories.

Swaminathan and Freedman (2012) recorded simultaneously from the dorsal PF cortex and parietal LIP after categories had been learned. The categories involved the direction in which dots moved coherently in a visual display. The coding for the motion categories was stronger in area LIP, and it was also evident earlier during the trial in LIP than in the PF cortex. This is consistent with the proposal that once the task has been learned, the categories are established in posterior areas.

In a review paper, Freedman and Assad (2016) concluded that the posterior parietal cortex can represent various categories, but they rely on physical similarity. This is unlike the findings for the ventral PF cortex where cells have been reported that code for categories that are purely arbitrary, cat/dogs or cat/limousines.

The distinction is made by an fMRI study. Garcin and Levy (2012) compared the two situations using a matching task. In one condition the correct match was physically similar; for example, a mouse as viewed from a different angle. In the other condition, the correct choice depended on an abstract relationship; for example, the subjects had to choose a banana if the sample was an orange, the reason being that both belong to the category 'fruit'. For the physical similarity match, the activation was in the posterior parietal cortex, whereas for the abstract categories the activation was in the cortex of the inferior frontal sulcus.

There is a final distinction to be made, that is between categorization and the responses that are appropriate. Cromer et al. (2011b) reported cell activity in the dorsal premotor cortex on categorization tasks (Cromer et al., 2011b). This lacked information about the visual categories but coded for the response.

Summary

When monkeys learn to choose objects on the basis of the category to which they belong, cell activity in the ventral PF cortex encodes that category. Furthermore, the cells encode these categories as monkeys learn to categorize or re-categorize stimuli into any subsets that the experimenters choose to impose. Although monkeys with lesions that include the ventral PF cortex can still recognize stimuli as being similar, they do so on the basis of physical similarity via stimulus generalization. Category information can also be found in the inferior temporal and posterior parietal cortex, but it is the ventral PF cortex that supports learning; the categorization in posterior areas results from overtraining.

Changing Responses

As the previous sections mentions, many experiments on associative learning and categorization use a matching design. The task is therefore a conditional task, meaning that the response depends on the sample presented. As a consequence, the animal has to behave in a flexible way, now doing this, now that. Changing associations requires the inhibition of the prior response and the activation of the new one.

Inhibiting and Activating Responses

On the categorization task studied by Freedman et al. (2002), the animals had to respond if the probe picture was in the same category as the sample but to withhold their response if it was not. Thus, the task was like a go no-go task in which the animals have to respond given one cue and withhold their response given another.

Sakagami et al. (2001) tested monkeys on a go no-go task. In one condition, a pattern of green dots meant go irrespective of the direction in which they moved, while red meant no go; in the other condition, downward motion of the dots meant go irrespective of the colour, whereas upward motion meant no go.

Because the recordings were taken in the ventral PF cortex (area 45A), most of the cells responded selectively to the association with colour cues rather than the motion cues. But the activity in this area was not related to withholding a response. Of the colour cells, 64% coded for going and 36% for not going. In other words, this area is not specialized for withholding a response but for learning the meaning of cues.

In the experiment by Sakagami et al. on no-go trials there was a decrease in the activity of the 'go' cells which showed a sensory response to the green cue, whereas there was an increase in the activity of the 'no-go' cells which had a sensory response to the red cue. It is probably this difference that accounts for the activation in the ventral PF cortex in an fMRI experiment on monkeys by Morita et al. (2004). The animals were rewarded for going when a green cue was presented and for withholding a response when a red one was presented. When no-go trials were contrasted with go trials, there was an activation in area 45A.

This area could influence withholding a response in one of two ways. One is via indirect connections with the dorsal premotor cortex, via the preSMA (Takahara et al., 2012). The other is via direct connections to the inferior caudal PF cortex (areas 45B and 44) which projects to both dorsal and ventral premotor cortex (Frey et al., 2014).

Evidence for the latter route comes from an fMRI study on human subjects by Chikazoe et al. (2009). They compared no-go with go trials and found activations in the right inferior caudal PF cortex (area 44), as well as in the ventral PF cortex.

Aron et al. (2003) devised the 'stop-signal' task to provide a purer measure of the ability to withhold a response. The subjects are instructed to press a key, but to withhold their response when a stop-signal is presented. The later this signal, the more difficult it is for the subjects to withhold their response.

Aron and Poldrack (2006) compared stop trials with go trials, and found a peak in the right inferior caudal PF cortex (area 44). However, a later meta-analysis of fMRI studies involving motor inhibition indicate that the activation is actually in the cortex of the IFJ. According to Amunts and Zilles (1999), this is the dorsal division of area 44 (44d), distinct from the opercular division of area 44 (44v) that is associated with speech (Kuhn et al., 2013). One possibility is that area 44d was activated because the instruction to stop was signalled by a sound.

In the study by Aron and Poldrack (2006), there were also activations in three other areas. These were the subthalamic nucleus, the preSMA and the dorsal paracingulate sulcus. When Aron and Poldrack looked for activations that were greater for subjects who were quick to stop compared with those who were slow, there was an activation that correlated with performance in the right subthalamic nucleus.

In order to study the pathway involved in stopping, Aron et al. (2007) combined fMRI with diffusion weighted images (DWI). They were able to show that the right inferior caudal PF cortex (area 44d) was connected anatomically with the subthalamic nucleus, as well as with the pre-SMA.

However, it was not clear from this study whether this system was involved in inhibition alone or in changing responses in general. So, with other colleagues (Verbruggen et al., 2010), Aron designed an experiment with several conditions. On some trials, a colour cue instructed the subjects to respond with one finger on a keyboard: green meant that the subjects should press one key and yellow another, using different fingers of the right hand. This was the 'primary task'.

On 'dual-signal' trials, a different cue appeared which instructed the subjects to respond with a finger at the same time as pressing the space bar with their right thumb. The experiment also included no-go trials in which a third cue instructed the subjects to do nothing.

Theta burst stimulation was applied over the right inferior caudal PF cortex (area 44). The effect was to interfere with both change and no-go trials. However, Aron et al. (2014) suggested that the interference on change trials could have been due to current spread to the ventral premotor cortex.

The alternative is that on change trials it is necessary to inhibit the previous response. There is evidence that the representation of the previous response can persist. This was demonstrated in a study by Akaishi et al. (2010) in which the subjects switched between saccades and antisaccades.

Zhang et al. (2012) used fMRI to scan subjects on a free choice task, especially on those trials in which the subjects spontaneously suppressed the tendency to repeat the previous finger movement. There was an activation which Zhang et al. attributed to the inferior caudal area 45B, though reference to the MNI atlas by Petrides (2018) suggests that it probably lay in area 44.

This activation provides direct evidence for suppression during the change: it correlated with a reduction in the rate of the accumulator for the previous action as estimated from the expected accumulator metabolic activity.

However as already mentioned, changing responses involves not only inhibiting the old one but also activating the new one. Neubert et al. (2010) compared the routes for these two operations. They taught two tasks to human subjects. On one, a red cue instructed a response with the right hand and green one a response with the left. On the other, a red cue instructed a response with the left hand and green with the right. The switch between tasks was marked by the colour of a central cue.

So, as to study the interactions between areas, the paired pulse TMS technique was used. In this, a pulse is applied to area A before another pulse is applied to area B, the timing between the two pulses being varied on a millisecond time scale.

When Neubert et al. applied a pulse to the right IFJ (44d) on switch trials, there was a decrease in the amplitude of the motor excitatory potential (MEP) that could be evoked by stimulation of the motor cortex. This suggests that the effect of TMS to the inferior caudal PF cortex was to suppress the routine task.

Yet in a parallel experiment, Mars et al. (2009) applied a TMS pulse to the preSMA before stimulating motor cortex on switch trials, and the result was that there was an increase in the MEP that was evoked by a pulse to the motor cortex. This suggests that the effect of TMS to the preSMA was to activate the new task.

Neubert et al. (2010) investigated whether these effects involved a cortico-cortical route, or a route from cortex to subcortex and thence back to the cortex. The distinction could be made by examining the latency for which the effects were found. Effects found with short latencies were taken to suggest a cortico-cortical route. But there were also effects at the longer latency of 12 seconds, and these were taken to suggest a cortico-subcortical-cortical route. The amplitude of these effects was correlated with individual differences in the size of the white matter tracts that ran near the subthalamic nucleus (STN) as measured using DWI.

However, there is a need to delineate the whole network of areas that are involved in switching between tasks. To do this Premereur et al. (2018) used fMRI to scan monkeys. The animals were trained to fixate a central point. If the target was green, the monkeys had to make a saccade to it; if the target was blue, the monkeys had to retract the hand that was ipsilateral to the target. Thus, switches were instructed by the colour of the target.

As is typical for task switching, there was a switch cost when the reaction times for switch and stay trials were compared. The analysis compared the activations for switch with stay trials. For this contrast, there were activations in the caudal inferior PF cortex, as well as in areas 45A and 9/46v and the anterior cingulate sulcus (sACC).

Johnston et al. (2007) compared the contribution of the sACC and caudal PF cortex by recording simultaneously in the two areas while monkeys switched between prosaccades and antisaccades. The two responses were rewarded in blocks of trials. However, there was no external cue to tell the animals when to switch. The cue for switching was the failure to obtain the predicted reward.

While the monkey prepared its response after an error, the cells in the sACC showed a higher level of task selectivity than cells in the caudal PF cortex. On subsequent trials on the same task, the task selectivity of the cells declined in the sACC, whereas it remained constant in the caudal PF cortex. These results suggest that the switch was marked by a signal for prediction error in the sACC (Chapter 3).

In an earlier paper, Shima and Tanji (1998) had previously showed that there were cells in the sACC that increased their activity when the amount of reward was less than predicted. Furthermore, when Shima and Tanji inactivated the cells with muscimol, the monkeys failed to switch between two responses, pushing or turning a handle. The same results have been reported when recordings were taken in a human patient during surgery (Williams et al., 2004). There were cells that increased their activity after a reduction in reward, and the subsequent surgical removal of tissue in the ACC impaired the ability of the patient to switch between responses.

The effect of the prediction error signal in the ACC is to engage the PF cortex. There is evidence for this from a study by Kerns et al. (2004) who scanned human subjects while they were tested on the Stroop task. In each condition the subjects had to read a colour word such as 'red', but in the Stroop condition the word was printed in a font of a different colour, for example the word 'red' in blue font.

The reaction time for naming was increased if a 'font' trial followed after a 'word' trial', the reason being that the subjects had to suppress the tendency to say the colour word itself. There was a corresponding increase in activation in the sACC on these trials. And the greater this activation, the more that the dorsal PF cortex was re-activated.

As in the study on monkeys by Johnston et al. (2007), these results suggest that activity in the sACC is involved in recruiting the PF cortex for the attentive selection of the goal on the subsequent trial. If so, a cortical lesion that invades the sACC should impair the ability to adjust on a trial that follows one that involves conflicting cues.

Maier et al.(2015) tested this prediction by testing human subjects on the Eriksen flanker task (1974). On this the subject has to press a key on the left or right, depending on the direction of a central arrowhead. However, there are also conflict trial they have to do so while ignoring arrowheads in the periphery which point in the opposite direction. Maier et al. tested patients with large cingulate lesions, and unlike the control subjects the patients failed to show an increase in accuracy on the trials that followed an error.

It is evident that there is a common network that supports the ability of subjects to change flexibly from routine to non-routine behaviour, whether on the stop-signal task, when switching between response, or when switching between tasks. The cortex in the sACC signals that a change is needed, and the PF cortex is recruited as a result. Via its connections with the premotor and thus motor cortex via the STN (Haynes & Haber, 2013), the PF cortex suppresses the routine response or task. Via its connections with the preSMA (Takahara et al., 2012), it activates the new response or task.

Summary

When monkeys or human subjects are required to change associations and inhibit the previous response, there is a common network of areas that are involved. These are the inferior caudal PF cortex, the subthalamic nucleus, the preSMA and the tissue in the sACC. A prediction error signal in the sACC recruits the PF to select the new goal. In turn the PF cortex influences action via two routes, one other running via the STN and the other via the preSMA. The subcortical path suppresses the routine response whereas the cortico-cortical path activates the new one.

Changing Cues

Flexible behaviour requires the subject not only to change responses but also to change the aspect of the stimuli to which they attend. The problem in the world outside the laboratory is that at any one time the environment is full of potential cues, but for consistent behaviour most of them have to be filtered out. As Allport (1986) pointed out, this is not because the brain cannot handle them but because only some of the information that is available is relevant for the task in hand.

The Wisconsin card sorting task has been used to test the ability of patients with frontal lobectomies to discover which of several cues or dimensions are currently relevant and to change between them (Milner, 1963). The patients are required to sort a series of cards into four separate piles according to coloured shapes on each card. On any one trial, the patient can sort a card according to the colour, shape, or number of the designs on that card.

The patients are told on each trial whether their choice had been correct or incorrect. At first, they must learn to sort the cards by a given feature, for example colour. The experimenter later changes the relevant feature, for example to shape. In the standard version of the task, the patient learns about the change only because of feedback: correct or incorrect. In a modified version of the task, the experimenter specifically instructs the patient to change category (Nelson, 1976).

In the original paper Milner (1963) stressed the fact that patients with large frontal lobectomies made perseverative errors, continuing to sort according to the previous category. However, Barcelo and Knight (2002) also found that patients with large frontal lesions made random errors as well as perseverative ones. In other words, they were impaired not only in sorting according to the previous dimension but also in continuing to sort according to the new one.

The lesions in these studies were very large, and this means that the studies are of no help in delineating the areas that are critical for the change. This led Mansouri et al. (2006) to devise a simplified version of the Wisconsin card sorting task for monkeys. The animals learned to match stimuli either on the basis of colour or shape, and there was a shift between categories on each day, with reward or non-reward as feedback. Buckley et al. (2009) showed that on this task lesions of the central and medial sectors of the orbital PF cortex caused an impairment in learning to stay after rewarded trials. As Chapter 4 describes, Rudebeck et al. (2008) had previously found that monkeys with an orbital PF lesion are impaired at using reward to guide a subsequent choice.

Buckley et al. also made lesions in the ventral PF cortex, testing the monkeys on a task in which the shifts occurred between days. The monkeys were unable to

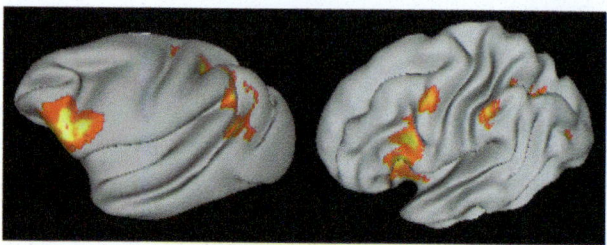

Figure 7.11 Activations for shifting between categories, shown on inflated surface reconstructions of the macaque monkey brain (left) and human brain (right)
The brains are expanded so as to show the activations in sulci.
Reproduced from Nakahara, K., Hayashi, T., Konishi, S., & Miyashita, Y. Functional MRI of macaque monkeys performing a cognitive set-shifting task. *Science*, 295 (5559), 1532–6, Figure 3c, Doi: 10.1126/science.1067653 Copyright © 2002, The American Association for the Advancement of Science

relearn either colour or shape matching at all after surgery. This is consistent with the findings reported earlier in this chapter for the effects of ventral PF lesions on simultaneous matching. It also supports the claim that it is the ventral PF cortex that learns the behavioural meaning of cues.

Finally, Buckley et al. also made lesions in the dorsal PF cortex. After the animals had reached criterion on a rule, the interval between trials was lengthened by five seconds. The result was that the performance of the monkeys fell to chance after this delay. Buckley et al. suggested that the reason was that they had failed to recall the current rule. But it is equally possible that they failed to code prospectively for the current rule during the delay.

Rather than making lesions, Nakahara et al. (2002) used fMRI to scan monkeys as they matched by colour or shape, and as they shifted between categories. When the monkeys shifted between categories, there was an activation in the inferior caudal PF cortex. The anatomical results are shown in Figure 7.11 (left). The surface of the brain has been inflated so as to show that in the macaque monkey the activation was centred on the descending limb of the arcuate sulcus.

For comparison, Nakahara et al. also scanned human subjects. Figure 7.11 (right) shows that there were peaks in the caudal inferior PF cortex (areas 45B and 44). In a related study on human subjects, Konishi et al. (1999) reported a peak in the same area for shifting between dimensions and for withholding a response on a go no-go task.

Mechanisms

Though the activations for shifting between cues lie in the inferior caudal PF cortex, a previous section shows that it is the ventral PF cortex that supports the ability of monkeys to learn the behavioural meaning of cues. Lauwereyns et al. (2001)

trained monkeys that green meant 'go' and red meant 'no-go', whereas downward motion meant 'no-go' and upward motion meant 'go'. Cells that responded to colour were concentrated in the ventral PF cortex (area 45A), whereas cells that responded to motion were found dorsally in areas 8Ad and 9/46. Surprisingly, about a third of the cells carried information about the irrelevant as well as the relevant cues, and the majority of these were found in the ventral PF cortex.

This suggests a top-down influence that allows monkeys to attend selectively to one cue or another. To study this Siegel et al. (2015) presented them with coloured moving dots. The animals had to make a saccade to the left or right, either in response to different shades of green and red, or to variants of upwards and downwards motion. Visual instruction cues indicated the relevant visual submodality, for example shades of colour. Because of this experimental design, the animals also had to divide the feature space in an arbitrary way, thus establishing categories.

By recording simultaneously from six sites, Siegel et al. were able to demonstrate that the order of the neural information about the colours could be detected first in the visual areas (V4) and the inferior temporal cortex, and information about motion in area MT and the parietal area LIP with which it is connected. Activity about the cues occurred earlier in these areas than in the PF area 8 or the more anterior PF cortex, including areas 9/46v and 45A.

By contrast, information about the relevant category was found earlier in the PF cortex and LIP than in the visual areas V4, MT, and the inferior temporal cortex. This result is consistent with the proposal that the PF cortex sends a top-down signal that encodes the cue or category that is currently relevant

To follow up these results, Brincat et al. (2018) showed that the PF cortex, including area 9/46 and 45A, maximally represented the abstract category, whereas the neural representations in areas V4 and MT were tightly linked to the sensory inputs. The representations in the caudal PF area 8, LIP, and inferior temporal cortex were intermediate, coding both for the current task rule and the sensory features. The categorical representations in the PF area 8, as well as the inferior temporal cortex and LIP, can be explained as being derived top-down, with the sensory representations occurring bottom-up.

However to prove a top-down influence, it is necessary to intervene in the system. So, Morishima et al. (2009) applied single-pulse TMS over the caudal PF cortex (area 8) of human subjects while recording with EEG over posterior visual areas. The subjects viewed faces composed of moving dots and had to make judgements either about the motion of the dots or about the gender of the face. Thus, they had to attend either to motion or to faces.

There were EEG sources in the MT complex and the fusiform face area. When the subjects prepared to make a judgement about motion, TMS over the PF cortex enhanced the activity in the MT complex. By contrast, when they prepared to make judgements about gender, TMS over the PF cortex enhanced activity in the fusiform face area (FFA).

Heinen et al. (2013) repeated this experiment, with the difference that they combined TMS with fMRI. They presented faces with moving dots; in one condition the subjects had to judge the gender of the face and in the other the direction of the moving dots. The fusiform face area (FFA) and the MT complex (MT+) were chosen as regions of interest. When a TMS pulse was applied over the caudal PF cortex (area 8), it led to an increase in the activation in the FFA on gender trials and in MT+ on motion trials. These results are shown in Figure 7.12 (next page) which also shows the null result for a passive condition in which there was no task to perform.

Though TMS was applied in these experiments over the caudal PF cortex, it is the PF cortex anterior to the caudal PF cortex that encodes the current task rule.

Buschman et al. (2012) studied the mechanisms for encoding the current task rule by teaching monkeys two rules. According to the colour rule, red meant saccade to the left and blue meant saccade to the right; according to the orientation rule, horizontal meant saccade to the left and vertical meant saccade to the right. The same rule applied for at least 20 trials before a switch. Thus, the animals had some forewarning that a switch was likely to occur.

Buschman et al. recorded local field potentials from pairs of electrodes that they describe as being placed in the 'dorsolateral prefrontal cortex'. In their usage, this refers to both the dorsal and ventral PF cortex. One ensemble of potentials showed coherent firing for the colour rule and another ensemble for the orientation rule. However, the orientation rule was dominant.

As the animals prepared to switch from the orientation rule to the weaker colour rule, there was an increase in synchrony in the alpha band. By contrast, beta synchronization was evident around the time of the presentation of the stimuli. Buschman et al. suggested that the implication was that alpha was involved in deselecting the current dominant rule, and beta in selecting the new rule.

Summary

PF lesions impair the ability to switch between cues and dimensions. The reason is that the PF cortex sends a top-down signal to the sensory areas to indicate which cues are currently relevant. This leads to enhancement in the relevant area. There is a suggestion that an increase in alpha power is involved in deselecting the current rule, and an increase in beta power in selecting the new rule.

Conclusions

Chapter 6 discusses the claim that the dorsal PF cortex is specialized for working memory. It reviews the evidence that this area does not in fact support short-term memory per se, but rather that it is critical for order memory.

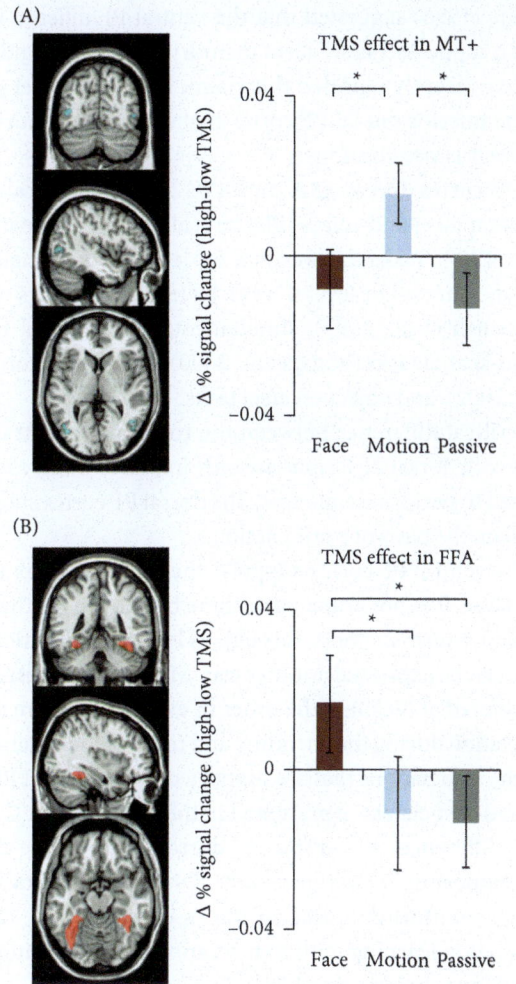

Figure 7.12 (A) The effect of a TMS pulse over the FEF on the activation in the MT complex when motion is relevant. (B) The effect of a TMS pulse over the FEF on the activation in the fusiform face area when the shape of the face is relevant
The ordinate shows the change in activation when comparing a high intensity with low intensity pulse over the FEF. MT+, MT complex; FFA, fusiform face area
Reproduced from Heinen, K., Feredoes, E., Weiskopf, N., Ruff, C.C., & Driver, J. Direct evidence for attention-dependent influences of the frontal eye-fields on feature-responsive visual cortex. *Cereb Cortex*, 24 (11), 2815–21, Figure 3, Doi: 10.1093/cercor/bht157 Copyright © 2013, Oxford University Press.

Goldman-Rakic (1996) suggested that the ventral PF differed from the dorsal PF cortex in that it supported short-term memory for objects. And Constantinidis and Qi (2018) have recently repeated this claim. But this chapter reviews the evidence that it is the inferior caudal PF cortex that supports memory for objects including faces, as well as vocalizations.

The ventral PF cortex (areas 45A and 47/12) is not critical for short-term memory, whether for faces and voices (Plakke et al., 2015) or for patterns or colours (Rushworth et al., 1997). Instead it is critical for learning or changing associations. This has been demonstrated in several ways, by testing monkeys on simultaneous matching (Rushworth et al., 1997), simultaneous visual-visual tasks (Gutnikov et al., 1997), visuo-spatial tasks (Wang et al., 2000), matching according to category (Freedman et al., 2002), and object reversal tasks.

So there is a critical difference between the roles of the ventral and dorsal PF cortex. Monkeys with dorsal PF lesions are only impaired if there is a delay and the animals are trained to perform sequences. The dorsal PF cortex supports the ability to generate a goal on the basis of a *prior* action.

By contrast, the ventral PF cortex supports the ability to generate a goal on the basis of a current cue. It is not critical that there be a delay. Furthermore, the goal is retrieved from long-term memory through its learned association with the cue.

The difference can be expressed another way. Monkeys with lesions of the dorsal PF cortex are impaired at recalling the order of events, and there is cell activity as well as fMRI activation during the encoding and recall of the order. However, order is irrelevant for the associations that are supported by the ventral PF cortex. It does not matter if paired-associates are presented in the order A–C or C–A.

In spite of this difference, the ventral and dorsal PF cortex are alike in that they are involved in supporting conditional tasks. On conditional tasks, the goal depends on the cue, and thus the response varies over time. The tasks differ from sensory discrimination learning in that they cannot be solved simply by the accumulative reward of one choice over trials.

It is intrinsic to a conditional task that there must be a transformation from the action or cue to the goal. Chapter 6 reviews evidence for cell activity that supports such a transformation from one action to another. The present chapter mentions evidence that there is a change in beta power and synchrony in the ventral PF cortex that reflects the transformation during associative learning (Brincat & Miller, 2016).

Selective Advantage

The matching and paired-associate tasks are laboratory tasks, and non-human anthropoids do not learn these in their natural surroundings. Yet, the ventral PF cortex (areas 47/12 and 45A) evolved in anthropoid primates as their brain and bodies expanded, after the evolution of the primate fovea and along with the development of trichromatic vision. So it is necessary to provide an account of the selective advantage that these areas conferred.

The ventral PF cortex allows anthropoids to learn which types of fruit and leaves should be the target for search. It supports the ability to identify different food items as being similar in kind in term of their taste or nutritional value. Since inputs to the ventral PF cortex also provide information about faces and vocalizations, the system may also support the ability to detect similarities between conspecific temperament or differences between predators the way in which they hunt.

More generally, the ventral PF cortex allows anthropoids to grasp abstract concepts such as shape, colour and motion. It is for this reason that macaque monkeys can learn task rules that involve these in laboratory experiments. The ability to do so is an indicator of the ability to make sense of the world. In other words, it is an aspect of intelligence.

Connectivity

Among its many connections, the ventral PF cortex has five that, collectively, make it unique. It has (1) connections with the inferior and superior temporal cortex as well as the perirhinal cortex; (2) interconnections with the orbital PF cortex and amygdala; (3) outputs to the caudal PF cortex (area 8Av and the FEFv); (4) direct outputs to the ventral premotor cortex and indirect outputs to the dorsal premotor cortex via the preSMA; (5) connections with the inferior caudal PF cortex (areas 45B and 44) which are connected in turn with the dorsal and ventral premotor cortex.

Neural Operation

Like previous chapters, this one offers a proposal concerning the transformations that are performed by the area.

There are visual and auditory inputs to the ventral PF cortex from the temporal lobe. These carry information about objects and cues as well as about faces and their associated vocalizations. The conjunction of inputs from the temporal lobe and the orbital PF cortex specifies the current goal on the basis of the relevant cue. This involves a transformation from cue to response, and there are changes in beta power and synchrony that may reflect this transformation.

The cue can be a specific item or the member of an abstract class since the ventral PF cortex can learn associations between objects even when they are not visually similar. The goal is retrieved from long-term memory via connections with the hippocampal system, including the perirhinal and entorhinal cortex.

Outputs to the ventral FEF transform the goal into the target for search, and outputs to the premotor areas transform it into the target for manual retrieval. Top-down connections from the ventral PF cortex to the ventral visual stream, either directly or via the FEF, enhance the representation of those aspects of the stimulus inputs that are relevant given the current objective.

References

Akaishi, R., Morishima, Y., Rajeswaren, V.P., Aoki, S., & Sakai, K. (2010). Stimulation of the frontal eye field reveals persistent effective connectivity after controlled behavior. *J Neurosci*, 30, 4295–305.

Allport, A. (1986) Selection for action: Some behavioral and neurophysiological considerations of attention and action. In Heuer, H., Sanders, A.F. (eds) *Perspectives on Perception and Action*. Erlbaum, Hillsdale, pp. 395–417.

Amaral, D.G. & Price, J.L. (1984) Amygdalo-cortical projections in the monkey ({IMacaca fascicularis}). *J Comp Neurol*, 230, 465–96.

Amunts, K., Lenzen, M., Friederici, A.D., Schleicher, A., Morosan, P., Palomero-Gallagher, N., & Zilles, K. (2010) Broca's region: Novel organizational principles and multiple receptor mapping. *PLoS Biol*, 8.

Amunts, K., Schleicher, A., Burgel, U., Mohlberg, H., Uylings, H.B.M., & Zilles, K. (1999) Broca's region revisited: Cytoarchitecture and intersubject variability. *J Comp Neurol*, 412, 319–41.

Amunts, K. & Zilles, K. (2012) Architecture and organizational principles of Broca's region. *Trends Cogn Sci*, 16, 418–26.

Andreau, J.M. & Funahashi, S. (2011) Primate prefrontal neurons encode the association of paired visual stimuli during the pair-association task. *Brain Cogn*, 76, 58–69.

Aron, A.R., Behrens, T.E., Smith, S., Frank, M.J., & Poldrack, R.A. (2007) Triangulating a cognitive control network using diffusion-weighted magnetic resonance imaging (MRI) and functional MRI. *J Neurosci*, 27, 3743–52.

Aron, A.R., Fletcher, P.C., Bullmore, E.T., Sahakian, B.J., & Robbins, T.W. (2003) Stop-signal inhibition disrupted by damage to right inferior frontal gyrus in humans. *Nat Neurosci*, 6, 115–16.

Aron, A.R. & Poldrack, R.A. (2006) Cortical and subcortical contributions to stop signal response inhibition: Role of the subthalamic nucleus. *J Neurosci*, 26, 2424–33.

Aron, A.R., Robbins, T.W., & Poldrack, R.A. (2014) Inhibition and the right inferior frontal cortex: One decade on. *Trends Cogn Sci*, 18, 177–85.

Asaad, W.F., Rainer, G., & Miller, E.K. (1998) Neural activity in the primate prefrontal cortex during associative learning. *Neuron*, 21, 1399–407.

Barcelo, F. & Knight, R.T. (2002) Both random and perseverative errors underlie WCST deficits in prefrontal patients. *Neuropsychol*, 40, 349–56.

Boettiger, C.A. & D'Esposito, M. (2005) Frontal networks for learning and executing arbitrary stimulus-response associations. *J Neurosci*, 25, 2723–32.

Brass, M., Derrfuss, J., Forstmann, B., & von Cramon, D.Y. (2005) The role of the inferior frontal junction area in cognitive control. *Trends Cogn Sci*, 9, 314–16.

Brincat, S.L. & Miller, E.K. (2015) Frequency-specific hippocampal-prefrontal interactions during associative learning. *Nat Neurosci*, 18, 576–81.

Brincat, S.L. & Miller, E.K. (2016) Prefrontal cortex networks shift from external to internal modes during learning. *J Neurosci*, 36, 9739–54.

Brincat, S.L., Siegel, M., von Nicolai, C., & Miller, E.K. (2018) Gradual progression from sensory to task-related processing in cerebral cortex. *Proc Natl Acad Sci USA*, 115, E7202–11.

Buckley, M.J., Mansouri, F.A., Hoda, H., Mahboubi, M., Browning, P.G., Kwok, S.C., Phillips, A., & Tanaka, K. (2009) Dissociable components of rule-guided behavior depend on distinct medial and prefrontal regions. *Science*, 325, 52–8.

Buckley, M.J. & Sigala, N. (2010) Is top-down control from prefrontal cortex necessary for visual categorization? *Neuron*, 66, 471–3.

Buschman, T.J., Denovellis, E.L., Diogo, C., Bullock, D., & Miller, E.K. (2012) Synchronous oscillatory neural ensembles for rules in the prefrontal cortex. *Neuron*, 76, 838–46.

Bussey, T., Wise, S., & Murray, E. (2001) The role of ventral and orbital prefrontal cortex in conditional visuomotor learning and strategy use in rhesus monkeys ({Macaca mulatta}). *Behav Neurosci*, 115, 971–82.

Bussey, T.J., Wise, S.P., & Murray, E.A. (2002) Interaction of ventral and orbital prefrontal cortex with inferotemporal cortex in conditional visuomotor learning. *Behav Neurosci*, 116, 703–15.

Chikazoe, J., Jimura, K., Asari, T., Yamashita, K., Morimoto, H., Hirose, S., Miyashita, Y., & Konishi, S. (2009) Functional dissociation in right inferior frontal cortex during performance of go/no-go task. *Cereb Cortex*, 19, 146–52.

Constantinidis, C. & Qi, X.L. (2018) Representation of spatial and feature information in the monkey dorsal and ventral prefrontal cortex. *Front Integr Neurosci*, doi. 10.3389/fnint.2018.00031.

Coudé, G., Ferrari, P.F., Roda, F., Maranesi, M., Borelli, E., Veroni, V., Monti, F., Rozzi, S., & Fogassi, L (2011) Neurons controlling voluntary vocalization in the macaque ventral premotor cortex. *Plos One*, 6, e26822.

Cromer, J.A., Machon, M., & Miller, E.K. (2011a) Rapid association learning in the primate prefrontal cortex in the absence of behavioral reversals. *J Cogn Neurosci*, 23, 1823–8.

Cromer, J.A., Roy, J.E., Buschman, T.J., & Miller, E.K. (2011b) Comparison of primate prefrontal and premotor cortex neuronal activity during visual categorization. *J Cogn Neurosci*, 23, 3355–65.

Cromer, J.A., Roy, J.E., & Miller, E.K. (2010) Representation of multiple, independent categories in the primate prefrontal cortex. *Neuron*, 66, 796–807.

DeGutis, J. & D'Esposito, M. (2009) Network changes in the transition from initial learning to well-practiced visual categorization. *Front Hum Neurosci*, 3, 44.

Distler, C., Boussaoud, D., Desimone, R., & Ungerleider, L.G. (1993) Cortical connections of inferior temporal area TEO in macaque monkeys. *J Comp Neurol*, 334, 125–50.

Eriksen, B.A. & Eriksen, C.W. (1974) Effect of noise letters upon the identification of a target letter in a nonsearch task. *Percep Psychophys*, 16, 143–9.

Fitzgerald, J.K., Swaminathan, S.K., & Freedman, D.J. (2012) Visual categorization and the parietal cortex. *Front Integr Neurosci*, 6, 18.

Folloni, D., Sallet, J., Khrapitchev, A.A., Sibson, N., Verhagen, L., & Mars, R.B. (2019) Dichotomous organization of amygdala/temporal-prefrontal bundles in both humans and monkeys. *Elife*, 8.

Freedman, D.J. & Assad, J.A. (2016) Neuronal mechanisms of visual categorization: An abstract view on decision making. *Annu Rev Neurosci*, 39, 129–47.

Freedman, D.J., Riesenhuber, M., Poggio, T., & Miller, E.K. (2001) Categorical representation of visual stimuli in the primate prefrontal cortex. *Science*, 291, 312–16.

Freedman, D.J., Riesenhuber, M., Poggio, T., & Miller, E.K. (2002) Visual categorization and the primate prefrontal cortex: Neurophysiology and behavior. *J Neurophysiol*, 88, 929–41.

Freedman, D.J., Riesenhuber, M., Poggio, T., & Miller, E.K. (2003) A comparison of primate prefrontal and inferior temporal cortices during visual categorization. *J Neurosci*, 23, 5235–46.

Frey, S., Mackey, S., & Petrides, M. (2014) Cortico-cortical connections of areas 44 and 45B in the macaque monkey. *Brain Lang*, 131, 36–55.

Funahashi, S. & Andreau, J.M. (2013) Prefrontal cortex and neural mechanisms of executive function. *J Physiol Paris*, 107, 471–82.

Fuster, J.M. & Jervey, J.P. (1982) Neuronal firing in the inferotemporal cortex of the monkey in a visual memory task. *J Neurosci*, 2, 361–75.

Gaffan, D. & Harrison, S. (1991) Auditory-visual associations, hemispheric specialization and temporal-frontal interaction in the rhesus monkey. *Brain*, 114 (Pt 5), 2133–44.

Garcin, B., Volle, E., Dubois, B., & Levy, R. (2012) Similar or different? The role of the ventrolateral prefrontal cortex in similarity detection. *PLoS One*, 7, e34164.

Gavrilov, N., Hage, S.R., & Nieder, A. (2017) Functional specialization of the primate frontal lobe during cognitive control of vocalizations. *Cell Rep*, 21, 2393–406.

Genovesio, A., Wise, S.P., & Passingham, R.E. (2014) Prefrontal-parietal function: From foraging to foresight. *Trends Cogn Sci*, 18, 72–81.

Gerbella, M., Belmalih, A., Borra, E., Rozzi, S., & Luppino, G. (2010) Cortical connections of the macaque caudal ventrolateral prefrontal areas 45A and 45B. *Cereb Cortex*, 20, 141–68.

Gerbella, M., Borra, E., Rozzi, S., & Luppino, G. (2016) Connections of the macaque granular frontal opercular (GrFO) area: A possible neural substrate for the contribution of limbic inputs for controlling hand and face/mouth actions. *Brain Struct Funct*, 221, 59–78.

Ghazizadeh, A., Hong, S., & Hikosaka, O. (2018) Prefrontal cortex represents long-term memory of object values for months. *Curr Biol*, 28, 2206–17 e2205.

Goldman-Rakic, P.S. (1996) The prefrontal landscape: implications of functional architecture for understanding human mentation and the central executive. *Philos Trans R Soc Lond B Biol Sci*, 351, 1445–53.

Gutnikov, S.A., Ma, Y.Y., & Gaffan, D. (1997) Temporo-frontal disconnection impairs visual-visual paired association learning but not configural learning in Macaca monkeys. *Eur J Neurosci*, 9, 1524–9.

Hartstra, E., Kuhn, S., Verguts, T., & Brass, M. (2011) The implementation of verbal instructions: An fMRI study. *Hum Brain Mapp*, 32, 1811–24.

Hasegawa, I., Fukushima, T., Ihara, T., & Miyashita, Y. (1998) Callosal window between prefrontal cortices: Cognitive interaction to retrieve long-term memory. *Science*, 281, 814–18.

Haxby, J.V., Horwitz, B., Ungerleider, L.G., Maisog, J.M., Pietrini, P., & Grady, C.L. (1994) The functional organization of human extrastriate cortex: A PET-rCBF study of selective attention to faces and locations. *J Neurosci*, 14, 6336–53.

Haynes, W.I. & Haber, S.N. (2013) The organization of prefrontal-subthalamic inputs in primates provides an anatomical substrate for both functional specificity and integration: Implications for Basal Ganglia models and deep brain stimulation. *J Neurosci*, 33, 4804–14.

Hebb, D.O. (1949) *The Organization of Behavior*. Wiley, New York.

Heinen, K., Feredoes, E., Weiskopf, N., Ruff, C.C., & Driver, J. (2013) Direct evidence for attention-dependent influences of the frontal eye-fields on feature-responsive visual cortex. *Cereb Cortex*, 24, 2815–21.

Higuchi, S. & Miyashita, Y. (1996) Formation of mnemonic neuronal responses to visual paired associates in inferotemporal cortex is impaired by perirhinal and entorhinal lesions. *Proc Natl Acad Sci USA*, 93, 739–43.

Huxlin, K.R., Saunders, R.C., Marchionini, D., Pham, H.A., & Merigan, W.H. (2000) Perceptual deficits after lesions of inferotemporal cortex in macaques. *Cereb Cortex*, 10, 671–83.

Hwang, J. & Romanski, L.M. (2015) Prefrontal neuronal responses during audiovisual mnemonic processing. *J Neurosci*, 35, 960–71.

Jackson, J.B. & Woolgar, A. (2018) Adaptive coding in the human brain: Distinct object features are encoded by overlapping voxels in frontoparietal cortex. *Cortex*, 108, 25–34.

Johnston, K., Levin, H.M., Koval, M.J., & Everling, S. (2007) Top-down control-signal dynamics in anterior cingulate and prefrontal cortex neurons following task switching. *Neuron*, 53, 453–62.

Kaskan, P.M., Costa, V.D., Eaton, H.P., Zemskova, J.A., Mitz, A.R., Leopold, D.A., Ungerleider, L.G., & Murray, E.A. (2017) Learned value shapes responses to objects in frontal and ventral stream networks in macaque monkeys. *Cereb Cortex*, 27, 2739–57.

Kerns, J.G., Cohen, J.D., MacDonald, A.W., 3rd, Cho, R.Y., Stenger, V.A., & Carter, C.S. (2004) Anterior cingulate conflict monitoring and adjustments in control. *Science*, 303, 1023–6.

Klein, A., Andersson, J., Ardekani, B.A., Ashburner, J., Avants, B., Chiang, M.C., Christensen, G.E., Collins, D.L., Gee, J., Hellier, P., Song, J.H., Jenkinson, M., Lepage, C., Rueckert, D., Thompson, P., Vercauteren, T., Woods, R.P., Mann, J.J., & Parsey, R.V. (2009) Evaluation of 14 nonlinear deformation algorithms applied to human brain MRI registration. *Neuroimage*, 46, 786–802.

Konishi, S., Nakajima, K., Uchida, I., Kikyo, H., Kameyama, M., & Miyashita, Y. (1999) Common inhibitory mechanism in human inferior prefrontal cortex revealed by event-related functional MRI. *Brain*, 122, 981–91.

Kostopoulos, P. & Petrides, M. (2008) Left mid-ventrolateral prefrontal cortex: Underlying principles of function. *Eur J Neurosci*, 27, 1037–49.

Kowalska, D.M., Bachevalier, J., & Mishkin, M. (1991) The role of the inferior prefrontal convexity in performance of delayed nonmatching-to-sample. *Neuropsychol*, 29, 583–600.

Kriegeskorte, N., Mur, M., & Bandettini, P. (2008a) Representational similarity analysis – connecting the branches of systems neuroscience. *Front Syst Neurosci*, doi:10.3389/neuro.06.004.2998.

Kriegeschorte, N., Mur, M., Ruff, D.A., Kiani, R., Bodurka, J., Esteky, H., Tanaka, K., & Bandettini, P.A. (2008b) Matching categorical object *representations* in inferior temporal cortex of man and monkey. *Neuron*, 60, 1126–47.

Ku, S.P., Tolias, A.S., Logothetis, N.K., & Goense, J. (2011) fMRI of the face-processing network in the ventral temporal lobe of awake and anesthetized macaques. *Neuron*, 70, 352362.

Kuhn, S., Brass, M., & Gallinat, J. (2013) Imitation and speech: Commonalities within Broca's area. *Brain Struct Funct*, 218, 1419–27.

Kurata, K. & Hoffman, D.S. (1994) Differential effects of muscimol microinjection into dorsal and ventral aspects of the premotor cortex of monkeys. *J Neurophysiol*, 64, 1151–64.

Lauwereyns, J., Sakagami, M., Tsutsui, K., Kobayashi, S., Koizumi, M., & Hikosaka, O. (2001) Responses to task-irrelevant visual features by primate prefrontal neurons. *J Neurophysiol*, 86, 2001–10.

Liu, Y., Brincat, S.L., Miller, E.K., & Hasselmo, M.E. (2020) A Geometric characterization of population coding in the prefrontal cortex and hippocampus during a paired-associate learning task. *J Cogn Neurosci*, 32, 1455–65.

Lorenc, E.S., Lee, T.G., Chen, A.J., & D'Esposito, M. (2015) The effect of disruption of prefrontal cortical function witranscranial magnetic stimulation on visual working memory. *Front Syst Neurosci*, 9, 169.

Maier, M.E., Di Gregorio, F., Muricchio, T., & Di Pellegrino, G. (2015) Impaired rapid error monitoring but intact error signaling following rostral anterior cingulate cortex lesions in humans. *Front Hum Neurosci*, 9, 339.

Mansouri, F.A., Matsumoto, K., & Tanaka, K. (2006) Prefrontal cell activities related to monkeys' success and failure in adapting to rule changes in a Wisconsin Card Sorting Test analog. *J Neurosci*, 26, 2745–56.

Mars, R.B., Klein, M.C., Neubert, F.X., Olivier, E., Buch, E.R., Boorman, E.D., & Rushworth, M.F. (2009) Short-latency influence of medial frontal cortex on primary motor cortex during action selection under conflict. *J Neurosci*, 29, 6926–31.

McAndrews, M.P. & Milner, B. (1991) The frontal cortex and memory for temporal order. *Neuropsychol*, 29, 849–59.

Miller, B.T., Deouell, L.Y., Dam, C., Knight, R.T. & D'Esposito, M. (2008) Spatio-temporal dynamics of neural mechanisms underlying component operations in working memory. *Brain Res*, 1206, 61–75.

Miller, E.K. & Desimone, R. (1994) Parallel neuronal mechanisms for short-term memory. *Science*, 263, 520–2.

Miller, E.K., Erickson, C.A., & Desimone, R. (1996) Neural mechanisms of visual working memory in prefrontal cortex of the macaque. *J Neurosci*, 16, 5154–67.

Milner, B. (1963) Effects of different brain lesions on card sorting. *Archiv Neurol*, 9, 90–100.

Milner, B., Corsi, P., & Leonard, G. (1991) Frontal-lobe contribution to recency judgements. *Neuropsychol*, 29, 601–18.

Minamimoto, T., Saunders, R.C. & Richmond, B.J. (2010) Monkeys quickly learn and generalize visual categories without lateral prefrontal cortex. *Neuron*, 66, 501–7.

Miyashita, Y., Morita, M., Naya, Y., Yoshida, M., & Tomita, H. (1998) Backward signal from medial temporal lobe in neural circuit reorganization of primate inferotemporal cortex. *Comptes rendus de l'Academie des sciences. Serie III, Sciences de la vie*, 321, 185–92.

Morecraft, R.J., Stilwell-Morecraft, K.S., Cipolloni, P.B., Ge, J., McNeal, D.W. & Pandya, D.N. (2012) Cytoarchitecture and cortical connections of the anterior cingulate and adjacent somatomotor fields in the rhesus monkey. *Brain Res Bull*, 87, 457–97.

Morishima, Y., Akaishi, R., Yamada, Y., Okuda, J., Toma, K., & Sakai, K. (2009) Task-specific signal transmission from prefrontal cortex in visual selective attention. *Nat Neurosci*, 12, 85–91.

Morita, M., Nakahara, K., & Hayashi, T. (2004) A rapid presentation event-related functional magnetic resonance imaging study of response inhibition in macaque monkeys. *Neurosci Lett*, 356, 203–6.

Muakkassa, K.F. & Strick, P.L. (1979) Frontal lobe inputs to primate motor cortex: Evidence for four somatotopically organized 'premotor areas'. *Brain Res*, 177, 176–82.

Murray, E.A. & Richmond, B.J. (2001) Role of perirhinal cortex in object perception, memory, and associations. *Curr Opin Neurobiol*, 11, 188–93.

Mushiake, H., Saito, N., Sakamoto, K., Itoyama, Y., & Tanji, J. (2006) Activity in the lateral prefrontal cortex reflects multiple steps of future events in action plans. *Neuron*, 50, 631–41.

Nakahara, K., Hayashi, T., Konishi, S., & Miyashita, Y. (2002) Functional MRI of macaque monkeys performing a cognitive set-shifting task. *Science*, 295, 1532-6.

Naya, Y., Sakai, K., & Miyashita, Y. (1996) Activity of primate inferotemporal neurons related to a sought target in pair-association task. *Proc Natl Acad Sci USA*, 93, 2664-9.

Naya, Y., Yoshida, M., & Miyashita, Y. (2001) Backward spreading of memory-retrieval signal in the primate temporal cortex. *Science*, 291, 661-4.

Nee, D.E. & D'Esposito, M. (2018) The representational basis of working memory. *Curr Top Behav Neurosci*, 37, 213-30.

Nelson, H.E. (1976) A modified card sorting test sensitive to frontal lobe defecits. *Cortex*, 12, 313-24.

Neubert, F.X., Mars, R.B., Buch, E.R., Olivier, E., & Rushworth, M.F. (2010) Cortical and subcortical interactions during action reprogramming and their related white matter pathways. *Proc Natl Acad Sci USA*, 107, 13240-5.

Neubert, F.X., Mars, R.B., Thomas, A.G., Sallet, J., & Rushworth, M.F. (2014) Comparison of human ventral frontal cortex areas for cognitive control and language with areas in monkey frontal cortex. *Neuron*, 81, 700-13.

O.Scalaidhe, P.E., Wilson, F.A.W., & Goldman-Rakic, P.S. (1997) Area segregation of face-processing neurons in prefrontal cortex. *Science*, 278, 1135-8.

O.Scalaidhe, S.P., Wilson, F.A.W., & Goldman-Rakic, P.S. (1999) Face-selective neurons during passive viewing and working memory performance of rhesus monkeys: Evidence for intrinsic specialization of neuronal coding. *Cer Cort*, 9, 484-96.

Passingham, R.E. (1975) Delayed matching after selective prefrontal lesions in monkeys. *Brain Res*, 92, 89-102.

Passingham, R.E. & Wise, S.P. (2012) *The Neurobiology of Prefrontal Cortex*. Oxford University Press, Oxford.

Perry, V.H. & Cowey, A. (1985) The ganglion cell and cone distributions in the monkey's retina: Implications for central magnification factors. *Vision Res*, 25, 1795-810.

Petrides, M. (2018) *Atlas of the Morphology of the Human Cerebral Cortex on the Average MNI Brain*. Academic Press, New York.

Petrides, M., Cadoret, G., & Mackey, S. (2005) Orofacial somatomotor responses in the macaque monkey homologue of Broca's area. *Nature*, 435, 1235-8.

Petrides, M. & Pandya, D.N. (2002) Comparative cytoarchitectonic analysis of the human and macaque ventrolateral prefrontal cortex and corticocortical connection pattern in the monkey. *Eur J Neurosci*, 16, 291-310.

Plakke, B., Hwang, J., & Romanski, L.M. (2015) Inactivation of primate prefrontal cortex impairs auditory and audiovisual working memory. *J Neurosci*, 35, 9666-9675.

Premereur, E., Janssen, P., & Vanduffel, W. (2018) Functional MRI in macaque monkeys during task switching. *J Neurosci*, 38, 10619-30.

Preuss, T.M. & Goldman-Rakic, P.S. (1991) Myelo- and cytoarchitecture of the granular frontal cortex and surrounding regions in the strepsirhine primate Galago and the anthropoid primate Macaca. *J Comp Neurol*, 310, 429-474.

Puig, M.V., Antzoulatos, E.G., & Miller, E.K. (2014) Prefrontal dopamine in associative learning and memory. *Neurosci*, 282, 217-29.

Rainer, G. & Miller, E.K. (2002) Timecourse of object-related neural activity in the primate prefrontal cortex during a short-term memory task. *Eur J Neurosci*, 15, 1244-54.

Rainer, G., Rao, S.C., & Miller, E.K. (1999) Prospective coding for objects in primate prefrontal cortex. *J Neurosci*, 19, 5493-505.

Rama, P. & Courtney, S.M. (2005) Functional topography of working memory for face or voice identity. *Neuroimage*, 24, 224-34.

Rama, P., Poremba, A., Sala, J.B., Yee, L., Malloy, M., Mishkin, M., & Courtney, S.M. (2004) Dissociable functional cortical topographies for working memory maintenance of voice identity and location. *Cereb Cortex*, 14, 768–780.

Rodu, J., Klein, N., Brincat, S.L., Miller, E.K., & Kass, R.E. (2018) Detecting multivariate cross-correlation between brain regions. *J Neurophysiol*, 120, 1962–72.

Romanski, L.M., Averbeck, B.B., & Diltz, M. (2005) Neural representation of vocalizations in the primate ventrolateral prefrontal cortex. *J Neurophysiol*, 93, 734–47.

Romanski, L.M., Bates, J.F., & Goldman-Rakic, P.S. (1999) Auditory belt and parabelt projections to the prefrontal cortex in the rhesus monkey. *J Comp Neurol*, 403, 141–57.

Roy, J.E., Riesenhuber, M., Poggio, T., & Miller, E.K. (2010) Prefrontal cortex activity during flexible categorization. *J Neurosci*, 30, 8519–28.

Rudebeck, P.H. & Murray, E.A. (2008) Amygdala and orbitofrontal cortex lesions differentially influence choices during object reversal learning. *J Neurosci*, 28, 8338–43.

Rushworth, M., Nixon, P.D., Eacott, M.J., & Passingham, R.E. (1997) Ventral prefrontal cortex is not essential for working memory. *J Neurosci*, 17, 4829–38.

Sakagami, M., Tsutsui, K., Lauwereyns, J., Koizumi, M., Kobayashi, S., & Hikosaka, O. (2001) A code for behavioral inhibition on the basis of color, but not motion, in ventrolateral prefrontal cortex of macaque monkey. *J Neurosci*, 21, 4801–08.

Sakai, K. & Miyashita, Y. (1991) Neural organization for the long-term memory of paired associates. *Nature*, 354, 152–5.

Sala, J.B., Rama, P., & Courtney, S.M. (2003) Functional topography of a distributed neural system for spatial and nonspatial information maintenance in working memory. *Neuropsychol*, 41, 341–56.

Saleem, K.S., Kondo, H., & Price, J.L. (2008) Complementary circuits connecting the orbital and medial prefrontal networks with the temporal, insular, and opercular cortex in the macaque monkey. *J Comp Neurol*, 506, 659–93.

Schon, K., Tinaz, S., Somers, D.C., & Stern, C.E. (2008) Delayed match to object or place: An event-related fMRI study of short-term stimulus maintenance and the role of stimulus pre-exposure. *Neuroimage*, 39, 857–72.

Scott, B.H., Leccese, P.A., Saleem, K.S., Kikuchi, Y., Mullarkey, M.P., Fukushima, M., Mishkin, M., & Saunders, R.C. (2017) Intrinsic connections of the core auditory cortical regions and rostral supratemporal plane in the macaque monkey. *Cereb Cortex*, 27, 809–40.

Sharma, S., Mantini, D., Vanduffel, W., & Nelissen, K. (2019) Functional specialization of macaque premotor F5 subfields with respect to hand and mouth movements: A comparison of task and resting-state fMRI. *Neuroimage*, 191, 441–56.

Shen, L. & Alexander, G.E. (1997) Preferential representation of instructed target location versus limb trajectory in dorsal premotor cortex. *J Neurophysiol*, 77, 1195–212.

Shima, K. & Tanji, J. (1998) Role for cingulate motor area cells in voluntary movement selection based on reward. *Science*, 282, 335–8.

Siegel, M., Buschman, T.J., & Miller, E.K. (2015) Cortical information flow during flexible sensorimotor decisions. *Science*, 348, 1352–5.

Sobotka, S., Diltz, M.D. & Ringo, J.L. (2005) Can delay-period activity explain working memory? *J Neurophysiol*, 93, 128–36.

Sprung-Much, T. & Petrides, M. (2018) Morphological patterns and spatial probability maps of two defining sulci of the posterior ventrolateral frontal cortex of the human brain: The sulcus diagonalis and the anterior ascending ramus of the lateral fissure. *Brain Struct Funct*, 223, 4125–52.

Sreenivasan, K.K., Curtis, C.E., & D'Esposito, M. (2014a) Revisiting the role of persistent neural activity during working memory. *Trends Cogn Sci*, 18, 82–9.
Sreenivasan, K.K., Gratton, C., Vytlacil, J. & D'Esposito, M. (2014b) Evidence for working memory storage operations in perceptual cortex. *Cogn Affect Behav Neurosci*, 14, 117–28.
Sreenivasan, K.K., Vytlacil, J., & D'Esposito, M. (2014c) Distributed and dynamic storage of working memory stimulus information in extrastriate cortex. *J Cogn Neurosci*, 26, 1141–53.
Stanley, D.A., Roy, J.E., Aoi, M.C., Kopell, N.J., & Miller, E.K. (2018) Low-beta oscillations turn up the gain during category judgments. *Cereb Cortex*, 28, 116–30.
Sugihara, T., Diltz, M.D., Averbeck, B.B. & Romanski, L.M. (2006) Integration of auditory and visual communication information in the primate ventrolateral prefrontal cortex. *J Neurosci*, 26, 11138–47.
Swaminathan, S.K. & Freedman, D.J. (2012) Preferential encoding of visual categories in parietal cortex compared with prefrontal cortex. *Nat Neurosci*, 15, 315–20.
Takahara, D., Inoue, K., Hirata, Y., Miyachi, S., Nambu, A., Takada, M. & Hoshi, E. (2012) Multisynaptic projections from the ventrolateral prefrontal cortex to the dorsal premotor cortex in macaques – anatomical substrate for conditional visuomotor behavior. *Eur J Neurosci*, 36, 3365–75.
Takeuchi, D., Hirabayashi, T., Tamura, K., & Miyashita, Y. (2011) Reversal of interlaminar signal between sensory and memory processing in monkey temporal cortex. *Science*, 331, 1443–7.
Tomita, H., Ohbayashi, M., Nakahara, K., Hasegawa, I., & Miyashita, Y. (1999) Top-down signal from prefrontal cortex in executive control of memory retrieval. *Nature*, 401, 699–703.
Toni, I., Ramnani, N., Josephs, O., Ashburner, J., & Passingham, R.E. (2001) Learning arbitrary visuomotor associations: Temporal dynamic of brain activity. *Neuroimage*, 14, 1048–57.
Ungerleider, L.G., Gaffan, D., & Pelak, V.S. (1989) Projections from inferior temporal cortex to prefrontal cortex via the uncinate fascicle in rhesus monkeys. *Exp Brain Res*, 76, 473–484.
Verbruggen, F., Aron, A.R., Stevens, M.A., & Chambers, C.D. (2010) Theta burst stimulation dissociates attention and action updating in human inferior frontal cortex. *Proc Natl Acad Sci USA*, 107, 13966–71.
Wang, M., Zhang, H., & Li, B.M. (2000) Deficit in conditional visuomotor learning by local infusion of bicuculline into the ventral prefrontal cortex in monkeys. *Eur J Neurosci*, 12, 3787–96.
Williams, Z.M., Bush, G., Rauch, S.L., Cosgrove, G.R., & Eskandar, E.N. (2004) Human anterior cingulate neurons and the integration of monetary reward with motor responses. *Nat Neurosci*, 7, 1370–5.
Wilson, F.A., Scalaidhe, S.P., & Goldman-Rakic, P.S. (1993) Dissociation of object and spatial processing domains in primate prefrontal cortex. *Science*, 260, 1955–8.
Wutz, A., Loonis, R., Roy, J.E., Donoghue, J.A., & Miller, E.K. (2018) Different levels of category abstraction by different dynamics in different prefrontal areas. *Neuron*, 97, 716–26 e718.
Xu, Y. (2017) Reevaluating the sensory account of visual working memory storage. *Trends Cogn Sci*, 21, 794–815.
Zhang, J., Hughes, L.E., & Rowe, J.B. (2012) Selection and inhibition mechanisms for human voluntary action decisions. *Neuroimage*, 63, 392–402.

PART III
THE PREFRONTAL CORTEX WITHIN THE SYSTEM AS A WHOLE

8
Prefrontal Cortex
Abstract Rules and Attentional Performance

Introduction

Chapters 6 and 7 argue that the dorsal and ventral PF cortex support the ability to learn conditional rules. But that cannot be a sufficient description of their function because other mammals can learn conditional tasks. For example, rats can learn the delayed response task (Kolb et al., 1974), cats can learn delayed matching (Okuava et al., 2005), and dogs can learn go no-go tasks (Lawicka, 1969). All these tasks have a conditional logic: if context A, then choice X is rewarded; if context B, then choice Y is rewarded (Passingham, 1993).

It could be argued that it is not surprising if there is a similarity between the functions of the granular PF cortex in primates and the agranular PF areas in other mammals. In primates the granular and agranular areas are neighbours; and the assumption is that the granular areas differentiated out from the agranular ones. The result might well be that they share many connections (Krubitzer & Huffman, 2000), and thus show similarities in function.

Nonetheless, the new areas must have provided some selective advantage. One possibility is that this involved greater processing power. The effect of lesions of the dorsal PF cortex on delayed response tasks in monkeys (Goldman et al., 1971) is much more devastating than the effect of lesions in the agranular medial PF cortex in rats (Kolb et al., 1974). So, it is clear that in anthropoids the dorsal PF cortex confers a critical capacity. In evolution, even small improvements can provide a critical advantage.

The alternative is that the new areas supported abilities that were not possible without them. This chapter suggests that this is true of the dorsal and ventral PF cortex in anthropoid primates. It proposes, specifically, that these areas can encode abstract rules, and that this is the key to the PF cortex.

As in previous chapters, to understand how the PF cortex can do this, it is necessary to review its cortico-cortical connections. The basic insight is that the PF cortex lies at the top of the neocortical processing hierarchy (Fuster, 2004). In the neocortex, but not the basal ganglia or cerebellum, lower order representations are re-represented, stage by stage, at higher levels of the hierarchy. It is this process of re-representation that provides the means via which the representations can become more abstract.

Cortico-Cortical Connections

There is a clear organization of the cortico-cortical pathways. Sensory information is relayed via the thalamus to different sensory-specific areas, and the analysis of the information occurs in parallel streams. For example, the processing of colour and form occurs at the same time as the processing of sounds. Similarly, the processing of visual information from the fovea occurs at the same time as the processing of visual information concerning the periphery. The advantage of this parallel organization is that it increases the speed of processing.

However, the animal has to respond, not to individual cues, but to the world as a whole; and this means that it has to take into account the context or surroundings. This means that it is necessary for the information from the individual sensory areas to be integrated in multimodal areas. It has been suggested that these multimodal areas serve as a global workspace (Dehaene et al., 1998). By this is meant a core integrated system for solving general problem solving.

Integration between and within Areas

Markov et al. (2012) visualized this core by injecting tracers into 29 of the 91 areas that they identified in the macaque monkey brain. By applying graph theory, they were able to show that within this system some areas are more highly interconnected than others. They argued that these serve as the core or hub for processing. The core is shown by the ring in Figure 8.1 (next page).

The core is illustrated in Figure 8.1. It includes the superior temporal polysensory area (STP), the posterior parietal cortex and its projections into the frontal eye field (FEF) and dorsal prefrontal (PF) cortex. The dorsal premotor cortex (F2) and ventral premotor cortex (F5) are also included; these are interconnected because there is a need to integrate reaching with grasping.

However, other areas that we know to be multisensory do not appear in Figure 8.1. The reason is that there were no injections into either the ventral or orbital PF cortex (Markov et al., 2012). Yet, as Chapters 4 and 7 review, these areas also have multimodal inputs, and are closely interconnected (Petrides & Pandya, 2002).

In a parallel study, Harriger et al. (2012) collated data from three sources on the connections of the macaque monkey neocortex (Young, 1993; Honey et al., 2007; Modha & Singh, 2010), and analyzed it by using graph theory. They identified fourteen 'rich clubs' of areas that were closely interconnected. They also looked for the club that had the tightest and most dense interconnections. It comprised the entire PF cortex, the medial, dorsal, ventral, and orbital cortex. It also included the inputs, the intraparietal area (LIP) and inferior parietal area (PFG), and the outputs, the anterior cingulate cortex (ACC) and parahippocampal cortex (Chapters 3–7).

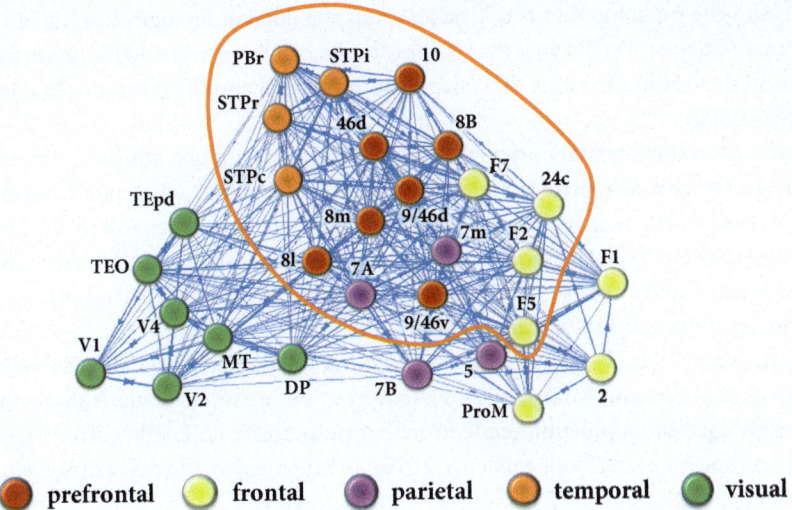

Figure 8.1 The central neocortical hub or core as shown by graph theory
See the text for the abbreviations

Reproduced from Ercsey-Ravasz, M., Markov, N.T., Lamy, C., Van Essen, D.C., Knoblauch, K., Toroczkai, Z., & Kennedy, H. A predictive network model of cerebral cortical connectivity based on a distance rule. *Neuron*, 80 (1), 184–97, Figure 7d, Doi: 10.1016/j.neuron.2013.07.036 Copyright © 2013, Elsevier Inc. All rights reserved.

Reasons for Integration

Though there are multimodal cells in the posterior parietal cortex, the *only* area that receives inputs from all modalities is the PF cortex. It processes information from vision, hearing, touch, smell, and taste (Carmichael & Price, 1995; Petrides & Pandya, 2002). The evidence that these inputs are integrated is that there are multimodal cells in the PF cortex (Bruni et al., 2015). There are also cells that integrate the inputs from the dorsal and ventral visual stream. Rao et al. (1997) taught monkeys two tasks, one to choose the matching object and the other to make a saccade to a remembered location. Of the cells with delay activity, 52% had conjunctive properties, responding both for shape (ventral stream) and location (dorsal stream).

As previous chapters stress, the consequence of the close integration between different PF areas (Harriger et al., 2012) is that it is the PF cortex that is in a position is to learn to generate the goal that is appropriate given the current context and the desired outcome. To put it another way, it is in a position to learn conditional rules.

The dorsal PF cortex encodes the prior context and the order of events, and it generates the spatial goal that is appropriate (Chapter 6). The ventral PF cortex encodes the current context, whether the presence of an object, animal or sound, and generates the choice that is appropriate (Chapter 7). Finally, the orbital PF cortex encodes the outcome that is desirable, given the current biological needs of the animal (Chapter 4). The goal that is appropriate can be generated via conjunctive cells in the dorsal PF cortex that encode both the goal and the outcome (Wallis & Miller, 2003).

The posterior parietal cortex also has multisensory inputs, but for a different reason. The specification of a spatial location is not dependent on a single modality. Thus, the priority map in area LIP is specified both by visual cues and by the location of sounds (Cohen et al., 2004). The area VIP has visual, auditory, vestibular, and tactile inputs because it represents movement through the surroundings in head-centred coordinates (Chen et al., 2011).

The area VIP is also involved in representing the numerosity of dots as well as of lengths (Tudusciuc & Nieder, 2009). However, a more recent analysis shows that intermingled subpopulations code for either numerosity or length, with very few cells coding for both (Eiselt & Nieder, 2016). In other words, the cells are not genuinely conjunctive.

The parietal cortex also differs from the PF cortex in that it has few fibres from the amygdala (Amaral & Price, 1984); and it also lacks a direct input from the orbital PF cortex (Petrides & Pandya, 2002). It is not, therefore, in a position to learn conditional rules, because these are learned on the basis of the rewards that the animals receive.

It is true that it is possible to record cell activity in the posterior parietal cortex that represents goals (Bonini et al., 2012) and the behavioural relevance of cues (Brincat et al., 2018), but this activity is probably derived. The evidence comes from the timing: the activity occurs after the dorsal premotor cortex represents the goal, and not before (Westendorff et al., 2010). It is likely that the posterior parietal cortex would no longer encode either goals or the relevance of cues if the recordings were taken in monkeys with prefrontal lesions.

Feedback Connections

The previous section describes the feedforward connections of the neocortex starting with the primary sensory areas. The inputs from area A' terminate in layer 4 of area A". But the return or feedback projections from area A" to A' avoid layer 4, terminating mostly in the supragranular or infragranular layers instead. By recording using electrodes with multiple contacts, Takeuchi et al. (2011) could distinguish between the feedforward and feedback signals by recording in the inferior temporal cortex during the retrieval of paired associates. The immediate feedback signal comes from the perirhinal cortex (Hirabayashi et al., 2013a).

The feedforward and feedback connections can achieve contact. For example, the feedback connections from layers 3A terminate in layer 3A; and this means that they come into contact with feedforward connections that originate in layers

3B (Markov & Kennedy, 2013). The bulk of the feedback connections from layer 6 terminate in layer 6, and this may mean that they can come into connect with the feedforward connections that originate in layer 5 (Markov & Kennedy, 2013). This arrangement is illustrated in Figure 8.2.

There have been various attempts to account for the role of feedback connections, and these are reviewed in Markov and Kennedy (2013). It has been suggested, for example, that feedforward connections are 'driving' whereas feedback connections are weak and 'modulatory' (Crick & Koch, 1998). More specifically Desimone and Duncan (1995) proposed that that the modulatory effect is to bias competition between items or features.

A more fundamental account of the role of the feedback connections from the PF cortex is that they support both the selection of actions and attentional selection. Shipp (2005) has commented that the feedback connections from the PF cortex to posterior cortical areas are like the connections to the premotor and motor cortex in that they avoid layer 4, the reason being that these areas lack a layer 4. He therefore suggested that both sets of connections could be regarded as being feedback connections.

Thus, activity in the PF cortex not only specifies the behavioural goal but also carries the sensory predictions that the animal makes. More specifically, it influences cells that represent the error between the prediction and the actual input (Bastos et al., 2012).

Evidence for a role of the PF cortex in these predictions comes from a study by Yamada et al. (2010). They taught monkeys that stimuli from one set predicted juice whereas stimuli from another set predicted saline. The associations were repeatedly reversed, and the animals were able to learn the opposite associations after just one error. The animals quickly learned what to expect, and even before presentation of the stimuli there was anticipatory activity in cells in the PF cortex. This activity could not have been driven bottom-up; instead, it is likely it represented a top-down predictive signal.

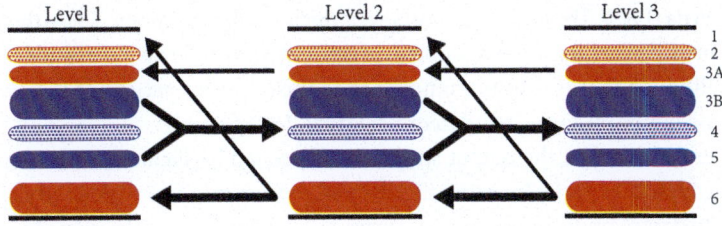

Figure 8.2 The organization of the feedforward (blue) and feedback (red) connections in the neocortex

Reproduced from Markov, N.T. & Kennedy, H. The importance of being hierarchical. *Current Opin Neurobiol*, 23 (2), 187–94, Figure 3d, Doi: 10.1016/j.conb.2012.12.008 Copyright © 2013 Elsevier Ltd. All rights reserved.

The top-down projections from the PF cortex are also involved in attentional selection. Chapters 5 and 7 review the evidence that feedback signals from the PF cortex to the more caudal sensory areas enhance the representation of the relevant visual cues in the inferior temporal (Zhou & Desimone, 2011) and posterior parietal cortex (Siegel et al., 2015).

Thus, the output pathways from the PF cortex, whether directed to the motor system or to sensory areas, are concerned with the *effect* that the animal itself has on the environment. They support the attentional selection of actions and cues, as well as predictions.

Summary

When graph theory is applied to the neocortical connectome of the macaque monkey, the integrated core consists of the PF cortex and its inputs from the posterior parietal cortex and outputs to the anterior cingulate and parahippocampal cortex. Integration within the PF cortex is achieved by cells with conjunctive properties, and these support the ability to learn conditional rules. There is integration within the posterior parietal cortex because it encodes different metrics. The feedforward organization of the neocortex provides information about the world and the state of the animal, whereas the feedback organization reflects the interaction of the animal with that world.

Cortico-Subcortical Connections

Basal Ganglia

Like most of the cerebral cortex, the PF cortex sends a heavy projection to the basal ganglia, targeting its input structure, the striatum. In turn, the striatum sends an inhibitory GABAergic projection to the globus pallidus pars externa (GPe) and pars interna (GPi). The GPi targets different nuclei of the thalamus. These thalamic nuclei then project back to the motor, premotor, and PF areas (Alexander et al., 1986), although there are minor outputs go to the posterior parietal (Clower et al., 2005) and temporal cortex (Middleton & Strick, 1996), as well.

The organization of the output streams is illustrated by Figure 8.3. The connections were established by injecting the various neocortical areas with a virus that is transmitted retrogradely and across synapses, so enabling the visualization of a multisynaptic pathway. Figure 8.3A (next page) shows the different areas of the neocortex, whereas Figures 8.3B and D indicate the streams within the globus pallidus and substantia nigra pars reticulata that send projections to the cortex via the thalamus.

Though Figures 8.3B and D show that the basal ganglia outputs to the different neocortical are relatively independent, the inputs to the striatum converge.

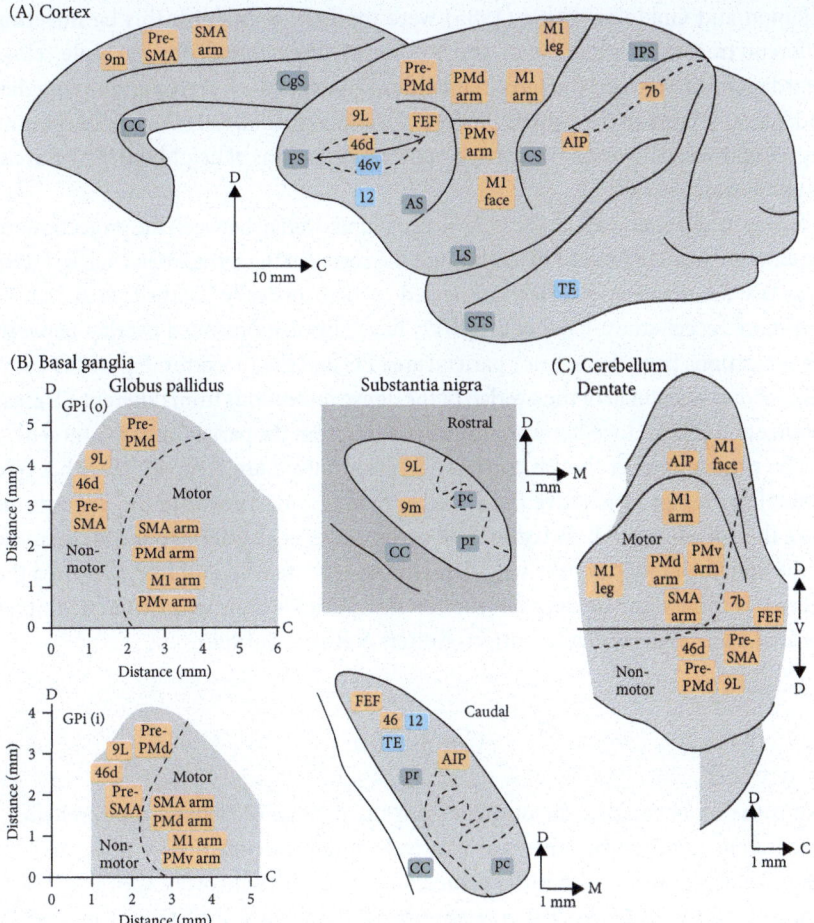

Figure 8.3 Organization of basal ganglia and cerebellar outputs to the cerebral cortex
The cortical targets of basal ganglia and cerebellar outputs are indicated on medial and lateral views of the brain of the cebus monkey (*Cebus apella*). (A) Neocortical target areas. (B & D) Summary maps of topography in the globus pallidus, pars interna, and the substantia nigra, pars reticulata. (C) Summary maps of topography in the dentate nucleus of the cerebellum. The dashed line illustrates the division between the projections to motor and non-motor areas. In all panels, orange labels indicate areas of the cerebral cortex that are the targets of both basal ganglia and cerebellar outputs, whereas blue labels indicate areas of the cerebral cortex that are the targets of basal ganglia, but not cerebellar, output. The numbers refer to cytoarchitectonic areas. AIP, anterior intraparietal area; AS, arcuate sulcus; C, caudal; CC, corpus callosum; CgS, cingulate sulcus; CS, central sulcus; D, dorsal; FEF, frontal eye field; I, the inner portion of the internal segment of the globus pallidus; IPS, intraparietal sulcus; LS, lateral sulcus; M, medial; M1, primary motor cortex; M1 arm, arm area of M1; M1 face, face area of M1; M1 leg; leg area of M1; O, the outer portion of the internal

(continued in footnote of next page)

Selemon and Goldman-Rakic (1985) were the first to establish this by injecting different tracers in subareas of the PF cortex They found, for example, close interdigitation between the terminal fields of the dorsal PF cortex and the medial and lateral sectors of the orbital PF cortex. Averbeck et al. (2014) showed that the degree of overlap between the cortico-striatal projections of neighbouring PF areas can be as much as 80%.

Selemon and Goldman-Rakic (1985) also found overlap between the projections to the striatum from the dorsal PF cortex and the cortex in the intraparietal sulcus (IPS), areas that are themselves closely connected cortico-cortically (Gerbella et al., 2013). In a more recent study, Choi et al. (2017) have also demonstrated overlap between the projections from the inferior parietal area PG and the PF cortex. Figure 8.4 (next page) shows a heatmap of the overlap of the dense projections from different PF areas (orange and yellow) together with the projections from the parietal area PG (green).

The connections of the PF cortex with the striatum are compatible with a role in the learning of associative tasks, since there is convergence in the projections from the dorsal, ventral, and orbital PF cortex. Levy et al. (1997) found an increase in the uptake of 2-deoxyglucose (2-DG) in the caudate when monkeys learned the delayed response and delayed alternation tasks. And lesions in the caudate impair performance on delayed alternation (Butters & Rosvold, 1968).

Cerebellum

As for the basal ganglia, the outputs from the dentate nucleus of the cerebellum to different neocortical areas are relatively independent. This is illustrated in Figure 8.3C. However, again the projections from the neocortex appear to show a degree of convergence. These terminate in the pons (Glickstein et al., 1985). The overlap is evident, for example, by comparing the results of injections in the dorsomedial PF cortex (areas 9 and 10) and the dorsal PF cortex (area 9/46) (Schmahmann & Pandya, 1997). Although no one has quantified the degree of corticopontine overlap, as Averbeck et al. (2014) have done for corticostriatal projections, the neuroanatomical literature amply documents the extensive overlap of

Figure 8.3 Continued

segment of the globus pallidus; pc, pars compacta; PMd arm, arm area of the dorsal premotor area; PMv arm, arm area of the ventral premotor area; pr, pars reticulata; Pre- PMd, predorsal or rostral dorsal premotor area; Pre- SMA, presupplementary motor area; PS, principal sulcus; SMA arm, arm area of the supplementary motor area; STS, superior temporal sulcus; TE, area of inferotemporal cortex.

Reproduced from Bostan, A.C. & Strick, P.L. The basal ganglia and the cerebellum: Nodes in an integrated network. *Nat Rev Neurosci*, 19 (6), 338–50, Figure 1, Doi: 10.1038/s41583-018-0002-7 Copyright © 2018, Springer Nature.

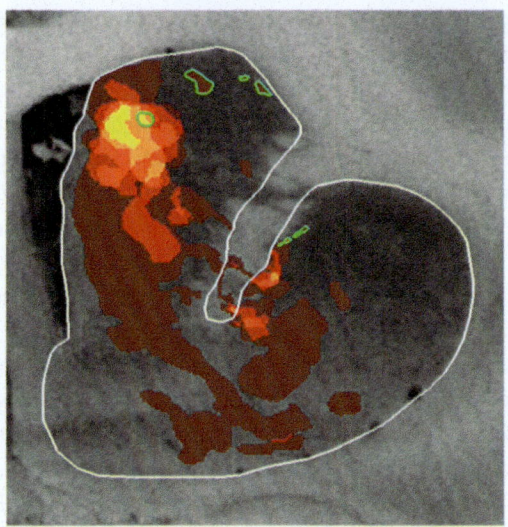

Figure 8.4 The overlap of the projections to the striatum from different PF areas is shown in orange. The projections from the inferior parietal area PG are shown in green

Reproduced from Choi, E.Y., Tanimura, Y., Vage, P.R., Yates, E.H., & Haber, S.N. Convergence of prefrontal and parietal anatomical projections in a connectional hub in the striatum. *Neuroimage*, 146, 821–32, Figure 3C, Doi: 10.1016/j.neuroimage.2016.09.037 © 2016, Published by Elsevier Inc.

corticopontine terminations from different cortical areas (Brodal & Bjaalie, 1992). In turn, the cells in the basilar pontine nuclei send mossy fibres to the cerebellar cortex (Brodal & Bjaalie, 1992).

However, there is an important difference between the cortical projections to the cerebellum and basal ganglia. Whereas most of the neocortical areas project to the striatum, the connections to the pons and thus to the cerebellum do not come from all areas of the neocortex. There are projections from the dorsal and medial cortical areas, but not from the inferior temporal cortex and the ventral and orbital PF cortex to which it projects (Schmahmann & Pandya, 1997; Glickstein & Doron, 2008). Furthermore, there are no return connections to these ventral areas (shown in blue in Figure 8.3A).

The absence of these connections indicates that the cerebellum is fundamentally a sensorimotor structure. It lacks direct information about colour and form. Thus, it is not in a position to learn object goals on the basis of the outcomes.

Thalamus

Like all other cortical areas, the PF cortex has reciprocal connections with the thalamus. Its main thalamic connections are with the mediodorsal (MD) nucleus. The multiform part of MD receives an input from the superior colliculus (Russchen et al., 1987; Erickson et al., 2004) and projects to the caudal PF cortex, which projects in turn back to the superior colliculus (Fries, 1984). This system is involved in directing eye movements (Chapter 5).

The parvocellular division of the MD nucleus projects to the medial and dorsal PF cortex, whereas the magnocellular divisions projects to the ventral and orbital PF cortex (Goldman-Rakic & Porrino, 1985). These divisions of the MD nucleus receive no direct sensory inputs. Roughly 20% of the layer 5 cells of the PF cortex project back to the MD nucleus (Xiao et al., 2009).

Like the ventral and orbital PF cortex, the magnocellular division of the MD nucleus receives an input from the amygdala (Russchen et al., 1987). Izquierdo and Murray (2010) disconnected the magnocellular division and the orbital PF cortex by making crossed unilateral lesions in the two areas. The effect was that, when a particular food was devalued by feeding it to the animal to satiety, the monkeys failed to shift to the object that was associated with the alternative food. This same effect can be found from bilateral lesions of the amygdala or orbital PF cortex.

The connections of the MD nucleus as a whole are compatible with a role in learning since it is linked to the medial, dorsal, ventral, and orbital PF cortex. Chakraborty et al. (2019) trained monkeys on the object-in scenes task (Figure 3.8). Bilateral lesions of the parvocellular and magnocellular MD had a severe effect on new learning, and so did crossed unilateral lesions of these two sectors of the nucleus.

Dopaminergic Midbrain

The midbrain dopamine cells fire to unexpected rewards (Fiorillo et al., 2003), thus coding for a quantitative prediction error signal. As learning proceeds, this phasic response transfers to sensory cues that predict reward (Schultz, 1998), and this provides a mechanism for associative learning through reinforcement (Arsenault et al., 2014). There is a heavy projection from the ventral tegmental area and substantia nigra pars compacta to both the dorsal and ventral striatum (Lynd-Balta & Haber, 1994).

There are also dopaminergic projections to the neocortex. The level of D1 receptors is consistently much higher than the level of D2 receptors. Whereas D2 receptors are at their highest in layer 5, D1 receptors are distributed in both the supragranular and infragranular layers (Lidow et al., 1991). Though both D1 and D2 receptors are found throughout the neocortex, the density is highest in the PF cortex, intermediate in the posterior parietal cortex, and at its lowest in the occipital cortex (Lidow et al., 1991).

This distribution is consistent with the hypothesis that dopamine receptors in the neocortex play a role in learning. Chapter 7 mentions a study by Puig et al. (2014) that compared the role of D1 and D2 receptors. The application of a D1 antagonist impaired new learning while leaving familiar associations intact. The application of a D2 antagonist led to a decrease in cognitive flexibility.

Summary

There is convergence of the projections of different PF areas to the striatum and pons and thus the cerebellum. The cerebellum differs from the striatum in that there are no inputs from the inferior temporal cortex or the ventral and orbital PF cortex to which it projects. The return outputs from the globus pallidus and dentate nucleus target the frontal lobe, with the different streams being relatively independent.

Hierarchical Organization

Though both the cortico-cortical and cortico-striatal connections of the PF cortex are involved in learning, there is a critical difference between the structure of the neocortex and the basal ganglia. The basal ganglia lack the layered structure of the neocortex and also the long-range connections that run in the white matter below the neocortical layers. It is this structural arrangement that supports hierarchical processing.

As brains enlarge, the number of neocortical areas increases (Changizi & Shimojo, 2005). The reason is probably that there is a need to maintain an efficient connectivity and to minimize long connections. This leads to the differentiation of the neocortex into interconnected subareas, a process that is referred to as arealization (Krubitzer & Huffman, 2000). However, these areas are not simply organized randomly: a hierarchical organization develops such that when area A differentiates, the outputs from area A' become the inputs to the new area A".

The consequence of this hierarchical organization is that the cells in area A" can combine inputs from different cells in area A'. That this combination occurs is suggested by comparing the receptive fields in the visual areas V1, V2, and V4. For example, the receptive fields in V4 are much larger than those in V2 (Desimone & Schein, 1987). A plausible explanation is that the larger receptive fields are achieved by combining inputs from cells in V2 that have smaller receptive fields.

In correspondence, the optimal stimulus for the cell becomes more complex. A line representation in V1 (Hubel & Wiesel, 1968) can contribute to a stopped line representation in V4 (Desimone et al., 1985) or the representation of a simple shape in the inferior temporal cortex (Tanaka, 1997). Cadieu et al. (2007) produced a model that was able to achieve invariance for position or shape by combining units in a hierarchical fashion.

Invariance involves a process of abstraction. It allows, for example, the identification of an object as being the same when viewed from different angles. Wang et al. (1998) used *in vivo* optical imaging to demonstrate that when a monkey is presented with a face as it turns, the activity that is evoked in the inferior temporal cortex travels across cortical columns. The inferior temporal cortex projects to the perirhinal cortex (Webster et al., 1991), and lesions of the perirhinal cortex impair the ability of macaque monkeys to recognize an object independently of the angle of view (Buckley & Gaffan, 1998). One way of achieving the transition from view dependent to view independent representations is via hierarchical processing.

That the inputs converge was demonstrated by Hirabayashi et al. (2013b). They taught monkeys to associate pairs of objects. They found some cells in area TE that responded selectively to one object of a pair and others that responded to both. By applying cross-correlation analyses to identify functional connectivity, they were able to show that there was convergence between cells representing single objects onto cells representing pairs. Pair-coding was found earlier in the inferior temporal area TE than in the perirhinal cortex; but the strength of the pair-coding was greater in the perirhinal cortex, suggesting that the associations were built up there over time.

The functional progression in the ventral visual system is accompanied by a neuroanatomical progression. Elston (2007) counted the spines on the basal dendrites of layer 3 pyramidal cells. He found that the number of spines in the ventral visual system increased from V1 through the early visual areas to the inferior temporal and perirhinal cortex, stage by stage. Elston also demonstrated that the maximum number of spines was found in the inferior convexity PF area 47/12 which has converging inputs from the inferior temporal and perirhinal cortex (Webster et al., 1994; Kondo et al., 2005) as well as from the superior temporal association cortex and the second somatosensory area S2 (Petrides & Pandya, 2002).

These findings support the claim that the PF cortex should be regarded as lying at the top of the sensory hierarchy. But it also sits at the top of the motor hierarchy. It generates goals irrespective of the actual movements that are needed to achieve them (Mushiake et al., 2006) (Figure 6.14). It is lesions of the motor cortex that interfere with the kinematics of movements of the hand and arm themselves (Hoffman & Strick, 1995).

The PF cortex also sits at the top of the outcome hierarchy. There are also cells in the orbital PF cortex that encode abstract value (Padoa-Schioppa & Assad, 2008). The existence of these cells could be explained if there is a hierarchy of processing within the orbital PF cortex, from cells that specify outcomes to cells that encode abstract value.

The PF cortex of monkeys and apes differs from the rodent PF cortex in that new granular areas have been added. These include the granular medial PF cortex (areas 9 and 10), the dorsal PF cortex (areas 9/46 and 46), and the ventral PF cortex (areas 45A and 47/12). It is the addition of these layers that mean that monkeys can learn abstract conditional rules, meaning rules that apply irrespective of the actual material.

Summary

The structure of the neocortex differs from that of the basal ganglia in that it is layered and there are long-range connections that run in the white matter under the cortex. It is this structure that makes hierarchical processing possible, thus enabling the generation of invariant or abstract representations. The PF cortex lies at the top of the sensory, motor, and outcome hierarchies. This means that uniquely it is in a position to support the learning of abstract rules.

Abstract Rules

A specific conditional rule requires that response R1 is appropriate given cue A, and response R2 given cue B. A matching task, for example, obeys the rule that given A, the animals should choose A', but given B, the animal should choose B'. Non-matching differs in that given A, response B' is appropriate, but given B, response A' is appropriate.

An abstract matching or non-matching task differs in that the rule is the same whatever the material. In other words, the matching rule applies whether the sample is A, B, C, or D. Wallis et al. (2001) taught monkeys matching and non-matching, and on each trial an instruction cue told the monkeys which rule currently applied. After a delay, a probe picture appeared. If the current rule was matching, the animal was taught to release the handle if the probe picture matched the sample; whereas if the current rule was non-matching the animal had to release the handle if the probe picture differed.

To ensure that any cell activity recorded after presentation of the instruction cue reflected the rule rather the sensory properties of the cue, Wallis et al. used two very different kinds of cues to specify the rule. A low-pitched tone or the delivery of juice instructed the monkey to apply the matching rule, and a high-pitched tone or the absence of juice delivery instructed the nonmatching rule. Before the delay-period between the sample and choice stimuli, one of these instructions appeared at the same time as the sample picture.

There were cells in the ventral and dorsal PF cortex that encoded the matching rule during the delay, and others that encoded the non-matching rule. Figure 8.5 (next page) shows the data for a cell that coded for the matching rule. As the figure shows, the cell coded for the abstract rule, responding irrespective of the object used as the sample (objects 1–4). These results could not be due to the nature of the instruction cue because they were the same whether the rule was specified by a tone or juice.

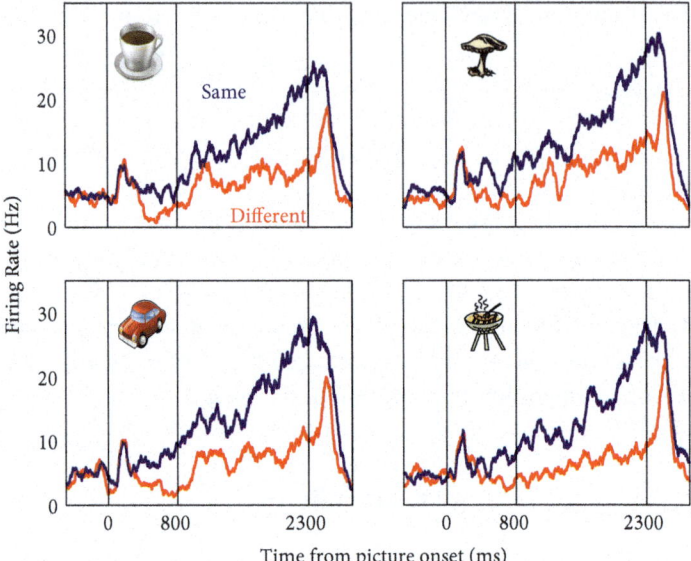

Figure 8.5 PF cell encoding the abstract matching rule
It is shown for four different sample stimuli (Objects 1, 2, 3, and 4). The blue colours indicate the activity for objects that matched, the red for objects that did not match

Reproduced from Wallis J.D., Anderson K.C., & Miller E.K. Single neurons in prefrontal cortex encode abstract rules. *Nature*, 411 (6840), 953–6, Figure 2, Doi: 10.1038/35082081 Copyright © 2001, Springer Nature.

To find out if cells that encoded an abstract rule were confined to the PF cortex, Muhammad et al. (2006) recorded from the inferior temporal cortex as monkeys performed the same task. They found only a few cells that encoded the rule. The activity of these cells was probably derived from the PF cortex; this suggestion could be tested by recording in monkeys with PF lesions.

Eiselt and Nieder (2014) used the same overall design, but with different rules. They taught monkeys to discriminate both the number of spots and the length of lines. In one case a sample number of dots were presented, and after a delay the monkeys had to respond to a probe array of dots. In the other case a sample line was presented, and after a delay the monkeys had to respond to a probe line.

There were two rules, 'greater than' and 'less than', but the animal did not know which rule was in operation until after the sample had been presented. The rules were cued visually, and as in the experiment by Wallis et al. (2001) two different instruction cues were used in the experiment to specify the current rule.

If cued that the current rule was 'greater than', the animal had to release the lever if the probe display had more dots or a longer line. If cued that the current rule was 'less than', the animal had to release the lever if the probe display had fewer dots or a shorter line.

Cell recordings were taken once the animals had been over-trained on these tasks. Sixteen percent of the cells in the ventral and caudal PF cortex fired differentially according to a rule, some on the number task and others on the length task. The comparable figures for the medial PF cortex and the rostral cingulate motor area (CMAr) were roughly 6%.

However, there were a further 8% of cells in the ventral and caudal PF that coded for a rule, irrespective of whether the displays consisted of dots or lines. These were referred to as 'generalists'. No such cells were found in either the medial PF cortex or CMAr.

Figure 8.6 shows the distribution of rule-selective cells and generalists in the PF cortex for one monkey.

Cells that code for 'greater than' or 'less than' for number have also been recorded in the intraparietal area VIP (Vallentin et al., 2012). However, the proportion of such cells was double in the PF cortex compared with VIP. These rule cells in area VIP may have been derived, and this could be tested by recording in monkeys with PF lesions.

The cells that coded for abstract rules in this experiment did so for decisions concerning the sensory input. But there are also cells in the PF cortex that fire before the generation of spatial sequences and that do encode the abstract structure of the sequences. Chapter 6 describes an experiment by Shima et al. (2007) who taught monkeys sequences of movements using a handle. They found cells were in the dorsal PF cortex (areas 46 and 9/46) that fired similarly for sequences that

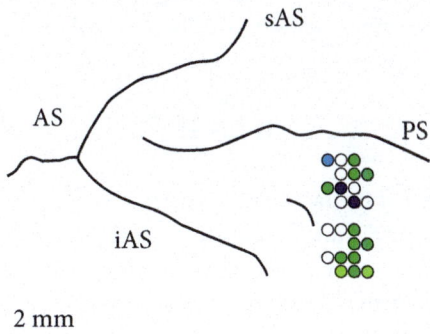

Figure 8.6 The location of rule selective and generalist cells in the PF cortex that coded for the general rule 'greater than' or 'lesser than'
These are shown for one monkey White circles, rule-selective cells, green, blue and black circles, generalist cells. AS = arcuate sulcus, sAS = superior arcuate sulcus, iAS = inferior arcuate sulcus, PS = principal sulcus.

Reproduced from Anne-Kathrin Eiselt & Andreas Nieder Rule activity related to spatial and numerical magnitudes: Comparison of prefrontal, premotor, and cingulate motor cortices. *J Cog Neurosci*, 26, 5 (May 2014), 1000–12. © 2014 by the Massachusetts Institute of Technology.

had the same abstract structure, even though the sequences differed in the actual movements themselves (Figure 6.11B, page 222).

The addition of new layers to the information-processing hierarchy means that the PF can represent abstract conditional rules that obtain in spite of variation in the sensory input or motor output. The tasks used by Shima et al. (2007) involved sequences, and this explains why the cells were found in the dorsal PF cortex (Chapter 6). The task used by Eiselt and Nieder (2014) involved associating the probe picture with the sample, and this explains why generalist cells were found in the ventral PF cortex (Chapter 7).

Though it is possible to find conjunctive cells in the PF cortex, the rules are supported by a subpopulation of cells, not single cells. Chapter 7 describes a study by Roy et al. (2010) in which monkeys were taught either to distinguish between cats and dogs or to distinguish between categories that paired different morphs of cats and dogs even though there were not alike in appearance. Each category distinction was represented in the ventral PF cortex by subpopulations that were largely, though not completely, independent.

Summary

The PF cortex supports conditional rules, but the addition of new layers to the processing hierarchy means that these rules can be abstract. This means that they apply whatever the material used. Conjunctive cells can be found that code for abstract properties, and it assumed that they achieve these properties because they have diverse inputs from lower-order areas. However, the representation of different abstract rules depends on subpopulations of cells, not single cells.

Transfer

There is a consequence of the fact that PF cortex can support the learning of abstract rules. Because the rules are abstract, they obtain whatever the sensory inputs or motor outputs. This means that, once the animals have learned the abstract rule, they can solve new problems very rapidly if the same rule applies. In other words, there is rapid transfer with few errors.

Harlow and Warren (1952) introduced the term 'learning set' to describe this transfer. They took, as one example, the learning of visual discrimination tasks. The animals were taught a series of problems, each for just six trials. On each problem the animals were faced with the choice between two objects. If they chose correctly, they obtained a reward. The term learning set referred to the improvement in learning that was shown across problems.

When macaque monkeys were tested on these discrimination problems, they showed a rapid improvement; and they were able at nearly 90% on trial two after they have solved a few hundred problems (Harlow & Warren, 1952) (Figure 8.7B).

The data in Figure 8.7 are shown in two ways. Figure 8.7A shows a cladogram with the number of trials to 60% correct performance plotted as a circle to the right. Macaque monkeys reached 60% correct choices on trial two very quickly, as indicated by the small size of the circle. Other mammals do much less well on trial two, as indicated by the proportionately larger circles.

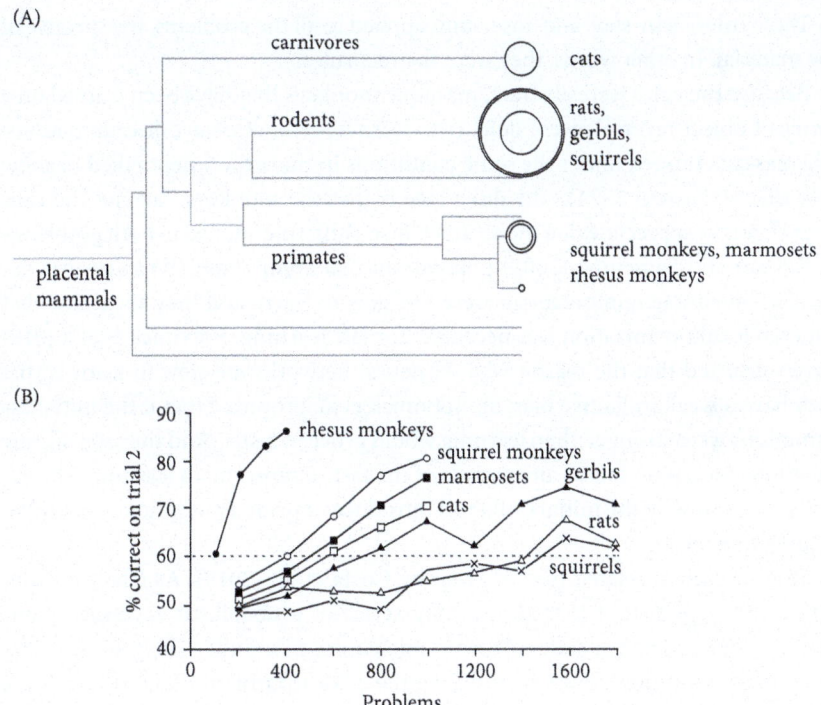

Figure 8.7 Development of a learning set for visual discrimination problems, in a selection of mammalian species
(A) Cladogram for selected mammals on a linear timescale. The diameter of the circles at the right shows the number of problems needed to reach 60% correct on trial two, interpolated from the data presented in (B). (B) Percentage of correct choices on trial two of a novel visual discrimination problem, as a function of the number of such problems previously solved. The dashed horizontal line shows the reference for the diameter plots shown in (A).

A: Reproduced from Passingham, R.E. & Wise, S.P. *The Neurobiology of Prefrontal Cortex*, p. 236, Figure 8.2 © 2012, Oxford University Press.

B: Adapted from Passingham *The Human Primate*, Figure 5.8 © 1982, W. H Freeman and Co., Ltd.

Figure 8.7B plots the improvement across blocks of trials, showing that, unlike the non-primate mammals tested, the macaque monkeys rapidly learned to achieve 90% correct performance on trial two.

Though there are data on learning set for other primates and other mammals, the results cannot be compared with those in Figure 8.7. The reason is that the conditions of testing were different: the animals were not been tested on each problem for just six trials.

It was the outcome of their choice on the first trial that told the animals which object was associated with food. If their choice was rewarded, the animals should repeat it on the next trial; if it was not rewarded, they could shift their choice to the other object on trial two.

These rules, 'win-stay' and 'lose-shift', applied to all the problems, irrespective of the material. In other words, they were abstract rules.

Because the rules were abstract, macaque monkeys that have been trained on a series of object reversals were quicker to learn a series of object discriminations, whereas cats trained under the same conditions in this experiment failed to show this effect (Warren, 1974). The difference is that the monkeys, but not the cats, could learn to appreciate that the abstract 'lose-shift' rule applied to both problems.

Kawato and Samejima (2007) pointed out that computational model that depended on simple reinforcement were too slow to learn; and they suggested that a hierarchical organization was necessary for fast learning. Botvinick et al. (2019) have suggested that the reason why AI neural networks are slow to learn is that they have a weak inductive bias. So Botvinick et al. proposed that if the initial assumptions were stronger, then learning would proceed faster. And they specifically mentioned learning sets as an example of the fast reinforcement learning. The experience of solving the initial problems introduces a prior assumption concerning future problems.

This account is related to that given by Costa et al. (2015). As Chapter 4 describes, they point out that after a long series of probabilistic reversals, monkeys are able to change suddenly from one choice to another. They suggest that this can be accounted for by considering the prior information that reversals are going to occur. In other word, there is a Bayesian prior that there will be reversals. Figure 4.12 (page 144) presents data on reversal learning set and an estimate of the Bayesian prior or evidence.

There are signals in the PF cortex that estimate the probability of reward (Roiser et al., 2010). This means that anthropoid primates can develop Bayesian priors concerning the location of rewards.

However, the data shown in Figure 8.7 do not prove that the improvement shown by macaque monkeys depends on the PF cortex. To do this, Browning et al. (2007) taught four monkeys object discrimination problems, presenting them on a screen in an automatic apparatus. Each problem was taught for just 11 trials, with

10 problems per session. Training continued until the animals had achieved a clear learning set, as shown by stable performance within a session.

Browning et al. then removed the inferior temporal cortex in one hemisphere of two monkeys and the PF cortex together with the premotor areas in the other hemisphere of the other two monkeys. The animals were tested for a further 8 sessions, of which the last 5 were used as measure of performance. Finally, the frontal lobe and inferotemporal cortex were disconnected by adding the other lesion in the previously intact hemisphere.

The results are shown in Figure 8.8. Before surgery there was a clear improvement from the first 10 preoperative problems (stage 1) through 10 problems taken from the middle of training (stage 2) to the last 10 problems (stage 3). There was no significant effect of the unilateral lesions. However, the dark grey bar shows that after the disconnection the performance of the animals was no better than it had been on the first 10 problems given before surgery (stage 1). The animals showed no retention of the learning set at all.

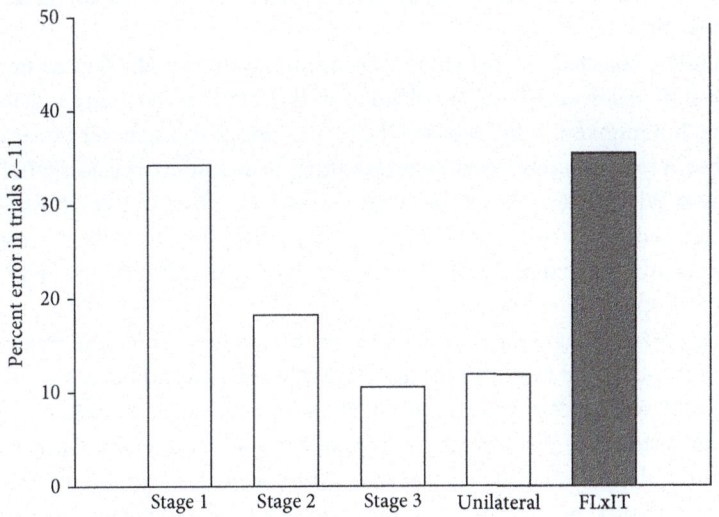

Figure 8.8 Histograms showing mean percent error in trials 2–11 at each performance test on discrimination learning set
(Stage 1) The first 10 problems of preoperative learning set formation; (Stage 2) 10 problems from the middle of preoperative learning; (Stage 30 the last 10 problems of preoperative learning. Unilateral = unilateral frontal or inferotemporal lesion. FL x IT refers to the disconnection of frontal lobe and inferior temporal cortex, with the percent errors on the postoperative tests.

Reproduced from Browning, P.G., Easton, A., & Gaffan, D. Frontal-temporal disconnection abolishes object discrimination learning set in macaque monkeys. *Cereb Cortex*, 17 (4), 859–64, Figure 4, https://doi.org/10.1093/cercor/bhk039 Copyright © 2006, Oxford University Press.

In this experiment the frontal lesion included not only the PF cortex but also the premotor areas. So, it needs to be repeated with lesions that are restricted to the granular PF cortex. Fortunately, as Chapter 7 describes, Bussey et al. (2001) have made PF lesions in monkeys that were trained monkeys on a series of problems until they could solve them in just a few trials. The difference is that the problems involved visual-spatial associations.

Rather than showing the improvement across sessions as in Figure 8.7, Figure 7.6 (page 254) plots the decrease in errors across trials within a session before surgery. It shows improvement within sessions averaged over problems.

The figure also shows that after the ventral an orbital PF cortex had been removed bilaterally, the monkeys failed to show any improvement at all within a session. They performed at a level of 70% correct throughout. This is the chance level given that there were either three or four directions.

The trials were divided into ones on which the cue was the same as on the previous trial (repeat) and trials on which the cue was different from the one on the previous trial (change). It therefore paid the animals to learn the abstract rules 'repeat-stay' and 'change-shift'. Before surgery, the monkeys showed a clear performance benefit for repeat-stay trials. However, the operated monkeys failed to show this effect.

It could be objected that the failure of the animals to learn the 'repeat-stay' and 'change-shift' rules was simply a reflection of the fact that they were impaired at learning. But monkeys with lesions of the hippocampus and subjacent cortex have been shown to be severely impaired on retention of conditional visuo-spatial tasks (Murray & Wise, 1996), and yet they were able to learn abstract rules. Bussey et al. (2001) showed this by computing a 'strategy' score. This measured the difference in performance between repeat and change trials. It was calculated for the first 8 trials of each problem.

Figure 8.9 (next page) shows the results for this strategy score. After removal of the ventral and orbital PF cortex the monkeys showed no retention as measured on this strategy score. Yet, monkeys with bilateral removal of the hippocampus and subjacent cortex were able to learn the 'repeat-stay' and 'change-shift' strategies.

There is independent evidence that it is the PF cortex that learns to represent the strategies 'repeat-stay' and 'change-shift. Genovesio et al. (2008) also taught monkeys visual-spatial associations, the difference being that the cues were associated with saccades in one of three directions. Cell recordings were taken in the dorsal and ventral PF cortex. Figure 8.10 (page 308) shows data for the subpopulations of cells in the PF cortex that coded for the 'change-shift' and 'repeat-stay' rules.

There is a critical difference between conditional tasks that depend on an abstract strategy and conditional tasks that do not. In the lesion experiment by Bussey et al. (2001), two of the operated animals were tested on conditional visuo-spatial problems for very many sessions. The result was that in the end they were able to learn them. In other words, the removal of the ventral and orbital PF cortex

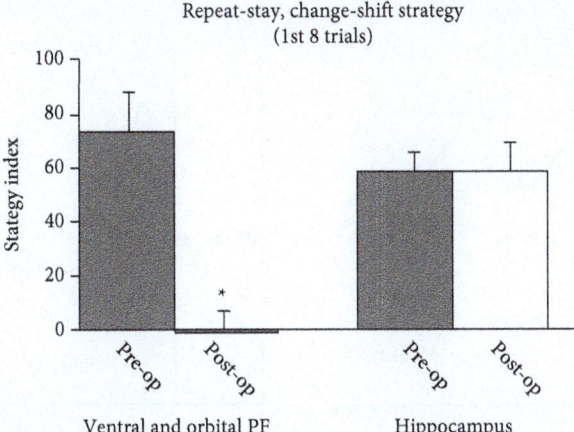

Figure 8.9 Strategy score (performance on repeat compared with change trials) before (black) and after (white) lesions
The data on the left are for monkeys with bilateral lesions of the ventral and orbital PF cortex, and the data on the right for monkeys with bilateral lesions of the hippocampus and subjacent tissue (H+).

Adapted from Bussey T.J., Wise S.P., & Murray E.A. The role of ventral and orbital prefrontal cortex in conditional visuomotor learning and strategy use in rhesus monkeys (*Macaca mulatta*), *Behav Neurosci*, 115 (5), 971–82, Doi: 10.1037//0735-7044.115.5.971 Copyright © 2001, American Psychological Association.

prevented fast learning that depends on abstract rules; but it did not prevent slow learning which depends on accumulative reinforcement.

Summary

Macaque monkeys show improvement when tested on a series of problems that have the same abstract structure or logic. The reason is that they can learn abstract rules, and these act as Bayesian priors concerning where food will be found or how to obtain it on future problems. There is evidence that it is the PF cortex that learns these rules and represents these priors. With experience, macaque monkeys can learn new problems with the same rules very rapidly. However, this ability is abolished by disconnection of the PF cortex from the inferior temporal cortex.

Prefrontal-Basal Ganglia Interactions

The ability to learn associations via reinforcement depends on connections between the PF cortex and the basal ganglia and dopaminergic midbrain. An earlier

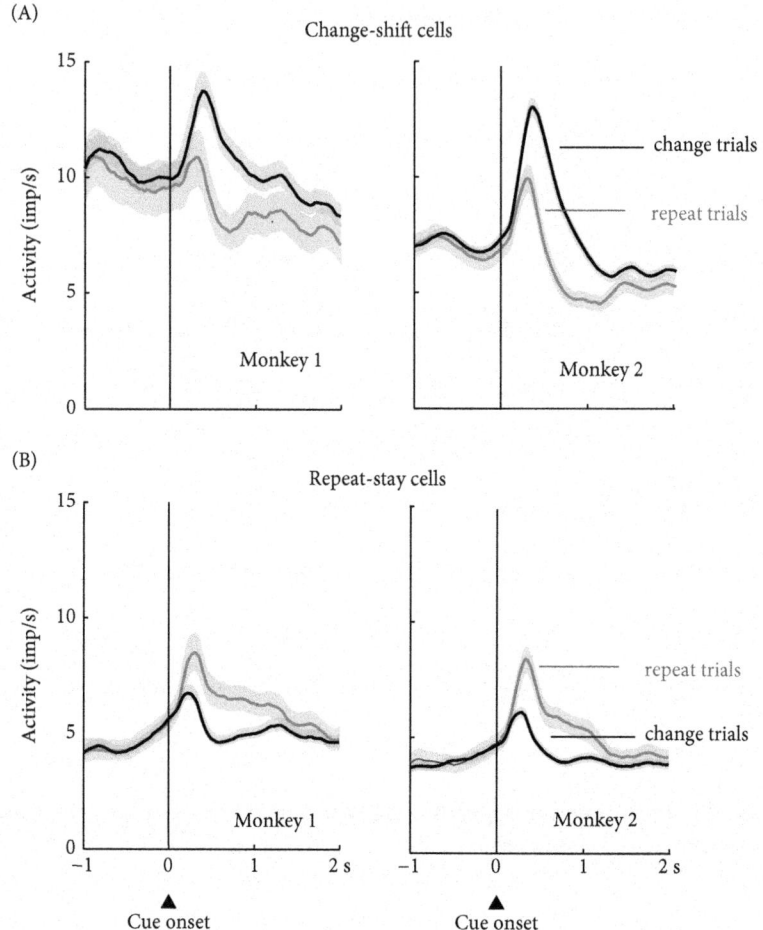

Figure 8.10 Population coding for abstract rules. (A) Average population activity for cells encoding the change-shift rule for two monkeys (left and right). (B) Average population activity for cells encoding the repeat-stay rule.

On change trials, there was a change in cue from the previous trial indicated that the monkey should choose a different goal. On repeat trials the cue was the same cue as on the previous trial. Shading: SEM.

Reproduced from Genovesio A. & Wise S.P. The neurophysiology of abstract strategies, in S.A. Bunge & J. Wallis (ed.), *Neurosc Rule Guided Behav*, pp. 81–105, Figure 7.11, Copyright © 2008, Oxford University Press.

section mentions that, unlike the cerebellum, the striatum receives a projection from the ventral and orbital PF cortex, and there is convergence between the projections from the dorsal and ventral PF cortex to the striatum.

Associative Learning

If recordings are taken simultaneously in the PF cortex and basal ganglia, it is possible to compare the contribution that they make to learning. Pasupathy and Miller (2005) presented monkeys with object cues, and each cue instructed the animal to make a saccade to one side or the other. Changes in the cell activity in the dorsal PF cortex tracked the slow improvement of performance. As learning of each pair progressed, cell activity encoding the direction of the saccade occurred earlier in the trial, but it did so earlier in the striatum than in the PF cortex.

However, this last result is not consistent with the findings of a later paper by Seo et al. (2012) who taught monkeys visual-visual associations. If the cue had more blue than red elements, the correct response was to make a saccade to the blue spot in the periphery; whereas if the cue had more red than blue elements, the correct response was to make a saccade to the red spot in the periphery.

Each trial consisted of a sequence of three saccades. In the fixed condition, 8 trials were presented with the same sequence, and this meant that the animal could learn a sequence of eye movements. In the random condition, the trials were varied so that the animals had to perform different sequences, and thus had to depend on the cue to tell them what to do.

Seo and Averbeck recorded both in the dorsal PF cortex and the dorsal striatum. The data were shown aligned to each movement. The selected goal or action was represented more strongly and occurred earlier in the PF cortex than in the striatum, and these differences were statistically significant.

Hoshi et al. (2013) recorded cell activity in the GPe rather than the striatum, and recorded simultaneously in the ventral and dorsal PF cortex and the dorsal premotor cortex. Rather than making a saccade, the monkeys were taught to reach to a box on the left or the right of a screen depending on the cues. There were two cues: one told them which arm to use and the other told them the side to which they were to respond. There was then a delay before a trigger stimulus that allowed the animals to respond.

Figure 8.11 (next page) presents the cumulative percent of cells in each area that showed selectivity for various aspects of the task. There were cells in the ventral PF cortex that showed a rise in response selectivity earlier than cells in the dorsal premotor cortex or the GPe (Figure 8.11B). However, when the cells for all these areas were considered together, the differences were not statistically significant.

Figure 8.11A shows that object selectivity was found early in the GPe as well as the ventral PF cortex. This is not surprising because the inferior temporal cortex

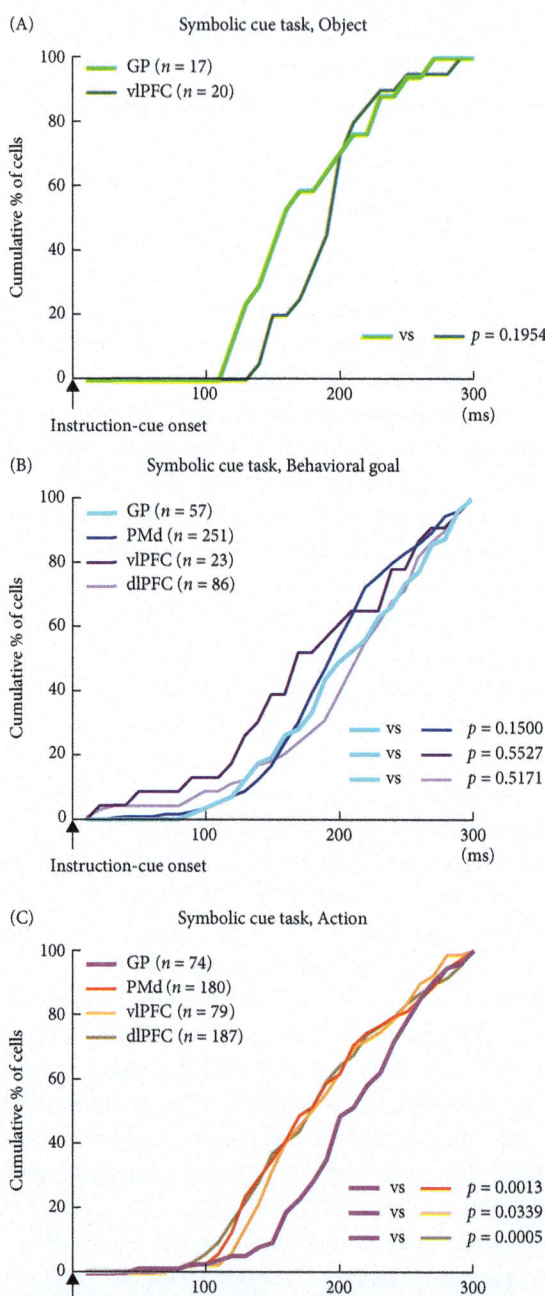

Figure 8.11 Timing of the development of selectivity for visual objects, behavioral goals, and actions
(A) Cumulative fractions of selectivity onset for visual objects in the GP (light green) and ventral PF cortex (vlPFC) (dark green) after instruction-cue onset in the symbolic
(continued in footnote of next page)

projects to the middle and tail of the caudate (Webster et al., 1993), as well as to the ventral PF cortex (Webster et al., 1994). However, Figure 8.11C also shows that the activity after the trigger stimulus occurred significantly later in the GPe than in the cortical areas.

The role of the basal ganglia and dopaminergic midbrain is to reinforce learned associations. In their experiment, Seo et al. (2012) devised a reinforcement learning algorithm to estimate the learning related action value. This value was represented more strongly in the striatum than the PF cortex. There was also an interaction such that the reinforcement learning parameter increased after the movement in the fixed but not random condition. This suggests a role in the automation of performance.

Abstract Categories and Rules

As an earlier section discusses, the structure of the neocortex differs from that of the striatum in that it is able to support hierarchical processing. It is this that allows the representation of abstract categories and abstract rules.

In order to compare the roles of the PF cortex and striatum, Antzoulatos and Miller (2011) taught monkeys to categorize displays of dots. There were two prototypes, and the exemplars in the same category consisted of dots that bore a resemblance in their general arrangement. The monkeys had to make a saccade to the left for one category and to the right for the other.

Early in learning, the striatal cell activity predicted the response earlier than the activity in the ventral and dorsal PF cortex. But as the number of exemplars increased, this pattern was reversed. When the monkeys had learned the categories, the population of PF cell activity predicted the response earlier than the striatal cells. These results are compatible with the suggestion that the striatum plays a role in the learning of individual associations between cues and responses, but

Figure 8.11 Continued

cue task. (B) Cumulative fractions of selectivity onset for the behavioural goal in the GP (light blue), dorsal premotor cortex (PMd) (dark blue), ventral PF cortex (vlPFC) (dark purple), and dorsal PF cortex (dlPFC) (light purple) after instruction-cue presentation in the symbolic cue task. (C) Cumulative fractions of the onset of action selectivity in the GP (purple), dorsal premotor cortex (PMd) (red), ventral PF cortex (vlPFC) (orange), and dorsal PF cortex (dlPFC) (brown) after choice-cue onset in the symbolic cue task.

Reproduced from Hoshi, E. Cortico-basal ganglia networks subserving goal-directed behavior mediated by conditional visuo-goal association. *Front Neur Circ*, 7, 158, Figure 11, Doi: 10.3389/fncir.2013.00158 © 2013 Hoshi. Licensed under CC BY 4.0.

not categories. In other words, it can support the learning of individual but not abstract rules.

Antzoulatos and Miller (2014) recorded from pairs of cells in the PF cortex and striatum while monkeys learned dot categories. After learning, significant category-selective synchrony emerged between multi-unit activity in the PF cortex and local field potentials in the striatum. However, an analysis using Grainger causality suggested that the striatum had an influence on oscillations in the ventral and dorsal PF cortex in both the beta and delta bands. Unfortunately, it is not clear what this meant in terms of learning categories; there was no change in the polysynaptic influence of the striatum on the PF cortex during learning.

Villagrasa et al. (2018) have produced a neurocomputational model to account for the results of this study. They presented various stimuli for categorization by the model, including morphs of the faces of Bill Clinton and George Bush. The model suggested that the decrease in category selectivity in the striatum with learning was due to the fact that the striatum came to code for a subset of displays in one category and a subset of displays in the other. The model suggested that the basal ganglia provide a teaching signal that supports Hebbian learning in the pathway from the inferior temporal cortex to the PF cortex, and this is sufficient to develop category specific signals in the PF cortex.

Antzoulatos and Miller (2014) were able to show that as monkeys learned new categories there was an increase in the synchronization in the beta band between the local field potentials in the PF cortex and striatum. Furthermore, different pairs of electrodes in the two areas showed stronger synchrony for one category rather than the other.

New Learning versus Automatic Performance

When monkeys are trained on a new task, they are typically over-trained. This means that there is a transition from attentional performance to automatic performance. This transition involves interactions between the PF cortex, the striatum and the cerebellum.

The degree to which a task can be performed automatically can be assessed by using the dual task paradigm. This is most easily arranged in studies with human subjects (Baddeley et al., 1998). In an early study, human subjects were taught spatial sequences eight moves long, performed with the fingers (Passingham, 1996). The subjects learned a new sequence by trial and error, with auditory feedback after each move to tell them whether the move was correct at that point in the sequence.

One of two secondary tasks was given at the same time as the subjects learned the sequence. One was easy: the subjects were instructed to repeat nouns while learning a new sequence. The other task was more difficult: the subjects had to generate nouns that were appropriate for verbs. The subjects made more errors on the new than the over-trained sequences while performing either of these secondary tasks. However, after overtraining they made very few errors when the secondary task was easy, repeating nouns.

In a PET study by Jueptner et al. (1997a; 1997b), the subjects learned a spatial sequence by trial and error during the scanning sessions, and they also performed a sequence that they had overlearned before scanning.

During new learning, as opposed to overlearned performance, there were activations reflecting attentional performance in the dorsal PF cortex (areas 46 and 9/46), the presupplementary motor area (preSMA), and the dorsal premotor cortex, as well as in the cortex in the IPS and the inferior parietal area PFG which project to the dorsal PF cortex. Subcortically there were activations in the caudate and MD nucleus of the thalamus with which the dorsal PF cortex is connected.

However, when the subjects performed a sequence automatically, there were no significant activations in the PF cortex. Instead, the activations were confined to the dorsal premotor cortex and supplementary motor area (SMA), the cortex in the IPS, the putamen and the cerebellar cortex and nuclei. These results indicate a shift from a system generating spatial goals to a sensorimotor system.

Rowe et al. (2002a) used fMRI to confirm that when subjects perform a simple automatic sequence there was no significant activation in the PF cortex. However, the lack of any activation could simply be due to the insensitivity of the method. Rainer and Miller (2000) taught monkeys a delayed-matching task with either novel or familiar objects, and recorded from cells in the dorsal and ventral PF cortex. When the objects were novel there was significant preparatory activity throughout the delay. But when the object were familiar, fewer cells responded to them, and cell activity was only found at the end of the delay, just before the response (Rainer & Miller, 2002).

These results suggest that once a spatial sequence has been overlearned, the cell activity in the PF cortex may be limited to the period just before movement. Functional brain imaging may be sensitive to preparatory activity between movements, but less so to a phasic signal.

In order to study preparation, Rowe et al. (2002a) specifically instructed their subjects to prepare or think about the next movement of a simple sequence, even though they were able to perform the sequence automatically. The result was that the dorsal PF cortex (areas 46 and 9/46) was re-activated, as well as the rostral part of the dorsal premotor cortex and the cortex in the IPS. In a parallel experiment, Rowe et al. (2002b) also found re-activation in the preSMA, though in the paper this was wrongly attributed to the paracingulate cortex.

In these experiments, structural equation modelling was used to compare the influence of the PF cortex and the tissue in the IPS. This method looks for the solution that minimizes the difference between the observed covariance and that implied by the anatomical model. The results of the analysis indicated that, when the subjects prepared their movements, it was the PF cortex, and not the posterior parietal cortex, that influenced the motor cortex via the dorsal premotor cortex and SMA.

However, when the subjects had to perform a secondary task at the same time, the subjects were no longer able to prepare (Rowe et al., 2002b). The result was that there was a significant decrease rather than an increase in the path coefficient between the PF cortex and the premotor cortex (Rowe et al., 2002a).

Floyer-Lea and Matthews (2004) also used a dual task paradigm. The primary task was to learn to press a sensor with a predictable sequence of forces so as to track the variation in the height of a histogram on a screen. The secondary task was to count backwards from 100, 101, or 99 in steps of three. The subjects made many fewer errors in tracking and counting when the task had been over-trained because the presses could be run off as a repeating motor sequence.

As the subjects had experience with the task, there was a learning-related decrease in activation in the dorsal PF cortex which reached baseline when the task was automatic. As learning progressed, there was also a decrease in activation with learning in the preSMA which is interconnected with the dorsal PF cortex (Wang et al., 2005), as well in the head of the caudate to which the PF cortex projects (Selemon & Goldman-Rakic, 1985).

At the same time, as the task became automatic there were corresponding increases in activation in the putamen around the level of the anterior commissure. The dorsal premotor cortex and SMA send projections to the putamen at this level (Selemon & Goldman-Rakic, 1985).

By recording from cells in the preSMA and SMA, Nakamura et al. (1998) have been able to show an anterior to posterior shift in the cell activity as monkeys learn sequences. Nakamura et al. presented five displays with a four by four array of LEDs in sequence, and on each display two of the squares were lit. During training, rewards were provided after each pair; but when the whole sequence had been learned, the reward was only presented at the end of the ten moves. The monkey had to learn the order in which to touch each of the squares. This design meant that the monkeys could learn new ten move sequence very rapidly, and thus the animals could be trained on a large set of different sequences.

Nakamura et al. compared new learning of these sets with the performance of sets on which the monkeys had already been over-trained. Whereas 31% of the cells in the preSMA responded during new learning, only 10% of the cells in the SMA did so. Correspondingly, whereas 3.6% of the cells in the preSMA responded during overlearned performance, 7.7% of the cells in the SMA did so.

Nakamura et al. (1999) followed these results up by injecting muscimol. Injections in the preSMA caused a marked increase in errors during new learning, whereas injections into the SMA caused an impairment that was less severe. However, inactivation of both areas had little effect on overlearned performance. The reason may be that, as the next section discusses, the cerebellum can support the automatic performance of sequences. The dentate nucleus of the cerebellum can influence the motor cortex via direct projections from the ventrolateral nucleus of the thalamus (Dum et al., 2002).

There is a similar anterior to posterior progression in the striatum with learning. Miyachi et al. (2002) compared the activity of cells in the striatum while monkeys performed sequences that they had learned and sequences on which they had been over-trained. There were more cells that fired preferentially for new sequences in the anterior striatum. By contrast, there were more cells that fired preferentially for over-trained sequences in the putamen at a level posterior to the anterior commissure. Gerbella et al. (2016) refer to the putamen at this level as 'the motor putamen' because it receives connections from the ventral premotor cortex.

These results account for the findings of an earlier study by Miyachi et al. (1997) in which they injected muscimol. Injections into the rostral striatum severely impaired new learning. By contrast, injections into the more caudal putamen impaired the performance of over-trained sequences.

Summary

The PF cortex interacts with the basal ganglia. Whereas the PF cortex specifies the goal and supports the learning of abstract categories and rules, the basal ganglia plays a role in reinforcing associations. During the automation of spatial sequences, there is a decrease in activity in the PF cortex and preSMA as well as the caudate nucleus. By contrast, there is a corresponding increase in the activity of cells in the posterior putamen. The reason is that there is a change from generating spatial goals to generating movements.

Prefrontal-Cerebellar Interactions

Automatic Performance

The previous section mentions that inactivation of the posterior putamen has been shown to cause an impairment in the performance of over-trained sequences; however the effect was small (Miyachi et al., 1997). The reason is that the cerebellum has the internal structure that allows it to support the performance of sequences, once learned.

In his classical description of the workings of the cerebellum, Marr (1969) suggested that the inputs to the olivary cells were elemental movements and that the Purkinje cells could initiate elemental movements. The proposal was that this arrangement allows the cerebellum to learn the motor contexts of movements, with one movement providing the cue for the next one.

Direct evidence that the cerebellum is involved in the performance of overlearned sequences of movements comes from a study by Lu et al. (1998). The monkeys were over-trained on sequences using the same design as in the study by Nakamura et al. (1998). Liu et al. (1998) then injected muscimol into the dorsal part of the dentate nucleus, and this had a severe effect on retention. The monkeys made errors on a third or so of the moves in the sequences, and they were slower to perform them. They also made fewer anticipatory saccades, meaning that they were less likely to look towards the next location in the sequence. Yet the animals were able to learn new sequences at a normal rate.

Nixon and Passingham (2000) made permanent lesions in the dentate on both sides by injecting an excitotoxin. They taught monkeys a sequence task in which four targets lit up for 300 milliseconds in turn in a repeating sequence. On the comparison task the targets lit up in a random sequence. In 2,000 trials the monkeys with total dentate lesions failed to learn the repetitive task to automaticity, as assessed in terms of the time to complete the sequence.

It is critical to note, however, that in the study by Lu et al. (1998) an impairment was only found when the hand ipsilateral to the inactivated dentate was used. There was no impairment if the other hand was used. This suggests that once the sequence had been overlearned, the task becomes a motor skill that was specific to the hand tested. In other words, it is performed in movement rather than spatial coordinates (Hikosaka et al., 1999; Hikosaka et al., 2002). During learning, location A is the context for the spatial goal B which is next in the sequence. But once the task has become automatic, the contexts are proprioceptive, so that movement A' is the context for movement B'.

Though monkeys with dentate lesions are impaired at performing automatic sequences, they can still learn a conditional visuo-spatial task. Nixon and Passingham (2000) trained monkeys with bilateral dentate lesions. At first the animals were clumsy in using the handle after surgery. However, once they had been given enough practice to improve their movements, they learned a new visual-spatial association as quickly as unoperated control animals. The reason is that visual associative learning involves an interaction between the PF cortex and the basal ganglia. The cerebellum is not involved because it has no interconnections with the ventral and orbital PF cortex (Schmahmann & Pandya, 1997).

The Cerebellum and Cognition

Marr's (1969) theory concerning the cerebellum accounts for its role in the automation of motor skills as well as the maintenance of postural reflexes. One aspect of motor skill involves exact timing, and there is cell activity in the cerebellar dentate nucleus that correlates with trial by trial variation in timing (Kunimatsu et al.,

2018). Thus, it is easy to see why the cerebellum should also be involved in the automation of rhythms (Ramnani and Passingham, 2001).

The theory can also be extended to account for the role of the cerebellum in eye-blink conditioning (Ramnani et al., 2000) and changes in skin conductance or heart rate during fear conditioning (Maschke et al., 2002). The reason is that in classical conditioning the conditioned stimulus predicts the exact time when the puff of air or electric shock will occur.

But it has also been claimed that the cerebellum is involved in verbal working memory (Desmond & Fiez, 1998), verb generation (Desmond & Fiez, 1998), set shifting (Lie et al., 2006) and problem solving (Kim et al., 1994). More generally it has been said to be involved in cognition (Schmahmann, 1997). And computational theories have even been produced to explain how it might do this (Honda & Ito, 2017).

Kelly and Strick (2003) suggested that the fact that there is a loop that connects the PF cortex with the cerebellum indicates a role in cognitive processing. But as Figure 8.3 illustrates, there are also loops with the motor cortex and premotor areas (Bostan et al., 2013). These may account for some of the findings since verbal working memory involves subvocal articulation (Baddeley, 1986), and this is true whether the words are read or heard (Conrad, 1973). The ability to articulate depends not only on the motor cortex but also the frontal opercular cortex amongst other areas (Ripamonti et al., 2018). Articulation is a learned skill that has been automated over very many years of practice. So, it is not surprising that the cerebellum is activated when subjects engage in verbal working memory (Desmond & Fiez, 1998); or that patients with cerebellar lesions can be impaired on verbal working memory (Peterburs et al., 2010). Verbal working memory involved the articulatory or phonological loop (Baddeley & Hitch, 2019), and cerebellar lesions cause dysarthria (Bodranghien et al., 2016).

Other findings have been taken to implicate the cerebellum in semantic processing as well. There is activation in the cerebellum when subjects generate verbs that are appropriate for nouns (Desmond & Fiez, 1998). But though both the PF cortex and the lateral cerebellar cortex (lobules VII and VIII) are activated when subjects are first tested on this task (Stoodley et al., 2012), the activation in the PF cortex decreases when the subjects repeat the same associations until they are automatic (Raichle et al., 1994). It is likely that the contribution of the lateral cerebellar cortex is to automate these associations.

The cerebellum is also not necessary for verb generation. It has been shown to be unaffected in patients with well-defined lesions of the cerebellum (Richter et al., 2004).

Nonetheless, Moberget et al. (2014) have argued that the cerebellum plays a role in language. In an fMRI study, they found that it was activated when the subjects learned to predict the end of a sentence depending on how it started. But the subjects were adults with years of experience with hearing language, and this means that they would have been able to make these predictions automatically. The initial part of the sentence serves as the context for the end.

The same explanation will not, however, account for the claims that the cerebellum is involved in performance of the Wisconsin Card Sorting Task (Schmahmann & Sherman, 1998). But Lie et al. (2006) used fMRI to scan subjects while they engaged in set shifting on this task, and they compared shifts that were or were not instructed. The cerebellar activation was only associated with shifts that were made on the basis of negative feedback. The central nucleus of the amygdala is connected with the pons (Price & Amaral, 1981) and it could be this connection that provided the negative feedback.

None of these examples are incompatible with the hypothesis that the cerebellum is involved in the automation of associations. But this is not true of claims that the cerebellum is involved in problem solving. To contribute to problem solving, the cerebellum would need to be able to generate novel goals. Yet, in an fMRI study Kim et al. (1994) found activation in the dentate nucleus while subjects solve a pegboard problem. They had to move pegs through a maze to achieve a goal. This experimental condition was contrasted with a control condition in which the subjects merely moved the pegs.

But the two conditions varied in several ways, for example in the eye movements that are involved (Glickstein & Doron, 2008). When trying to solve a problem, the subjects would look ahead whereas they would not need to do this when simply moving pegs. Furthermore, the critical test is whether cerebellar lesions impair performance. Patients with lesions that are confined to the cerebellum have been tested on the Tower of Hanoi, a planning task in which the subjects have to rearrange rings on pegs to achieve a specified goal arrangement (Daum et al., 1993). Yet, in spite of the lesions these patients were not impaired.

The best way to check whether the cerebellum is like the PF cortex in being able to generate novel goals is to test monkeys with lesions in the cerebellum on a task on which it is well established that the PF cortex plays a critical role. So, Nixon and Passingham (1999) trained macaque monkeys on the delayed alternation task, and then removed the dorsal prefrontal cortex (areas 46 and 9/46) in three animals, while making complete excitotoxic lesions of the dentate nuclei in three others. As expected, the animals with dorsal PF lesions failed to relearn the task in 2,000 trials. But the animals with lesions of the dentate nucleus were unimpaired.

It could be argued that it is odd that monkeys with total lesions of the dentate nuclei are unimpaired on the spatial delayed alternation task since it involves sequences. But there is a critical difference between delayed alternation and the sequences studied by Hikosaka et al. (1999). Film of the animals performing the sequences taught by Hikosaka et al. shows that, when the animals had learned the sequences, they performed them very rapidly and skillfully, with little time between the movements. But on delayed alternation a screen is interposed between movements, and the delay between movements is 5 seconds. So, the animals have to encode the next goal prospectively during the delay (Chapter 6). This is a task that can never become a simple motor skill.

Summary

The cerebellum is involved in the automation of motor sequences such that they become motor skills. These sequences are learned initially in spatial coordinates but when overlearned they are performed in motor-coordinates. Though there are connections between the dorsal PF cortex and the cerebellum, there is no convincing evidence that the cerebellum can generate novel goals. In experiments where it has been claimed that the cerebellum is involved in 'cognitive' performance, its role remains that of automating skills. Whereas monkeys with lesions of the dorsal PF cortex fail to relearn the spatial delayed alternation task in 2,000 trials, monkeys with total lesions of the dentate nuclei are unimpaired. The reason is that there is a 5-second interval between movements during which a screen is interposed, and so the task can never be run off as a motor skill.

Conclusions

Abstract Rules

The dorsal and ventral PF cortex are alike in that they support conditional behaviour Chapters 6 and 7). In this the goal depends on a single event, the current context, and thus conditional behaviour involves attentional performance.
But primates are not alone in being able to learn conditional tasks or respond flexibly. For example, it has been shown that rats can learn probabilistic reversals, and inactivation of either the medial frontal cortex or the orbital cortex impairs performance (Dalton et al., 2016).

But there is a critical difference between the abilities of macaque monkeys and rats. Unlike rats, macaque monkeys can learn to show rapid improvement when tested on tasks in which the visual cues differ from problem to problem. For example, they can achieve a level of 90% on trial 2 when trained on a series of visual discrimination problems (Passingham, 1982); and disconnection of the PF cortex from its inferior temporal input abolishes the learning set for visual-spatial problems (Browning et al., 2007). Macaques can also learn to solve novel visuo-spatial tasks rapidly, and removal of the ventral and orbital PF cortex abolishes this ability (Bussey et al., 2001).

Monkeys can show rapid learning because they are capable of learning abstract rules such as 'repeat-stay' and 'change-shift'. Though Dalton et al. (2016) argue that rats can learn the rules 'win-stay' and 'lose-shift', these are not abstract rules because they only apply to a single problem, probabilistic reversals between two levers.

Macaque monkeys can learn abstract rules because their PF cortex lies at the top of the sensory, motor, and outcome hierarchies. Hierarchical processing supports allows the generation of invariance or abstraction. And it is the neocortex and not the striatum or cerebellar cortex that has the appropriate structure.

Whereas the bottom-up connections carry information about the current state of the world, the connections that originate in the PF cortex carry information about the animal's goals. It is these that determine what the animal is interested in and therefore attends to.

Top-down pathways enhance the sensory representations that are relevant for the task in hand. Thus, these feedback pathways can also be regarded as output paths. They are involved in what the animal itself does. So, the PF cortex supports both response selection and attentional selection.

Conjunctive Cells

When the PF cortex is considered as a whole, it can be regarded as a mechanism for the rapid learning and solving of problems based on experience with previous ones. This is not to suggest that there are no functional distinctions within the PF cortex. These are easily demonstrated by removing discrete areas. However, they are much less obvious in the intact PF cortex. The reason is that the dense interconnections within it mean that much of the cell activity is derived.

However, the key is not that the cells are more adaptable than other cells, as suggested by Duncan (Duncan, 2001; Stokes et al., 2013). Cells in the inferotemporal cortex are also adaptable (Freedman et al., 2006). Instead, the key is that the cells in the PF cortex can develop conjunctive properties, because of the position of the PF cortex in the processing hierarchy. It is these conjunctive cells that are able to support the learning of behavioural rules.

Thus, interactions with the orbital PF cortex lead to the development of conjunctive cells in the dorsal PF cortex that encode both the spatial goal and the outcome (Wallis & Miller, 2003). Interactions with the ventral PF cortex lead to the development of cells in the dorsal PF cortex that encode both the cue and the goal (Asaad et al., 1998). Interactions with the hippocampus allow the development of conjunctive cells in the ventral PF cortex that code for the association between objects pairs over the long term (Brincat & Miller, 2015).

Conjunctive cells also provide the basis for learning abstract rules. For example, there cells in the PF cortex that code for matching or non-matching, irrespective of the object pairs that are used (Wallis et al., 2001). In other words, these cells show activity that is conjunctive across objects. There are also cells that code for the 'win-stay' or 'lose-shift' rules and do so irrespective of the cues that are presented (Genovesio & Wise, 2008). In the dorsal PF cortex, there are cells that code for the categorization of sequences, even when the actions are not alike (Shima et al., 2007), and in the ventral PF cortex cells that code for members of object categories even when the members do not look alike (Cromer et al., 2010).

Reinforcement and Skill

The interactions between the PF cortex and the basal ganglia support the ability to learn associative tasks via rewards or reinforcement. This ability does not depend on learning abstract rules, and it can be thought of as slow learning. When either monkeys or human subjects learn sequences, there is a shift from generating spatial goals to motor goals, and this shift can be detected in the striatum.

When sequences have been over-learned, they become a matter of skill. One movement serves as the cue for the next one. The cerebellum has the structure that allows it to learn the contexts for movements, in the absence of rewards. However, there is no evidence that the cerebellum can generate novel goals.

Selective Advantage

The problems that anthropoid primates face in the wild involve the need to feed rapidly and efficiently. This can be because of heat stress because descending into a more open environment makes them easier for predators to pick off. Furthermore, the volatility of the resources means that they need to be able to adapt quickly as resources deplete. So, any increase in the speed of learning and the reduction of costly errors is of selective advantage.

The ability of monkeys in the laboratory to learn abstract rules implies that in their natural environment they are also able to benefit from experience when tackling new problems. The ability to do this is one aspect of intelligence. But another aspect concerns the ability to understand the relations between things in the world. If one resource depletes, it is helpful to know which other foods have similar properties. It also pays to understand the similarities between the ways in which certain predators stalk and hunt.

The most impressive example of the ability to categorize objects concerns the use of tools by chimpanzees. Chimpanzees in the Ivory Coast use branches and stones to pound open nuts (Boesch & Boesch, 1990). In other words, they can understand that branches and stones have similar properties, even though they do not look alike. Similarly, chimpanzees in the Gombe stream reserve use grasses or twigs to fish for termites (Whiten et al., 1999). The modification of the branches or twigs so as to make them more appropriate can be treated as an index of the ability to innovate, a further aspect of intelligence (Reader & Laland, 2002).

However, once a problem has been solved, performance can become automatic. Chimpanzees learn to become more skillful in their use of tools. The advantage is that automatic tasks are less attentionally demanding. This means that the animal has the attentional resources for processing information about other aspects of the environment, such as the presence of potential predators.

References

Alexander, G.E., DeLong, M.R., & Strick, P.L. (1986) Parallel organization of functionally segregated circuits linking basal ganglia and cortex. *Annu Rev Neurosci*, 9, 357–81.

Amaral, D.G. & Price, J.L. (1984) Amygdalo-cortical projections in the monkey ({IMacaca fascicularis}). *J Comp Neurol*, 230, 465–96.

Antzoulatos, E.G. & Miller, E.K. (2011) Differences between neural activity in prefrontal cortex and striatum during learning of novel abstract categories. *Neuron*, 71, 243–9.

Antzoulatos, E.G. & Miller, E.K. (2014) Increases in functional connectivity between prefrontal cortex and striatum during category learning. *Neuron*, 83, 216–25.

Arsenault, J.T., Rima, S., Stemmann, H., & Vanduffel, W. (2014) Role of the primate ventral tegmental area in reinforcement and motivation. *Curr Biol*, 24, 1347–53.

Asaad, W.F., Rainer, G., & Miller, E.K. (1998) Neural activity in the primate prefrontal cortex during associative learning. *Neuron*, 21, 1399–407.

Averbeck, B.B., Lehman, J., Jacobson, M., & Haber, S.N. (2014) Estimates of projection overlap and zones of convergence within frontal-striatal circuits. *J Neurosci*, 34, 9497–505.

Baddeley, A. (1986) *Working Memory*. Oxford University Press, Oxford.

Baddeley, A., Emslie, H., Kolodny, J., & Duncan, J. (1998) Random generation and the executive control of working memory. *Q J Exp Psychol A*, 51, 819–52.

Baddeley, A.D. & Hitch, G.J. (2019) The phonological loop as a buffer store: An update. *Cortex*, 112, 91–106.

Bastos, A.M., Usrey, W.M., Adams, R.A., Mangun, G.R., Fries, P., & Friston, K.J. (2012) Canonical microcircuits for predictive coding. *Neuron*, 76, 695–711.

Bodranghien, F., Bastian, A., Casali, C., Hallett, M., Louis, E.D., Manto, M., Marien, P., Nowak, D.A., Schmahmann, J.D., Serrao, M., Steiner, K.M., Strupp, M., Tilikete, C., Timmann, D., & van Dun, K. (2016) Consensus paper: Revisiting the symptoms and signs of cerebellar syndrome. *Cerebellum*, 15, 369–91.

Boesch, C. & Boesch, H. (1990) Tool use and tool making in wild chimpanzees. *Folia Primatol*, 54, 86–99.

Bonini, L., Ugolotti Serventi, F., Bruni, S., Maranesi, M., Bimbi, M., Simone, L., Rozzi, S., Ferrari, P.F., & Fogassi, L. (2012) Selectivity for grip type and action goal in macaque inferior parietal and ventral premotor grasping neurons. *J Neurophysiol*, 108, 1607–19.

Bostan, A.C., Dum, R.P., & Strick, P.L. (2013) Cerebellar networks with the cerebral cortex and basal ganglia. *Trends Cogn Sci*, 17, 241–54.

Bostan, A.C. & Strick, P.L. (2018) The basal ganglia and the cerebellum: Nodes in an integrated network. *Nat Rev Neurosci*, 19, 338–50.

Botvinick, M., Ritter, S., Wang, J.X., Kurth-Nelson, Z., Blundell, C., & Hassabis, D. (2019) Reinforcement learning, fast and slow. *Trends Cogn Sci*, 23, 408–22.

Brincat, S.L. & Miller, E.K. (2015) Frequency-specific hippocampal-prefrontal interactions during associative learning. *Nat Neurosci*, 18, 576–81.

Brincat, S.L., Siegel, M., von Nicolai, C., & Miller, E.K. (2018) Gradual progression from sensory to task-related processing in cerebral cortex. *Proc Natl Acad Sci USA*, 115, E7202–211.

Brodal, P. & Bjaalie, J.G. (1992) Organization of the pontine nuclei. *Neurosci Res*, 13, 83–118.

Browning, P.G., Easton, A., & Gaffan, D. (2007) Frontal-temporal disconnection abolishes object discrimination learning set in macaque monkeys. *Cereb Cortex*, 17, 859–64.

Bruni, S., Giorgetti, V., Bonini, L., & Fogassi, L. (2015) Processing and integration of contextual information in monkey ventrolateral prefrontal neurons during selection and execution of goal-directed manipulative actions. *J Neurosci*, 35, 11877–90.

Buckley, M.J. & Gaffan, D. (1998) Perirhinal cortex ablation impairs visual object identification. *J Neurosci*, 18, 2268–75.

Bussey, T., Wise, S. & Murray, E. (2001) The role of ventral and orbital prefrontal cortex in conditional visuomotor learning and strategy use in rhesus monkeys (Macaca mulatta). *Behav Neurosci*, 115, 971–82.

Butters, N. & Rosvold, H.E. (1968) Effect of caudate and septal nuclei lesions on resistance to extinction and delayed-alternation. *J Comp Physiol Psychol*, 65, 397–403.

Cadieu, C., Kouh, M., Pasupathy, A., Connor, C.E., Riesenhuber, M., & Poggio, T. (2007) A model of V4 shape selectivity and invariance. *J Neurophysiol*, 98, 1733–50.

Carmichael, S.T. & Price, J.L. (1995) Sensory and premotor connections of the orbital and medial prefrontal cortex of macaque monkeys. *J Comp Neurol*, 363, 642–64.

Chakraborty, S., Ouhaz, Z., Mason, S., & Mitchell, A.S. (2019) Macaque parvocellular mediodorsal thalamus: Dissociable contributions to learning and adaptive decision-making. *Eur J Neurosci*, 49, 1041–54.

Changizi, M.A. & Shimojo, S. (2005) Parcellation and area-area connectivity as a function of neocortex size. *Brain Behav Evol*, 66, 88–98.

Chen, A., DeAngelis, G.C., & Angelaki, D.E. (2011) Representation of vestibular and visual cues to self-motion in ventral intraparietal cortex. *J Neurosci*, 31, 12036–52.

Choi, E.Y., Tanimura, Y., Vage, P.R., Yates, E.H., & Haber, S.N. (2017) Convergence of prefrontal and parietal anatomical projections in a connectional hub in the striatum. *Neuroimage*, 146, 821–32.

Clower, D.M., Dum, R.P., & Strick, P.L. (2005) Basal ganglia and cerebellar inputs to 'AIP'. *Cereb Cortex*, 15, 913–20.

Cohen, Y.E., Cohen, I.S., & Gifford, G.W. 3rd. (2004) Modulation of LIP activity by predictive auditory and visual cues. *Cereb Cortex*, 14, 1287–301.

Conrad, R. (1973) Some correlates of speech coding in the short-term memory of the deaf. *J Speech Hear Res*, 16, 375–84.

Costa, V.D., Tran, V.L., Turchi, J., & Averbeck, B.B. (2015) Reversal learning and dopamine: A bayesian perspective. *J Neurosci*, 35, 2407–16.

Crick, F. & Koch, C. (1998) Constraints on cortical and thalamic projections: The no-strong-loops hypothesis. *Nature*, 391, 245–50.

Cromer, J.A., Roy, J.E., & Miller, E.K. (2010) Representation of multiple, independent categories in the primate prefrontal cortex. *Neuron*, 66, 796–807.

Dalton, G.L., Wang, N.Y., Phillips, A.G., & Floresco, S.B. (2016) Multifaceted contributions by different regions of the orbitofrontal and medial prefrontal cortex to probabilistic reversal learning. *J Neurosci*, 36, 1996–2006.

Daum, I., Ackermann, H., Schugens, M.M., Reimold, C., Dichgans, J., & Birbaumer, N. (1993) The cerebellum and cognitive functions in humans. *Beh Neurosci*, 107, 411–19.

Dehaene, S., Kerszberg, M., & Changeux, J.-P. (1998) A neuronal model of a global workspace in effortful cognitive tasks. *Proc Nat Acad Sci*, 95, 14529–34.

Desimone, R. & Duncan, J. (1995) Neural mechanisms of selective visual attention. *Ann Rev Neurosci*, 18, 193–222.

Desimone, R. & Schein, S.J. (1987) Visual properties of neurons in area V4 of the macaque: Sensitivity to stimulus form. *J Neurophysiol*, 57, 835–68.

Desimone, R., Schein, S.J., Moran, J., & Ungerleider, L.G. (1985) Contour, color and shape analysis beyond the striate cortex. *Vision Res*, 25, 441–52.

Desmond, J.E. & Fiez, J. (1998) Neuroimaging studies of the cerebellum: Language, learning and memory. *Trends Cogn Sci*, 2, 355–61.

Dum, R.P., Li, C., & Strick, P.L. (2002) Motor and nonmotor domains in the monkey dentate. *Ann NY Acad Sci*, 978, 289–301.

Duncan, J. (2001) An adaptive coding model of neural function in prefrontal cortex. *Nat Rev Neurosci*, 2, 820–29.

Eiselt, A.K. & Nieder, A. (2014) Rule activity related to spatial and numerical magnitudes: Comparison of prefrontal, premotor, and cingulate motor cortices. *J Cogn Neurosci*, 26, 1000–12.

Eiselt, A.K. & Nieder, A. (2016) Single-cell coding of sensory, spatial and numerical magnitudes in primate prefrontal, premotor and cingulate motor cortices. *Exp Brain Res*, 234, 241–54.

Elston, G.N. (2007) Specialization of the neocortical pyramidal cell during primate evolution. In Kaas, J., Preuss, T.M. (eds) *Evolution of Nervous Systems: A Comprehensive Reference*. Elsevier, New York. pp. 191–242.

Ercsey-Ravasz, M., Markov, N.T., Lamy, C., Van Essen, D.C., Knoblauch, K., Toroczkai, Z., & Kennedy, H. (2013) A predictive network model of cerebral cortical connectivity based on a distance rule. *Neuron*, 80, 184–97.

Erickson, S.L., Melchitzky, D.S., & Lewis, D.A. (2004) Subcortical afferents to the lateral mediodorsal thalamus in cynomolgus monkeys. *Neurosci*, 129, 675–90.

Fiorillo, C.D., Tobler, P.N., & Schultz, W. (2003) Discrete coding of reward probability and uncertainty by dopamine neurons. *Science*, 299, 1898–902.

Freedman, D.J., Riesenhuber, M., Poggio, T., & Miller, E.K. (2006) Experience-dependent sharpening of visual shape selectivity in inferior temporal cortex. *Cereb Cortex*, 16, 1631–44.

Fries, W. (1984) Cortical projections to the superior colliculus in the macaque monkey: A retrograde study using horseradish peroxidase. *J Comp Neurol*, 230, 55–76.

Fuster, J.M. (2004) Upper processing stages of the perception-action cycle. *Trends Cogn Sci*, 8, 143–5.

Genovesio, A. & Wise, S.P. (2008) The neurophysiology of abstract response strategies. In Bunge, S.A., Wallis, J. (eds) *Neuroscience of Rule-Guided Behavior*. Oxford University Press, Oxford, pp. 81–106.

Gerbella, M., Borra, E., Mangiaracina, C., Rozzi, S., & Luppino, G. (2016) Corticostriate projections from areas of the 'lateral grasping network': Evidence for multiple hand-related input channels. *Cereb Cortex*, 26, 3096–115.

Gerbella, M., Borra, E., Tonelli, S., Rozzi, S., & Luppino, G. (2013) Connectional heterogeneity of the ventral part of the macaque area 46. *Cereb Cortex*, 23, 967–87.

Glickstein, M. & Doron, K. (2008) Cerebellum: Connections and functions. *Cerebellum*, 7, 589–94.

Glickstein, M., May, J.G., & Mercier, R.E. (1985) Corticopontine projections in the macaque: The distribution of labelled cortical cells after large injections of horseradish peroxidase in the pontine nuclei. *J Comp Neurol*, 235, 343–59.

Goldman, P.S., Rosvold, H.E., Vest, B., & Galkin, T.W. (1971) Analysis of the delayed-alternation deficit produced by dorsolateral prefrontal lesions in the rhesus monkey. *J Comp Physiol Psychol*, 77, 212–20.

Goldman-Rakic, P.S. & Porrino, L.J. (1985) The primate mediodorsal (MD) nucleus and its projection to the frontal lobe. *J Comp Neurol*, 242, 535–60.

Harlow, H.F. & Warren, J.M. (1952) Formation and transfer of discrimination learning sets. *J Comp Physiol Psychol*, 45, 482–9.

Harriger, L., van den Heuvel, M.P., & Sporns, O. (2012) Rich club organization of macaque cerebral cortex and its role in network communication. *PLoS One*, 7, e46497.

Hikosaka, O., Nakamura, H., Rand, M.K., Sakai, K., Lu, X., Nakamura, K., Miyachi, S., & Doya, K. (1999) Parallel neural networks for learning sequential procedures. *TINS*, 22, 464–71.

Hikosaka, O., Nakamura, K., Sakai, K., & Nakahara, K. (2002) Central mechanisms of motor skill learning. *Curr Opin Neurobiol*, 12, 217–22.

Hirabayashi, T., Takeuchi, D., Tamura, K., & Miyashita, Y. (2013a) Functional microcircuit recruited during retrieval of object association memory in monkey perirhinal cortex. *Neuron*, 77, 192–203.

Hirabayashi, T., Takeuchi, D., Tamura, K., & Miyashita, Y. (2013b) Microcircuits for hierarchical elaboration of object coding across primate temporal areas. *Science*, 341, 191–5.

Hoffman, D.S. & Strick, P.L. (1995) Effects of a primary motor cortex lesion on step-tracking movements of the wrist. *J Neurophysiol*, 73, 891–5.

Honda, T. & Ito, M. (2017) Developments from Marr's theory of the cerebellum. In Vaina, L., Passingham, R.E. (eds) *Computational Theories and Their Implementation in the Brain*. Oxford University Press, Oxford.

Honey, C.J., Kotter, R., Breakspear, M., & Sporns, O. (2007) Network structure of cerebral cortex shapes functional connectivity on multiple time scales. *Proc Natl Acad Sci USA*, 104, 10240–5.

Hoshi, E. (2013) Cortico-basal ganglia networks subserving goal-directed behavior mediated by conditional visuo-goal association. *Front Neural Circuits*, 7, 158.

Hubel, D.H. & Wiesel, T.N. (1968) Receptive fields and functional architecture of monkey striate cortex. *J Physiol*, 195, 215–43.

Izquierdo, A. & Murray, E.A. (2010) Functional interaction of medial mediodorsal thalamic nucleus but not nucleus accumbens with amygdala and orbital prefrontal cortex is essential for adaptive response selection after reinforcer devaluation. *J Neurosci*, 30, 661–9.

Jueptner, M., Frith, C.D., Brooks, D.J., Frackowiak, R.S.J., & Passingham, R.E. (1997a) Anatomy of motor learning. II. Subcortical structures and learning by trial and error. *J Neurophysiol*, 77, 1325–37.

Jueptner, M., Stephan, K.M., Frith, C.D., Brooks, D.J., Frackowiak, R.S.J., & Passingham, R.E. (1997b) Anatomy of motor learning. I. Frontal cortex and attention to action. *J Neurophysiol*, 77, 1313–24.

Kawato, M. & Samejima, K. (2007) Efficient reinforcement learning: Computational theories, neuroscience and robotics. *Curr Opin Neurobiol*, 17, 205–12.

Kelly, R.M. & Strick, P.L. (2003) Cerebellar loops with motor cortex and prefrontal cortex of a nonhuman primate. *J Neurosci*, 23, 8432–44.

Kim, S.-G., Ugurbil, S., & Strick, P.L. (1994) Activation of cerebellar output nucleus during cognitive processing. *Science*, 265, 949–61.

Kolb, B., Nonneman, A.J. & Singh, R.K. (1974) Double dissociation of spatial impairments and perseveration following selective prefrontal lesions in rats. *J Comp Physiol Psychol*, 87, 772–80.

Kondo, H., Saleem, K.S., & Price, J.L. (2005) Differential connections of the perirhinal and parahippocampal cortex with the orbital and medial prefrontal networks in macaque monkeys. *J Comp Neurol*, 493, 479–509.

Krubitzer, L. & Huffman, K.J. (2000) Arealization of the neocortex in mammals: Genetic and epigenetic contributions to the phenotype. *Brain Behav Evol*, 55, 322–35.

Kunimatsu, J., Suzuki, T.W., Ohmae, S., & Tanaka, M. (2018) Different contributions of preparatory activity in the basal ganglia and cerebellum for self-timing. *Elife*, 7.

Lawicka, W. (1969) Differing effectiveness of auditory quality and location cues in two forms of differentiation learning. *Acta Biol Exper* (Warsz), 29, 83–92.

Levy, R., Friedman, H.R., Davachi, L., & Goldman-Rakic, P.S. (1997) Differential activation of the caudate nucleus in primates performing spatial and nonspatial working memory tasks. *J Neurosci*, 17, 3870–82.

Lidow, M.S., Goldman-Rakic, P.S., Gallager, D.W., & Rakic, P. (1991) Distribution of dopaminergic receptors in the primate cerebral cortex: Quantitative autoradiographic analysis using [3H]raclopride, [3H]spiperone and [3H]SCH23390. *Neurosci*, 40, 657–71.

Lie, C.H., Specht, K., Marshall, J.C., & Fink, G.R. (2006) Using fMRI to decompose the neural processes underlying the Wisconsin Card Sorting Test. *Neuroimage*, 30, 1038–49.

Lu, X., Hikosaka, O. & Miyachi, S. (1998) Role of monkey cerebellar nuclei in skill for sequential movement. *J Neurophysiol*, 79, 2245–54.

Lynd-Balta, E. & Haber, S.N. (1994) The organization of midbrain projections to the striatum in the primate: Sensorimotor-related striatum versus ventral striatum. *Neurosci*, 59, 625–40.

Markov, N.T., Ercsey-Ravasz, M.M., Ribeiro Gomes, A.R., Lamy, C., Magrou, L., Vezoli, J., Misery, P., Falchier, A., Quilodran, R., Gariel, M.A., Sallet, J., Gamanut, R., Huissoud, C., Clavagnier, S., Giroud, P., Sappey-Marinier, D., Barone, P., Dehay, C., Toroczkai, Z., Knoblauch, K., Van Essen, D.C., & Kennedy, H. (2012) A weighted and directed interareal connectivity matrix for macaque cerebral cortex. *Cereb Cortex*, 24, 17–36.

Markov, N.T. & Kennedy, H. (2013) The importance of being hierarchical. *Curr Opin Neurobiol*, 23, 187–94.

Marr, D. (1969) A theory of cerebellar cortex. *J Physiol*, 202, 437–70.

Maschke, M., Schugens, M., Kindsvater, K., Drepper, J., Kolb, F.P., Diener, H.C., Daum, I., & Timmann, D. (2002) Fear conditioned changes of heart rate in patients with medial cerebellar lesions. *J Neurol Neurosurg Psychiatry*, 72, 116–18.

Middleton, F.A. & Strick, P.L. (1996) The temporal lobe is a target of output from the basal ganglia. *Proc Natl Acad Sci USA*, 93, 8683–7.

Miyachi, S., Hikosaka, O., & Lu, X. (2002) Differential activation of monkey striatal neurons in the early and late stages of procedural learning. *Exp Brain Res*, 146, 122–6.

Miyachi, S., Hikosaka, O., Miyashita, K., Karadi, Z., & Rand, M.K. (1997) Differential roles of monkey striatum in learning of sequential hand movement. *Exp Brain Res*, 115, 1–5.

Moberget, T., Gullesen, E.H., Andersson, S., Ivry, R.B., & Endestad, T. (2014) Generalized role for the cerebellum in encoding internal models: Evidence from semantic processing. *J Neurosci*, 34, 2871–78.

Modha, D.S. & Singh, R. (2010) Network architecture of the long-distance pathways in the macaque brain. *Proc Natl Acad Sci USA*, 107, 13485–90.

Muhammad, R., Wallis, J.D., & Miller, E.K. (2006) A comparison of abstract rules in the prefrontal cortex, premotor cortex, inferior temporal cortex, and striatum. *J Cogn Neurosci*, 18, 974–89.

Murray, E.A. & Wise, S.P. (1996) Role of the hippocampus plus subjacent cortex but not amygdala in visuomotor conditional learning in rhesus monkeys. *Beh Neurosci*, 110, 1261–70.

Mushiake, H., Saito, N., Sakamoto, K., Itoyama, Y., & Tanji, J. (2006) Activity in the lateral prefrontal cortex reflects multiple steps of future events in action plans. *Neuron*, 50, 631–41.

Nakamura, K., Sakai, K., & Hikosaka, O. (1998) Neuronal activity in medial frontal cortex during learning of sequential procedures. *J Neurophysiol*, 80, 2671–87.

Nakamura, K., Sakai, K., & Hikosaka, O. (1999) Effects of local inactivation of monkey medial frontal cortex in learning of sequential procedures. *J Neurophysiol*, 82, 1063–8.

Nixon, P.D. & Passingham, R.E. (1999) The cerebellum and cognition: Cerebellar lesions do not impair spatial working memory or visual associative learning in monkeys. *Eur J Neurosci*, 11, 4070–80.

Nixon, P.D. & Passingham, R.E. (2000) The cerebellum and cognition: Cerebellar lesions impair sequence learning but not conditional visuomotor learning in monkeys. *Neuropsychology*, 38, 1054–72.

Okuava, V., Natishvili, T., Gurashvili, T., Chipashvili, S., Bagashvili, T., Andronikashvil, G., & Kvernadze, G. (2005) One-trial visual recognition in cats. *Acta Neurobiol Exp* (Wars), 65, 205–11.

Padoa-Schioppa, C. & Assad, J.A. (2008) The representation of economic value in the orbitofrontal cortex is invariant for changes of menu. *Nat Neurosci*, 11, 95–102.

Passingham, R.E. (1982) *The Human Primate*. Freeman, Oxford.

Passingham, R.E. (1993) *The Frontal Lobes and Voluntary Action*. Oxford University Press, Oxford.

Passingham, R.E. (1996) Attention to action. *Philos Trans R Soc Lond B Biol Sci*, 351, 1473–9.

Pasupathy, A. & Miller, E.K. (2005) Different time courses of learning-related activity in the prefrontal cortex and striatum. *Nature*, 433, 873–6.

Peterburs, J., Bellebaum, C., Koch, B., Schwarz, M., & Daum, I. (2010) Working memory and verbal fluency deficits following cerebellar lesions: Relation to interindividual differences in patient variables. *Cerebellum*, 9, 375–83.

Petrides, M. & Pandya, D.N. (2002) Comparative cytoarchitectonic analysis of the human and macaque ventrolateral prefrontal cortex and corticocortical connection pattern in the monkey. *Eur J Neurosci*, 16, 291–310.

Price, J.L. & Amaral, D.G. (1981) An autoradiographic study of the projections of the central nucleus of the monkey amygdala. *J Neurosci*, 1, 1242–59.

Puig, M.V., Antzoulatos, E.G., & Miller, E.K. (2014) Prefrontal dopamine in associative learning and memory. *Neurosci*, 282, 217–29.

Raichle, M.E., Fiez, J.A., Videen, T.O., MacLeod, A.K., Pardo, J.V., Fox, P.T., & Petersen, S.E. (1994) Practice-related changes in human brain functional anatomy during non-motor learning. *Cer Cort*, 4, 8–26.

Rainer, G. & Miller, E.K. (2000) Effects of visual experience on the representation of objects in the prefrontal cortex. *Neuron*, 27, 179–189.

Rainer, G. & Miller, E.K. (2002) Timecourse of object-related neural activity in the primate prefrontal cortex during a short-term memory task. *Eur J Neurosci*, 15, 1244–1254.

Ramnani, N. & Passingham, R.E. (2001) Changes in the brain during rhythm learning. *J Cogn Neurosci*, 13, 952–66.

Ramnani, N., Toni, I., Josephs, O., Ashburner, J., & Passingham, R.E. (2000) Learning- and expectation-related changes in the human brain during motor learning. *J Neurophysiol*, 84, 3026–35.

Rao, S.C., Rainer, G., & Miller, E.K. (1997) Integration of what and where in the primate prefrontal cortex. *Science*, 276, 821–24.

Reader, S.M. & Laland, K.N. (2002) Social intelligence, innovation, and enhanced brain size in primates. *Proc Natl Acad Sci USA*, 99, 4436–41.

Richter, S., Kaiser, O., Hein-Kropp, C., Dimitrova, A., Gizewski, E., Beck, A., Aurich, V., Ziegler, W., & Timmann, D. (2004) Preserved verb generation in patients with cerebellar atrophy. *Neuropsychology*, 42, 1235–46.

Ripamonti, E., Frustaci, M., Zonca, G., Aggujaro, S., Molteni, F., & Luzzatti, C. (2018) Disentangling phonological and articulatory processing: A neuroanatomical study in aphasia. *Neuropsychology*, 121, 175–85.

Roiser, J.P., Stephan, K.E., den Ouden, H.E., Friston, K.J., & Joyce, E.M. (2010) Adaptive and aberrant reward prediction signals in the human brain. *Neuroimage*, 50, 657–64.

Rowe, J., Friston, K., Frackowiak, R., & Passingham, R. (2002a) Attention to action: Specific modulation of corticocortical interactions in humans. *Neuroimage*, 17, 988–98.

Rowe, J., Stephan, K.E., Friston, K., Frackowiak, R., Lees, A., & Passingham, R. (2002b) Attention to action in Parkinson's disease: Impaired effective connectivity among frontal cortical regions. *Brain*, 125, 276–89.

Roy, J.E., Riesenhuber, M., Poggio, T., & Miller, E.K. (2010) Prefrontal cortex activity during flexible categorization. *J Neurosci*, 30, 8519–28.

Russchen, F.T., Amaral, D.G., & Price, J.L. (1987) The afferent input to the magnocellular division of the mediodorsal thalamic nucleus in the monkey, *Macaca fascicularis*. *J Comp Neurol*, 256, 175–210.

Schmahmann, J. (ed) (1997) *The Cerebellum and Cognition*. Academic Press, San Diego.

Schmahmann, J.D. & Pandya, D.N. (1997) Anatomic organization of the basilar pontine projections from prefrontal cortices in rhesus monkey. *J Neurosci*, 17, 438–58.

Schmahmann, J.D. & Sherman, J.C. (1998) The cerebellar cognitive affective syndrome. *Brain*, 121 (Pt 4), 561–79.

Schultz, W. (1998) Predictive reward signal of dopamine neurons. *J Neurophysio*, 80, 1–27.

Selemon, L.D. & Goldman-Rakic, P.S. (1985) Longitudinal topography and interdigitation of corticostriatal projections in the rhesus monkey. *J Neurosci*, 5, 776–94.

Seo, M., Lee, E., & Averbeck, B.B. (2012) Action selection and action value in frontal-striatal circuits. *Neuron*, 74, 947–60.

Shima, K., Isoda, M., Mushiake, H., & Tanji, J. (2007) Categorization of behavioural sequences in the prefrontal cortex. *Nature*, 445, 315–18.

Shipp, S. (2005) The importance of being agranular: A comparative account of visual and motor cortex. *Philos Trans R Soc Lond B Biol Sci*, 360, 797–814.

Siegel, M., Buschman, T.J., & Miller, E.K. (2015) Cortical information flow during flexible sensorimotor decisions. *Science*, 348, 1352–55.

Stokes, M.G., Kusunoki, M., Sigala, N., Nili, H., Gaffan, D., & Duncan, J. (2013) Dynamic coding for cognitive control in prefrontal cortex. *Neuron*, 78, 364–75.

Stoodley, C.J., Valera, E.M., & Schmahmann, J.D. (2012) Functional topography of the cerebellum for motor and cognitive tasks: An fMRI study. *Neuroimage*, 59, 1560–70.

Takeuchi, D., Hirabayashi, T., Tamura, K., & Miyashita, Y. (2011) Reversal of interlaminar signal between sensory and memory processing in monkey temporal cortex. *Science*, 331, 1443–7.

Tanaka, K. (1997) Mechanisms of visual object recognition. *Curr Opin Neurobiol*, 7, 523–9.

Tudusciuc, O. & Nieder, A. (2009) Contributions of primate prefrontal and posterior parietal cortices to length and numerosity representation. *J Neurophysiol*, 101, 2984–94.

Vallentin, D., Bongard, S., & Nieder, A. (2012) Numerical rule coding in the prefrontal, premotor, and posterior parietal cortices of macaques. *J Neurosci*, 32, 6621–30.

Villagrasa, F., Baladron, J., Vitay, J., Schroll, H., Antzoulatos, E.G., Miller, E.K., & Hamker, F.H. (2018) On the role of cortex-basal ganglia interactions for category learning: A neurocomputational approach. *J Neurosci*, 38, 9551–62.

Wallis, J.D., Anderson, K.C., & Miller, E.K. (2001) Single neurons in prefrontal cortex encode abstract rules. *Nature*, 411, 953–6.

Wallis, J.D. & Miller, E.K. (2003) Neuronal activity in primate dorsolateral and orbital prefrontal cortex during performance of a reward preference task. *Eur J Neurosci*, 18, 2069–81.

Wang, G., Tanifuji, M., & Tanaka, K. (1998) Functional architecture in monkey inferotemporal cortex revealed by in vivo optical imaging. *Neurosci Res*, 32, 33–46.

Wang, Y., Isoda, M., Matsuzaka, Y., Shima, K., & Tanji, J. (2005) Prefrontal cortical cells projecting to the supplementary eye field and presupplementary motor area in the monkey. *Neurosci Res*, 53, 1–7.

Warren, J.M. (1974) Possibly unique characteristics of learning by primates. *J Hum Evol*, 3, 445–54.

Webster, M.J., Bachevalier, J., & Ungerleider, L.G. (1993) Subcortical connections of inferior temporal areas TE and TEO in macaque monkeys. *J Comp Neurol*, 335, 73–91.

Webster, M.J., Bachevalier, J., & Ungerleider, L.G. (1994) Connections of inferior temporal areas TEO and TE with parietal and frontal cortex in macaque monkeys. *Cer Cort*, 4, 471–83.

Webster, M.J., Ungerleider, L.G., & Bachevalier, J. (1991) Connections of inferior temporal areas TE and TEO with medial temporal-lobe structures in infant and adult monkeys. *J Neurosci*, 11, 1095–116.

Westendorff, S., Klaes, C., & Gail, A. (2010) The cortical timeline for deciding on reach motor goals. *J Neurosci*, 30, 5426–36.

Whiten, A., Goodall, J., McGrew, W.C., Nishida, T., Reynolds, V., Sugiyama, Y., Tutin, C.E., Wrangham, R.W., & Boesch, C. (1999) Cultures in chimpanzees. *Nature*, 399, 682–5.

Xiao, D., Zikopoulos, B., & Barbas, H. (2009) Laminar and modular organization of prefrontal projections to multiple thalamic nuclei. *Neurosci*, 161, 1067–81.

Yamada, M., Pita, M.C., Iijima, T., & Tsutsui, K. (2010) Rule-dependent anticipatory activity in prefrontal neurons. *Neurosci Res*, 67, 162–71.

Young, M.P. (1993) The organization of neural systems in the primate cerebral cortex. *Proc Roy Soc Lond B*, 252, 13–18.

Zhou, H. & Desimone, R. (2011) Feature-based attention in the frontal eye field and area V4 during visual search. *Neuron*, 70, 1205–17.

PART IV
THE HUMAN PREFRONTAL CORTEX

9
Evolution of the Prefrontal Cortex in the Hominins

Introduction

Evidence from DNA indicates that our closest ancestors are the bonobos or pygmy chimpanzees (*Pan paniscus*) and the common chimpanzee (*Pan troglodytes*) (Prufer et al., 2012). The overall difference in DNA is of the order of 1% (Caswell et al., 2008). However, it is the proteins that are responsible for the phenotypic difference, and 80% of the proteins show some difference between humans and chimpanzees (Glazko et al., 2005). Since many of these code for transcription factors that influence the rates of development during ontogeny, there is no mystery about how humans could be so different from chimpanzees.

It is possible to produce estimates of the time at which the lines leading to humans and modern chimpanzees diverged by using the molecular clock, an estimate of the rate of mutations over time. This gives an estimate of ~7 Ma for the divergence (Langergraber et al., 2012), and this is consistent with the fossil evidence. It is controversial whether *Sahelanthropus* from Chad is an ancestral hominin, but *Orrorin* from Kenya has a better claim and it lived ~6Ma. The fossil remains of the first universally accepted hominin, *Ardipethecus ramidus*, date from ~4.4 Ma.

The different hominins left Africa in a series of waves (Stewart & Stringer, 2012), though it is not clear exactly when the first wave occurred. However, it must have been before 2 Ma. Fossil remains of *Homo erectus* have been found in Lantian county in China dated at ~2.1 Ma (Zhu et al., 2018).

Further waves followed. The remains of *Homo heidelbergensis* from a cave in Mauer in Germany are dated ~610 Ka (Wagner et al., 2010). This had traits that were similar to *Homo erectus* and *Homo sapiens*, leading to the designation 'archaic *Homo sapiens*'.

The earliest modern humans date from ~315 Ka. The evidence comes from a facial fragment and mandible found at Jebel Irhoud in Morocco that have features that appear to be more like *Homo sapiens* than *Homo erectus* or *Homo neanderthalensis* (Hublin et al., 2017). All the earliest evidence for *Homo sapiens*, derived from fossil bones and associated tools, comes from Africa (Richter et al., 2017).

The earliest evidence for *Homo sapiens* in Europe comes from a fossil cranium and tooth that are referred to as Apidima 1, and found in Greece. These have a mixture of primitive and more modern features, and are dated to over 210 Ka (Harvati et al., 2019). At the same site, the cranium Apidima 2 is Neanderthal like, dated to more than 170 Ka. *Homo sapiens* and *Homo neanderthalensis* probably moved out of Africa around this time as evidenced by finds in Israel, with an estimated date from 177–194 Ka (Hershkovitz et al., 2018).

The fossil remains of *Homo sapiens* have been found in association with Upper Paleolithic tools in the Bacho Kiro cave in Bulgaria (Fewlass et al., 2020). The collagen from the bone has allowed radio-carbon dating to between 45,8200 and 43,650 Ka.

The account given here will, of course, change The reason is that we have relatively few fossil remains of our ancestors, and new discoveries are being made all the time, as for example the unexpected finding of *Homo floresiensis* (Baab, 2016). This means that our ancestral tree has become more and more 'bushlike', with many side branches and with different hominins being present at the same time.

Encephalization

The skull of *Sahelanthropus tchadensis* suggests that it had an endocranial volume of 360–370 cm^3, roughly that of a small chimpanzee (Guy et al., 2005). The later *Ardipithecus ramidus* from ~4.4 Ma had a similar cranial capacity of 300–350 cm^3 (Suwa et al., 2009), as did *Australopithecus anamensis* from ~3.8 Ma with a cranial capacity of 365–370 cc (Haile-Selassie et al., 2019).

Yet the foot of *Ardipethecus* indicates early adaptations for upright walking (Prang, 2019). The footprints from Laetoli from 3.6 Ma also suggest upright walking, even if it was less efficient than in modern humans (Crompton et al., 2012). Therefore, some form of bipedalism predated any significant change in brain size.

As judged by the encephalization quotient (EQ) (Chapter 2), the size of the hominin brain increased slowly from ~3.5 Ma (Figure 9.1, next page). However, this was followed by a much faster increase between the time of *Homo heidelbergensis* and ~100,000 years ago (Murray et al., 2017). Kappelman (1996) suggested that during the late evolution of *Homo sapiens* there was a selection from smaller body size, while keeping absolute brain size relatively constant, thus leading to a final increase in the EQ.

The increase in encephalization was associated with increasingly sophisticated behaviour. The evidence for this comes from the assemblages of tools that were associated with the different hominins.

The bottom of Figure 9.1 shows the dates for two tool making traditions. The earliest was the Oldowan, so named because of the stone tools collected by Mary Leakey (2009) from the Olduvai Gorge. The later Acheulian tradition included handaxes. This tradition lasted a very long time, perhaps because the handaxe was

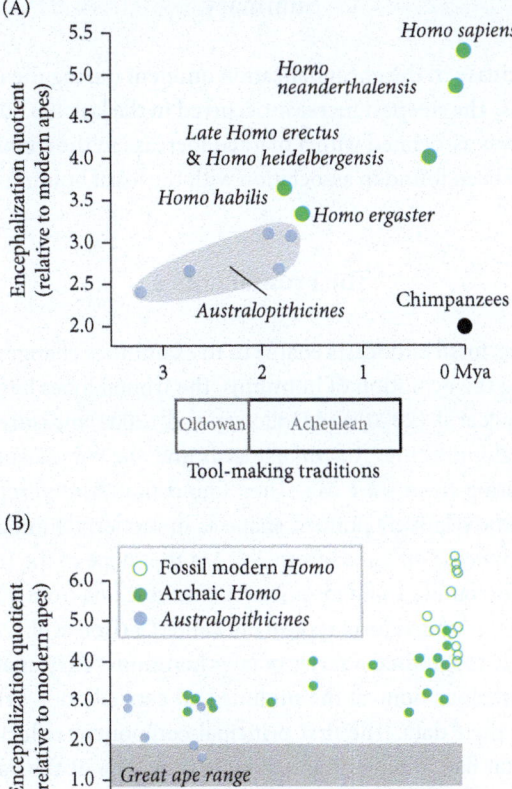

Figure 9.1 Encephalization quotients (EQs) for fossil hominins, modern humans, and modern chimpanzees
The Australopithecines are shown in blue and the different species of *Homo* in green.
EQs from Kappelman (1996)
The range of dates for two tool-making traditions appears at the bottom.
Reproduced from Murray, E.A., Wise, S.P., & Graham, K.S., *The Evolution of Memory Systems: Ancestors, Anatomy, and Adaptations*, Plate 6, Copyright © Oxford University Press 2017. Original data from Klein, RG (2009), *The Human Career*, University of Chicago Press.

an excellent all-purpose tool which served both for cutting and dismembering, and as a weapon. Handaxes could exhibit outstanding craftsmanship.

Not shown in Figure 9.1 is the later Levallois tradition from ~300 Ka which included points and blades, and the specialized Upper Paleolithic toolkit which had its antecedents in the Levant around 50 Ka (Bosch et al., 2015).

Summary

There was an increase in the encephalization quotient during the evolution of the various hominins. The steepest increase occurred in the last 400,000 years. During the last 50 Ka it was associated with a marked increase in the sophistication of the toolkits that have been found in association with the fossil hominins.

The Frontal Lobes

In addition to size, fossil endocasts enable us to examine of changes in the shape of the brain. During the evolution of hominins, the frontal lobes became wider and more rounded. Falk et al. (2000) and Holloway et al. (2004) measured endocasts for the gracile australopithecine, *Australopithecus* and the robust australopithecine, *Paranthropus*, dating from ~4.1 Ma. They found that *Paranthropus* had frontal lobes with the same relatively pointed shape as in modern chimpanzees and gorillas. In *Australopithecus africanus* from 3–2 Ma the shape of the frontal lobes became slightly more rounded, and in *Australopithecus sediba* from ~2 Ma the frontal lobes had the more rounded form typical of *Homo* (Carlson et al., 2011).

Bruner et al. (2013) studied a series of later hominins. They analysed the shape of the frontal squamous bone at the midline and carried out a principal component analysis on their data. The first principal component reflected frontal bulging versus frontal flattening, and it accounted for 40% of the variance. Bruner et al. found that this component separated early and recent modern humans (blue circles) from *Homo erectus*, *Homo heidelbergensis* (orange squares), and *Homo neanderthalensis* (black circles). However, modern humans had a small amount of overlap with *Homo heidelbergensis*.

Figure 9.2 compared CT scans of the skulls of a chimpanzee with specimens of *Homo heidelbergensis*, *Homo neanderthalensis*, and *Homo sapiens*. The base of the

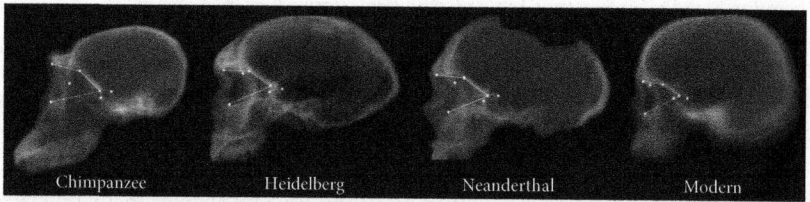

Figure 9.2 CT scans of skulls of a chimpanzee, *Homo heidelbergensis*, *Homo neanderthalensis*, and *Homo sapiens*

Adapted from Ana Sofia Pereira-Pedro, Michael Masters, & Emiliano Bruner. Shape analysis of spatial relationships between orbito-ocular and endocranial structures in modern humans and fossil hominids. *J Anat*, 231 (6), 947–60, Figure 6, Doi: 10.1111/joa.12693 Copyright © 2017 Anatomical Society.

frontal lobes is outlined with a white line, and it will be seen that in *Homo sapiens* the frontal lobes have extended further over the orbit. This could result from an increase in the size of the frontal lobes, mechanical constraints, or both.

When contrasted with the brains of *Homo neanderthalensis*, the brains of *Homo sapiens* showed a similar bulging of the parietal lobes. In an initial study, Bruner et al. (2004) analysed the shape of the midsagittal cranium in Neanderthals. More recently, Bruner et al. (2017) used magnetic resonance imaging (MRI) to obtain structural images of human and chimpanzee brains. Taken together, these studies indicate that, compared to Neanderthal and chimpanzee brains, human brains have an enlargement of the medial parietal cortex.

The assumption is that these changes in the cranium indicate changes in the brain. Alatorre Warren et al. (2019) have demonstrated that there can be changes in sulcal pattern that are not reflected in changes in the cranium. But there is no obvious reason why there would be an enlargement of the cranium over the medial parietal cortex if it bore no relation to a change in the brain.

It seems safe to conclude that, during the evolution of hominin brains, there was an increase in EQ in the later Australopithecines (Figure 9.1), and a rounding of the frontal lobe in *Australopithecus sediba* from ~2 Ma. There was a later rapid increase in EQ from ~400,000 years ago. During the evolution of *Homo sapiens*, this was associated with an increase in frontal and parietal cortex.

Summary

The frontal lobes were more rounded in *Australopithecus sediba* from ~2 Ma than in the earlier gracile Australopthecines. This was followed by the expansion of the frontal and the medial parietal cortex in *Homo sapiens*, when compared with either *Homo heidelbergensis* or *Homo neanderthalensis*.

Prefrontal Cortex as a Proportion of the Frontal Lobe

Fossil endocasts cannot distinguish the granular prefrontal (PF) cortex from other parts of the frontal lobe, but architectonic analysis in modern primates can. Brodmann (1912) measured the size of the granular PF cortex and the frontal lobe as a whole in a series of modern primates. Using his data, Elston et al. (2006) showed that the granular PF cortex takes up ~80% of the frontal lobe in humans, much more than in any other primate, including chimpanzees and gorillas (*Gorilla gorilla*). Figure 9.3 (next page) presents this analysis.

Passingham and Smaers (2014) also provided estimates for the size of the granular PF cortex, based on a dataset collected by Smaers and Zilles (2011). Figure 9.4A (page 339) shows the extent of PF cortex plotted against the rest of the frontal

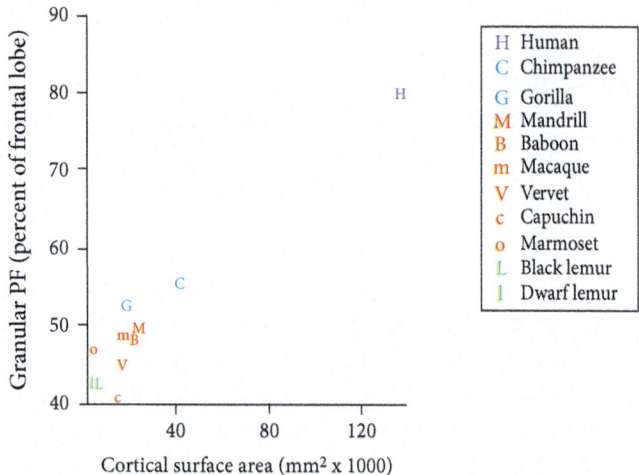

Figure 9.3 Granular PF cortex as a percentage of the frontal lobe, plotted versus function of cerebral extent, in modern primates

Reproduced from Guy N. Elston, Ruth Benavides-Piccione, Alejandra Elston, Bendan Zietsch, Javier Defelipe, Paul Manger, Vivien Casagrande, & Jon H. Kaas. Specializations of the granular prefrontal cortex of primates: Implications for cognitive processing. *Anatom Rec P A Discov Molec Cell Evol Biol*, 288 (1), 26–35, Doi: 10.1002/ar.a.20278, Copyright © 2005, Wiley-Liss, Inc.

lobe. The solid line shows the regression through the values for monkeys, together with the gibbon, and the dotted lines show the confidence limits. The grey dots in Figure 9.4A show the values for the great apes.

Three conclusions follow from the analysis shown in Figure 9.4. First, when related to the size of the frontal lobe as a whole, the granular PF cortex in great apes has undergone an upward grade shift relative to monkeys. The cladogram in Figure 9.4B marks the lineages for the great apes in dark blue.

Second, the granular PF cortex has undergone an additional and marked upward grade shift in humans (open circle) relative to the great apes (Passingham & Smaers, 2014). These differences are highly statistically significant (Smaers et al., 2017) and provide evidence that humans represent a new grade of frontal-lobe organization. Taken together with Figure 9.3, Figure 9.4A shows that the granular PF cortex dominates the frontal lobe of humans. This serves as an example of nonlinear scaling (Smaers & Soligo, 2013).

However the final observation is that the slope of the regression line for the monkeys and gibbon does not differ significantly from 1.0 (Passingham et al., 2017). This indicates that the proportion of the granular PF cortex within the frontal lobe remains constant. This is consistent with the claim made by Passingham et al. (2017) that in general the proportions of different brain areas remain constant within a grade.

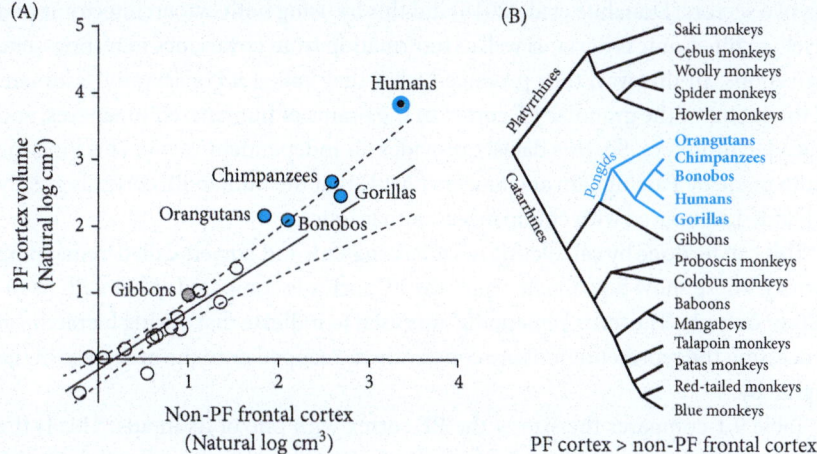

Figure 9.4 Log-log plot of the estimated volume of the granular PF cortex as a function of non-PF cortex in the frontal lobe using the data from Smaers et al. (2011) (A) The solid line shows the regression through the monkeys and gibbon, and the dotted lines show the confidence limits
The figure also shows the values for the great apes. Open circles, monkeys; grey circle, gibbons; blue circles, great apes; large blue circle with a dot, human.
(B) Cladogram of anthropoid primates, based on the same dataset as in A. Lineages in blue show a significant increase in the extent of granular PF cortex within the frontal lobe.
Reproduced from Smaers, J.B., Gomez-Robles, A., Parks, A.N., & Sherwood, C.C. Exceptional evolutionary expansion of prefrontal cortex in great apes and humans. *Cur Biol*, 27 (10), 1549, Figure 2, Doi: 10.1016/j.cub.2017.05.015 © 2017, Elsevier Ltd.

This claim can be illustrated by considering the neocortex of rodents. Krubitzer and Dooley (2013) showed that the proportion of the neocortex that is neither primary sensory nor motor is very similar in two caviomorph rodents of very different size. This is true, for example, of guinea pigs (*Cavia porcellus*) which weight just 0.7–1.2 kg and capybaras (*Hydrochoerus hydrochaeris*, family *Cavidae*) which weigh 35–66 kg. So, this pair of species serves an example of constant scaling *within* a grade.

By contrast, the same proportion can differ dramatically in two mammals of different grades, even though they have roughly similar body size. For example, Californian ground squirrels (*Otospermophilus beecheyi*) weigh 280–740 g and squirrel monkeys (*Saimiri sciureus*) can weigh as little as 750 g. Yet, there is much more non-primary cortex in squirrel monkeys. So this pair of species serve as an example of a change in scaling *between* grades.

Though Smaers and Zilles (2011) measured volumes of granular and agranular frontal cortex, unlike Brodmann (1912) they did not delineate the border between the two sectors. Donahue et al. (2018) did this by using both cytoarchitectonic and myeloarchitectonic criteria, as well as information from covariance in resting-state activations. In this way, they produced what they called a 'conservative estimate' of the extent of the granular PF cortex in the brains of humans, chimpanzees, and macaque monkeys. So, this dataset provides an independent way of checking the claim made by Passingham and Smaers (2014) that the human PF cortex is greatly expanded compared with chimpanzees or macaques.

This can be done by calculating what Passingham and Smaers called 'remapping factors'. These show how much bigger the PF cortex is compared either with its inputs or outputs. The term 'remapping' was used to indicate that during hierarchical processing the representations are constantly re-mapped onto the next stage in the hierarchy.

Table 9.1 compares the size of the PF cortex with one of its inputs. This is the primary visual cortex, although of course the visual inputs are relayed to the PF cortex via the ventral and dorsal visual systems. The remapping factor indicates how many times larger the PF cortex is than the primary visual cortex.

The table compares the results for two datasets. The first is that produced over a century ago by Brodmann (1912); the second is the recent one from Donahue et al. (2018) (Table 9.1).

The figures for the two datasets are remarkably similar, especially given the different definition of the PF cortex used by Donahue et al. (2018) for their 'conservative estimate'. It is clear that, compared with the PF cortex of chimpanzees or macaques, the human PF cortex is vastly expanded in relation to the indirect inputs from the primary visual cortex.

Table 9.2 (next page) presents the data for the remapping factors for the PF cortex compared with its outputs, the premotor and motor cortex.

The actual numbers differ for the two datasets because Brodmann (1913) presented data for the premotor areas plus motor cortex, whereas Donahue et al. (2018) presented data for the motor cortex alone. But if the remapping factors for the human and chimpanzee brain are compared, the human remapping factor is 4.1 times larger in the Brodmann data, and 3.2 times larger in the Donahue data. So, both datasets indicate a marked expansion in the human compared with the chimpanzee brain when the PF cortex is compared with its outputs.

Table 9.1 Remapping factors.

PF/primary visual V1	Human	Chimpanzee	Macaque monkey
Brodmann	8.9	2.1	0.9
Donahue	7.9	3.1	1.3

Table 9.2 Remapping factors.

PF/premotor + Motor	Human	Chimpanzee	Macaque monkey
Brodmann	4.9	1.2	1.0
PF/motor			
Donahue	9.4	2.9	3.1

The cerebellum has also expanded (Rilling & Insel, 1998), and the expansion cannot be put down simply to bipedalism. Balsters et al. (2010) compared the size of the different cerebellar lobules in humans, chimpanzees, and a Cebus monkey. Lobules Crus I and II project to the granular PF cortex via the thalamus, and others send information to the premotor and motor cortex (Kelly & Strick, 2003). Although Balsters et al. (2010) did not conduct an allometric analysis, their data indicate that it is the lobules Crus I and II that show an especial increase in the human brain. This is consistent with the conclusion that during hominin evolution the granular PF cortex expanded disproportionately compared with the premotor and motor cortex.

Summary

Several lines of evidence point to the same conclusion. The granular PF cortex is especially expanded in the human brain compared with either the primary visual cortex or the premotor areas and motor cortex.

Controversy

Three groups have disputed the claim that the PF cortex is especially expanded in the human brain. Therefore, it is necessary to consider carefully why they have reached a different conclusion.

The Frontal Lobe

First, Semendeferi et al. (2002) measured the size of the frontal lobe as a proportion of the neocortex as a whole. The mean value for humans was 38% compared with 35% for chimpanzees. Given variation in the values between individuals within each species, this difference was not statistically significant. Furthermore, Semendeferi and Damasio (2000) found that the proportions of the temporal and parieto-occipital cortex were also the same in the human and chimpanzee brain.

However, the claim made in the Table 9.2 relates to the proportion of the PF cortex *within* the frontal lobe. Semendeferi et al. (2002) did not carry out a Cytoarchitectural analysis. Instead, they measured the cortex anterior to the precentral sulcus as an estimate of the PF cortex. But Bailey et al. (1950) showed that in the chimpanzee brain there is extensive dysgranular cortex that lies anterior to this sulcus. This is less true for the human brain (Geyer, 2004), and this means that it is not valid to use the cortex anterior to the precentral cortex to make comparisons across species.

Cortex Rostral to the Corpus Callosum

Schoenemann et al. (2005) measured the amount of the frontal tissue that lay rostral to the genu of the corpus callosum in a large series of primate including humans. Barton and Venditti (2013) plotted this PF estimate against the size of the neocortex as a whole. As estimated in this way, the human PF cortex was no different from the size predicted for a brain with a necortex our size. So, Barton and Venditti concluded that size of the human PF cortex could be accounted for by simple scaling.

But the method assumes that the *same* proportion of the PF cortex lies rostral to the genu of the corpus callosum in the different brains. But the cytoarchitectonic and myeloarchitectonic study by Donahue et al. (2018) has shown this assumption to be incorrect. Much more of the granular PF cortex lies posterior to this limit in the human brain than in the brains of the chimpanzee and macaque monkey. So the estimate used by Barton and Venditti (2013) for the human PF cortex is a severe *underestimate*, and this invalidates the allometric analysis.

This effect can be seen in Figure 9.5 (next page). Glasser et al. (2014) compared the density of myelin in the frontal lobe in humans, chimpanzees, and macaques. The lightly myelinated cortex in the frontal lobe roughly corresponds to the granular PF cortex. In the human brain, much of the lightly myelinated cortex, shown in dark blue, lies caudal to the genu of the corpus callosum, marked with a red arrow. This is not the case in the chimpanzee and macaque brains.

Proportion of Neurones

Gabi et al. (2016) plotted the density of neurones as opposed to glial cells throughout the neocortex, from the back of the brain to the front. They found that the neuronal density decreased rostrally in humans in the same way as in other anthropoids. They also reported that eight percent of the neurones lay rostral to the genu of the corpus callosum, and that surprisingly this was true in the human brain

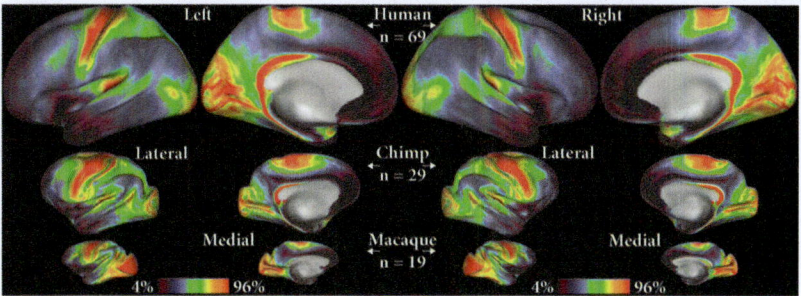

Figure 9.5 Myeloarchitectonics in the brains of humans, chimpanzees, and macaque monkeys
Red arrows indicate the genu of the corpus callosum. White arrows point to the central sulcus.

Reproduced from Glasser, M.F., Goyal, M.S., Preuss, T.M., Raichle, M.E. and Van Essen, D.C., Trends and properties of human cerebral cortex: Correlations with cortical myelin content. *Neuroimage*, 93 (Part 2), 165-75, Figure 3, Doi: 10.1016/j.neuroimage.2013.03.060 Copyright © 2013, Elsevier Inc. All rights reserved.

as well as in the brains of other primates. Herculano-Houzel (2016) concluded that the human PF cortex is not especially expanded.

There are two problems with this conclusion. First, the neurones were counted by distinguishing between neuronal and non-neuronal cells in a free-floating suspension (Herculano-Houzel et al., 2015). No attempt was made to distinguish PF neurones from premotor or motor neurones on the basis of morphology. Second, the neurones were counted rostral to the genu of the corpus callosum. But as the previous section demonstrates, this produces a severe underestimate of the PF cortex and therefore PF neurones in the human brain.

Allometry of Area 10

Semendeferi et al. (2001) measured the size of the polar PF cortex (area 10) in the brains of humans, four species of great apes (*Hominidae*), and gibbons (*Hylobatidae*). Figure 9.6A (next page) shows that Semendeferi et al. fitted a log-log regression line though the data for the five non-human primates, and claimed that the human value for area 10 was as predicted, given the size of the human brain. On this basis, Holloway (2002) concluded that area 10 did not expand significantly during human evolution.

There are three problems. First, Semendeferi et al. also plotted the data on a liner scale, and as Figure 9.6B shows the human value does not lie on or anywhere near this line.

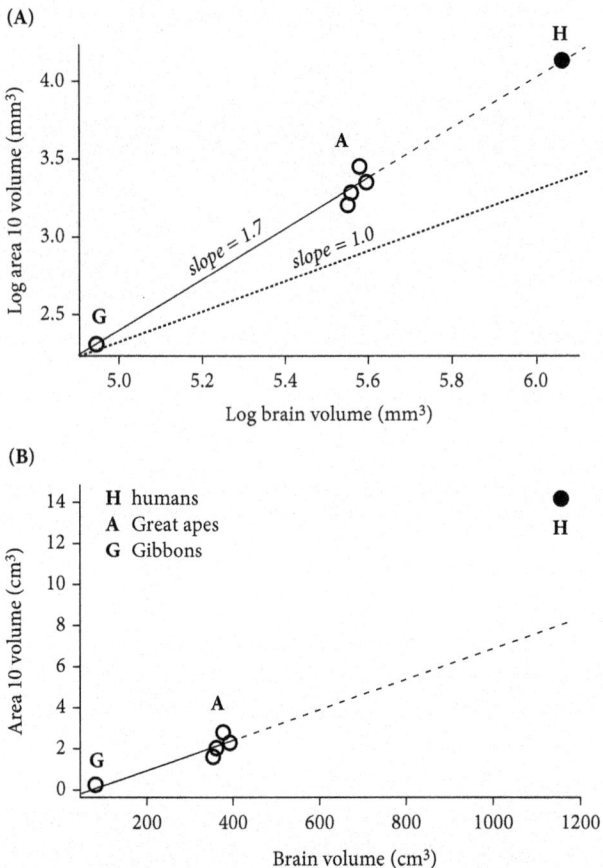

Figure 9.6 Regressions of the volume of the frontal pole cortex (area 10) as a function of brain volume in selected primates
Solid line, regressions from the ape data; dashed line, extrapolation to a brain the size of that in humans. (A) Log–log scale, with unity slope indicated by the dotted line. (B) Linear scale. Abbreviations: G, gibbons; A, great apes; H, humans.

Adapted from Semendeferi, K., Armstrong, E., Schleicher, A., Zilles, K., & Van Hoesen, G.W. Prefrontal cortex in humans and apes: A comparative study of area 10, *Am J Phys Anthropol*, 114 (3), 224–41, Figure 8a and b, https://doi.org/10.1002/1096-8644(200103)114:3 Copyright © 2001, Wiley-Liss, Inc.

Next, the data collected by Semendeferi et al. (2001) indicate that the absolute size of the human area 10 exceeds the size of area 10 in any other anthropoid and does so dramatically. Figure 9.7 (next page) presents these data in graphical form, with the size of each circle proportionate to the absolute volume of area 10. When the volume of the human polar PF cortex is compared with the brain as a whole, it is twice as large as in great apes.

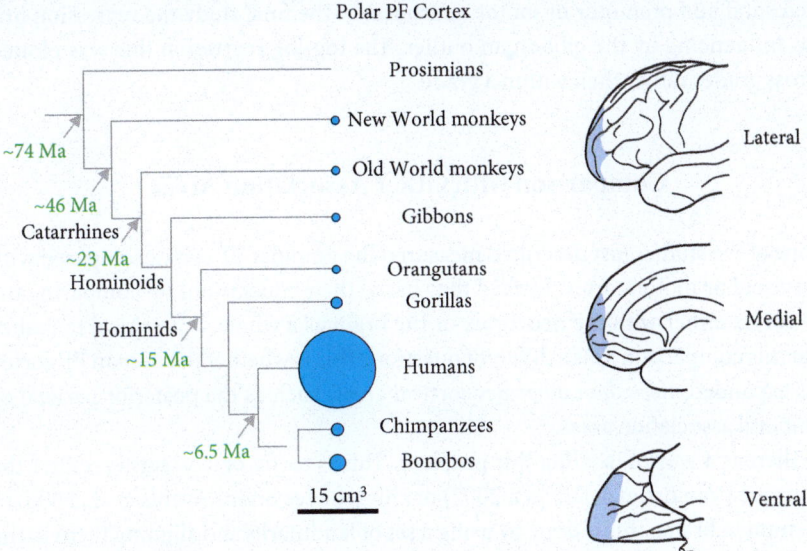

Figure 9.7 Expansion of the area 10 in humans
Cladogram for selected primates with approximate divergence times on the left (arrows) in millions of years ago (Ma). The diameter of each circle encodes the volume of area 10. The column shows the location of area 10 in the human brain. The figures on the right show area 10 in the human brain.

Adapted from Tsujimoto, S., Genovesio, A., & Wise, S.P. Frontal pole cortex: encoding ends at the end of the endbrain. *Trends Cog Sci*, 15 (4), 169–76, Figure 1, Doi: 10.1016/j.tics.2011.02.001 Copyright © 2011, Elsevier Ltd. All rights reserved.

The final problem is that in Figure 9.6A the value for gibbons is an outlier, and it is this value that almost wholly determines the slope of the line. The slope of the regression line in Figure 9.7A is 1.7. Yet, Passingham et al. (2017) showed that a log-log regression slope should not differ significantly from 1.0 when plotted *within* a grade. Figure 9.6 shows a slope of 1.0 for reference. It is clear that Figure 9.6A plots a regression line *across* grades, and not within a grade; and that is why the proportions differ between humans and the other primates, A regression through the great apes alone would establish the grade for the size of area 10 for great apes, but the limited variation in the volume of their brains precludes this approach.

Summary

There has been controversy about whether the human PF cortex is especially expanded. But in two studies the measure used for the human PF cortex was an underestimate. In another study no distinction was made between the number of prefrontal and premotor or motor neurones. In the final study the regression line was influencing by the gibbon an outlier. The log-log regression line was plotted across grades rather than within a grade.

Comparison with Other Association Areas

None of the studies just described measured the granular PF cortex compared with the visual or motor cortex. Instead they based their conclusions on comparing the PF cortex either with the neocortex or the brain as a whole. It is critical to realize that this comparison asks a different question. This is whether the human PF cortex has expanded *more than other* neocortical areas, such as the posterior parietal or temporal association areas.

There is a way of tackling this problem. This is to 'fit' or 'co-register' either the macaque (Van Essen & Dierker, 2007) or chimpanzee brain (Avants et al., 2006) to the human brain. This is done by using a set of landmarks and aligning them in the different brains. The distortions that are needed to achieve the fit can be used to visualize the areas that have expanded most in the human brain. The method allows differences in the expansion of particular areas to be measured irrespective of differences in overall brain size.

Figure 9.8A (next page) shows the results of fitting the macaque to the human brain (Van Essen & Dierker, 2007). The colour scale indicates the degree to which particular areas have to be expanded to achieve this fit. In the human brain the primary visual cortex lies on the medial surface, and the light blue shows that it is of the size that one would expect for a brain our size. This is consistent with the finding that if the values for striate cortex are plotted for non-human primates, the value for the human striate cortex is as predicted for a primate with a body our size (Passingham, 1973). The reason is that the body, in this case the eye, is mapped onto the striate cortex, though with greater expansion for foveal than central vision (Perry & Cowey, 1985).

The secondary visual areas are in a darker blue, indicatijng a degree of expansion. The tissue in the central sulcus and the premotor cortex are also in a darker blue. This probably reflects an increase in skilled performance with the hand, mouth and larynx. The same data are shown in more detail on the front cover.

The areas with greatest expansion are shown in reds and yellows, and these are the association areas. The relative expansion of these areas is not related to mapping of the body onto the brain.

Instead, as Chapter 2 explains, these reflect a difference in grades, and this is due to 'an adaptive factor' (Passingham et al., 2017). The expansion of the association cortex and in particular the PF cortex supports the need to solve new problems by transfer from previous problems (Chapter 8). In other words, it supports rapid learning.

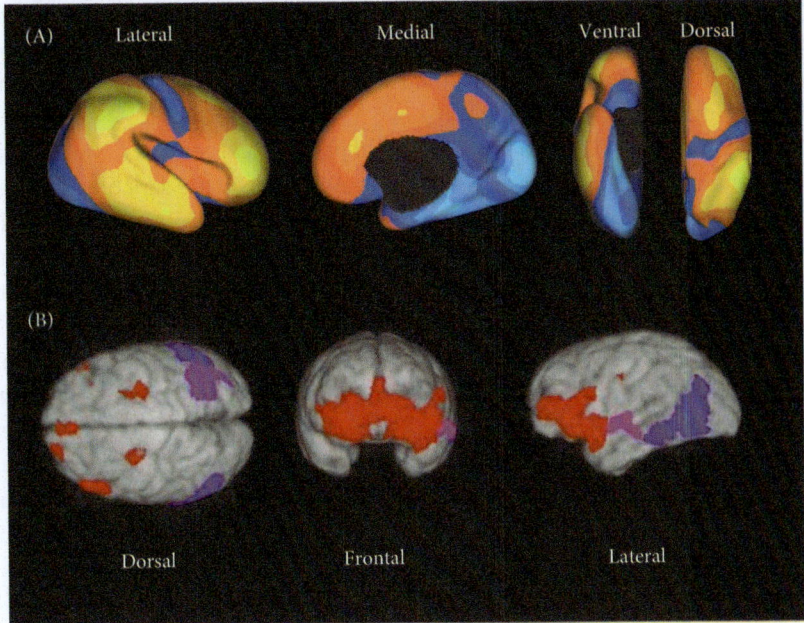

Figure 9.8 (A) Relative expansion of cortical regions from macaque to human brains, shown on the human brain. (B) Relative expansion of cortical regions from chimpanzee to human brain, shown on the chimpanzee brain

A: Reproduced from Hill, J., Inder, T., Neil, J., Dierker, D., Harwell, J., & Van Essen, D. Similar patterns of cortical expansion during human development and evolution. *Proc Nat Acad Scienc USA*, 107 (29), 13135–40, Figure 4a, https://doi.org/10.1073/pnas.1001229107 Copyright © 2010 National Academy of Sciences.

B: Reproduced from Avants, B.B., Schoenemann, P.T., & Gee, J.C. Lagrangian frame diffeomorphic image registration: Morphometric comparison of human and chimpanzee cortex. *Med Im Analysis*, 10 (3), 397–412, Figure 14, Doi: 10.1016/j.media.2005.03.005 Copyright © 2005, Elsevier B.V. All rights reserved.

Whether the fit involves macaque monkeys (Figure 9.8A) or chimpanzees (Figure 9.8B), expansion has occurred *within* each of the lobes of the human brain, the frontal, parietal, and temporal. This explains how it can be that the proportions of the lobes as a whole have remained constant during human evolution, in spite of changes in the overall size of the neocortex.

Passingham and Smaers (2014) tried to find out whether the human PF cortex had expanded *more* than other association areas. They did so by plotting the volume of the PF cortex against the volume of the non-frontal cortex minus the striate cortex. The idea was to derive an estimate of the rest of the association cortex. It is, however, a very imperfect measure as it includes the primary somatosensory and auditory cortices.

The result is shown in Figure 9.9A. The value for the human PF cortex proved to be greater than predicted using a regression line through the monkeys and gibbon (solid line). However, the value for the chimpanzee was also almost outside the confidence limits (dotted lines).

Smaers et al. (2017) exploited the power of phylogenetic statistics, which takes into account the relationships among species. As shown by the dark blue lineages in Figure 9.9B, the analysis showed that the PF cortex has expanded significantly more than the non-frontal association areas in the human–chimpanzee clade.

Because the measure of non-frontal cortex minus striate cortex is an imperfect measure of the posterior association areas, it is important to see if there is a better way of making the comparison. Fortunately, there is another way of tackling the issue. Rilling and Seligman (2002) have produced data for the size of the temporal lobe (mm^2) in a series of primates. Because they also provide data on the grey matter of the superior temporal gyrus (mm^2), it is possible to work out the grey matter of the middle and inferior gyrus (mm^2). All of this is homotypical or association cortex.

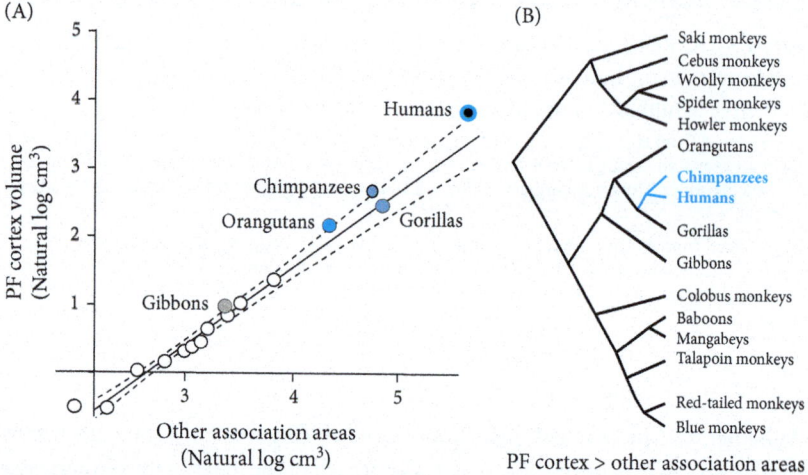

Figure 9.9 Log-log plot of volume of the PF cortex as a function of an estimate of the other association areas (see text)
(A) The solid line shows the regression through monkeys and the gibbon and the dotted lines show the confidence limits. Open circles, monkeys; grey circle, gibbons; blue circles, great apes; large blue circle with a dot, human.
(B) Cladogram of anthropoid primates, based on the same dataset as in A. Lineages in blue show a significant increase in the extent of granular PF cortex versus estimate of other association areas.

Reproduced from Smaers, J.B., Gomez-Robles, A., Parks, A.N., & Sherwood, C.C. Exceptional evolutionary expansion of prefrontal cortex in great apes and Humans. *Cur Biol*, 27 (10), 1549, Figure 2, Doi: 10.1016/j.cub.2017.05.015 © 2017, Elsevier Ltd.

Table 9.3 Remapping factors.

PF/inferior temporal	Human	Chimpanzee	Macaque monkey
Brodmann	2.1	1.4	0.9

So Passingham and Smaers (2014) compared the size of the PF cortex with the inferior temporal cortex. They used the value for the PF cortex (mm^2) from Brodmann (1912) to calculate how many times larger the PF cortex was than the visual temporal cortex. In the human and chimpanzee brain the measure included the middle and inferior temporal cortex, whereas in the macaque monkey there is no clear convexity cortex that corresponds to the middle temporal cortex. Table 9.3 presents the results.

They indicate that the human PF cortex is expanded relative to the temporal association cortex. However, there are currently no available measures of the posterior parietal cortex which also provides visual inputs. Figure 9.5 shows that there is an extensive area of lightly myelinated cortex in the inferior and medial parietal cortex. However, since no numerical measure has been provided for the posterior parietal cortex, we do not yet know whether the PF cortex is more expanded than the posterior parietal cortex.

Nonetheless, we know that the PF cortex lies at the top of the sensory processing hierarchy (Chapter 8). The PF cortex is expanded compared with the middle and inferior temporal cortex, and this means that there is more tissue in the human PF cortex for re-representing the input from the ventral visual stream.

The PF cortex also lies at the top of the motor hierarchy (Chapter 8). The consequence of the expansion of the PF cortex compared with the premotor and motor cortex is that there is relatively more tissue for generating goals, as opposed to directing the movements that are needed to achieve them.

Summary

During the evolution of the anthropoids, there has been an expansion in the homotypical or association cortex within each lobe, and not just within the frontal lobe. However, the human PF cortex has expanded more than the middle and inferior temporal cortex, and more than the premotor and motor cortex. The consequence of these differences is that there is more tissue in the human PF cortex for re-representing visual information from the ventral visual stream as well as for generating goals as opposed to directing the actions that achieve them.

Connectivity

The relative expansion of the human PF cortex has been associated with an expansion of the underlying white matter. Schoenemann et al. (2005) measured the extent of the grey matter and white matter in the frontal lobe anterior to the genu of the corpus callosum. Compared with the grey matter, there was significantly more PF white matter in the human brain than in the brains of great apes or monkeys.

This study could not distinguish between the white matter tracts entering the PF cortex from posterior areas from the white matter tracts that connect the different subareas of the PF cortex internally. However, Barrett et al. (2020) were able to draw this distinction by using DWI to visualize the major tracts in the human brain and the brains of macaque monkeys and vervet monkeys (*Cholorecubs aethiops*). However, there are two limitations of this study. Firstly, no data were collected for chimpanzees, and secondly no distinction was drawn between the PF cortex and the frontal cortex as a whole.

Barrett et al. calculated the proportions of the different tracts in relation to all the tracts visualized by tractography. On this proportional measure, there was significantly more white matter connecting the different frontal areas internally in the human brain than in the brains of the monkeys studied.

There was also significantly more white matter in the human brain that connected the frontal lobe with other areas. This was true for the three superior longitudinal tracts that connect the posterior parietal cortex with the premotor and dorsal PF cortex, as well as for the uncinate fasciculus that connects the temporal lobe with the ventral and orbital PF cortex.

Sneve et al. (2019) analysed resting-state covariance instead. They first co-registered the brains of marmosets (*Callithrix jacchus*) to the brains of cebus monkeys (*Cebus apella*), cebus monkey to macaque brains, and macaque brains to human brains. This enabled them to visualize the relative expansion of the homotypical areas that accompanies the expansion of the brain as a whole. Though the data were comparative, the expansion was taken to reflect 'evolutionary' expansion.

Sneve et al. applied graph theory to the patterns of connectivity as measured by resting-state covariance in the human brain. They compared what they called 'high expanding' regions with 'low expanding' regions. Compared with low expanding nodes, the high expanding nodes had a high 'participation coefficient', meaning that the nodes participated in many of the subnetworks of the brain. These nodes also had a high 'community density', a measure of the number of networks that are represented in a particular cortical location.

Finally, Sneve et al. scanned human subjects while they performed a visuomotor task. The high expanding regions had a greater 'closeness centrality' than low expanding regions. This measures the average shortest pathlength from one node to others in the network, and so the tightness of the coupling with the rest of the network. Figures 9.10A and 9.10B show that the PF cortex showed the highest community centrality during task performance as opposed to rest. The other area with high centrality was the cortex of the superior temporal sulcus. This includes the polymodal sensory area (Bruce et al., 1981) which is interconnected with the PF cortex (Seltzer & Pandya, 1989; Ercsey-Ravasz et al., 2013).

Figure 9.10C shows in orange and yellow the areas that had greater coupling with the high-expanding notes during task performance than rest. It will be seen that the PF cortex interacted with prestriate, posterior parietal, and temporal association areas during task performance. Figure 9.10C also shows in blue the nodes that had the highest coupling with high-expanding regions when rest was compared with task performance. They include the ventromedial PF cortex and ventral prelimbic area 32. Figure 9.10D shows the data for the areas in blue and yellow in Figure 9.10C.

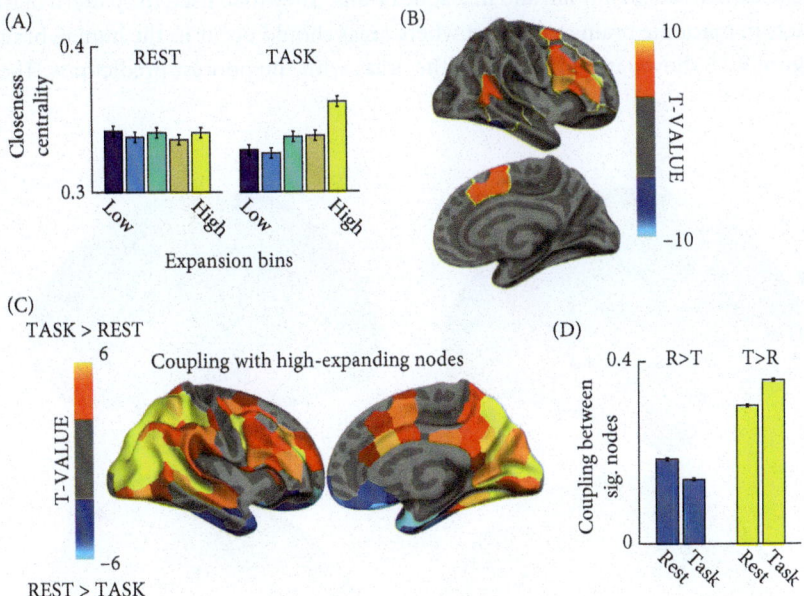

Figure 9.10 Degree of expansion and closeness centrality
(A) Closeness centrality for lowest expanding (dark blue) to highest expanding (yellow) area during task performance versus rest. (B) Nodes with high expansion hotspots outlined in yellow. (C) Areas that had greater coupling with high expanding areas during task performance than rest (oranges and yellow), and during rest than task performance. (D) Histograms showing the data for the areas shown in blue and yellow in C. R, rest; T, task.

Reproduced from Markus H. Sneve, Håkon Grydeland, Marcello G.P. Rosa, Tomáš Paus, Tristan Chaplin, Kristine Walhovd, & Anders M. Fjell, High-expanding regions in primate cortical brain evolution support supramodal cognitive flexibility. *Cerebral Cortex*, 29 (9), 3891–901, Figure 2, https://doi.org/10.1093/cercor/bhy268, Copyright © 2018, Oxford University Press.

These data are consistent with the claim that both the PF grey and white matter are especially expanded. They are also consistent with the evidence from tracer studies that the different areas of the PF cortex are highly interconnected (Petrides & Pandya, 1999; 2002; Saleem et al., 2014). It is these internal interconnections that allow the PF cortex to generate the goal that is appropriate, given the current context and the desired outcome (Chapter 8).

Changes in Connectivity

The previous section reviews evidence on the degree of the connectivity of the PF cortex. However, along with the expansion of the human brain, there have been changes in the origin and termination of some of the connections.

Mars et al. (2018b) used diffusion weighted imaging (DWI) to visualize the major fibre tracts in human and macaque brains. They then used the pattern of the tracts in macaque brains to predict where areas should occur in the human brain. Figure 9.11 shows, in red and blue, the areas with the poorest predictions. They

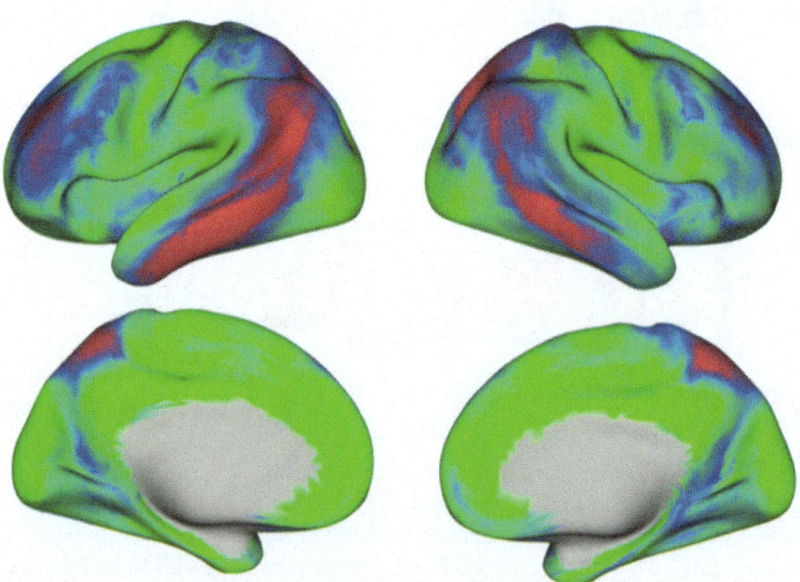

Figure 9.11 Areas of the brain that have changed most in their connectivity as estimated from visualizing the major tracts using diffusion weighted imaging (DWI) Red and blue areas differed the most in humans compared to macaque monkeys.

Reproduced from Mars, R.B., Sotiropoulos, S.N., Passingham, R.E., Sallet, J., Verhagen, L., Khrapitchev, A.A., Sibson, N., & Jbabdi, S. Whole brain comparative anatomy using connectivity blueprints. *Elife*, 7, e35237, Doi: 10.7554/eLife.35237 © 2018, Mars et al. Licensed under CC BY 4.0.

include the polar PF cortex (red) as well as the dorsal, ventral, and caudal PF cortex (blue), the inferior and medial parietal cortex (red), the middle temporal (blue), and inferior temporal cortex (red).

Bryant et al. (2019) used DWI to chart connections from the primary visual and auditory cortices in the macaque, chimpanzee and human brain. In chimpanzees and humans, but not macaques, connections from the primary visual cortex appear to project as far as the anterior temporal lobe. And in humans, but not chimpanzees, there appear to be connections with the superior temporal pole from both the primary auditory cortex and the primary visual cortex.

However, the most dramatic difference lies in the areas that contribute axons to the arcuate fascicle. In macaque monkeys this links the inferior parietal cortex and area Tpt at the back of the Sylvian fissure with the inferior part of the inferior part of the caudal PF cortex (Broca's areas 44 and 45B) (Frey et al., 2014). Figure 1.2 shows the location of the temporo-parietal area Tpt in a macaque brain.

The chimpanzee differs in that the arcuate fasciculus also originates from the superior temporal gyrus and the superior temporal sulcus (Rilling et al., 2012). And it differs again in the human brain in which it also originates from the middle temporal gyrus (Rilling et al., 2012; Mars et al., 2018a). There is little evidence for fibres originating from the middle temporal gyrus in chimpanzees (Mars et al., 2019).

Figure 9.12 (A and B) on the next page show that in the human brain the arcuate fascicle terminates in a greater area of the PF cortex than it does in macaque monkeys (Figure 9.12C and D). There are two general regions of termination (Eichert et al., 2019). One includes the inferior caudal PF cortex (area 44 and 45B), extending more rostrally into the ventral PF area 45A. The other involves the inferior frontal junction (IFJ), at the border between the precentral sulcus and the inferior frontal sulcus. As Chapter 7 describes, the IFJ was classified by Amunts et al. (2010) as being the dorsal part of area 44 (44d), as opposed to the opercular part of Broca's area (44v). Chapter 11 discusses the evidence that these changes in connectivity in the human brain support the ability to repeat sounds.

The change in connectivity has been accompanied by an expansion in all these areas. This can be appreciated from Figure 9.8B which shows the results of co-registering the chimpanzee to the human brain (Avants et al., 2006). There has been an expansion in the human inferior parietal cortex, middle temporal gyrus, the inferior caudal PF cortex (areas 44 and 45B), and the ventral PF cortex more rostrally.

Ardesch et al. (2019) also compared the connectivity of the human and chimpanzee brain using DWI. They found that the strength of the connections between Broca's area and the parietal and temporal lobe was significantly stronger in the human brain.

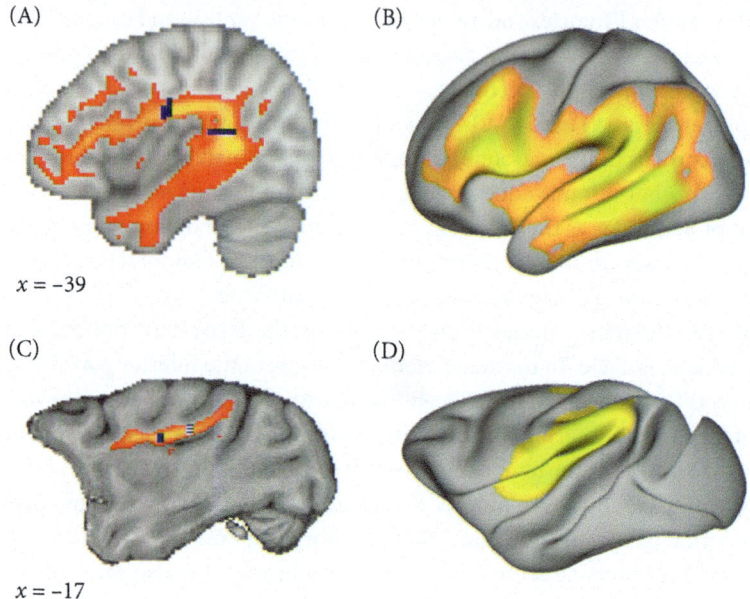

Figure 9.12 Arcuate fasciculus in the human and macaque monkey brain as visualized by diffusion weighted imaging
(A) Parasaggital section through human brain. (B) Arcuate fasciculus in human brain. (C) Parasaggital section through macaque monkey brain. (D) Arcuate fasciculus in macaque monkey brain.

Reproduced from Eichert, N., Verhagen, L., Folloni, D., Jbabdi, S., Khrapitchev, A.A., Sibson, N.R., Mantini, D., Sallet, J., & Mars, R.B. What is special about the human arcuate fasciculus? Lateralization, projections, and expansion. *Cortex*, 118, 107–15, Figure 1, Doi: 10.1016/j.cortex.2018.05.005 © The Authors. Licensed under CC BY 4.0.

Summary

The expansion of the PF cortex and other association areas has been accompanied by changes in connectivity. The PF cortex, in particular, is highly connected with other networks, as well as having dense internal interconnections. The arcuate fascicle also has a more extensive origin and termination in the human than chimpanzee or macaque monkey brain. It originates in the middle, as well as the superior temporal gyrus, as well as the inferior parietal cortex.

Microstructure

The previous sections discuss the area (Brodmann, 1912) or volume (Smaers et al., 2011; Donahue et al., 2018) of the PF cortex. But the cortex consists of neurones and glial cells, and we cannot simply assume that the expansion of the human PF cortex is due to an expansion in the number of neurones.

Fortunately, Herculano-Houzel (2009) has counted the ratio of neurones to glial cells in the neocortex of humans and other primates, as well as in other mammals. It turns out that the ratio in the human brain is as expected for a primate our size. The human brain is similar to the brains of other primates in having a significantly greater density of neurones per volume of tissue than the brains of other mammals (Herculano-Houzel, 2012).

The neurones in the PF cortex of primates differ from those in other cortical areas in being larger, having a greater complexity of the dendritic branching, and with more spines on the basal dendrites. The evidence comes from data on adult humans, chimpanzees, Old World monkeys, a New World monkey, and one species of bushbaby (Jacobs et al., 2001; Elston et al., 2006; Bianchi et al., 2013).

However, a correction factor is needed since differences between species in dendritic size or spine density could be due to the differences in the overall size of the neocortex or brain. This can be achieved by comparing the values for PF neurones with the values for neurones in the primary or secondary areas of the brain since these areas roughly scale according to body size.

Using the data from Elston et al. (2006), it is possible to compare the total number of basal dendritic spines on layer III pyramidal neurones in the PF cortex with the numbers in visual area V2 across a range of primate species. For the four monkey species studied, the mean for the PF cortex was 4.3 times greater than that for V2. The mean for the human PF cortex was 6.2 times greater than that for V2, but the value for the macaque monkey was similar, being 7.7 times greater than that for V2. Unfortunately, no data were collected in this study for the chimpanzee brain.

However, Jacobs et al (2001) and Bianchi et al. (2013) have produced data for the polar PF cortex (area 10) in the chimpanzee and human brains. For the chimpanzee the mean number of spines in PF area 10 was 2.3 times greater than for area V2. The mean for the human brain was similar, 2.1 times greater than for area V2.

It could, of course, be that what matters is not the dendritic spine number but the spacing between cells instead, that is the neuropil space. This is occupied by dendritic processes, axons, synapses, and microvasculature. Semendeferi et al. (2011) and Spocter et al. (2012) have found that the neuropil space is greater in the human PF area 10 than in brains of great apes. However, this difference may reflect a scaling rule such that the neuropil fraction increases as area 10 increases. We know that the neuropil fraction in areas V1 and V2 scales in correlation with the volumes of visual system structures across hominoids (de Sousa et al., 2010).

Nonetheless, it is important to recognize that even if the difference between humans and other primates fits a scaling rule, that does not mean that the difference is of no functional consequence. Given that the dendrites in the PF polar cortex are larger and have more spines in the human than chimpanzee brain (Bianchi et al.,

2013), it is likely that the human PF cortex has an increased capacity for integrating a wide range of inputs from diverse sources.

These inputs can derive from neocortical or subcortical areas. One source of local-circuit inputs comes from subtypes of GABAergic inhibitory interneurons. Sherwood et al. (2010) examined the GABAergic subtypes of inter-neurones in the PF cortex in a series of anthropoid primates. They found similar distributions of these inhibitory subtypes of inter-neurone across regions of the PF cortex of all anthropoid primates, including humans.

A significant source of extrinsic afferents that modulate the excitability of cortical networks comes from subcortical neurones that express neurotransmitters such as dopamine, serotonin, and acetylcholine. Humans and great apes differ from macaques in having relatively greater innervation of the PF cortex by axons that express these neuromodulators (Raghanti et al., 2008b; c; a). This is evidence of a grade shift in the common ancestor of humans and great apes.

Summary

There have been studies that indicate that the number of spines on PF pyramidal cells or the packing density differ in the human brain. But these differences are probably due to the greater brain size.

Grade Shifts and Ontogeny

The data in the previous sections are all taken from modern primates; in other words, the evidence is comparative. But chimpanzees are not our ancestors. Since the last common ancestor of chimpanzees and modern humans, there have been changes in the line leading to modern chimpanzees as well as in the line leading to modern humans.

Nonetheless, the comparative evidence indicates that there are different grades amongst modern primates. So, there must be a way in which *during evolution* ancestral species shifted from one grade to another. One mechanism for grade-shifts involves heterochrony in the timing of changes during development (Gould, 1977).

This can be illustrated by comparing the growth of the brain and body in chimpanzees, humans, and macaque monkeys. The adult human brain is 3.5 times larger than expected for an ape of our size, and 4.5 times larger than a monkey of our size (Passingham, 2008).

The size of the adult human brain is the result of extending the time for which it grows at the foetal rate (Passingham, 1975). It continues to grow at this rate for about a year after birth, and the rate then starts to decrease until the brain reaches its final size at the age of 20. In contrast, the chimpanzee brain no longer grows at the foetal rate after birth, reaching its adult size at the age of around 10 (Passingham, 1975).

However, though the rates of growth from foetus to adulthood differ for the human and chimpanzee brain, the difference can be accounted for by a simple mathematical transformation. If the chimpanzee curve is extended so as to match the length of development of the human brain, then the two curves superimpose (Rice, 2002). This is not it true for the curves for body weight (Rice, 1997). Instead, the curves for body weight can only be fitted via a mathematical transformation called 'sequential heterochrony'. This means that different discrete growth phases change rates or extend individually (McNamara, 2012).

The consequence of these changes is that the *relative* rates of the growth of the brain and body are not matched between modern humans and chimpanzees. And nor are they matched for macaque monkeys and chimpanzees (Rice, 1997, 2002).

These results mean that even if a chimpanzee or macaque monkey had as long a period of ontogeny as modern humans, they would not have the same ratio of brain to body weight as humans. They represent different grades. There was a change in the relative rates of the development of the brain and body in the ancestral hominins.

The same argument applies to the proportion of the neocortex that is taken up by homotypical or association cortex, including the PF cortex. Consider modern species first. The development of the primary sensory and motor areas of the neocortex parallels that of the growth of the body. Chugani et al. (1987) scanned foetuses and infants using PET with a fluorodeoxyglucose label (^{18}F). The study showed that, as judged by functional activation, the striate and sensorimotor cortex matured early. Other areas such as the PF and premotor cortex reached maturity much later.

The consequence is that, during the long period of human postnatal development, the areas that are neither primary, sensory, nor motor take up an increasing proportion of the brain. This has been demonstrated by Hill et al. (2010). They produced maps showing how much larger different areas are in the adult human brain than they were at birth. The areas that are most expanded are the association areas including the PF cortex.

These areas are shown in Figure 9.13 (next page). The areas that show the greater expansion during human ontogeny are those that can be shown to be expanded when either the macaque or chimpanzee brains are co-registered to the adult human brain (Figure 9.8). In other words, the changes during ontogeny are similar to the changes that occurred during evolution.

Sneve et al. (2019) collected imaging data for forty-six children and teenagers from the ages of 6 to 17 and compared the data with that for adults. Relative to the average closeness, the centrality closeness of the high expanding regions was higher in the adults than the young members of sample. Yet there was no difference between the groups in the closeness centrality for the low expanding areas. The high expanding areas reach their maximal size at adolescence (Amlien et al.,

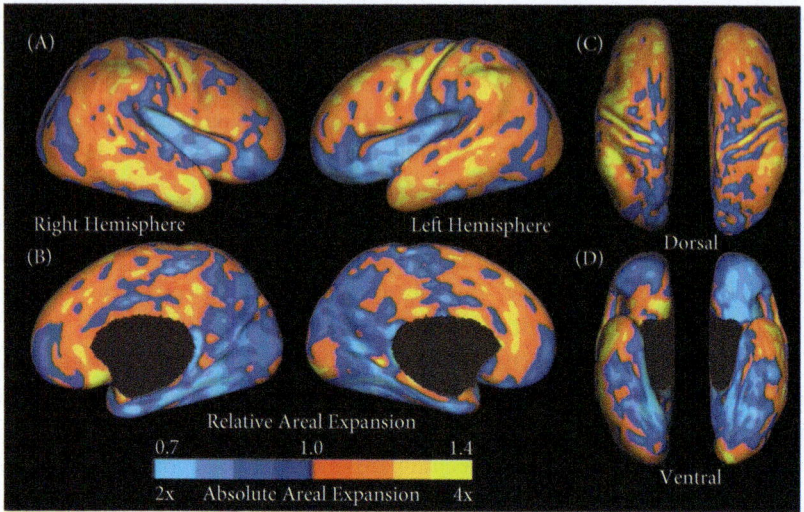

Figure 9.13 Postnatal cortical surface expansion
A = lateral, B = medial, C = dorsal, and D = ventral views. The absolute expansion scale indicates how many times larger the surface area of a given region is in adulthood relative to that region's area at term. The relative expansion scale indicates the difference in proportion of total surface area at birth and adulthood.

Reproduced from Hill, J., Inder, T., Neil, J., Dierker, D., Harwell, J., & Van Essen, D., Similar patterns of cortical expansion during human development and evolution. *Proc Nal Ac Sci USA*, 107 (29), 13135–40, Figure 1, https://doi.org/10.1073/pnas.1001229107 © The Authors.

2016). However, this does not take into account the further changes in the volume of the underlying white matter with age (Bartzokis et al., 2001). Finally, Sneve et al. (2019) used their data from marmosets, cebus monkeys, macaques, and humans to calculate a measure of the areas that showed the greatest 'evolutionary expansion'. The relation between the closeness centrality of high expanding regions and evolutionary expansion was only significant when the human brain had reached adulthood, at 18 years of age.

For evolution to produce a grade shift, there must be variation on which natural selection can act. It turns out that even in the brain of modern humans there are individual differences in the degree of expansion of the association areas (Reardon et al., 2018). Furthermore, this variation in expansion correlates with the size of the brain as a whole. So, even today there is variation on which in principle selection could act.

The evolutionary change between grades requires the operation of control genes. These control the rate and time at which DNA is transcribed to messenger RNA. Therefore, changes in transcription have the potential to influence the development of the brain by shifts in timing or heterochrony (Somel et al., 2014).

Berto and Nowick (2018) compared transcription factors in the PF cortex of human, chimpanzee, and macaque monkey brains. They looked for the species-specific transcription factors that are derived from genes with species-specific changes in the PF cortex. These were found to be over-represented in the human PF cortex.

However, there are also changes in other association areas. Sousa et al. (2017) looked at differences in gene expression between different brain areas in humans, chimpanzees and macaque monkeys. There were significant differences between the human and chimpanzee or macaque monkey brain in the PF cortex, inferior parietal cortex and inferior temporal cortex amongst other areas. The differences were much less in the primary auditory cortex, the primary somatosensory or motor cortex. In other words, it is the high expanding association areas that showed the greatest difference in gene regulation.

Summary

Comparative evidence between modern species suggests that evolutionary changes between grades could be achieve by changes in the relative rates of development of the brain compared to the body, and changes in the relative rates of development of the association areas compared to the primary sensory and motor areas. These changes could be achieved by transcription factors influencing the timing at which genes are expressed. When humans are compared with chimpanzees and macaque monkeys, the greatest difference in transcription factors are in the PF cortex and other association areas.

Selection Pressures

Given the expansion of the granular PF cortex in humans, it is important to establish what selective pressures were responsible. But there is a danger of spinning 'just so stories'. These are speculative explanations of what 'might have' happened during evolution. They are so called because, in his book for children called 'Just So Stories', Rudyard Kipling gave fanciful explanations as to how, for example, the elephant got its trunk or the camel its hump.

The problem is that many factors influenced the evolution of the hominins, including climate change, new environments, new predators, new foods, new prey, and new competitors. Table 9.4 (next page) presents a list of potential selective factors in the left column, along with a list of behavioural adaptations in the middle column and a list of functions that have been suggested for the PF cortex in the right column.

The entries in the three columns are in alphabetical order and are independent, meaning that the entries in a row are not meant to link up in any way. It would, of

Table 9.4 Factors that might account for the expansion of the PF cortex in hominins. Note that the entries in each column appear in alphabetical order, so there is no suggestion that entries in a row are related to each other.

Selective pressures	Behavioural responses	Prefrontal functions
Climate instability	Clothing	Analogical reasoning
Global cooling (ice ages)	Fire for cooking	Semantic associations
New environments	Fire for protection	Imagination or simulation
New foods	Group defence	Metacognition
New predators	Social cooperation	Social and moral rules
New prey	Spoken language	Planning and prospection
New social systems	Tools (e.g. hafted spears)	Theory of mind

course, be possible to try to associate specific selective pressures to specific functions that are performed by the human PF cortex; but this problem has too many degrees of freedom to reach any firm conclusions, see Table 9.4.

Rather than linking particular selection pressures to particular functions, it is safer to simply consider some of the factors that are most likely to have influenced the expansion of the granular PF cortex during the evolution of the hominins. They have in common that they involve changes in the environment in which our ancestors lived. It is changes in environments that promote the evolution of new grades due to adaptive change.

The reason is that new environments make demands on the ability to solve new problems rapidly. And, as Chapter 8 documents, it is the PF cortex that in a position to transfer prior knowledge to new situations.

Climate Change

After the dramatic global cooling event of the early Oligocene (~34–23 Ma), the climate again cooled in the late Miocene, which saw another 2°C decrease that occurred gradually from ~15 Ma until the end of the Miocene (~5 Ma). In Africa, this cooling regime led to increased aridity, deforestation and the development of savannas and grasslands (Kaya et al., 2018).

An earlier section suggests that the hominins diverged from our common ancestor with chimpanzees ~7 Ma, and that they did so in Africa. So these climate changes affected the foraging options of the common ancestor as well as the early hominins. The latest common ancestor of chimpanzees and hominins lived in tropical deciduous woodland, with more open habitat (Andrews, 2020). However, even as late as ~2.3–1.4 Ma, *Homo habilis* was adapted for life in the trees as well as on the ground (Andrews & Johnson, 2020). It is not clear why bipedalism developed, perhaps for wading through water while foraging (Niemitz, 2010) given that the early hominins lived near rivers (Gani & Gani, 2011). Bipedalism had the further advantage that it was easy to carry foods and tools.

There is evidence of more animal foods in the diet of the hominins ~2.5 Ma, but the assumption is that the meat was scavenged (Smith et al., 2015). This change in diet was associated with tools that enabled these hominins to exploit a large variety of tough foods that could be found at ground level.

The earliest tools date to ~3.3 Ma (Lewis & Harmand, 2016). This is before the earliest Oldowan tools associated with the Australopithecines (Leakey, 2009). The early Acheulean assemblage of hand axes, cleavers, and picks were associated with *Homo ergaster* in East Africa (Semaw et al., 2018). (Figure 9.1) shows that there was an increase in brain size, relative to the body, with the development of these assemblages.

The ability to stride required longer legs, and this developed with the evolution of *Homo erectus* (Andrews & Johnson, 2020). Striding locomotion is much more energetically efficient than quadrupedal walking (Pontzer et al., 2009), and this meant that there were less metabolic costs in travelling between food and water resources as they became more dispersed. With the increase in naturally occurring forest fires, *Homo erectus* discovered the use of fire for cooking (Parker et al., 2016). The majority of the animals that these hominins ate were small; and the larger animals that were probably scavenged rather than hunted (O'Connell et al., 2002).

New Environments

Because of the ability to move further distances, *Homo erectus* expanded its range within Africa. As the climate changed, *Homo erectus* migrated out of Africa (Stewart & Stringer, 2012). This exodus occurred in several waves, beginning more than 2 Ma. *Homo erectus* left first, followed by *Homo heidelbergensis*. These waves of migration compelled these people to adapt to new environments and find new refuges (Stewart & Stringer, 2012). The move to new climes meant that our ancestors had to exploit new plants and find new sources of meat either by scavenging or hunting. They also faced new predators. So, there were a myriad of selective pressures during the invasion of new ecosystems that might have provided the driving force for further encephalization (Figure 9.1).

The earliest modern humans emerged ~314,000 years ago (Richter et al., 2017). They moved out of Africa through narrow corridors into Europe and Asia (Hershkovitz et al., 2018). And it is modern rather than archaic humans that there was evidence of frontal bulging (Bruner et al., 2013). Since the frontal lobes of modern humans are not expanded relative to the neocortex (Semendeferi et al.,

2002), this frontal bulging is likely to indicate an expansion of the PF cortex specifically.

As modern humans migrated through Eurasia, they met new challenges for survival. An expansion of the PF cortex would have meant that they would have been especially equipped to transfer knowledge rapidly from one problem to another.

During the North African Middle Stone Age, which began ~150,000 years ago, evidence for the Aterian technology appears. This cultural tradition involved the flaking and the hafting of points on spears (Iovita, 2011). A bone knife found in a cave in Morocco dates to ~90,000 years ago (Bouzouggar et al., 2018). This represents the oldest specialized bone-derived tool yet discovered. Manufacturing it required sequential movements at a new level of complexity and sophistication. Shell beads of around the same date have also been found in a cave in Eastern Morocco (Bouzouggar et al., 2007). They are coloured with red ochre and the wear patterns suggest that they were worn as a necklace. These finding are from a time long after the PF cortex had ceased to expand. They are evidence for cultural rather than evolutionary change.

Communication

Modern chimpanzees often vocalize at the same time as they produce gestures (Heesen et al., 2019). It is therefore likely that the early hominins did the same. Clark and Henneberg (2017) have analysed the craniofacial morphology of *Ardipithecus ramidus* from ~4.4 Ma. They have suggested that the larynx lay deeper in the vocal tract than in modern chimpanzees, and that it has increased mobility. They interpreted this to mean that *Ardipethecus* would have been better able to modulate vocalizations.

Later, when the hominins hunted rather simply scavenging, there would have been a selection pressure towards better communication, whether by gesture or vocalization. There are chimpanzees in the Tai National Park that hunt colobus monkeys, and to do so they take different roles during the hunt. These chimpanzees show an increase in cooperation and communication compared with the chimpanzees in neighbouring areas of the Tai park that do not hunt (Luncz & Boesch, 2015). As *Homo neanderthalensis* and *Homo sapiens* depended increasingly for food on animals both big (Marin et al., 2017) and small (Wedage et al., 2019), there would have been a similar need to communicate.

Barney et al. (2012) have modelled the vocal tract and the position of the hyoid bone in series of Neanderthals and modern humans. Their analysis suggests Neanderthals could produce some of our vowels but not all of them; in other words, they had a restricted vowel space. Furthermore, genomic sequencing has revealed the modern human-like variant of the FOXP2 gene in Neanderthal DNA (Enard et al., 2002). This gene plays a critical role in the development of the cortical–basal ganglia networks that contribute to learning auditory-motor articulation patterns and other aspects of orofacial coordination (Enard, 2011; Watkins, 2011).

If there was a difference between communication by the Neanderthals and our direct ancestors, one possibility is that it could have related to the production of grammatical strings rather than sounds alone. It is not clear when the earliest grammatical language developed, but Bolhuis et al. (2014) suggested that it might have been ~70 Ka.

However, protolanguages may well have existed long before this time (Dunbar, 2014). This refers to the use of small clauses as opposed to sentences (Progovac, 2016). Thus, spoken language could be one factor that accounted for the rapid expansion of the brain of hominins from ~400Ka (Figure 9.1).

The development of grammatical language would have conferred a critical selective advantage. The reason is that it is possible to give specific instructions using clauses or whole sentences. In this way the teacher can pass on knowledge about the world, as well as skills such as how to make a complex tools (Kolodny & Edelman, 2018). Rather than individuals transferring their knowledge from one problem to another (Chapter 8), they can transfer it to other individuals. The result is that students can in principle avoid trial-and-error completely.

Social Factors

The environment also includes other individuals within the group, as well as other groups. An increase in the size of the groups would have been of advantage for defence and hunting. This would have promoted the abilities that were needed to adapt to living in such groups.

Dunbar (2014) has used information about the climate and environment to estimate the likely sizes of the groups of hominins at different stages during evolution. Few experts doubt that one way that hominins adapted to the problems they faced involved cooperation within their social group (Tomasello, 2016).

To cooperate, our ancestors would have gained an advantage if they were able to infer the intentions of other members of their group. Tomasello (2019) has referred to this as 'joint intentionality', meaning that different individuals have similar intentions on a joint enterprise. In an early and classical paper, Premack and Woodruff argued that the ability to infer the mental states of others involves a 'theory of mind'. Frith and Frith (Frith & Frith, 2004) have suggested the ugly but useful term 'mentalizing'.

The requirement to co-operate would also have promoted the need to understand social rules and conventions (Tomasello, 2016). However, it is not possible to have a system of social rules without imposing sanctions on those who disobey them. Those who obeyed would have had a better chance of integrating in the group, and thus surviving.

The next two chapters argue that the ability learn by instruction, as well as the ability to read the mental states of others and to learn social rules depend on mechanisms of the PF cortex.

Conclusions

Hominin evolution led to greater adaptability in a changing environment, and it seems likely that the expanding PF cortex of these species played a role in this capacity. The expansion of the PF cortex and other association areas was driven by changes in the environment. These were caused by climate change and new ecosystems.

The hominins responded by using and making tools, changing their diet to include animals, and the use of fire for cooking. As they migrated into Eurasia, they met new challenges; and there is evidence of frontal bulging, indicating expansion of the PF cortex in our direct ancestors. The need to hunt led to improved ways of communicating, and to the development of protolanguage and fully grammatical language.

This promoted cultural transmission and the development of intricate social systems that foster cooperativity. Spoken language allowed the accumulation of knowledge across generations. It is this that accounts for the acceleration in cultural innovation that characterizes the past 100,000 years or so of human history.

References

Alatorre Warren, J.L., Ponce de Leon, M.S., Hopkins, W.D., & Zollikofer, C.P.E. (2019) Evidence for independent brain and neurocranial reorganization during hominin evolution. *Proc Natl Acad Sci USA*, 116, 22115–21.

Amlien, I.K., Fjell, A.M., Tamnes, C.K., Grydeland, H., Krogsrud, S.K., Chaplin, T.A., Rosa, M.G.P., & Walhovd, K.B. (2016) Organizing principles of human cortical development thickness and area from 4 to 30 years: Insights from comparative primate neuroanatomy. *Cereb Cortex*, 26, 257–67.

Amunts, K., Lenzen, M., Friederici, A.D., Schleicher, A., Morosan, P., Palomero-Gallagher, N., & Zilles, K. (2010) Broca's region: Novel organizational principles and multiple receptor mapping. *PLoS Biol*, 8, doi: 10.1371/journal.pbio.1000489

Andrews, P. (2020) Last common ancestor of apes and humans: Morphology and environment. *Folia Primatol* (Basel), 91, 122–48.

Andrews, P. & Johnson, R.J. (2020) Evolutionary basis for the human diet: Consequences for human health. *J Intern Med*, 287, 226–37.

Ardesch, D.J., Scholtens, L.H., Li, L., Preuss, T.M., Rilling, J.K., & van den Heuvel, M.P. (2019) Evolutionary expansion of connectivity between multimodal association areas in the human brain compared with chimpanzees. *Proc Natl Acad Sci USA*, 116, 7101–6.

Avants, B.B., Schoenemann, P.T., & Gee, J.C. (2006) Lagrangian frame diffeomorphic image registration: Morphometric comparison of human and chimpanzee cortex. *Med Image Anal*, 10, 397–412.

Baab, K. (2016) The place of Homo floresiensis in human evolution. *J Anthropol Sci*, 94, 5–18.
Bailey, P., von Bonin, G., & McCullogh, W.S. (1950) *The Isocortex of the Chimpanzee*. University of Illinois Press, Urbana.
Balsters, J.H., Cussans, E., Diedrichsen, J., Phillips, K.A., Preuss, T.M., Rilling, J.K., & Ramnani, N. (2010) Evolution of the cerebellar cortex: The selective expansion of prefrontal-projecting cerebellar lobules. *Neuroimage*, 49, 2045–52.
Barney, A., Martelli, S., Serrurier, A., & Steele, J. (2012) Articulatory capacity of Neanderthals, a very recent and human-like fossil hominin. *Philos Trans R Soc Lond B Biol Sci*, 367, 88–102.
Barrett, R.L.C., Dawson, M., Dyrby, T.B., Krug, K., Ptito, M., D'Arceuil, H., Croxson, P.L., Johnson, P.J., Howells, H., Forkel, S.J., Dell'Acqua, F. Monay, & Catani, M. (2020) Differences in frontal network anatomy across primate species. *J Neurosci*, 40, 2094–107.
Barton, R.A. & Venditti, C. (2013) Human frontal lobes are not relatively large. *Proc Natl Acad Sci USA*, 110, 9001–6.
Bartzokis, G., Beckson, M., Lu, P.H., Nuechterlein, K.H., Edwards, N., & Mintz, J. (2001) Age-related changes in frontal and temporal lobe volumes in men: A magnetic resonance imaging study. *Arch Gen Psychiatry*, 58, 461–5.
Berto, S. & Nowick, K. (2018) Species-specific changes in a primate transcription factor network provide insights into the molecular evolution of the primate prefrontal cortex. *Genome Biol Evol*, 10, 2023–36.
Bianchi, S., Stimpson, C.D., Bauernfeind, A.L., Schapiro, S.J., Baze, W.B., McArthur, M.J., Bronson, E., Hopkins, W.D., Semendeferi, K., Jacobs, B., Hof, P.R., & Sherwood, C.C. (2013) Dendritic morphology of pyramidal neurons in the chimpanzee neocortex: Regional specializations and comparison to humans. *Cereb Cortex*, 23, 2429–36.
Bolhuis, J.J., Tattersall, I., Chomsky, N., & Berwick, R.C. (2014) How could language have evolved? *PLoS Biol*, 12, e1001934.
Bosch, M.D., Mannino, M.A., Prendergast, A.L., O'Connell, T.C., Demarchi, B., Taylor, S.M., Niven, L., van der Plicht, J., & Hublin, J.J. (2015) New chronology for Ksar 'Akil (Lebanon) supports Levantine route of modern human dispersal into Europe. *Proc Natl Acad Sci USA*, 112, 7683–88.
Bouzouggar, A., Barton, N., Vanhaeren, M., d'Errico, F., Collcutt, S., Higham, T., Hodge, E., Parfitt, S., Rhodes, E., Schwenninger, J.L., Stringer, C., Turner, E., Ward, S., Moutmir, A., & Stambouli, A. (2007) 82,000-year-old shell beads from North Africa and implications for the origins of modern human behavior. *Proc Natl Acad Sci USA*, 104, 9964–9.
Bouzouggar, A., Humphrey, L.T., Barton, N., Parfitt, S.A., Clark Balzan, L., Schwenninger, J.L., El Hajraoui, M.A., Nespoulet, R., & Bello, S.M. (2018) 90,000 year-old specialised bone technology in the Aterian Middle Stone Age of North Africa. *PLoS One*, 13, e0202021.
Brodmann, K. (1912) Neue ergebnisse uber die verleichende histologische localisation der grosshirnrinde mit besonderer berucksichtigung des stirnhirns. *Anat Anz spllp*, 41, 157–216.
Brodmann, K. (1913) Neue forchungsergebnisse der grosshirnrindeanatomische mit besonderer berucksichtung anthropologischer fragen. *Gesselch Deuts Naturf Artze*, 85, 200–40.
Bruce, C., Desimone, R.M., & Gross, C.G. (1981) Visual properties of neurons in a polysensory area in superior temporal sulcus of the macaque. *J Neurophysiol*, 46, 369–84.
Bruner, E., Athreya, S., de la Cuetara, J.M., & Marks, T. (2013) Geometric variation of the frontal squama in the genus homo: Frontal bulging and the origin of modern human morphology. *Am J Phys Anthropol*, 150, 313–23.

Bruner, E., Preuss, T.M., Chen, X., & Rilling, J.K. (2017) Evidence for expansion of the precuneus in human evolution. *Brain Struct Funct*, 222, 1053–60.

Bruner, E., Saracino, B., Ricci, F., Tafuri, M., Passarello, P., & Manzi, G. (2004) Midsagittal cranial shape variation in the genus Homo by geometric morphometrics. *Coll Antropol*, 28, 99–112.

Bryant, K.L., Glasser, M.F., Li, L., Jae-Cheol Bae, J., Jacquez, N.J., Alarcon, L., Fields, A., 3rd, & Preuss, T.M. (2019) Organization of extrastriate and temporal cortex in chimpanzees compared to humans and macaques. *Cortex*, 118, 223–43.

Carlson, K.J., Stout, D., Jashashvili, T., de Ruiter, D.J., Tafforeau, P., Carlson, K., & Berger, L.R. (2011) The endocast of MH1, australopithecus sediba. *Science*, 333, 1402–7.

Caswell, J.L., Mallick, S., Richter, D.J., Neubauer, J., Schirmer, C., Gnerre, S., & Reich, D. (2008) Analysis of chimpanzee history based on genome sequence alignments. *PLoS Genet*, 4, e1000057.

Chugani, H.T., Phelps, M.E., & Mazziotta, J.C. (1987) Positron emission tomography study of human brain functional development. *Ann Neurol*, 22, 487–97.

Clark, G. & Henneberg, M. (2017) Ardipithecus ramidus and the evolution of language and singing: An early origin for hominin vocal capability. *Homo*, 68, 101–21.

Crompton, R.H., Pataky, T.C., Savage, R., D'Aout, K., Bennett, M.R., Day, M.H., Bates, K., Morse, S., & Sellers, W.I. (2012) Human-like external function of the foot, and fully upright gait, confirmed in the 3.66 million year old Laetoli hominin footprints by topographic statistics, experimental footprint-formation and computer simulation. *J R Soc Interface*, 9, 707–19.

de Sousa, A.A., Sherwood, C.C., Schleicher, A., Amunts, K., MacLeod, C.E., Hof, P.R., & Zilles, K. (2010) Comparative cytoarchitectural analyses of striate and extrastriate areas in hominoids. *Cereb Cortex*, 20, 966–81.

Donahue, C.J., Glasser, M.F., Preuss, T.M., Rilling, J.K., & Van Essen, D.C. (2018) Quantitative assessment of prefrontal cortex in humans relative to nonhuman primates. *Proc Natl Acad Sci USA*, 115, E5183–92.

Dunbar, R. (2014) *Human Evolution*. Penguin, London.

Eichert, N., Verhagen, L., Folloni, D., Jbabdi, S., Khrapitchev, A.A., Sibson, N.R., Mantini, D., Sallet, J., & Mars, R.B. (2019) What is special about the human arcuate fasciculus? Lateralization, projections, and expansion. *Cortex*, 118, 107–15.

Elston, G.N., Benavides-Piccione, R., Elston, A., Zietsch, B., Defelipe, J., Manger, P., Casagrande, V., & Kaas, J.H. (2006) Specializations of the granular prefrontal cortex of primates: Implications for cognitive processing. *Anat Rec A Discov Mol Cell Evol Biol*, 288, 26–35.

Enard, W. (2011) FOXP2 and the role of cortico-basal ganglia circuits in speech and language evolution. *Curr Opin Neurobiol*, 21, 415–24.

Enard, W., Przeworski, M., Fisher, S.E., Lai, C.S., Wiebe, V., Kitano, T., Monaco, A.P., & Paabo, S. (2002) Molecular evolution of FOXP2, a gene involved in speech and language. *Nature*, 418, 869–72.

Ercsey-Ravasz, M., Markov, N.T., Lamy, C., Van Essen, D.C., Knoblauch, K., Toroczkai, Z., & Kennedy, H. (2013) A predictive network model of cerebral cortical connectivity based on a distance rule. *Neuron*, 80, 184–97.

Falk, D., Redmond, J.C., Jr., Guyer, J., Conroy, C., Recheis, W., Weber, G.W., & Seidler, H. (2000) Early hominid brain evolution: A new look at old endocasts. *J Hum Evol*, 38, 695–717.

Fewlass, H., Talamo, S., Wacker, L., Kromer, B., Tuna, T., Fagault, Y., Bard, E., McPherron, S.P., Aldeias, V., Maria, R., Martisius, N.L., Paskulin, L., Rezek, Z., Sinet-Mathiot, V.,

Sirakova, S., Smith, G.M., Spasov, R., Welker, F., Sirakov, N., Tsanova, T., & Hublin, J.J. (2020) A (14)C chronology for the middle to upper palaeolithic transition at Bacho Kiro Cave, Bulgaria. *Nat Ecol Evol*, 4, 794–801.
Frey, S., Mackey, S., & Petrides, M. (2014) Cortico-cortical connections of areas 44 and 45B in the macaque monkey. *Brain Lang*, 131, 36–55.
Frith, U. & Frith, C. (2004) Development and neurophysiology of mentalizing. In Frith, C., Wolpert, D.M. (eds). *The Neuroscience of Social Interaction*. Oxford University Press, Oxford, pp. 45–76.
Gabi, M., Neves, K., Masseron, C., Ribeiro, P.F., Ventura-Antunes, L., Torres, L., Mota, B., Kaas, J.H., & Herculano-Houzel, S. (2016) No relative expansion of the number of prefrontal neurons in primate and human evolution. *Proc Natl Acad Sci USA*, 113, 9617–22.
Gani, M.R. & Gani, N.D. (2011) River-margin habitat of Ardipithecus ramidus at Aramis, Ethiopia 4.4 million years ago. *Nature Communications*, 2, 602.
Geyer, S. (2004) The microstructural border between the motor and the cognitive domain in the human cerebral cortex. *Adv Anat Embryol Cell Biol*, 174, I-VIII, 1–89.
Glasser, M.F., Goyal, M.S., Preuss, T.M., Raichle, M.E., & Van Essen, D.C. (2014) Trends and properties of human cerebral cortex: Correlations with cortical myelin content. *Neuroimage*, 93, 165–75.
Glazko, G., Veeramachaneni, V., Nei, M., & Makalowski, W. (2005) Eighty percent of proteins are different between humans and chimpanzees. *Gene*, 346, 215–19.
Gould, S.J. (1977) *Ontogeny and Phylogeny*. Belknap, Cambridge.
Guy, F., Lieberman, D.E., Pilbeam, D., de Leon, M.P., Likius, A., Mackaye, H.T., Vignaud, P., Zollikofer, C., & Brunet, M. (2005) Morphological affinities of the Sahelanthropus tchadensis (Late Miocene hominid from Chad) cranium. *Proc Natl Acad Sci USA*, 102, 18836–41.
Haile-Selassie, Y., Melillo, S.M., Vazzana, A., Benazzi, S., & Ryan, T.M. (2019) An 3.8-million-year-old hominin cranium from Woranso-Mille, Ethiopia. *Nature*, 573, 214–19.
Harvati, K., Roding, C., Bosman, A.M., Karakostis, F.A., Grun, R., Stringer, C., Karkanas, P., Thompson, N.C., Koutoulidis, V., Moulopoulos, L.A., Gorgoulis, V.G., & Kouloukoussa, M. (2019) Apidima Cave fossils provide earliest evidence of Homo sapiens in Eurasia. *Nature*, 571, 500–4.
Heesen, R., Hobaiter, C., Ferrer, I.C.R., & Semple, S. (2019) Linguistic laws in chimpanzee gestural communication. *Proc Biol Sci*, 286, 20182900.
Herculano-Houzel, S. (2009) The human brain in numbers: A linearly scaled-up primate brain. *Front Hum Neurosci*, 3, 31.
Herculano-Houzel, S. (2012) Neuronal scaling rules for primate brains: The primate advantage. *Prog Brain Res*, 195, 325–40.
Herculano-Houzel, S. (2016) *The Human Advantage*. MIT Press, Cambridge, M.A.
Herculano-Houzel, S., von Bartheld, C.S., Miller, D.J., & Kaas, J.H. (2015) How to count cells: The advantages and disadvantages of the isotropic fractionator compared with stereology. *Cell Tissue Res*, 360, 29–42.
Hershkovitz, I., Weber, G.W., Quam, R., Duval, M., Grun, R., Kinsley, L., Ayalon, A., Bar-Matthews, M., Valladas, H., Mercier, N., Arsuaga, J.L., Martinon-Torres, M., Bermudez de Castro, J.M., Fornai, C., Martin-Frances, L., Sarig, R., May, H., Krenn, V.A., Slon, V., Rodriguez, L., Garcia, R., Lorenzo, C., Carretero, J.M., Frumkin, A., Shahack-Gross, R., Bar-Yosef Mayer, D.E., Cui, Y., Wu, X., Peled, N., Groman-Yaroslavski, I., Weissbrod, L., Yeshurun, R., Tsatskin, A., Zaidner, Y., & Weinstein-Evron, M. (2018) The earliest modern humans outside Africa. *Science*, 359, 456–9.

Hill, J., Inder, T., Neil, J., Dierker, D., Harwell, J., & Van Essen, D. (2010) Similar patterns of cortical expansion during human development and evolution. *Proc Natl Acad Sci USA*, 107, 13135–40.

Holloway, R.L. (2002) Brief communication: How much larger is the relative volume of area 10 of the prefrontal cortex in humans? *Am J Phys Anthropol*, 118, 399–401.

Holloway, R.L., Broadfield, D.C., & Yuan, M.S. (2004) *The Fossil Record: Brain Endocasts—The Paleoneurological Evidence*. Wiley-Liss, New York.

Hublin, J.J., Ben-Ncer, A., Bailey, S.E., Freidline, S.E., Neubauer, S., Skinner, M.M., Bergmann, I., Le Cabec, A., Benazzi, S., Harvati, K., & Gunz, P. (2017) New fossils from Jebel Irhoud, Morocco, and the pan-African origin of *Homo sapiens*. *Nature*, 546, 289–92.

Iovita, R. (2011) Shape variation in Aterian tanged tools and the origins of projectile technology: A morphometric perspective on stone tool function. *PLoS One*, 6, e29029.

Jacobs, B., Schall, M., Prather, M., Kapler, E., Driscoll, L., Baca, S., Jacobs, J., Ford, K., Wainwright, M., & Treml, M. (2001) Regional dendritic and spine variation in human cerebral cortex: A quantitative golgi study. *Cereb Cortex*, 11, 558–71.

Kappelman, J. (1996) The evolution of body mass and relative brain size in fossil hominids. *J Hum Evol*, 30, 242–76.

Kaya, F., Bibi, F., Zliobaite, I., Eronen, J.T., Hui, T., & Fortelius, M. (2018) The rise and fall of the Old World savannah fauna and the origins of the African savannah biome. *Nat Ecol Evol*, 2, 241–6.

Kelly, R.M. & Strick, P.L. (2003) Cerebellar loops with motor cortex and prefrontal cortex of a nonhuman primate. *J Neurosci*, 23, 8432–44.

Kolodny, O. & Edelman, S. (2018) The evolution of the capacity for language: The ecological context and adaptive value of a process of cognitive hijacking. *Philos Trans R Soc Lond B Biol Sci*, 373.

Krubitzer, L. & Dooley, J.C. (2013) Cortical plasticity within and across lifetimes: How can development inform us about phenotypic transformations? *Front Hum Neurosci*, 7, 620.

Langergraber, K.E., Prufer, K., Rowney, C., Boesch, C., Crockford, C., Fawcett, K., Inoue, E., Inoue-Muruyama, M., Mitani, J.C., Muller, M.N., Robbins, M.M., Schubert, G., Stoinski, T.S., Viola, B., Watts, D., Wittig, R.M., Wrangham, R.W., Zuberbuhler, K., Paabo, S., & Vigilant, L. (2012) Generation times in wild chimpanzees and gorillas suggest earlier divergence times in great ape and human evolution. *Proc Natl Acad Sci USA*, 109, 15716–21.

Leakey, M. (2009) *Olduvai Gorge*. Cambridge University Press, Cambridge, U.K.

Lewis, J.E. & Harmand, S. (2016) An earlier origin for stone tool making: Implications for cognitive evolution and the transition to Homo. *Philos Trans R Soc Lond B Biol Sci*, 371, doi: 10.1098/rstb.2015.0233.

Luncz, L.V. & Boesch, C. (2015) The extent of cultural variation between adjacent chimpanzee (Pan troglodytes verus) communities; a microecological approach. *Am J Phys Anthropol*, 156, 67–75.

Marin, J., Saladie, P., Rodriguez-Hidalgo, A., & Carbonell, E. (2017) Neanderthal hunting strategies inferred from mortality profiles within the Abric Romani sequence. *PLoS One*, 12, e0186970.

Mars, R.B., O'Muircheartaigh, J., Folloni, D., Li, L., Glasser, M.F., Jbabdi, S., & Bryant, K.L. (2019) Concurrent analysis of white matter bundles and grey matter networks in the chimpanzee. *Brain Struct Funct*, 224, 1021–33.

Mars, R.B., Passingham, R.E., & Jbabdi, S. (2018a) Connectivity fingerprints: From areal descriptions to abstract spaces. *Trends Cogn Sci.*, 22, 1026–37.

Mars, R.B., Sotiropoulos, S.N., Passingham, R.E., Sallet, J., Verhagen, L., Khrapitchev, A.A., Sibson, N., & Jbabdi, S. (2018b) Whole brain comparative anatomy using connectivity blueprints. *Elife*, 7.

McNamara, K.J. (2012) Heterochrony: The evolution of development. *Evo. Edu.Outreach*, 5, 203–18.
Murray, E.A., Wise, S.P., & Graham, K.S. (2017) *The Evolution of Memory Systems*. Oxford University Press, Oxford.
Niemitz, C. (2010) The evolution of the upright posture and gait—a review and a new synthesis. *Naturwissenschaften*, 97, 241–63.
O'Connell, J.F., Hawkes, K., Lupo, K.D., & Blurton Jones, N.G. (2002) Male strategies and Plio-Pleistocene archaeology. *J Hum Evol*, 43, 831–72.
Parker, C.H., Keefe, E.R., Herzog, N.M., O'Connell J, F., & Hawkes, K. (2016) The pyrophilic primate hypothesis. *Evol Anthropol*, 25, 54–63.
Passingham, R.E. (1973) Anatomical differences between the neocortex of man and other primates. *Brain Behav Evol*, 7, 337–59.
Passingham, R.E. (1975) Changes in the size and organization of the brain in man and his ancestors. *Brain Beh Evol*, 11, 73–90.
Passingham, R.E. (2008) *What is Special about the Human Brain*. Oxford University Press, Oxford.
Passingham, R.E. & Smaers, J.B. (2014) Is the prefrontal cortex especially enlarged in the human brain allometric relations and remapping factors. *Brain Behav Evol*, 84, 156–66.
Passingham, R.E., Smaers, J.B., & Sherwood, C.C. (2017) Evolutionary specializations of the human prefrontal cortex. In Preuss, T.M., Kaas, J. (eds) *Evolution of Nervous Systems*. Elsevier, New York.
Perry, V.H. & Cowey, A. (1985) The ganglion cell and cone distributions in the monkey's retina: Implications for central magnification factors. *Vision Res*, 25, 1795–810.
Petrides, M. & Pandya, D.N. (1999) Dorsolateral prefrontal cortex: Comparative cytoarchitectonic analysis in the human and the macaque brain and corticocortical connection patterns. *Eur J Neurosci*, 11, 1011–36.
Petrides, M. & Pandya, D.N. (2002) Comparative cytoarchitectonic analysis of the human and macaque ventrolateral prefrontal cortex and corticocortical connection pattern in the monkey. *Eur J Neurosci*, 16, 291–310.
Pontzer, H., Raichlen, D.A., & Sockol, M.D. (2009) The metabolic cost of walking in humans, chimpanzees, and early hominins. *J Hum Evol*, 56, 43–54.
Prang, T.C. (2019) The African ape-like foot of Ardipithecus ramidus and its implications for the origin of bipedalism. *Elife*, 8.
Progovac, L. (2016) A gradualist scenario for language evolution: Precise linguistic reconstruction of early human (and neandertal) grammars. *Front Psychol*, 7, 1714.
Prufer, K., Munch, K., Hellmann, I., Akagi, K., Miller, J.R., Walenz, B., Koren, S., Sutton, G., Kodira, C., Winer, R., Knight, J.R., Mullikin, J.C., Meader, S.J., Ponting, C.P., Lunter, G., Higashino, S., Hobolth, A., Dutheil, J., Karakoc, E., Alkan, C., Sajjadian, S., Catacchio, C.R., Ventura, M., Marques-Bonet, T., Eichler, E.E., Andre, C., Atencia, R., Mugisha, L., Junhold, J., Patterson, N., Siebauer, M., Good, J.M., Fischer, A., Ptak, S.E., Lachmann, M., Symer, D.E., Mailund, T., Schierup, M.H., Andres, A.M., Kelso, J., & Paabo, S. (2012) The bonobo genome compared with the chimpanzee and human genomes. *Nature*, 486, 527–31.
Raghanti, M.A., Stimpson, C.D., Marcinkiewicz, J.L., Erwin, J.M., Hof, P.R., & Sherwood, C.C. (2008a) Cholinergic innervation of the frontal cortex: Differences among humans, chimpanzees, and macaque monkeys. *J Comp Neurol*, 506, 409–24.
Raghanti, M.A., Stimpson, C.D., Marcinkiewicz, J.L., Erwin, J.M., Hof, P.R., & Sherwood, C.C. (2008b) Cortical dopaminergic innervation among humans, chimpanzees, and macaque monkeys: A comparative study. *Neurosci*, 155, 203–20.

Raghanti, M.A., Stimpson, C.D., Marcinkiewicz, J.L., Erwin, J.M., Hof, P.R., & Sherwood, C.C. (2008c) Differences in cortical serotonergic innervation among humans, chimpanzees, and macaque monkeys: A comparative study. *Cereb Cortex*, 18, 584–97.

Reardon, P.K., Seidlitz, J., Vandekar, S., Liu, S., Patel, R., Park, M.T.M., Alexander-Bloch, A., Clasen, L.S., Blumenthal, J.D., Lalonde, F.M., Giedd, J.N., Gur, R.C., Gur, R.E., Lerch, J.P., Chakravarty, M.M., Satterthwaite, T.D., Shinohara, R.T., & Raznahan, A. (2018) Normative brain size variation and brain shape diversity in humans. *Science*, 360, 1222–7.

Rice, S.H. (2002) The role of heterochrony in primate brain evolution. In Minugh-Purvis, N., MacNamara, K.J. (eds) *Human Evolution Through Developmental Change*. Johns Hopkins University Press, Baltimore.

Richter, D., Grun, R., Joannes-Boyau, R., Steele, T.E., Amani, F., Rue, M., Fernandes, P., Raynal, J.P., Geraads, D., Ben-Ncer, A., Hublin, J.J., & McPherron, S.P. (2017) The age of the hominin fossils from Jebel Irhoud, Morocco, and the origins of the Middle Stone Age. *Nature*, 546, 293–6.

Rilling, J.K., Glasser, M.F., Jbabdi, S., Andersson, J., & Preuss, T.M. (2012) Continuity, divergence, and the evolution of brain language pathways. *Front Evol Neurosci*, 3, 1–6.

Rilling, J.K. & Insel, T.R. (1998) Evolution of the cerebellum in primates: Differences in relative volume among monkeys, apes and humans. *Brain Behav Evol*, 52, 308–14.

Rilling, J.K. & Seligman, R.A. (2002) A quantitative morphometric comparative analysis of the primate temporal lobe. *J Hum Evol*, 42, 505–33.

Saleem, K.S., Miller, B., & Price, J.L. (2014) Subdivisions and connectional networks of the lateral prefrontal cortex in the macaque monkey. *J Comp Neurol*, 522, 1641–90.

Schoenemann, P.T., Sheehan, M.J., & Glotzer, L.D. (2005) Prefrontal white matter volume is disproportionately larger in humans than in other primates. *Nat Neurosci*, 8, 242–52.

Seltzer, B. & Pandya, D.N. (1989) Frontal lobe connections of the superior temporal sulcus in the rhesus monkey. *J Comp Neurol*, 281, 97–113.

Semaw, M.J., Rogers, M.J., Caceries, I., Stout, D., & Lewis, A.C. (2018) The Early Acheulian 1.6-1.2 Ma from Gona, Ethiopia. Issues related to the emergence of the Acheulean in Africa. In Gallotti, A., Mussi. (eds) *The Emergence of the Acheulean in East Africa and Beyond*. Springer, Berlin.

Semendeferi, K., Armstrong, E., Schleicher, A., Zilles, K., & Van Hoesen, G.W. (2001) Prefrontal cortex in humans and apes: A Comparative study of area 10. *Am J Phys Anthropol*, 114, 224–41.

Semendeferi, K. & Damasio, H. (2000) The brain and its main anatomical subdivisions in living hominoids using magnetic resonance imaging. *J Hum Evol*, 38, 317–32.

Semendeferi, K., Lu, A., Schenker, N., & Damasio, H. (2002) Humans and great apes share a large frontal cortex. *Nat Neurosci*, 5, 272–6.

Semendeferi, K., Teffer, K., Buxhoeveden, D.P., Park, M.S., Bludau, S., Amunts, K., Travis, K., & Buckwalter, J. (2011) Spatial organization of neurons in the frontal pole sets humans apart from great apes. *Cereb Cortex*, 21, 1485–97.

Sherwood, C.C., Raghanti, M.A., Stimpson, C.D., Spocter, M.A., Uddin, M., Boddy, A.M., Wildman, D.E., Bonar, C.J., Lewandowski, A.H., Phillips, K.A., Erwin, J.M., & Hof, P.R. (2010) Inhibitory interneurons of the human prefrontal cortex display conserved evolution of the phenotype and related genes. *Proc Biol Sci*, 277, 1011–20.

Smaers, J.B., Gomez-Robles, A., Parks, A.N., & Sherwood, C.C. (2017) Exceptional evolutionary expansion of prefrontal cortex in great apes and humans. *Curr Biol*, 27, 1549.

Smaers, J.B. & Soligo, C. (2013) Brain reorganization, not relative brain size, primarily characterizes anthropoid brain evolution. *Proc Biol Sci*, 280, 20130269.

Smaers, J.B., Steele, J., Case, C.R., Cowper, A., Amunts, K., & Zilles, K. (2011) Primate prefrontal cortex evolution: Human brains are the extreme of a lateralized ape trend. *Brain Behav Evol*, 77, 67–78.

Smith, A.R., Carmody, R.N., Dutton, R.J., & Wrangham, R.W. (2015) The significance of cooking for early hominin scavenging. *J Hum Evol*, 84, 62–70.

Sneve, M.H., Grydeland, H., Rosa, M.G.P., Paus, T., Chaplin, T., Walhovd, K., & Fjell, A.M. (2019) High-expanding regions in primate cortical brain evolution support supramodal cognitive flexibility. *Cereb Cortex*, 29, 3891–901.

Somel, M., Rohlfs, R., & Liu, X. (2014) Transcriptomic insights into human brain evolution: Acceleration, neutrality, heterochrony. *Curr Opin Genet Dev*, 29, 110–19.

Sousa, A.M.M., Zhu, Y., Raghanti, M.A., Kitchen, R.R., Onorati, M., Tebbenkamp, A.T.N., Stutz, B., Meyer, K.A., Li, M., Kawasawa, Y.I., Liu, F., Perez, R.G., Mele, M., Carvalho, T., Skarica, M., Gulden, F.O., Pletikos, M., Shibata, A., Stephenson, A.R., Edler, M.K., Ely, J.J., Elsworth, J.D., Horvath, T.L., Hof, P.R., Hyde, T.M., Kleinman, J.E., Weinberger, D.R., Reimers, M., Lifton, R.P., Mane, S.M., Noonan, J.P., State, M.W., Lein, E.S., Knowles, J.A., Marques-Bonet, T., Sherwood, C.C., Gerstein, M.B., & Sestan, N. (2017) Molecular and cellular reorganization of neural circuits in the human lineage. *Science*, 358, 1027–32.

Spocter, M.A., Hopkins, W.D., Barks, S.K., Bianchi, S., Hehmeyer, A.E., Anderson, S.M., Stimpson, C.D., Fobbs, A.J., Hof, P.R., & Sherwood, C.C. (2012) Neuropil distribution in the cerebral cortex differs between humans and chimpanzees. *J Comp Neurol*, 520, 2917–29.

Stewart, J.R. & Stringer, C.B. (2012) Human evolution out of Africa: The role of refugia and climate change. *Science*, 335, 1317–21.

Suwa, G., Kono, R.T., Simpson, S.W., Asfaw, B., Lovejoy, C.O., & White, T.D. (2009) Paleobiological implications of the Ardipithecus ramidus dentition. *Science*, 326, 94–9.

Tomasello, M. (2016) *A Natural History of Human Morality*. Harvard University Press, Cambrdige, MA.

Tomasello, M. (2019) *Becoming Human*. Harvard University. Press, Cambridge, MA.

Van Essen, D.C. & Dierker, D.L. (2007) Surface-based and probabilistic atlases of primate cerebral cortex. *Neuron*, 56, 209–25.

Wagner, G.A., Krbetschek, M., Degering, D., Bahain, J.J., Shao, Q., Falgueres, C., Voinchet, P., Dolo, J.M., Garcia, T., & Rightmire, G.P. (2010) Radiometric dating of the type-site for *Homo heidelbergensis* at Mauer, Germany. *Proc Natl Acad Sci USA*, 107, 19726–30.

Watkins, K. (2011) Developmental disorders of speech and language: From genes to brain structure and function. *Prog Brain Res*, 189, 225–38.

Wedage, O., Amano, N., Langley, M.C., Douka, K., Blinkhorn, J., Crowther, A., Deraniyagala, S., Kourampas, N., Simpson, I., Perera, N., Picin, A., Boivin, N., Petraglia, M., & Roberts, P. (2019) Specialized rainforest hunting by *Homo sapiens* ~45,000 years ago. *Nat Comm*, 10, 739.

Zhu, Z., Dennell, R., Huang, W., Wu, Y., Qiu, S., Yang, S., Rao, Z., Hou, Y., Xie, J., Han, J., & Ouyang, T. (2018) Hominin occupation of the Chinese Loess Plateau since about 2.1 million years ago. *Nature*, 559, 608–12.

10
Human Prefrontal Cortex
Reasoning, Imagination, and Planning

Introduction

Evolution is both conservative and innovative. There are two principles. The first is that it retains mechanisms that have worked in previous generations. For example, there are homologues in the human genome of the 'wee' genes that act as protein kinases as in fission yeast (Moris et al., 2016). Yet, the lines leading to yeast and humans diverged hundreds of millions of years ago.

The second principle is that evolution modifies what it already has. For example, the ossicles of our middle ear are thought to have evolved from the gill arches of ancestral fish. Thus, evolution co-opts structures and adapts them for new purposes.

Most of what we know about the organization of the human prefrontal (PF) cortex comes from the use of fMRI. The reason is that studies of patients with brain lesions have the disadvantage that the damage includes the white matter. As Chapter 1 points out, fMRI activations are in the grey matter, and they have the spatial resolution that is adequate for the purposes of functional anatomy.

However, it is not enough to simply report that a particular area or set of areas was activated. We need to understand *why* it was this area that was activated, and to do this we need to know why the human brain is organized in the way that it is. This means that we have to ask questions concerning evolution and selective advantage. The evolutionary reason why an area of the human brain is activated is that, either it performed the same function in the common ancestors of humans and other primates, or that that function has been co-opted and then modified in the human brain.

However, fossil evidence does not allow us to study the functional organization of the brains of ancestral primates. Furthermore, for reasons that Chapter 1 explains, we know very little of the PF cortex of the great apes, our nearest ancestors. So, to gain indirect clues as to the ancestral organization we have to resort to studying modern macaques. As Chapters 3 to 8 document, we have a detailed knowledge of the neuroanatomy and neurophysiology of their brains. One might think that given that the last common ancestors of humans and macaques lived between 25 and 30 million years ago (Stevens et al., 2013), the enterprise would be hopeless. In fact, as this chapter and the next one document, the fundamental organization of the neocortex is remarkably similar in humans and macaque monkeys.

Multiple-Demand System

It has been suggested that human intelligence depends on a network that includes the PF cortex and the parietal association cortex, and that this acts as a 'multiple-demand system' (Duncan, 2010b) or 'global workspace' (Dehaene et al., 1998; Mashour et al., 2020). There are two strands of evidence that are taken to support this view. The first comes from studying the network of connections, the second from studying the range of tasks that lead to activation in this system.

Connectivity

The interconnectivity of the PF cortex can be studied by measuring resting state or task covariance. Osada et al. (2015) measured psychophysical interactions so as to study the functional connectivity of the PF cortex in monkeys while they performed a task which tested their memory for the order in which objects had been presented. They estimated the 'betweenness centrality' of different areas in the neocortex; this measured the degree to which a node interacted with all the other nodes. The PF area 9/46 had the highest betweenness centrality.

In a study of the human brain, Cole et al. (2015) used a graph-theoretical measure to compare the connectivity of the PF cortex with that of other regions. They computed three measures. The first was the overall global connectivity. However, this did not distinguish between high connectivity within the region and high connectivity with other regions. They described the former as 'provincial hubs' and the latter as 'connector hubs'.

So, they computed two further measures. One was 'betweenness'; this is high for regions with high connectivity with other areas. The other was 'participation'; this is the proportion of the connections of a region that are with other areas.

These measures were computed for the dorsal and ventral PF cortex combined. In terms of global connectivity, the PF cortex ranked in the top 3.4% of areas of the neocortex. In terms of 'betweenness', it ranked in the top 10.6%, and in terms of 'participation' in the top 1.5%.

These results show that the human dorsal and ventral PF cortex receive information from a very wide variety of sources indeed. The dorsal PF cortex receives visuospatial information from the cortex of the intraparietal sulcus (IPS), and information about objects and words from the ventral PF cortex (Chapter 6). This explains how there can be activation in the human dorsal PF cortex on the n-back task, irrespective of whether the material consists of locations, objects, numbers, or letters.

Tasks

The richness of its connectivity means that the PF cortex can operate on many different inputs. So, it was a major advance when Duncan and Owen (2000) pointed out that the human PF cortex is activated during the performance of a wide variety of tasks.

Duncan and his colleagues (Fedorenko et al., 2013) followed up this finding by claiming that the same voxels were activated during six different tasks. Yet, the material was varied, consisting of numbers, spatial locations, words, and coloured adjectives. And the operations were also varied; two involved remembering a supra-span sequence, one adding a series of numbers, and the remaining three tasks required the subjects to inhibit a prepotent response.

The analysis contrasted hard and easy versions of each of these tasks. Fedorenko et al. (2013) presented an activation map showed common voxels not only in the PF cortex but also in the cortex of the IPS, dorsal and ventral frontal eye fields (FEFd, FEFv), and the cortex in the anterior cingulate sulcus (sACC).

Furthermore, the data are replicable. Hugdahl et al. (2015) gave subjects five tasks, the Wisconsin Card Sorting Task, a spatial working memory task, a go–no-go task, a left–right discrimination task, and the Stroop task. Of these, only the spatial working memory task and the Stroop task were included in the battery used by Fedorenko et al. (2013). Yet, the maps showing areas of common activation across tasks looked similar in the two studies.

However, there are two concerns. The first relates to the statistics. In these studies, the maps were shown at a very low significance level. For example, in the paper by Fedorenko et al. (2013) a level of $t < 1.5$ was used for the activation map presented. This is far from t value that is needed to reach the standard significance level of $p < 0.05$. In the analysis of voxels, the statistical analysis used regions of interest from a prior study, and this is unremarkable; but the significance level was only for $p < 0.001$ without correction for multiple comparisons.

The second concern is that it was assumed that the tasks were unrelated. But in fact, the operations fell into two classes. One group of tasks used by Fedorenko et al. (2013) and Hugdahl et al. (2015) involved inhibiting prepotent response. These were the interference tasks, the Stroop and the go–no-go task. Cieslik et al. (2015) have carried out a meta-analysis of 173 studies that included the Stroop task, spatial interference tasks, the stop-signal task, and the 'go–no-go task'. There were activations that were common to all tasks in the frontal eye field (FEF), the inferior frontal junction (IFJ), the anterior insula and the anterior cingulate cortex

(ACC). But, when standard significance levels were used, there were no activations in the dorsal PF cortex (areas 46 and 9/46).

Cieslik et al. (2015) suggested that the FEF was involved in 'supervisory attentional control'. It is necessary to attend and take care when a prepotent response must be inhibited. The anterior insula and ACC are interconnected (Hutchison et al., 2012), and are also activated when subjects must attend because they are aware of making errors by failing to inhibit a prepotent response (Orr & Hester, 2012). So, it is not surprising that there was activation in these areas in the study by Fedorenko et al. (2013) because they compared hard with easy versions of each task, and the hard versions differed in their attentional demands.

The other group of tasks used by Fedorenko et al. (2013) included spatial and verbal short-term memory tasks and calculation. At first sight these appear to differ in the operations that they require; but in fact, there is a commonality. The memory tasks used by Fedorenko et al. required the subjects to distinguish between the order in which the words or locations had been presented. And calculation involves the manipulation of numbers that are arranged in an ordinal sequence.

Calculation is similar to the task devised by Postle et al. (1999) in which subjects had to manipulate letters by rearranging them in alphabetic order. The ability to do this was disrupted by repetitive transcranial magnetic stimulation (rTMS) applied over the dorsal PF cortex (Postle et al., 2006). The cognitive operation has often been referred to as 'manipulation in memory' or 'transformation in memory' (Nee & D'Esposito, 2018). However, the critical factor is that it involves re-ordering sequences.

So as to characterize the multiple-demand system with high statistical power, Duncan collaborated with those running the Human Connectome Project. A paper by Assem et al. (2020) analysed the data from 449 subjects who had been scanned with fMRI for this project. Assem et al. selected three cognitive tasks, again contrasting difficult with easy versions of the tasks. These were the n-back memory task, an arithmetic task, and a task that involved relational reasoning. On the reasoning task, the subjects viewed two stimuli at the top and two at the bottom. The pairs could differ in shape or texture or both, and the task was to say whether the bottom pair of stimuli differed in the same way as the top pair.

The conjunction of the activations for these tasks were superimposed on the parcellation of the human brain by Glasser et al. (2016) from the Human Connectome Project. Figure 10.1 (next page) shows the results. The areas in yellow showed significant activation for three of the contrasts, and these were taken to be the core network. These were dorsal PF areas 46 (a9–46v), area 9/46 extending into the inferior frontal sulcus (p9–46v), the cortex in the IPS (IP1), the supplementary eye field (8Bm) and the anterior insula (AVI). There were only activations in the caudal PF area 8 (8C), and the IFJ (IFJp) for two of the tasks.

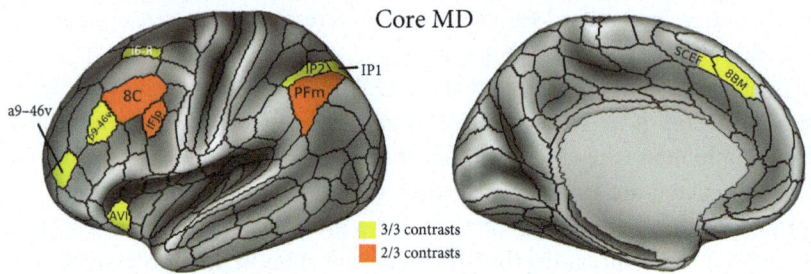

Figure 10.1 The multiple-demand system as visualized on the parcellation of the human brain by Glasser et al.
For areas see text.

Reproduced from Assem, M., Glasser, M.F., Van Essen, D.C., & Duncan, J. A domain-general cognitive core defined in multimodally parcellated human cortex. *Cereb Cort*, 30 (8), 4361–80, Figure 2b, Doi: 10.1093/cercor/bhaa023 Copyright © 2020, Oxford University Press.

The activations for relational reasoning were greater in the ventral than the dorsal PF cortex. The reason is that, whereas the n-back task and arithmetic involved sequences, the relational reasoning task was an associative task.

These results indicate that the fundamental organization of the PF cortex is the same in the human and macaque monkey brains. Previous chapters on macaque monkeys show that the caudal PF cortex sends a top-down signal that leads to attentional enhancement (Chapter 5). The dorsal PF cortex is involved in the generation of ordered sequences (Chapter 6). And the ventral PF cortex is involved in associating items (Chapter 7).

However, there are two difference in the human brain. One is that these PF areas are vastly expanded (Chapter 9). The remapping factors indicate that there is very much more tissue for analysing the inputs and generating the goals. The second difference is that there are new inputs to the system, such as letters, words, and numbers.

Summary

The dorsal and ventral PF cortex rank highest amongst the areas of the brain in terms of their connectivity with other areas. The result is that they can operate on a very wide variety of material. However, the fundamental operations of these areas are the same in the human as in the macaque monkey brain. The difference is that there are new inputs, such that the material can be verbal or numerical.

Fluid Intelligence

The study by Assem et al. (2020) included a test of non-verbal relational reasoning. Non-verbal IQ tests assess the ability to reason by deliberately presenting novel material. The aim is that the tests should be 'culture fair' because they do not depend on the acquisition of semantic knowledge.

The intelligence tests that have the highest loading on g or 'general intelligence' are those that use diagrams or symbols (Duncan, 2010a). One such battery is the Cattell tests of fluid intelligence (Cattell et al., 1973). Fluid intelligence is distinguished from crystallized intelligence; the latter is often assessed by the vocabulary subtest of the Wechsler Adult Intelligence Scale (Wechsler, 2008) because this tests knowledge of language.

In an early PET study by Duncan et al. (2000), the subjects attempted problems that involved spatial diagrams or letters. Duncan et al. reported that the PF cortex was activated irrespective of whether the material was spatial or not. However, though the dorsal PF cortex was activated on the spatial problems, the ventral PF cortex was activated on the problems which involved letters because these consist of shapes.

A later study by Liang et al. (2014) used fMRI instead. The subjects completed either a letter series or number series. Just as with the diagrams on the Cattell tests, these completion tests involve detecting the rule governing a sequence. For example, the letters in the series 'a, c, e' are separated by two in the alphabet. Irrespective of whether the material was verbal or numerical, Liang et al. found activation in common in the inferior frontal sulcus, the border between the PF areas 9/46v and 45A.

The Raven's Progressive Matrices (2003) differs in that it is a non-verbal IQ test that uses a series of figures. This also has a high loading on g (Rao & Baddeley, 2013). Figure 10.2 shows a typical problem used in a study by (Crone et al., 2009). The task for the subject was to complete the third line by picking between the answers 1, 2, and 3 to the right.

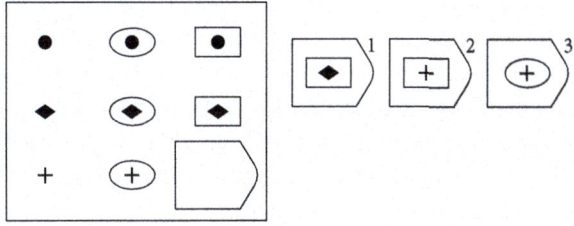

Figure 10.2 A typical problem on the Raven's Progressive Matrices

Reproduced from Passingham, RE, *Cognitive Neuroscience: A Very Short Introduction*, p. 56, Figure 14 Copyright © 2016, The Author.

Crone et al. (2009) scanned subjects with fMRI while they solved problems of this sort. The comparison conditions differed in that they could be solved simply by matching the figures. The contrast of the experimental with the control conditions revealed activations in the dorsal PF cortex (areas 9/46 and 46) extending into the polar PF cortex, as well as in the IPS.

Of course, tests of fluid intelligence have been developed to assess the abilities of human subjects, not macaque monkeys. But the human ability to solve these problems indicates an extension of the fundamental function of the dorsal PF cortex. This can be illustrated by considering items on the Raven's Matrices (Figure 10.2). To solve the problems the subject has to find the abstract rule that applies to the first two lines, and this rule concerns the order or sequence of the items. As Chapter 6 describes, there are cells the dorsal PF cortex of macaque monkeys that represent abstract sequences (Figure 6.11, page 222).

However, the fundamental function of the dorsal PF cortex has been extended in the human brain in two ways. The first is that the area is greatly expanded in relation to its inputs (Chapter 9). Cole et al. (2015) found that the degree of across-network connectivity of the PF cortex predicted differences in fluid intelligence between individuals. This correlation remained statistically significant when controlling for within-network connectivity.

There is a second way in which the fundamental function has been extended in the human brain. Whereas there are cells in the dorsal PF cortex of macaque monkeys that encode abstract structure, when subjects solve problems such as the one in Figure 10.3, they have to *compare* two sequences to see if they have the same abstract structure. It is this comparison that allows the subjects to apply the rule to the material in line three, and so complete the series.

The Relation to Short-Term Memory

Because the three lines are visible throughout problems such as the one illustrated in Figure 10.2, there should be a minimal load on short-term memory. However, tasks of this sort involve shifts of attention, what Duncan and his colleagues (2013; Kadohisa et al., 2020) have called 'attentional episodes'. It is, therefore, important to establish whether short-term memory is required across episodes.

One way of finding out is to use the dual task paradigm. Rao and Baddeley (2013) gave the Raven's Matrices to subjects and required them to continuously repeat three numbers while solving the problems. There was no interference, and Rao and Baddeley took this as evidence that solving the Raven's Matrices does not place a major demand on short-term memory.

However, in another dual task condition the subjects were required to count backwards in twos. This did lead to an increase in the response times when solving the Raven's Matrices. But the reason is that the ability to calculate and the ability to solve the matrices problems both depended on the dorsal PF cortex because they both involved ordered sequences. In other words, the interference was caused because the same operation was required in both cases.

The Effect of Lesions

Given that the dorsal PF cortex is activated when subjects attempt problems that assess fluid intelligence, we might expect that lesions that include the PF cortex to have a detrimental effect on performance. Therefore, Woolgar et al. (2010) studied patients with frontal, parietal, and temporal lobe lesions. The patients were tested on the Cattell Culture Fair Test, Scale 2, Form A.

Figure 10.3 shows the results. The plots under the brain diagrams show the relation between performance on the Cattell test and the volume of the lesion. The value 0 on the ordinate denotes no impairment and negative values denote an impairment.

There was a significant relation between an impairment and the volume of the lesions in the parietal and frontal cortex. The reason why this correlation was slightly lower for the frontal than the parietal lobe was because there was an impairment even with a small degree of damage.

By contrast, the volume of the lesions in the temporal lobe showed no relation with performance. The reason is that the test was specifically devised so as not to

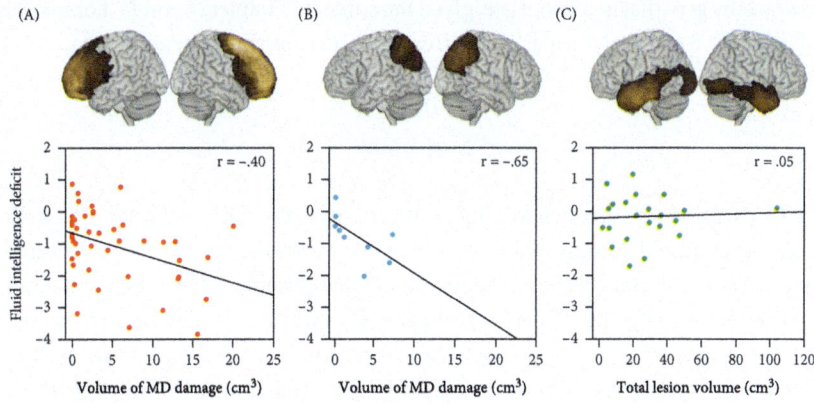

Figure 10.3 The relation between fluid intelligence and the volume of damage to the frontal, parietal, or temporal lobe
The value 0 on the ordinate means no impairment in intelligence, whereas negative values mean an impairment.

Reproduced from Woolgar, A., Parr, A., Cusack, R., Thompson, R., Nimmo-Smith, I., Torralva, T., Roca, M., Antoun, N., Manes, F., & Duncan, J., Fluid intelligence loss linked to restricted regions of damage within frontal and parietal cortex. *Proc Nat Academy of Sciences of the United States of America*, 107 (3), 14899–902, Figure 3, https://doi.org/10.1073/pnas.1007928107 © 2008 The Authors.

require semantic or associative knowledge. Chapter 11 reviews the evidence that semantic knowledge depends on the temporal lobe and its extension into the ventral PF cortex. Woolgar et al. found no relation between the degree of impairment and amount of damage to the ventral PF cortex.

The impairment after lesions of the frontal and parietal cortex was also not related to the ability of the patients to use language. Woolgar et al. (2018) scanned patients frontal and parietal lesions while they read a sentence. Woolgar et al. then weighted the lesion maps by the activation maps for reading. This enabled them to see if the resulting maps predicted the decline in fluid intelligence from before to after the lesion; and they failed to do so. Woolgar et al. argued that this demonstrated that solving problems which assess fluid intelligence need not involve the use of language.

This claim is supported by a study by Levine et al. (1982) who described a patient who was totally aphasic and lacked inner speech. Yet, he was able to solve problems on the Raven's Matrices. It is possible to solve the problem using visual imagery or an image code, since the material consists of diagrams.

Though both parietal and PF lesions lead to an impairment on tests of fluid intelligence, it is likely that they do so for different reasons. It is the posterior parietal cortex that represents metrics (Genovesio et al., 2014), and these can include diagrams (Na et al., 2000; Jordan et al., 2001), numbers (Dehaene et al., 2004), and letters (Fias et al., 2007). Therefore, damage to the posterior parietal cortex will impair the representation of the material itself.

But it is the PF cortex that manipulates this information (Postle et al., 2006) and that represents abstract rules (Chapter 8). It is the PF cortex that is in a position to generate the goal that is appropriate, given the context (Chapters 6 and 7). For reasons that Chapter 8 explains, the parietal cortex is not in a position to generate goals.

Summary

The PF cortex is activated when subjects try to solve problems that test fluid intelligence. As expected, lesions of dorsal also impair performance these tests. Whereas the posterior parietal cortex represents the metrics used, such as diagrams, numbers, and letters, it is the PF cortex that generates the final goal item in accordance with the abstract rule. Furthermore, there is a significant correlation between individual differences in the connectedness of the PF cortex and fluid intelligence.

Solving Problems with Imagery

When human subjects attempt problems on tests of fluid intelligence, they view the displays and consider the various alternative solutions. The alternatives can be represented using either visual or motor imagery. Trying out the various solutions allows subjects to derive the abstract rule that applies, and so generate the last item in the series or goal item.

It is not clear to what extent macaque monkeys and chimpanzees have access to visual imagery. Inoue and Matsuzawa (2007) tested an infant chimpanzee, Ayumu, on a task in which the subject had to remember the locations on the screen of the numbers 1–9. Even though the locations were varied from trial to trial, and the display was presented as briefly as 210ms, Ayumu achieve a performance of around 80% correct. This suggests that Ayumu had access to eidetic imagery.

However, even if this suggestion is accepted, it does not prove that chimpanzees retain this capacity into adulthood, or indeed that other chimpanzees and other species have the same ability. Nor does it prove that they can use imagery for planning the solution of problems.

Tulving (2001) suggested a way of finding out whether subjects can plan, involving the 'spoon test'. The idea is that when children think of a party that they are going to, they bring a spoon in the knowledge that they might need this to eat ice cream. The ability to do this could imply visual imagery of the party or motor imagery of using the spoon.

The proposal led Mulcahy and Call (2006) to try the spoon test on bonobos and orangs (*Pongo pygmaeus*). The animals learned to open an apparatus using a particular tool. They then waited in an area from which they could see the testing apparatus. When they returned to the room, they were more likely to bring the tool that was suitable for that specific problem rather than a tool that was unsuitable.

However, this test did not require the apes to imagine a future situation, because they could see the room throughout. It is true that apes have been shown to transport stone tools (Proffitt et al., 2018), but that does not prove that they use motor imagery to represent using them in the future. Furthermore, the distance the stones are transported is short. It does not compare with the 10 Km estimated for the transport of stone tools by the hominins at Olduvai ~2.6 Ma (Wynn et al., 2011).

Visual Imagery

It is possible to prove that modern humans can use a visual code for imagination by using fMRI (Ishai, 2010). In an experiment by Ishai et al. (2000), when the subjects imagined faces the activation was in the fusiform face area (FFA), whereas when they imagined houses the activation was more medial. It has even been shown that during imagination the activation extends to the primary visual cortex (Pearson & Kosslyn, 2015).

When subjects imagine objects as opposed to perceiving them, there is also activation in the ventral PF cortex (Ishai et al., 2002). Mechelli et al. (2004) used dynamic causal modelling to show that during visual imagination, both the PF and posterior parietal cortex exert a top-down influence. However, whereas the influence from the parietal cortex was non-specific, the influence from the PF cortex was specific. It led to the enhancement of activation in the FFA for faces but more medially for houses.

Human subjects can also imagine a future scene, where for example they will be using the spoon. In an fMRI study, Zeidman et al. (2015) presented words such as 'jungle' and required the subjects to imagine immersing themselves in the scenes. There were activations in the parahippocampal cortex and the anterior medial part of the hippocampus, both when subjects actually perceived scenes and when they imagined them.

Motor Imagery

Carrying a spoon to a party can also involve imagining what to do with it, in other words a motor code. In an fMRI study, Ehrsson et al. (2003) instructed their subjects to imagine moving their hand, toe, or tongue. The location of the activations in the anterior part of the motor cortex (area 4a) differed depending whether the subjects were imagining moving their tongue, hand, or foot. This somatotopic mapping for imagined hand and foot actions has been replicated (Lorey et al., 2013).

Imagining actions is like imagining objects in that there is also activation in the PF cortex. However, the difference is that for imagining actions it is in the dorsal as opposed to the ventral PF cortex (Gerardin et al., 2000; Ehrsson et al., 2003). It is the dorsal PF cortex that has direct outputs to both the dorsal and ventral premotor cortex (Borra et al., 2019).

Imagining the Outcome

There is also an outcome of taking a spoon, namely eating the ice cream. Gerlach et al. (2014) used fMRI to scan subjects while they imagined the series of steps that they would need to achieve a goal that they desired. They reported activations in the amygdala as well as the central and medial sectors of the orbital PF cortex that were associated with achievement of the goal.

Covert Trial and Error

The previous sections show that humans can imagine or simulate objects, actions, and the outcome, as well as scenes. This result is that, when faced with a problem, they can imagine alternative courses of action and the outcomes before committing themselves to action. This can be described as 'covert trial and error', in contrast to actual or overt trial and error.

Boorman et al. (2011) have shown that human subjects can represent choices that they do not take. In an fMRI study, the subjects were given the opportunity to decide between three alternatives. These were pictures of a face, a house, and a body, each of which was associated with an independent reward probability. These reward probabilities varied across trials according to a fixed volatility. Boorman et al. used a Bayesian model to infer the probabilities of the outcomes.

There were activations in the ventrolateral polar PF cortex, in the area that Neubert et al. (2014) suggested might be unique to the human brain (Chapter 7). These activations correlated positively with the probability of the reward for the best of the three options that was not chosen. In other words, it represented a counterfactual choice. This was one that could have been chosen but was not.

Laboratory Tasks to Assess Planning

The ability to engage in covert trial and error can also be assessed on by presenting tasks such as the 'Tower of Hanoi'. In the standard version, the subjects are allowed to move coloured disks from one peg to another so as to achieve a particular goal arrangement. The aim is to do so in the minimum number of moves. In order to succeed, it pays to plan potential moves, considering the consequence of moving a particular disk before actually doing so.

Shallice (1982) devised a simplified version of this task for testing patients. It consists of three pegs and three coloured balls which must be moved from peg to peg. A computerized version of this task has been named 'The Tower of Cambridge': the display is shown in Figure 10.4 (next page). Instead of slotting onto pegs, the balls are shown in pockets as in a snooker table.

The arrangement in the top row shows the starting position and the arrangement in the bottom row the goal or target arrangement. The task is to decide how many moves are needed to rearrange the balls into the target position, moving one ball at a time into any of the pockets. The problem on the left in Figure 10.4 can be solved in a minimum of two moves, and the one on the right in a minimum of five moves.

Rowe et al. (2001) used fMRI to scan subjects while they planned the moves in their head. The analysis looked for activations that were related to planning rather than actually moving the balls. There were activations in the dorsal PF cortex (area 9/46) as well as in the IPS when subjects planned in this way.

However, Rowe et al. also included a control condition in which the start position was the same as the target arrangement. In this case the task was to plan any series of four moves, irrespective of the target arrangement. When planning so as to achieve a goal arrangement was compared with this control condition, there were activations in the IPS and FEF but not in the dorsal PF cortex. The implication is that the activations in the IPS and FEF probably reflected comparing the

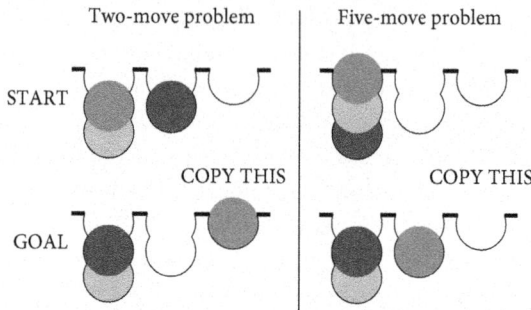

Figure 10.4 Computerized version of the 'Tower of London'
The figure shows a problem that can be solved in a minimum of two moves and a problem that can be solved in a minimum of five moves.
Reproduced from Passingham, R.E. *Cog Neurosci: A V Short Intro*, p. 75, Figure 18 Copyright © 2016, The Author.

moves with the goal arrangement. Subjects make saccadic eye movements when they make this comparison (Hodgson et al., 2000).

However, Rowe et al. (2001) were able to show that it was the dorsal PF cortex that was activated when the subjects generate a sequence of moves. There was also a significant correlation between the time it took the subjects to plan their moves and the degree of activation in the polar PF cortex. This was the only area that showed such a correlation.

Unterrainer et al. (2004) compared the percentage of six-move problems that different subjects were able to solve. The better the performance, the greater the activation in the dorsal PF cortex (areas 9/46 and 46) in the right, but not left hemisphere. The activation in the right hemisphere probably reflected right hemisphere dominance for visuospatial attention (Shulman et al., 2010).

However, the fact that there was a correlation between performance and activation does not prove a causal connection. So Kaller et al. (2013) applied theta-burst TMS to either the right or the left PF cortex, using the activation peak from the study by Unterrainer et al. (2004) to guide positioning of the TMS coil. Appling rTMS over the right, but not left, PF cortex led to a significant increase the time that the subject took to plan their solution.

Goel and Grafman (1995) studied patients with frontal lesions. They found that they tended to make errors when counter-intuitive movements were required. These are moves that take a ball away further away from its final position than appears necessary; in this case the more obvious move causes problems later.

The Tower of London is, of course, a relatively simple task compared with games such as chess and Go. These require the player to plan very many moves ahead and to react to the moves that the other player makes. Unterrainer et al. (2006) looked to see whether chess players would be better at solving problems on the Tower of

London than subjects who did not play chess. The subjects were tested on problems that required four, five, six, or seven moves. There was a very significant effect such that the more difficult the problems, the better the chess players did, compared with subjects who did not play chess.

In other words, there was transfer from one type of problem to another. This could occur because both chess and the Tower of London require the subject to generate moves in their head and evaluate the effect that any particular move will have. Chapter 8 reviews the evidence from macaque monkeys that the PF cortex supports the ability to show transfer across problems.

Planning in Everyday Life

The extended period of human development means that people have abundant opportunity to learn how to plan, and to do so whatever the type of problem they face. Shallice and Burgess (1991) tested three patients with PF lesions on the sorts of tasks that people are required to perform in everyday life. They sent them on a series of errands in a shopping centre.

The errands included, for example, buying a brown loaf of bread, finding out the price of tomatoes, and being in a certain place 15 minutes after starting. The patients were also instructed that they should not go into the same shop twice, and this meant that they needed to be careful in planning the sequence of errands. The subjects carried a card around with them that had the instructions on it, and this meant they did not have to remember what was required.

All three patients were significantly less efficient in carrying out these tasks than control subjects. Furthermore, even though they had the card, they also broke the rules. This has been described by Duncan and his colleagues (Roca et al., 2018) as 'goal neglect', that is a failure to act correctly in spite of being aware of the instructions.

Unfortunately, the lesion of only one of these patients has been described, and this was a large bilateral lesion that involved the orbital, medial, and polar PF cortex (Shallice & Cooper R, 2011). So Tranel et al. (2007) repeated the study with a series of nine patients, together with a description of their lesions. The patients had bilateral damage to the medial PF cortex, extending from the ventral prelimbic area to the polar PF cortex, as well as the orbital PF cortex. Figure 11.12 illustrates similar lesions from a later paper with a larger sample of patients.

Tranel et al. devised their own version of the multiple errands test involving a series of tasks to be performed in a shopping centre. Again, the patients with medial and orbital lesions made more errors than control patients or healthy subjects.

However, it was not clear whether the patients failed to produce adequate plans or failed to execute properly the plans that they had made. It is possible to find out by simply asking them to describe the series of steps that are needed to shop for groceries or carry out other tasks. Godbout and Doyon (1995) asked eight patients with frontal lobe lesions to generate scripts, meaning a verbal account of how they would go about planning the tasks. The task was also given to control subjects. The patients with PF lesions generated fewer steps and made more sequence errors.

Goel and Grafman (2000) also studied an architect with a meningioma that affected the right medial PF cortex. After the development of the tumour, he was no longer able to hold down a job. There could, of course, been many reasons for this that had nothing to do with his ability to plan: he could, for example, have simply lacked the motivation. To pinpoint the reason, Goel and Grafman asked him to redesign the laboratory. His plans for doing so turned out to be sketchy and lacking in detail, consisting of three unrelated ideas. He also failed to develop the proposals that he had, and was completely unable to redesign the laboratory as required.

However, these lesions studies have the problem that the lesions were large and variable, and the white matter was damaged. To find out which areas are involved in planning in everyday life, Spreng et al. (2010) used fMRI to scan subjects while they generated scripts, for example planning how they could achieve academic success. The various steps are working hard, going to lectures, passing exams, and so on. The subjects were also tested a simplified version of the Tower of Hanoi.

Figure 10.5 (next page) gives the results. This shows in red the areas that were activated when the subjects planned solutions on the Tower of Hanoi, and these included the dorsal PF cortex (area 46). The areas in blue were activated when subjects generated plans as to how to achieve their own goals, but de-activated when they tackled problems on the Tower of Hanoi. The PF activations for this condition included the medial PF cortex extending from the prelimbic area 32 forwards to areas 9 and 10, as well as the retrosplenial cortex. Kolling et al. (2018) also reported activation in the prelimbic area 32 and PF area 9 when human subjects were tested on a task designed to mimic planning on the job market.

One difference between the two types of planning task is that the Tower of Hanoi involves planning how to solve a given problem whereas the scripts involved plans as to how to achieve the subject's own goals. The multiple-errands tasks is similar in that the subjects has to plan how to shop, so achieving the goals of everyday life.

However, there is another difference between the two planning tasks. Solving problems on the Tower of Hanoi involves maintaining sequences in short-term memory. By contrast, planning how to achieve academic success involves maintaining sequences in long-term memory, planning into the distant future. The multiple errands task is similar in that it involves long term memory.

Whatever the critical difference, the study by Spreng et al (2010) suggests that the impairment of patients on the multiple-errands tests (Tranel et al., 2007) was due to the damage to the medial and polar PF cortex.

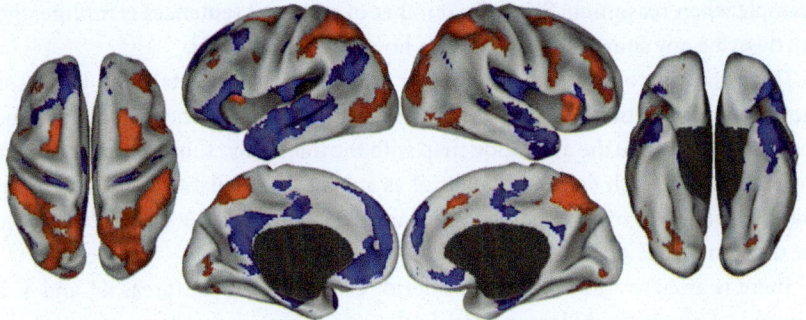

Figure 10.5 Areas that are activated when subjects plan
The areas shown in red were activated when subjects planned how to solve problems on the Tower of Hanoi. The areas shown in blue were activated when the subjects planned the series of steps needed to achieve their goals in everyday life.

Reproduced from Spreng, R.N., Stevens, W.D., Chamberlain, J.P., Gilmore, A.W., & Schacter, D.L. Default network activity, coupled with the frontoparietal control network, supports goal-directed cognition, *Neuroimage*, 53 (1), 303–17, Figure 2a, Doi: 10.1016/j.neuroimage.2010.06.016 Copyright © 2010 Elsevier Inc. All rights reserved.

Summary

Human subjects can solve problems in their heads. One way that they can do this is by imaging the alternatives using both visual and motor imagery. Damage to the medial and polar PF cortex severely impairs the ability of patients to plan a series of tasks in everyday life.

Solving Problems with Symbolic Thought

There is a disadvantage of using an image code. This is that visual imagery is faint and lacking in detail. Though activations have been demonstrated in the ventral visual stream when subjects imagine objects, they were less extensive than the activations when the subjects actually viewed the objects (Ishai et al., 2000; Pearson & Kosslyn, 2015).

However, in the human brain there has been an expansion of the inferior caudal PF cortex (areas 44 and 45B) and the ventral PF cortex (Avants et al., 2006). This has made possible the development of grammatical language, and the result is that human subjects have access to a propositional code as well as an image code (Pearson & Kosslyn, 2015). A propositional code involves the gist of what can be expressed in a verbal statement (Pearson et al., 2015). It has the advantage over a visual code in that it is easy to specify multiple steps towards a goal in detail, for example when reasoning. Since the number of potential sentences is infinite, they can describe any situation or possibility, however hypothetical.

There is a way of finding out whether when human subjects think or reason, they use a propositional code. This is to scan subjects while they engage in inner speech and compare the activation map with the map when subjects engage in reasoning. Inner speech can be thought of as subvocal articulation, and this means that it is related to overt articulation (Martinez-Manrique & Vicente, 2015), with the difference that it does not involve full sentences.

There is an activation in the left inferior caudal PF cortex (areas 44 and 45B) (Simons et al., 2010; Alderson-Day et al., 2016) when subjects engage in inner speech. The distinction between these two areas is that area 45 is engaged during semantic processing whereas area 44 is specialized for syntactic processing (Goucha & Friederici, 2015). Thus, activation in these areas can be used as a marker for inner speech when subjects engage in reasoning.

Goel et al. (2000) used tests of deductive reasoning. All involved syllogisms such as 'all poodles are pets; all pets have names; so all poodles have names'. The three statements were presented on the screen, one by one. The task was to judge whether the conclusion followed from the premises, and the subjects reported their judgement by pressing one of two buttons on a keypad.

To control for activations that were associated with reading, the experimental design included a condition in which the last sentence was irrelevant for the argument. In the example given previously, a typical sentence was 'no napkins are white'. In this condition it is also necessary to understand the words, but the syllogism is clearly invalid; we might call it a 'pseudo-syllogism'. Thus, by contrasting syllogisms with pseudo-syllogisms, it was possible to ensure that the activations for reasoning did not simply reflect reading and understanding the words.

Syllogisms hold because they obey the rules of logic. So in another condition, the subjects evaluated syllogisms in the form 'All P are B, all B are C, so all P are C'. And again there was a control condition with a pseudo-syllogism.

The data were analysed for the period after the last statement had been presented, that is while the subjects were evaluating whether the syllogism was valid. Figure 10.6 (next page) shows the whole brain map for the activations when the data for syllogisms with content and without content were combined. There were activations in the left inferior caudal PF cortex (areas 44 and 45B) as well as in the superior temporal sulcus (STS) which is interconnected via the arcuate fasciculus (Eichert et al., 2019). There was also an activation in the cerebellum, and this is additional evidence that reasoning involved covert articulation (Chapter 8). Thus, the results indicate that inner speech is involved when the subjects evaluate verbal syllogisms or formal logic.

Figure 10.6B presents the activations that were greater when the content was familiar. There were activations in the PF area 45 as well as the STS and middle

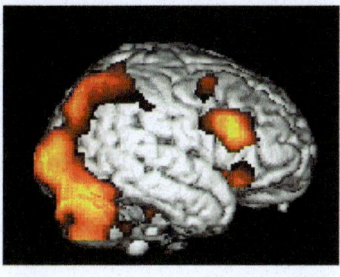 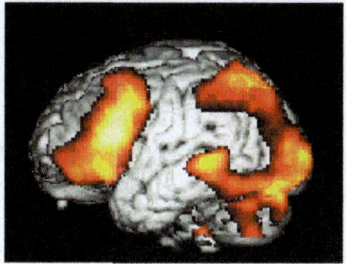

Deductive reasoning (words and symbolic logic)

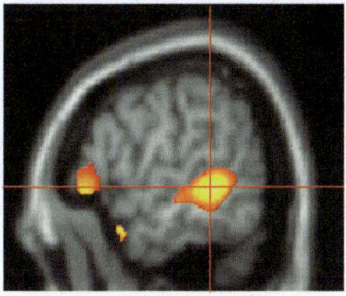 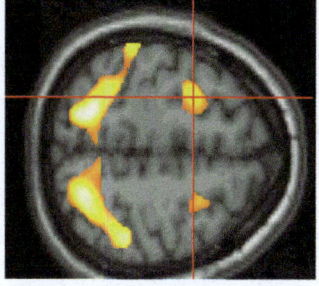

Familiar material   Symbolic logic

Figure 10.6 Activations in the left inferior caudal PF cortex (areas 44 and 45B) and the STS during deductive reasoning
(A) Activations irrespective of whether the syllogisms were presented in words or formal logic. (B) Activations for syllogisms that were expressed in words, with content that was familiar. (C) Activations for syllogisms that were presented in formal logic.
Reproduced from Goel, V. Anatomy of deductive reasoning. *Trends Cog Sci*, 11 (10), 435–41, Figure 1, Doi: 10.1016/j.tics.2007.09.003. Crown copyright © 2007 Published by Elsevier Ltd. All rights reserved.

temporal gyrus. Chapter 11 reviews evidence that these are part of the system that supports semantic knowledge.

Figure 10.6C presents the activations that were greater when the content was unfamiliar, as in formal logic. In this condition there was also an activation in the cortex of the IPS. Activations in the parietal association cortex, but not the inferior caudal PF cortex (areas 44 and 45B), have also been reported when subject evaluated syllogisms that involved spatial relations, as in 'K is in front of L, L is in front of J, so is K is in front of J?' (Goel & Dolan, 2001). A similar pattern has been reported when subjects evaluated simple 'if, then' conditionals with shapes as the material, as in 'If Black square, Yellow Circle'; 'Black square, so Yellow circle?' (Noveck et al., 2004). Both of these studies involve spatial relations.

Johnson-Laird (2001) proposed that reasoning involves mental models with a spatial code, and a spatial code is adequate for problems such as these. But a propositional code has the advantage that when the premises are complex, as in logical syllogisms, working out what follows requires an ordered flow of thought, even if this is not expressed in full syntactic detail. And Figure 10.6A shows that reasoning about logical syllogisms involves a propositional as well as a spatial code.

In these experiments the subjects were set problems to solve. But Spreng et al. (2010) also reported activation in the left inferior caudal PF cortex (area 45B) when subjects planned a series of steps towards a personal goal such as academic success. This indicates that planning for goals of this sort involve not only visual imagery but also symbolic thought.

Summary

The left inferior caudal PF cortex has expanded in the human brain, supporting both syntax and semantics. The consequence is that humans can solve complex problems by representing the alternative solutions using inner speech. This conveys the gist of what can be expressed in a verbal statement. The left inferior caudal PF cortex (areas 44 and 45B) is activated when subjects evaluate logical syllogisms and plan for their future.

Self-Generated Thoughts

Humans are aware of having a spontaneous inner life. It is not clear how much of a typical day is taken up by spontaneous thoughts; the reported proportions vary depending on how mind wandering is defined (Seli et al., 2018). While human subjects lie at rest in the scanner, there are activations in the medial PF cortex as well as other parts of the medial network (Mason et al., 2007). And when they are asked at random intervals about their thoughts, they frequently report spontaneous thoughts about the past, present, or future, as well as brief periods when there were no thoughts (Kawagoe et al., 2019).

Retrieving Events and Imagining the Future

As Chapter 3 describes, Rilling et al. (2007) reported that the same areas of the medial network including the medial PF cortex are active when chimpanzees rest, either lying down or sitting quietly. However, there was a striking difference between the activations on the lateral surface for chimpanzees and human subjects. In the human brain there were extensive activations in the resting scans in the left hemisphere; these included the inferior caudal PF cortex (areas 44 and 45B), the ventral PF cortex (area 45A) and the middle and inferior temporal cortex. These areas are interconnected in the human brain via the arcuate fasciculus (Eichert et al., 2019). These activations indicate that humans use a propositional code when thinking.

By contrast, the activations on the medial surface may indicate the use of imagery. The same medial network has been shown to be activated when human subjects are asked to 're-experience' memories from the past or to imagine future scenarios (Addis et al., 2007; Hassabis et al., 2007). In a meta-analysis, Chase et al. (2020) found that the area that was most associated with episodic memory was the subgenual cortex, area 25.

Because the same core medial network is involved for episodic memory and imagination, Schacter et al. (2008) suggested the term 'episodic simulation'. They have also proposed that imaging future scenarios involves using details from past memories and re-arranging them, thus leading to the term 'constructive episodic simulation' (Thakral et al., 2020).

It is possible that the activation of the medial PF cortex when chimpanzees are at rest reflects spontaneous cognition. However, the best guess is that, if so, the activations reflect experiencing the current situation in which the animals find themselves. At most the mental horizons of chimpanzees extend to the previous day (Schwartz et al., 2005) or next day (Janmaat et al., 2014).

We cannot rule out the possibility that the activations also reflect the replay of memories. But even if this is the case, it is unlikely that they resemble human memories in either the detail or temporal distance. The reasons for this judgement are given in the following paragraphs.

There is evidence on both detail and temporal distance from fMRI studies. In an experiment by Addis et al. (2012), the subjects were scanned while they retrieved memories such as losing a dog in a pond. After retrieving each memory, the subjects filled out ratings on a scale of 1–5 for the level of detail, the degree of emotional involvement, and the level of personal significance. The mean scores for detail were 3.0 for detail, 2.1 for emotional significance, and 2.3 for personal significance 2.3. The subjects also described the scene as if they were viewing it: 95% of the memories were rated as being viewed from a field perspective. It is clear that human subjects can remember past events from their life in considerable detail.

They can also remember events from the distant past. Bonnici et al. (2012a; Bonnici & Maguire, 2018) required subjects to remember memories that were either from 2 or 10 years ago. They then used multivoxel pattern analysis (MVPA) to show that it is possible to distinguish between memories from the different periods; this was possible both in the hippocampus and in medial PF cortex (ventral prelimbic 32). Figure 10.7 (next page) shows the accuracy of the classification of memories that were either 2 or 10 years old across a series of brain areas. As the figure illustrates, there was a significant difference in that the accuracy of decoding;

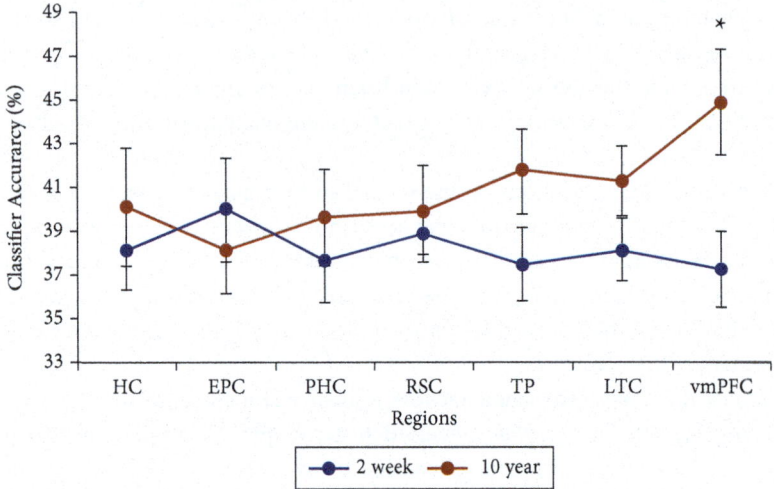

Figure 10.7 The accuracy of decoding using a multivoxel pattern analysis for memories that were of events that occurred either 2 years ago or 10 years ago HC, hippocampus; EPC, entorhinal and perirhinal cortex; RSC, retrosplenial cortex; TP, temporal pole; LTC, lateral temporal cortex; vmPFC, ventromedial PF cortex (ventral prelimbic area 32). The star indicates a statistically significant difference.

Reproduced from Bonnici, H.M. and Maguire, E.A. Two years later—Revisiting autobiographical memory representations in vmPFC and hippocampus. *Neuropsychologia*, 110, 159–69, Figure 2a, Doi: 10.1016/j.neuropsychologia.2017.05.014 Copyright © The Authors, 2018. Licensed under CC BY 4.0.

in the medial PF cortex it was significantly greater for the most distant events, that is the ones that had occurred 10 years ago. Bonnici et al. took this to be suggestive of a process of consolidation.

Given these results, one would expect lesions that include the medial PF cortex to impair the ability of patients to re-experience scenes from the past or imagine ones in the future. And Philippi et al. (2014) have reported that bilateral lesions of the medial PF cortex, including the ventral prelimbic area 32 and medial polar PF cortex, impair the ability of patients to retrieve autobiographical memories.

Bertossi et al. (2016) also carried out a detailed analysis of the memories of patients with lesions of these PF areas and healthy controls. The memories of the patients were very much less detailed, and this was especially true for events that were distant in the past. The same patients were also required to rate the degree to which their mind wandered, either while they carried out tasks or while they simply passively viewed a series of numbers (Bertossi & Ciaramelli, 2016). The patients reported many fewer spontaneous thoughts. Furthermore, compared with those of healthy controls, the thoughts were much more likely to be restricted to the present than the past or future.

Rather than depending on ratings in the laboratory, McCormick et al. (2018) accompanied patients with bilateral hippocampal damage and healthy controls for two days. At random intervals the subjects were asked to report what they were thinking of. Whereas the control subjects reported spontaneous thoughts about the past, present, and future, the patients with hippocampal lesions only reported thoughts about the present or that day. Furthermore, their thoughts differed in that they concerned facts rather than scenes or events; in other words, their semantic knowledge was intact whereas their episodic memory was severely impaired.

Taken together these results suggest that the activation of the medial network in chimpanzees at rest is unlikely to reflect detailed memories or memories at a temporal distance. As in patients with lesions in this network, it is much more likely that the spontaneous cognitions are related to the present.

The Interplay of Brain Expansion and Culture

In the experiment by Bonnici and Maguire (2018), the subjects were first asked before scanning for memories that were recent or remote. And, of course, this was only possible because there are words in the language to draw the distinctions between days, weeks, and years. This suggests that the human ability to remember events from the distant past depends in part on the fact that children learn how to read the time and tell the days of the week, as well as being read stories about the past and future.

This is not to deny that the ability to think about the remote past also depends on the degree to which the PF cortex has expanded in the human brain. An earlier section introduced the concept of remapping factors which relate the size of the PF cortex to its inputs or outputs. Table 10.1 present remapping factors for the volume of the PF cortex in relation to the volume of the hippocampus. The data for the PF cortex are taken from the conservative estimate in Donahue et al. (2018), and the data for the hippocampus from Stephan et al. (1981). The figures refer to how many times greater the PF cortex is than the hippocampus.

The table shows that, relative to the hippocampus, there is much more PF tissue in the human than chimpanzee brain. The importance of this is that it is the PF cortex that is responsible for retrieving events and constructing scenarios. The evidence comes from two studies.

The first is an MEG study by Barry et al. (2019). As an earlier section mentions, the subjects were presented with a word such as 'jungle', and their task was

Table 10.1 Remapping factors

Prefrontal/hippocampus	Human	Chimpanzee	Macaque monkey
	101.9	61.0	33.8

to immerse themselves in the scene in their imagination. Imagining scenes was compared with imagining objects. In both cases there were changes in theta oscillations in the hippocampus. However, the coherence between activity in the medial PF cortex and the hippocampus was also significantly greater for imagining scenes compared with objects.

Barry et al. used dynamic causal modelling to analyse the MEG data. The advantage of MEG is that its temporal resolution meant that it was possible to distinguish the relative timing of activity in the medial PF cortex and hippocampus. The analysis indicated that it was the medial PF cortex that drove the activity in the anterior hippocampus, thus supporting the claim that it is the medial PF cortex that is responsible for retrieving the details used in imagining scenes.

The second study applied similar methods to data on the retrieval of personal memories from the past. McCormick et al (2020). Just as for imagining the future, there were theta oscillations in the medial PF cortex and hippocampus Figure 10.8. And dynamic causal modelling indicated that it was the medial PF cortex that drove activity in the hippocampus.

The ventral prelimbic cortex is connected with the hippocampus indirectly via the retrosplenial cortex (Kobayashi & Amaral, 2003; 2007), and directly with the parahippocampal cortex including the entorhinal cortex (Kondo et al., 2005). As Chapter 3 discusses, both areas are involved in representing scenes. Using intracranial recording Woolnough et al. (2020) were able to distinguish the scenes that their human subjects were viewing by recording from cells in the posterior parahippocampal gyrus. In an earlier fMRI study Bonnici et al. (2012b) used MVPA to analyse activations in the human hippocampus, and they were also able to discriminate between activations that were associated with different scenes.

The conclusion is that the ability to visualize scenes and events in detail, whether relating to the past or future, depends on the medial PF cortex and other areas in the medial network. Within this network, it is the medial PF cortex that generates

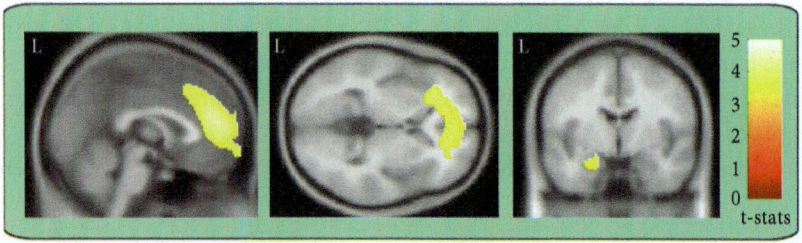

Figure 10.8 The area in yellow shows the source of the theta oscillations as measured by MEG while the subjects retrieved personal memories from the past

Reproduced from Cornelia McCormick, Daniel N. Barry, Amirhossein Jafarian, Gareth R. Barnes, & Eleanor A Maguire. VmPFC drives hippocampal processing during autobiographical memory recall regardless of remoteness. *Cereb Cortex*, 172, Figure 2, https://doi.org/10.1093/cercor/bhaa172 Copyright © The Authors, 2020. Licensed under CC BY 4.0.

the content of spontaneous thoughts, and it does so via its connections with the hippocampal system. However, the ability to retrieve events from the distant past also depends on the ability to tell the time and distinguish the days of the week and months of the year. In other words, though the brain mechanisms are necessary, they are not sufficient. Hearing stories about the past and learning about history may help children to think about the past themselves.

Co-opting the Mechanisms for Self-Generation

Chapter 3 reviews the evidence that in macaque monkeys the prelimbic cortex supports the ability to generate actions spontaneously (Khalighinejad et al., 2019). The existence of a paracingulate sulcus in the chimpanzee and human brain is evidence of expansion of this area (Amiez et al., 2019). There is a further expansion of the granular medial PF cortex in the human compared with the chimpanzee brain (Avants et al., 2006).

It is these areas that have become co-opted in the human brain for the generation of spontaneous thoughts, as opposed to actions alone. This has allowed the extension of memory and imagination into the distant past and future. At the same time, the development of language has allowed the addition of detail via a phonological code.

Summary

The medial network is activated in chimpanzees and human subjects at rest. However, it is likely that in chimpanzees this activation relates to their immediate experiences. In the human brain the PF cortex has expanded relative to the hippocampus, and this may be one reason why the human mind can re-generate scenes and events from the distant past and imagine events into the distant future. Another reason is that humans are brought up in a cultural environment in which they are read stories and taught history. The detail with which events can be characterized is further enhanced because of the availability of a phonological code.

Perceptual Awareness

Unlike monkeys or chimpanzees, human subjects can report on their perceptual awareness by describing the contents of what they see. However, there a way to probing for awareness without requiring a verbal account. This is to require the subject to press one key if they are aware and another if they are not. The key is referred to as a 'commentary key', and it was introduced by Weiskrantz (2009) in his studies of patients with 'blindsight'.

Cowey and Stoerig (1997) tested monkeys with the equivalent of a commentary key. They taught the animals to initiate each trial by pressing a start light. On some trials a target then appeared on the screen, and on other trials there was no target. The monkeys were trained to press the target if one appeared but press a rectangle on the screen if no target appeared.

Cowey and Stoerig then removed the primary visual cortex in one hemisphere. So as to check that the monkeys still knew the rule, they first presented the target lights to the intact hemisphere. The monkeys behaved as they had before surgery. But when the target lights were presented to the hemisphere with the lesion of V1, the monkeys almost always pressed the rectangle on the screen. In other words, they reported that no target had appeared.

Yet we know from the work of Humphrey (1974) that a monkey with a bilateral lesion of V1 can still pick up small objects from the floor. Cowey and Stoerig (1997) therefore suggested that monkeys with a lesion of VI are like patients with 'blindsight' who are able to point to stimuli even though they are not aware of seeing them (Weiskrantz, 2009).

There is another possible implication from the study by Cowey and Stoerig (1997). This is that the monkeys experienced visual sensations when the target was presented to the intact field but no such sensations when it was presented to the 'blind' field. A later section suggests that, for the purposes of mental hygiene, a distinction should be drawn between the ability to experience visual sensations and consciousness as it is commonly used in the human literature. The reason is that consciousness is usually taken to imply self-consciousness.

Because a commentary key was used in the human and monkeys studies, Weiskrantz (1997) suggested that the critical experiment was to find out where the 'commentary stage' was in the cortical processing of visual stimuli.

In search of this commentary stage, Lau and Passingham (2006) used fMRI to scan human subjects while they judged whether a square or a rectangle had been presented. Backward masking was used with a metacontrast mask so as to make the judgement difficult.

Lau and Passingham varied the stimulus onset asynchrony (SOA) and found that with SOAs of 33 and 104ms the subjects were equally accurate at distinguishing which of the two targets had been presented. However, they were significantly more likely to judge that they had seen the target at the long compared with the short SOA.

By comparing the scans for the long versus the short SOA, it was therefore possible to visualize the activations that were associated with the greater, as opposed to lesser, probability of awareness. Strikingly, there was only one activation for this contrast, and this was in the cortex in the inferior frontal sulcus, the border between the ventral and dorsal PF cortex. Lau and Passingham assumed that this activation could not reflect the response itself, because the subjects responded in both conditions.

If this area is critical for awareness, disruption of activity in the region should lead to a decrease in the percentage of trials on which the subjects report seeing the stimulus. So, Rounis et al. (2010) applied rTMS at theta frequency over the dorsal PF cortex in both hemispheres before testing the subjects on the task. As in the earlier experiment by Lau and Passingham (2006), the perceptual task was made difficult by using backward masking, with the difference that there was a fixed SOA of 100ms. The stimulation had no significant effect on performance, that is on the percent of the trials on which the subjects judged the identity of the stimulus correctly. However, compared with their performance before the rTMS, the subjects reported being aware of a smaller percentage of the targets on correct trials.

Rounis et al. also computed a measure of metacognitive accuracy. This is based on the matrix of high confidence hits, false alarms, low confidence misses, and correct rejections (Fleming & Lau, 2014). Receiver operating characteristic (ROC) curves are then plotted that relate the probability of confident judgments that are correct against the probability of confident judgements that are incorrect. Differences between these curves provide a measure of metacognitive sensitivity or metacognitive d'. Rounis et al. (2010) were able to show that theta-burst rTMS led to a decrease in metacognitive sensitivity as shown by the ROC curves.

Bor et al. (2017) reported that they were unable to replicate this result. However, as Ruby et al. (2018) have pointed out, Bor et al. (2017) initially found an effect of rTMS on metacognition, but they then discarded the results for twenty-seven of the ninety subjects because of stringent exclusion criteria.

The controversy is best settled by tackling the problem in a different way. Del Cul et al. (2009) collected data for fifteen patients with lesions that included the PF cortex. The greatest area of overlap was in the anterior ventral PF cortex. The task was to detect a digit, and it was made difficult by using backward masking with a variable SOA.

Overall, the patients performed less well than control subjects, but they also reported that they were aware of seeing fewer of the digits. However, there was no difference in the level of performance once the trials were matched for awareness. Yet, when the trials were matched for performance, the patients still reported seeing fewer of the stimuli than the control subjects. Unfortunately, ROC curves were not plotted to see if this was the result of a change in metacognitive d'.

However, in a later study Fleming et al. (2014) did perform such an analysis. There were seven patients with lesions of the PF cortex, with the greatest degree of overlap being in the medial, polar, and ventral PF cortex. These patients were compared with patients with lesions of the temporal lobe as well as a group of healthy control subjects.

The task was to judge whether the display on the left or the right had more dots in it. And the judgement was difficult because the stimuli were only shown for 700ms. The subjects were also asked to rate their degree of confidence after each judgement by moving a cursor along a six-point scale.

As in the study by Del Cul et al. (2009) the patients with PF lesions were as accurate in their performance as the control patients and healthy control subjects. However, Fleming et al. (2014) calculated meta-d' or type 2 sensitivity in the same units as type 1 sensitivity (d'). On this comparative measure of metacognitive accuracy, called meta-d'/d', the patients with the PF lesions scored 50% below the value for the control groups. This is a major effect.

By recording with intracranial electrodes in patients, it is possible to detect other indicators of perceptual awareness. Gaillard et al. (2009) presented words that were either masked or visible, and looked for activity that was associated with seeing the words. There were sustained voltage changes in the evoked potentials for words that were seen, especially in the PF cortex. Visible words were also accompanied by an increase in synchrony in the activity across different cortical areas, including the PF cortex and posterior parietal cortex. Moutard et al. (2015) refer to this increase in synchrony as 'ignition'.

Perceptual Awareness in Monkeys

Given all these results, it is possible to see whether the same neural markers can be found in the brains of non-human primates. In a neurophysiological study, van Vugt et al. (2018) taught monkeys to detect whether a circle had or had not been presented. The level of difficulty was varied by changing the contrast of the circle against a background square. The monkeys reported detection of the circle by making a saccade to where it had been presented; and they reported a failure to detect the circle by making a saccade to a dot.

Van Vugt et al. specifically analysed trials on which a circle had been presented, and the monkey had either detected the circle or failed to detect it. On both trial types a stimulus had been presented. The difference relates to factors that were internal to the animals, such as the momentary degree of vigilance.

Multiunit activity was recorded in the visual areas V1 and V4, and in the ventral and dorsal PF cortex. The activity in V1 and V4 was phasic, but in the PF cortex it was sustained.

Van Vugt et al. then computed an index to measure the degree to which the activity differed for misses and hits. As Figure 10.9 (next page) shows, the miss fraction was calculated as (the activity [miss] over the activity [hit] x 100%). If the index was 100, there was no difference between the activity for hits and misses, whereas if the index is 0 there was no activity at all for misses.

Figure 10.9 plots the indices for the three areas, and for three levels of difficulty or contrast.

There was a striking difference in the PF cortex which was much greater for hits than misses. In other words, it was in the PF cortex that the activity was most closely related to the animal's report.

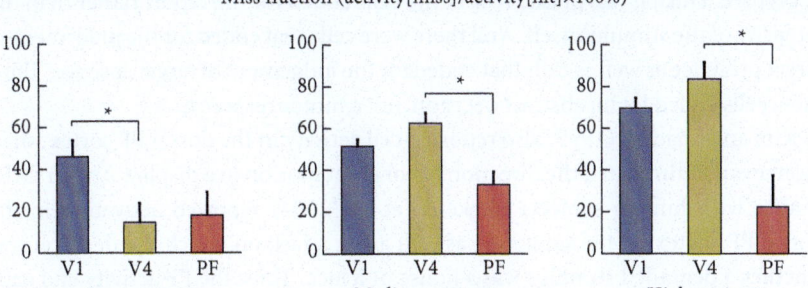

Figure 10.9 Indices relating the activity for hits and misses
These are shown for three levels of difficulty depending on the contrast.
Reproduced from van Vugt, B., Dagnino, B., Vartak, D., Safaai, H., Panzeri, S., Dehaene, S., & Roelfsema, P.R. The threshold for conscious report: Signal loss and response bias in visual and frontal cortex. *Science*, 360 (6388), 537–42, Figure 2b, Doi: 10.1126/science.aar7186 Copyright © 2018 The Authors.

Unfortunately, no recordings were taken in the inferior temporal cortex. This would have allowed a comparison with the inferior temporal cortex and the PF areas to which it projects (Webster et al., 1994). However, in an MEG study with human subjects, Wyart et al. (2011) found a correlate of reportability in the inferior temporal and PF cortex 220ms after stimulus presentation. Further work is needed to see if the activity in the inferior temporal cortex occurred bottom-up or was the result of a top-down signal from the PF cortex. In a study that used backward masking, Thompson et al. (1999) reported activity in the FEF 58ms after the presentation of the mask, and the activity was greater for hits than misses.

In these experiments, the monkeys reported whether they had detected the circle or not by making different saccadic eye movements. This means that it is necessary to show that the cell activity in the PF cortex reflected the report rather than the response itself. Merten and Nieder (2012) did this in an experiment in which they recorded from single cells in the dorsal PF cortex while macaque monkeys reported whether a square had been presented or not. As in the experiment by van Vugt et al. (2018), the contrast of the stimulus was varied to compare judgements of different difficulty.

On trials on which a target was presented, a red square instructed the monkeys to release the lever to report detection and a blue square instructed them to continue to hold to report a lack of detection. On trials on which no target was presented, the rules were reversed, so that, for example, the red square instructed the monkeys to continue to hold in order to report a lack of detection. This arrangement meant that it was possible to dissociate the report from a specific response.

Recordings were taken while the animal waited before the instruction cue for the task rule was presented, in other words before the animals knew what response to prepare. During this phase, 15% of the cells coded for the report rather than the intensity of the stimulus itself. And there were cells that coded for the judgement of target presence as well as cells that coded for the judgement of target absence. Thus, these cells coded for an abstract decision, not a motor response.

Kim and Shadlen (1999) also reported cell activity in the dorsal PF cortex when monkeys discriminated the direction of coherent motion in a display. And in fMRI studies with human subjects Heekeren et al. (2004) reported activation in the dorsal PF cortex on the same task as well as on a task on which the decision was whether a degraded stimulus was a house or a face. Thus, the PF activity and activation reflect the decision.

Studies with No Report

Some have argued the PF cortex is not essential for perceptual awareness per se, just for the decision (Koch et al., 2016). The way to find out is to carry out studies in which the subjects are not required to make a decision. These are referred to as 'no report studies'.

In one such study, Frassle et al. (2014) presented a green sinusoidal grating moving to the left to one eye, and a red grating moving to the right to the other eyes. Under these conditions there is binocular rivalry, and subjects typically report alternating between seeing one stimulus and then the other. The use of this task meant that it was possible to detect the times at which the percept switched by measuring optokinetic nystagmus, since the eyes drift to the left with one stimulus and to the right with the other. Frassle et al. validated this measure by comparing it with transitions as reported by button presses.

They then compared binocular rivalry with simulated rivalry by using gratings that actually changed; they called this control condition 'replay'. They found activation in the ventral and dorsal PF cortex when the subjects had to report the switch, the active condition; but they found no activation when the switches were measured by optokinetic nystagmus, the passive condition. On this basis they concluded that the activation in the PF cortex reflected the report, since they found no activation in the PF cortex in the passive condition.

But this conclusion is not warranted. Their own data showed that there was activation in the ventral and dorsal PF cortex when the passive condition was contrasted with baseline. The reason why there was no activation in the contrast of the passive condition with replay is presumably that the PF cortex was activated in both conditions. The explanation for why it was more activated in the active condition than replay is presumably that it was more difficult for the subjects to decide on their report during rivalry.

The lesson is that the results of no report studies are not valid unless the images are compared against baseline. In too many studies the comparison is with another condition, and this means that there is a danger that activation of the PF cortex is subtracted out. This applies, for example, to the fMRI study by Nunn et al. (2002). They claimed that there was no activation in the PF cortex when subjects with synaesthesia heard words and thus saw colours as opposed to simply hearing sounds. The study is of interest because it showed that there was activation in the human homologue of V4 (area V8) in the synaesthetic condition. But it is likely that the PF cortex was activated in both conditions, and thus that it was subtracted out in the contrast.

Total Prefrontal Lesions

There is another way of trying to find out whether the PF cortex is essential for perceptual awareness. This is to study the effect of PF lesions. Koch et al. (2016) claimed that a patient described in the literature by Brickner (1952) had a total bilateral PF lobectomy, and yet was perceptually aware. But Koch et al. (2016) had misread the pattern of the sulci in the photograph of the brain, and the lobectomy was not total at all (Odegaard et al., 2017).

The famous patient Phineas Gage had a massive PF lesion due to damage as the result of an explosion in which a tamping iron penetrated his skull; yet there is no evidence that he was perceptually unaware. But the actual damage was almost entirely to the left PF cortex (Van Horn et al., 2012). Similarly, the patient EVR had a large bilateral PF ablation, but the damage was mainly to the medial and orbital PF cortex (Eslinger & Damasio, 1985).

The conclusion is that it is likely that the PF cortex is essential for perceptual awareness (Odegaard et al., 2017). The findings in the literature so far from fMRI and lesions are consistent with higher-order theories of awareness (Lau & Rosenthal, 2011; Shea & Frith, 2019). These theories assume that there is a meta-representation of first-order states (Brown et al., 2019). The position of the PF cortex in the sensory hierarchy means that it is in a position to support these meta-representations. (Chapter 8).

Summary

In an fMRI study, the only area that was associated with perceptual awareness was the PF cortex. Inactivation or permanent lesions of this area leads to a decrease in the number of trials on which human subjects report seeing a target. As in human subjects, there is sustained activity in the PF cortex when monkeys detect targets. Furthermore, this activity is much greater for hits than misses. Neither 'no report' studies nor reports of the effects of large PF lesions rule out the claim that the PF cortex supports the meta-representations of lower-order states that are necessary for awareness.

Awareness of Agency

The PF lies at the top of the motor as well as the sensory hierarchy (Chapter 8). In a classical experiment, Libet et al. (1983) introduced a way of finding out if human subjects were aware of their intention to act. Rather than asking them to describe their intention, Libet et al. simply asked them to identify the time at which they were first aware.

The subjects were instructed to flex a finger at whatever time they chose after the presentation of a tone. The action was to be as spontaneous as possible. While they did this, they viewed a rapidly moving clock face, and they were later asked at what time on the clock they had first been aware that they were going to move their finger.

To study the neural correlates of awareness of intention, Lau et al. (2004) used fMRI to scan subjects while they pressed a mouse button at self-paced intervals during the scans. As in the experiment by Libet et al. (1983), the subjects viewed a rapidly rotating clock. There were two conditions. In one the subjects reported when they were first aware of the urge to move; Libet et al. referred to this as the willed condition (W). In the other condition, the subjects reported when they were first aware of actually pressing the button; Libet et al. referred to this as the movement condition (M).

When the first of these conditions was contrasted with the second, there was a difference in activation on the border between the presupplementary motor area (preSMA) and the supplementary motor area (SMA). So as to examine this difference, Rigoni et al. (2013) recorded evoked potentials (ERPs) in the same two conditions. As Figure 10.10 (next page) shows, that the amplitude of the ERPs recorded from the preSMA and SMA were significantly larger in the W than M condition.

Rigoni et al. assumed that the enhancement of the ERPs in the W condition was driven by a top-down modulation, but they did not prove this. However, in their fMRI study Lau et al. (2004) found evidence for a top-down effect. There was more activation in the dorsal PF cortex and the cortex of the IPS in the W than M condition.

To see which area exerted a top-down effect, Lau et al. looked for a 'psychophysiological interaction' (Friston et al., 1997); this uses a regression method to look for 'effective connectivity', meaning that the relation between the activations in two areas differs as a function of the condition or context. Lau et al. found a significant psychophysiological interaction between the dorsal PF cortex and the preSMA, but there was no such interaction between the cortex of the IPS and the preSMA. This result suggests that it is the PF cortex that drives the enhancement in the preSMA in the W condition.

The conclusion is that, just as the PF cortex is involved in driving the enhancement of activation in sensory areas (Heinen et al., 2013), it is also involved in driving the enhancement of activation in motor areas. Both are top-down effects.

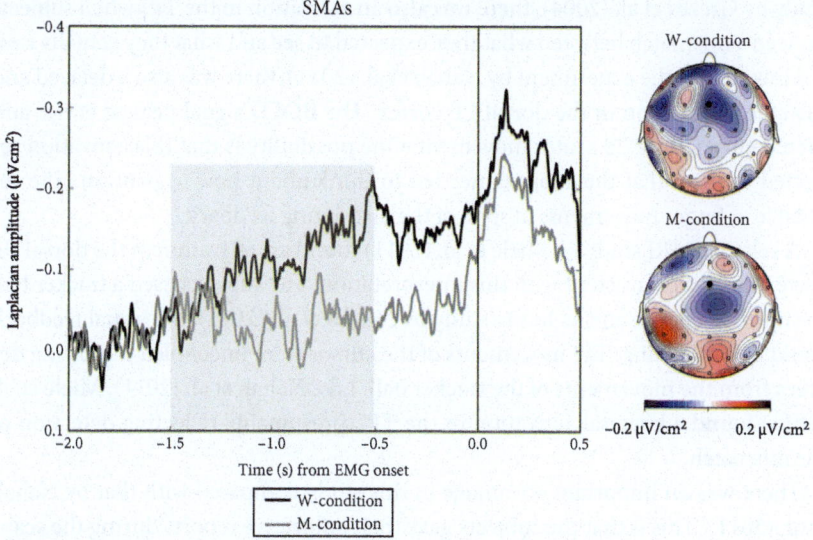

Figure 10.10 ERPs recorded on the Libet task for the conditions in which the subjects timed their intention to move (W) or the actual movement itself (M)
The EEG sources are shown on the right.

Reproduced from Rigoni, D., Brass, M., Roger, C., Vidal, F., & Sartori, G. Top-down modulation of brain activity underlying intentional action and its relationship with awareness of intention: An ERP/Laplacian analysis. *Exp Brain Res*, 229, 347–57, Figure 2, https://doi.org/10.1007/s00221-013-3400-0 Copyright © 2013, Springer Nature.

When human subjects make a movement, they are aware that they intended it, in other words that they are the agent. As the philosopher Searle (2007) pointed out, there is a difference between raising your arm and your arm rising, for example because someone else knocked it.

One way of testing for awareness of agency is to vary the extent to which the subjects have control over the visual feedback about their actions. This procedure allows one to measure the subject's awareness of when their actions as intended or not.

A study by Nahab et al. (2011) used this procedure. The subjects viewed a virtual reality image of their hand while performing a simple sequence of finger movements. These movements were recorded with a glove with sensors. The virtual image either moved in synchronization with the movements that the subjects made or in various degrees of desynchronization.

Nahab et al. used time-resolved fMRI. In most areas, the BOLD signals were of greater amplitude the greater the disparity between the intended movement and the visual feedback. There were early peaks in the dorsal PF cortex and the temporoparietal junction (TPJ). The activation in the TPJ probably reflected detection of the mismatch between the movement as intended and the visual feedback. In a study by Grezes et al. (2004), there was also an activation in the TPJ when subjects noticed a mismatch between what they expected to see and what they actually saw.

However, in the experiment by Nahab et al. (2011), there was also a delayed and sustained activation in the dorsal PF cortex. The BOLD signal peaked at 8 s, and then continued for 20 s after movement. One possibility is that this activation reflected the fact that the subjects needed to think about how to continue the sequence of finger movements in spite of the distracting feedback.

A related fMRI study by Miele et al. (2011) found an activation in the dorsal PF cortex that was consistent with this interpretation. The subjects used a tracker ball to control a cursor, and as in the study by Nahab et al. (2011), the visual feedback could be misleading. The movements of the cursor were uncoupled to varying degrees from the movements of the tracker ball. Like Nahab et al. (2011), Miele et al. (2011) found a bilateral activation in the TPJ, presumably reflecting detection of the mismatch.

There was an important advantage in this study compared with that by Nahab et al. (2011). This is that the subjects gave their subjective reports during the scans rather than in a separate behavioural study outside the scanner. There were two ratings, one on the extent to which they were in control of the cursor (performance), and the other on the extent to which they believed themselves to be agents of the movements (agency).

Miele et al. (2011) then computed a measure of agency that was independent of the perceived degree of control. This controlled for differences in the subjects' perception of their performance. There was only one activation that was related to this measure of agency, and this was in the dorsal PF cortex (area 46). In other words, the PF cortex is activated both when human subjects report that 'I did it' and when they report that 'I saw it' (Passingham & Lau, 2019).

Summary

Human subjects are aware of the time at which they intend to act, and there is enhancement in the preSMA and SMA when they attend to that time. This enhancement is driven by a top-down effect from the dorsal PF cortex. Subjects are aware that they are the agent, and activation in the dorsal PF cortex is associated with this awareness. This area is activated both when subjects report 'I did it' and 'I saw it' (Passingham and Lau, 2019).

Representation of the Self

The concept of the self includes the representation of what individuals see, hear, and feel; what they do; what they recall from their past history; what they plan for their future; the emotions they experience; and the personality they possess. Just as human subjects can think about the minds of others (Chapter 11), so they can think about their own minds. They see themselves as distinct from other individuals.

Confidence in Monkeys

The ability to represent the self requires metacognition. One way of studying metacognition in monkeys is to require them to report their confidence in their own judgements. This can be done by giving them the opportunity to bet on what they are seeing or remembering. This method was introduced by Persaud et al. (2011) in a study of the patient GY who exhibits blindsight.

The confidence with which the monkeys make their judgements can be tested by requiring them, for example, to choose between a red square for a high bet and a green square for a low bet.

Miyamoto et al. (2017) investigated metacognition in monkeys by showing them a series of pictures and later testing their recognition memory. Confidence was measured by requiring the monkeys to bet on their judgement. Correct judgements were rewarded with three juice drops when the bet was high and with two drops when the bet was low. Incorrect judgements led to a long time-out when the bet was high.

Miyamoto et al. used fMRI to scan the monkeys while they performed this task. There was an activation in the medial PF cortex on the border between areas 9 and 10 when the monkeys recognized a picture that was early in the series and did so with high confidence.

Miyamoto et al. went on to test whether the activation in the medial PF cortex was causal for the meta-cognitive judgement by inactivating the area using muscimol. The inactivation led to a significant impairment in metacognitive accuracy.

Miyamoto et al. also reported an activation in the supplementary eye field (SEF) when the monkeys recognized a picture that was late in the series. Stuphorn et al. (2000) had previously suggested that the SEF was involved in monitoring, and Middlebrooks and Sommer (2012) had reported that cell activity in the SEF was associated with metacognitive accuracy on a perceptual task.

However, the studies by Miyamoto et al. (2017) and Middlebrooks and Sommer (2012) cannot be taken as evidence that monkeys can attain an explicit representation of themselves. Their metacognitive ability is probably implicit. Chapter 11 discusses the difference between implicit and explicit representations in more detail.

Furthermore, the representation of the self requires more than just confidence in one's judgements. For example, the individual must be able to distinguish their own thoughts from those of others.

Metacognition in Human Subjects

Bang and Fleming (2018) argued that it is necessary to separate metacognition for perception and the sensory evidence on which the judgement was based. They designed a task on which the subjects had to decide the direction in which the dots in a display were moving; the difficulty of the task was also varied by altering the degree to which the dots moved coherently.

After the presentation of the dots, a reference point was shown, and the subjects had to judge whether the dots were moving clockwise or counter-clockwise compared with this point. Finally, the subjects also rated their degree of confidence on a scale from 50% to 100%.

Both the degree of coherence and the distance from the reference point had an effect on the confidence with which the subjects formed their judgement. Bang and Fleming then looked for activations that showed an effect of both coherence and distance, and the analysis revealed an activation in the medial PF cortex (prelimbic area 32). There was a high correlation between this activation and the confidence that the subjects expressed in their judgements.

So as to compare judgement of confidence for perception and memory, Morales et al. (2018) presented words or shapes. On the perceptual task, the stimuli were presented briefly. and the subjects had to detect which of the two stimuli was the brighter. On the memory task, the subjects had to indicate which of two probe stimuli had been presented previously.

The subjects rated their confidence on a four-point scale. Meta-cognitive accuracy could therefore be measured for both the perceptual and memory task. MVPA was used to look for domain general activity predicting confidence and accuracy. The analysis revealed common activations in the medial PF cortex (area 9) and the preSMA.

An earlier section reviews the evidence that there is activation in the medial PF cortex when subjects retrieve autobiographical memories (Addis et al., 2007; Hassabis et al., 2007) or plan how to achieve personal goals (Spreng et al., 2010). These activations reflect episodic memory and simulation as opposed to semantic memory.

However, human individuals also have personal names and surnames, as well as having a date of birth and place at which they live. This is semantic knowledge, but it relates to the person themselves. So, in an fMRI study Tacikowski et al. (2017) presented words referring to the subjects themselves or words referring to others. The task was to indicate whether the words did or did not refer to the subjects themselves, and it was made difficult by using visual masking.

To look for the activations that related to the self, Tacikowski used the BOLD adaptation method. This makes use of the fact that there is a decrease in activation trials when the same stimulus is presented twice as opposed to trials on which the stimuli on the two trials differ.

There were activations in the anterior middle temporal gyrus, and Chapter 11 reviews the evidence that this is part of the semantic system. However, there were also activations in the medial PF cortex (ventral prelimbic area 32 extending into the polar PF cortex), as well as the retrosplenial cortex and hippocampus. In other words, semantic information that related to the self is also represented in the medial network.

Another aspect of the self relates to individual differences in emotion, such as the tendency to be anxious or to get depressed. Taschereau-Dumouchel et al. (2019) presented human subjects with photos of animals that can cause phobias, such as spiders and snakes. The degree to which each individual was fearful was in two ways. The first involved subjective ratings; the second was an objective measure, the skin conduction response (SCR) that results a change in skin conductance due to sweat. MVPA was then used to decode the areas in which it was possible to distinguish photos of feared objects from photos of neutral ones, both animals and objects.

Figure 10.11A (next page) shows the lateral and medial views of the brain. The areas in which the activations predicted the subjective ratings better than skin conductance are shown in reds and orange. These included the medial and dorsal PF cortex.

The areas in which the activations predicted the physiological responses (SCRs) better are shown in blue. The histograms in the right hand panel of Figure 10.11B show the degree of these differences. The areas included the amygdala and the anterior insula.

The medial PF cortex is also activated when subjects brood about themselves. Berman et al. (2011) looked at the role of the medial network in rumination in subjects who were depressed. Depression was assessed on the Beck scale (1996). This measures negative thoughts about oneself, as well feelings of hopelessness concerning the future. When the subjects were just resting, there was more activation in the subgenual area 25 in the subjects who were ruminating than in those who were not. And rumination was associated with an increase in the connectivity of this area with the posterior cingulate and retrosplenial cortex.

Summary

The concept of the self includes the representation of what individuals see, hear, or feel, what they do, what they recall from their past history, what they plan for their future and their emotions and personality. Many studies point to the role of the medial PF cortex in representing the self.

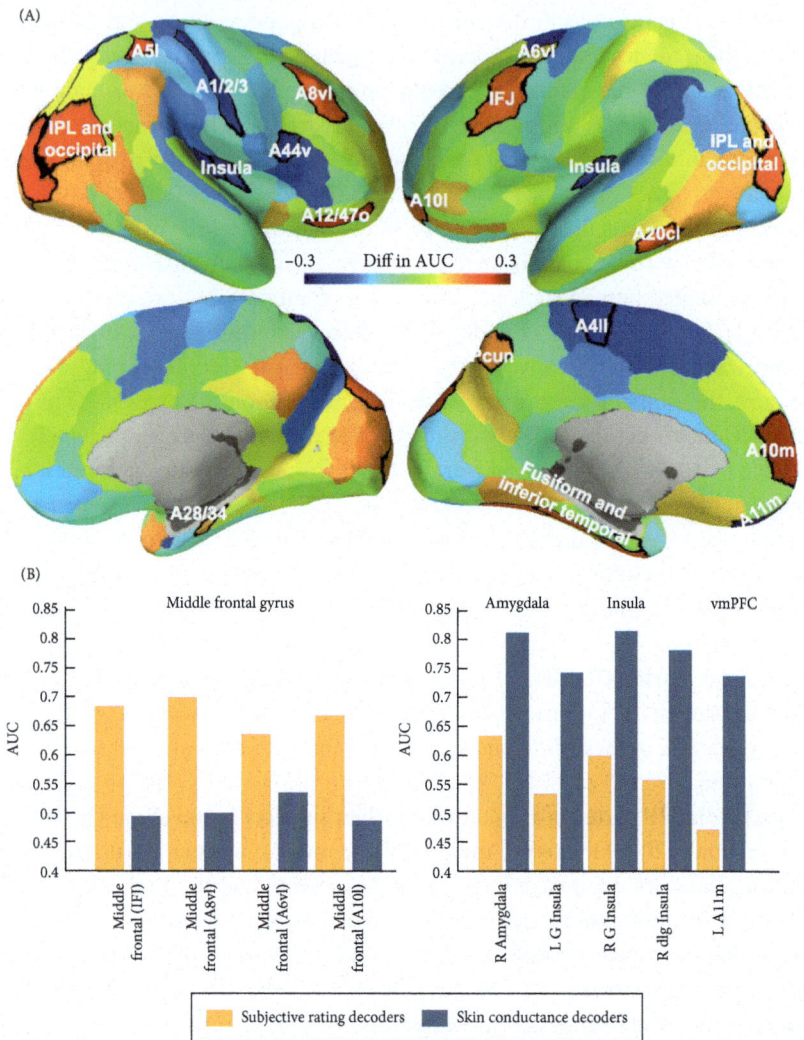

Figure 10.11 Brain regions with a significant difference between the prediction of the subjective ratings and skin conductance reactivity
(A) A positive difference in the area under the curve indicates a better prediction of the subjective ratings (red–orange regions) whereas a negative difference indicates a better prediction of skin conductance reactivity (blue regions). (B, Left) Significant regions of the middle frontal gyrus associated with subjective feelings. (B, Right) significant activations in the amygdala, insula, and ventral medial prefrontal cortex (vmPFC) associated with skin conductance reactivity. Dashed lines represent the critical value ($p = 0.05$).

Reproduced from Taschereau-Dumouchel, V., Kawato, M., & Lau, H. Multivoxel pattern analysis reveals dissociations between subjective fear and its physiological correlates. *Mol Psychiat*, Figure 4, https://doi.org/10.1038/s41380-019-0520-3 Copyright © 2019, The Authors. Licensed under CC BY 4.0.

Conclusions

The term consciousness is usually used to imply self-consciousness. There is no evidence that other primates can form an explicit representation the self.

The fact that humans can represent the self is due in part to the expansion of the human PF cortex since it is PF areas that are activated when subjects reflect on themselves. But it is also due to the cultural environment in which children are brought up.

Expansion of the Human Prefrontal Cortex

The expansion of the human PF cortex in relation to its sensory inputs has the consequence that there is more tissue available for analysing and re-representing those inputs. The expansion of the inferior caudal PF cortex (areas 44 and 45B) also means that human subjects are unique in having access to a propositional code. This is used when subjects solve verbal problems, such as evaluating logical syllogisms. In the human brain the arcuate fasciculus connects with middle and inferior temporal cortex with the inferior caudal PF cortex in the human brain, and Chapter 11 reviews the evidence that these areas support the semantic system.

The expansion of the dorsal PF cortex in relation to its outputs (Chapter 9) has the consequence that more tissue is also available for manipulating and re-ordering items. The ancestral mechanisms for doing this have been co-opted so as to allow subjects to tackle problems on tests of non-verbal or fluid intelligence. Solving these problems also involves discovering the abstract rule, and it is the PF cortex that supports the ability to represent abstract rules (Chapter 8).

The dorsal PF is highly interconnected with other areas of the brain, both through its connections with the ventral PF cortex and connections with posterior association areas. It therefore has access to a very wide variety of material. This means that in the human brain there are new inputs, for example letters, words, and numbers. The degree of the connectivity of the PF cortex with other areas is related to individual differences in intelligence.

There has also been an expansion of the ventral and medial PF cortex including the polar PF cortex. Thus, there is also more tissue for generating visual imagery. The medial PF cortex supports the ability to retrieve personal memories from the past and to imagine the future. It is also activated when subjects reflect on themselves.

Development in a Cultural Environment

The data on human subjects reported in this chapter comes from fMRI studies of adults. But the ability to re-represent depends not only on the expansion of the PF cortex but also on the extended period of childhood during which children are taught in language (Chapter 11). They hear stories and then read them for themselves.

This is likely to promote the ability to engage in the internal monologue or narrative that Brown et al. (2019) take to be characteristic of humans as they think about themselves and their position in the world. When human subjects are at rest there is activation in the inferior caudal PF cortex of the left hemisphere as well as the middle temporal gyrus with which it is interconnected (Rilling et al., 2007). This suggests that this internal monologue depends on the language system.

Karmiloff-Smith (1992) has documented the role of the repeated re-representation of ideas and information from babyhood on. This allows the development from implicit to explicit verbal representations. Chapter 11 discusses the long period during which children learn to read the minds of others, from implicit to explicit representations (Heyes & Frith, 2014).

The same progression is likely for the re-representation of one's own mental states. Consciousness of the self depends on an inner awareness of these states (Brown et al., 2019), and that awareness is probably learned (Cleeremans et al., 2020). Only humans have an explicit awareness of the self.

Selective Advantage

The hominins evolved at a time of very rapid climactic and environmental change. They adapted in several ways, but they can be summed up by saying that the expansion of the PF cortex and other association areas led to an astonishing increase in intelligence as assessed by the ability to solve novel problems.

First, our ancestors developed the ability to engage in covert trial and error. By simulating the various alternatives, it was possible to eliminate the ones that would be unsuccessful. The ability to do this was aided by the development of grammatical language which could represent any alternatives and do so in great detail. Thus, it was possible to solve a new problem with no error at all.

And errors mattered. Genetic evidence suggests that, between 60 and 10 Ka, there was a severe bottleneck in the population size of the hominins that had left Africa for Europe and China (Li & Durbin, 2011). And the rapid changes in the environment during the Ice Ages put severe pressure on the hominins in Europe, as the food resources became less dependable, and large predators posed a danger.

The development of grammatical language, an estimated 70,000 Ka (Bolhuis et al., 2014), also had the effect that it allowed teaching. Whereas other primates can transfer their knowledge from one problem to another, our ancestors were able to transfer knowledge and skills from one individual to another. This also had the effect that the student could solve a problem without error.

Chapter 11 also argues that the ability to transmit knowledge and skills from one generation to another makes possible cumulative change across the generations as individuals make discoveries of their own. These can then be passed on to others. To take a recent example, Newton invented the calculus. But it can now be mastered at senior school.

The result is that humans are unique amongst animals in the ways in which they communicate, teach, think, remember, and cooperate. We are the only animal that knows itself and has a scientific understanding of the world. It is this that accounts for our extraordinary success as a species.

References

Addis, D.R., Knapp, K., Roberts, R.P., & Schacter, D.L. (2012) Routes to the past: Neural substrates of direct and generative autobiographical memory retrieval. *Neuroimage*, 59, 2908–22.

Addis, D.R., Wong, A.T., & Schacter, D.L. (2007) Remembering the past and imagining the future: Common and distinct neural substrates during event construction and elaboration. *Neuropsychol*, 45, 1363–77.

Alderson-Day, B., Weis, S., McCarthy-Jones, S., Moseley, P., Smailes, D., & Fernyhough, C. (2016) The brain's conversation with itself: Neural substrates of dialogic inner speech. *Soc Cogn Affect Neurosci*, 11, 110–20.

Amiez, C., Sallet, J., Hopkins, W.D., Meguerditchian, A., Hadj-Bouziane, F., Ben Hamed, S., Wilson, C.R.E., Procyk, E., & Petrides, M. (2019) Sulcal organization in the medial frontal cortex provides insights into primate brain evolution. *Nat Comm*, 10, 3437.

Amunts, K., Lenzen, M., Friederici, A.D., Schleicher, A., Morosan, P., Palomero-Gallagher, N., & Zilles, K. (2010) Broca's region: Novel organizational principles and multiple receptor mapping. *PLoS Biol*, 8.

Assem, M., Glasser, M.F., Van Essen, D.C., & Duncan, J. (2020) A domain-general cognitive core defined in multimodally parcellated human cortex. *Cereb Cort*, 30, 4361–80.

Avants, B.B., Schoenemann, P.T., & Gee, J.C. (2006) Lagrangian frame diffeomorphic image registration: Morphometric comparison of human and chimpanzee cortex. *Med Image Anal*, 10, 397–412.

Bang, D. & Fleming, S.M. (2018) Distinct encoding of decision confidence in human medial prefrontal cortex. *Proc Natl Acad Sci USA*, 115, 6082–7.

Barry, D.N., Barnes, G.R., Clark, I.A., & Maguire, E.A. (2019) The Neural Dynamics of Novel Scene Imagery. *J Neurosci*, 39, 4375–86.

Beck, A.T., Steer, A.R., & Brown, G.K. (1996) *Manual for the Beck Depression Inventory II*. Psychological Corporation, San Antonio.

Berman, M.G., Peltier, S., Nee, D.E., Kross, E., Deldin, P.J., & Jonides, J. (2011) Depression, rumination and the default network. *Soc Cogn Affect Neurosci*, 6, 548–55.

Bertossi, E. & Ciaramelli, E. (2016) Ventromedial prefrontal damage reduces mind-wandering and biases its temporal focus. *Soc Cogn Affect Neurosci*, 11, 1783–91.

Bertossi, E., Tesini, C., Cappelli, A., & Ciaramelli, E. (2016) Ventromedial prefrontal damage causes a pervasive impairment of episodic memory and future thinking. *Neuropsychol*, 90, 12–24.

Bolhuis, J.J., Tattersall, I., Chomsky, N., & Berwick, R.C. (2014) How could language have evolved? *PLoS Biol*, 12, e1001934.

Bonnici, H.M., Chadwick, M.J., Lutti, A., Hassabis, D., Weiskopf, N., & Maguire, E.A. (2012a) Detecting representations of recent and remote autobiographical memories in vmPFC and hippocampus. *J Neurosci*, 32, 16982–91.

Bonnici, H.M., Kumaran, D., Chadwick, M.J., Weiskopf, N., Hassabis, D., & Maguire, E.A. (2012b) Decoding representations of scenes in the medial temporal lobes. *Hippocampus*, 22, 1143–53.

Bonnici, H.M. & Maguire, E.A. (2018) Two years later - Revisiting autobiographical memory representations in vmPFC and hippocampus. *Neuropsychol*, 110, 159–69.

Boorman, E.D., Behrens, T.E., & Rushworth, M.F. (2011) Counterfactual choice and learning in a neural network centered on human lateral frontopolar cortex. *PLoS Biol*, 9, e1001093.

Bor, D., Schwartzman, D.J., Barrett, A.B., & Seth, A.K. (2017) Theta-burst transcranial magnetic stimulation to the prefrontal or parietal cortex does not impair metacognitive visual awareness. *PLoS One*, 12, e0171793.

Borra, E., Ferroni, C.G., Gerbella, M., Giorgetti, V., Mangiaracina, C., Rozzi, S., & Luppino, G. (2019) Rostro-caudal connectional heterogeneity of the dorsal part of the macaque prefrontal area 46. *Cereb Cortex*, 29, 485–504.

Brickner, R.M. (1952) Brain of patient A after bilateral frontal lobectomy; status of frontal-lobe problem. *AMA Arch Neurol Psychiatry*, 68, 293–313.

Brown, R., Lau, H., & LeDoux, J.E. (2019) Understanding the higher-order approach to consciousness. *Trends Cogn Sci*, 23, 754–68.

Cattell, R.B., Krug, S.E., & Barton, K. (1973) *Technical Supplement for the Culture Fair Intelligence Test, Scales 2 and 3*. Champaign, Illinois.

Chase, H.W., Grace, A.A., Fox, P.T., Phillips, M.L., & Eickhoff, S.B. (2020) Functional differentiation in the human ventromedial frontal lobe: A data-driven parcellation. *Hum Brain Mapp*, 41, 3266–83.

Cieslik, E.C., Mueller, V.I., Eickhoff, C.R., Langner, R., & Eickhoff, S.B. (2015) Three key regions for supervisory attentional control: Evidence from neuroimaging meta-analyses. *Neurosci Biobehav Rev*, 48, 22–34.

Cleeremans, A., Achoui, D., Beauny, A., Keuninckx, L., Martin, J.R., Munoz-Moldes, S., Vuillaume, L., & de Heering, A. (2020) Learning to Be Conscious. *Trends Cogn Sci*, 24, 112–3.

Cole, M.W., Ito, T., & Braver, T.S. (2015) Lateral prefrontal cortex contributes to fluid intelligence through multinetwork connectivity. *Brain Connect*, 5, 497–504.

Cowey, A. & Stoerig, P. (1997) Visual detection in monkeys with blindsight. *Neuropsychol*, 35, 929–39.

Crone, E.A., Wendelken, C., van Leijenhorst, L., Honomich, R.D., Christoff, K., & Bunge, S.A. (2009) Neurocognitive development of relational reasoning. *Dev Sci*, 12, 55–66.

Dehaene, S., Kerszberg, M., & Changeux, J.-P. (1998) A neuronal model of a global workspace in effortful cognitive tasks. *Proc Nat Acad Sci*, 95, 14529–34.

Dehaene, S., Molko, N., Cohen, L., & Wilson, A.J. (2004) Arithmetic and the brain. *Curr Opin Neurobiol*, 14, 218–24.

Del Cul, A., Dehaene, S., Reyes, P., Bravo, E., & Slachevsky, A. (2009) Causal role of prefrontal cortex in the threshold for access to consciousness. *Brain*, 132, 2531–40.

Donahue, C.J., Glasser, M.F., Preuss, T.M., Rilling, J.K., & Van Essen, D.C. (2018) Quantitative assessment of prefrontal cortex in humans relative to nonhuman primates. *Proc Natl Acad Sci USA*, 115, E5183–92.

Duncan, J. (2010a) *How Intelligence Happens*. Yale University Press, New Haven.

Duncan, J. (2010b) The multiple-demand (MD) system of the primate brain: Mental programs for intelligent behaviour. *Trends Cogn Sci*, 14, 172–9.

Duncan, J. (2013) The structure of cognition: Attentional episodes in mind and brain. *Neuron*, 80, 35–50.

Duncan, J. & Owen, A.M. (2000) Common regions of the frontal lobe recruited by diverse cognitive demands. *TINS*, 23, 475–82.

Duncan, J., Seitz, R.J., Kolodny, J., Bor, D., Herzog, H., Ahmed, A., Newell, F.N., & Emslie, H. (2000) A neural basis for general intelligence. *Science*, 289, 457–60.

Ehrsson, H.H., Geyer, S., & Naito, E. (2003) Imagery of voluntary movement of fingers, toes, and tongue activates corresponding body-part-specific motor representations. *J Neurophysiol*, 90, 3304–16.

Eichert, N., Verhagen, L., Folloni, D., Jbabdi, S., Khrapitchev, A.A., Sibson, N.R., Mantini, D., Sallet, J., & Mars, R.B. (2019) What is special about the human arcuate fasciculus? Lateralization, projections, and expansion. *Cortex*, 118, 107–15.

Eslinger, P.J. & Damasio, A.R. (1985) Severe disturbance of higher cognition after bilateral frontal lobe ablation: Patient EVR. *Neurol*, 35, 1731–41.

Fedorenko, E., Duncan, J., & Kanwisher, N. (2013) Broad domain generality in focal regions of frontal and parietal cortex. *Proc Natl Acad Sci USA*, 110, 16616–21.

Fias, W., Lammertyn, J., Caessens, B., & Orban, G.A. (2007) Processing of abstract ordinal knowledge in the horizontal segment of the intraparietal sulcus. *J Neurosci*, 27, 8952–6.

Fleming, S.M. & Lau, H.C. (2014) How to measure metacognition. *Front Hum Neurosci*, 8, 443.

Fleming, S.M., Ryu, J., Golfinos, J.G., & Blackmon, K.E. (2014) Domain-specific impairment in metacognitive accuracy following anterior prefrontal lesions. *Brain*, 137, 2811–22.

Frassle, S., Sommer, J., Jansen, A., Naber, M., & Einhauser, W. (2014) Binocular rivalry: Frontal activity relates to introspection and action but not to perception. *J Neurosci*, 34, 1738–47.

Friston, K.J., Buechel, C., Fink, G.R., Morris, J., Rolls, E., & Dolan, R.J. (1997) Psychophysiological and modulatory interactions in neuroimaging. *Neuroimage*, 7, 218–29.

Gaillard, R., Dehaene, S., Adam, C., Clemenceau, S., Hasboun, D., Baulac, M., Cohen, L., & Naccache, L. (2009) Converging intracranial markers of conscious access. *PLoS Biol*, 7, e61.

Genovesio, A., Wise, S.P., & Passingham, R.E. (2014) Prefrontal-parietal function: From foraging to foresight. *Trends Cogn Sci*, 18, 72–81.

Gerardin, A., Sirigu, A., Lehericy, S., Poline, J.-B., Gaymard, B., Marsault, C., Agid, Y., & Le Bihan, D. (2000) Partially overlapping neural networks for real and imagined hand movements. *Cer Cort*, 10, 1093–104.

Gerlach, K.D., Spreng, R.N., Madore, K.P., & Schacter, D.L. (2014) Future planning: Default network activity couples with frontoparietal control network and reward-processing regions during process and outcome simulations. *Soc Cogn Affect Neurosci*, 9, 1942–51.

Glasser, M.F., Coalson, T.S., Robinson, E.C., Hacker, C.D., Harwell, J., Yacoub, E., Ugurbil, K., Andersson, J., Beckmann, C.F., Jenkinson, M., Smith, S.M., & Van Essen, D.C. (2016) A multi-modal parcellation of human cerebral cortex. *Nature*, 536, 171–8.

Godbout, L. & Doyon, J. (1995) Mental representation of knowledge following frontal-lobe or postrolandic lesions. *Neuropsychol*, 33, 1671–96.

Goel, V., Buchel, C., Frith, C., & Dolan, R.J. (2000) Dissociation of mechanisms underlying syllogistic reasoning. *Neuroimage*, 12, 504–14.

Goel, V. & Dolan, R.J. (2001) Functional neuroanatomy of three-term relational reasoning. *Neuropsychol*, 39, 901–9.

Goel, V. & Grafman, J. (1995) Are the frontal lobes implicated in 'planning' functions? Interpretating data from the Tower of Hanoi. *Neuropsychol*, 33, 623–42.

Goel, V. & Grafman, J. (2000) Role of the right prefrontal cortex in ill-structured planning. *Cogn Neuropsychol*, 17, 415–36.

Goucha, T. & Friederici, A.D. (2015) The language skeleton after dissecting meaning: A functional segregation within Broca's Area. *Neuroimage*, 114, 294–302.

Grezes, J., Frith, C., & Passingham, R.E. (2004) Brain mechanisms for inferring deceit in the actions of others. *J Neurosci*, 24, 5500–5.

Hassabis, D., Kumaran, D., & Maguire, E.A. (2007) Using imagination to understand the neural basis of episodic memory. *J Neurosci*, 27, 14365–74.

Heekeren, H.R., Marrett, S., Bandettini, P.A., & Ungerleider, L.G. (2004) A general mechanism for perceptual decision-making in the human brain. *Nature*, 431, 859–62.

Heinen, K., Feredoes, E., Weiskopf, N., Ruff, C.C., & Driver, J. (2013) *Direct Evidence for Attention-Dependent Influences of the Frontal Eye-Fields on Feature-Responsive Visual Cortex*. Cereb Cortex.

Heyes, C.M. & Frith, C.D. (2014) The cultural evolution of mind reading. *Science*, 344, 1243091.

Hodgson, T.L., Bajwa, A., Owen, A.M., & Kennard, C. (2000) The strategic control of gaze direction in the Tower-of-London task. *J Cogn Neurosci*, 12, 894–907.

Hugdahl, K., Raichle, M.E., Mitra, A., & Specht, K. (2015) On the existence of a generalized non-specific task-dependent network. *Front Hum Neurosci*, 9, 430.

Humphrey, N.K. (1974) Vision in a monkey without striate cortex: A case study. *Perception*, 3, 241–55.

Hutchison, R.M., Womelsdorf, T., Gati, J.S., Leung, L.S., Menon, R.S., & Everling, S. (2012) Resting-state connectivity identifies distinct functional networks in macaque cingulate cortex. *Cereb Cortex*, 22, 1294–308.

Inoue, S. & Matsuzawa, T. (2007) Working memory of numerals in chimpanzees. *Curr Biol*, 17, R1004–5.

Ishai, A. (2010) Seeing with the mind's eye: top-down, bottom-up, and conscious awareness. *Biol Rep*, 34. Doi: 10.3410/B2-34.

Ishai, A., Haxby, J.V., & Ungerleider, L.G. (2002) Visual imagery of famous faces: Effects of memory and attention revealed by fMRI. *Neuroimage*, 17, 1729–41.

Ishai, A., Ungerleider, L.G., & Haxby, J.V. (2000) Distributed neural systems for the generation of visual images. *Neuron*, 28, 979–90.

Janmaat, K.R.L., Polasnky, L. Ban, S.D. & Boesch, C. (2014). Wild chimpanzees plan their breakfast time, type, and location. *Proc Nat Acad Sci*, 111, 16343–8.

Johnson-Laird, P.N. (2001) Mental models and deduction. *Trends Cogn Sci*, 5, 434–42.

Jordan, K., Heinze, H.J., Lutz, K., Kanowski, M., & Jancke, L. (2001) Cortical activations during the mental rotation of different visual objects. *Neuroimage*, 13, 143–52.

Kadohisa, M., Watanabe, K., Kusunoki, M., Buckley, M.J., & Duncan, J. (2020) Focused Representation of Successive Task Episodes in Frontal and Parietal Cortex. *Cereb Cortex*, 30, 1779–96.

Kaller, C.P., Heinze, K., Frenkel, A., Lappchen, C.H., Unterrainer, J.M., Weiller, C., Lange, R., & Rahm, B. (2013) Differential impact of continuous theta-burst stimulation over left and right DLPFC on planning. *Hum Brain Mapp*, 34, 36–51.

Karmiloff-Smith, A. (1992) *Beyond Modularity*. MIT Press, Cambridge.

Kawagoe, T., Onoda, K., & Yamaguchi, S. (2019) The neural correlates of 'mind blanking': When the mind goes away. *Hum Brain Mapp*, 40, 4934–40.

Khalighinejad, N., Bongioanni, A., Verhagen, L., Folloni, D., Attali, D., Aubry, J.F., Sallet, J., & Rushworth, M.F.S. (2019) A basal forebrain-cingulate circuit in macaques decides it is time to act. *Neuron*, 105, 370–84.

Kim, J.-N. & Shadlen, M.N. (1999) Neural correlates of a decision in the dorsolateral prefrontal cortex of the macaque. *Nat Neurosci*, 2, 176–85.

Kobayashi, Y. & Amaral, D.G. (2003) Macaque monkey retrosplenial cortex: II. Cortical afferents. *J Comp Neurol*, 466, 48–79.

Kobayashi, Y. & Amaral, D.G. (2007) Macaque monkey retrosplenial cortex: III. Cortical efferents. *J Comp Neurol*, 502, 810–33.

Koch, C., Massimini, M., Boly, M., & Tononi, G. (2016) Neural correlates of consciousness: Progress and problems. *Nat Rev Neurosci*, 17, 307–21.

Kolling, N., Scholl, J., Chekroud, A., Trier, H.A., & Rushworth, M.F.S. (2018) Prospection, perseverance, and insight in sequential behavior. *Neuron*, 99, doi: 10.1016/jneuron.2018.08.018.

Kondo, H., Saleem, K.S., & Price, J.L. (2005) Differential connections of the perirhinal and parahippocampal cortex with the orbital and medial prefrontal networks in macaque monkeys. *J Comp Neurol*, 493, 479–509.

Lau, H. & Rosenthal, D. (2011) Empirical support for higher-order theories of conscious awareness. *Trends Cogn Sci*, 15, 365–73.

Lau, H.C. & Passingham, R.E. (2006) Relative blindsight in normal observers and the neural correlate of visual consciousness. *Proc Natl Acad Sci USA*, 103, 18763–8.

Lau, H.C., Rogers, R.D., Haggard, P., & Passingham, R.E. (2004) Attention to intention. *Science*, 303, 1208–10.

Levine, D.N., Calvanio, R., & Popovics, A. (1982) Language in the absence of inner speech. *Neuropsychol*, 20, 391–409.

Li, H. & Durbin, R. (2011) Inference of human population history from individual whole-genome sequences. *Nature*, 475, 493–6.

Liang, P., Jia, X., Taatgen, N.A., Zhong, N., & Li, K. (2014) Different strategies in solving series completion inductive reasoning problems: An fMRI and computational study. *Int J Psychophysiol*, 93, 253–60.

Libet, B., Gleason, C.A., Wright, E.W., & Pearl, D.K. (1983) Time of conscious intention to act in relation to onset of cerebral activity (readiness-potential). The unconscious initiation of a freely voluntary act. *Brain*, 106, 623–42.

Lorey, B., Naumann, T., Pilgramm, S., Petermann, C., Bischoff, M., Zentgraf, K., Stark, R., Vaitl, D., & Munzert, J. (2013) How equivalent are the action execution, imagery, and observation of intransitive movements? Revisiting the concept of somatotopy during action simulation. *Brain Cogn*, 81, 139–50.

Martinez-Manrique, F. & Vicente, A. (2015) The activity view of inner speech. *Front Psychol*, 6, 232.

Mashour, G.A., Roelfsema, P., Changeux, J.P., & Dehaene, S. (2020) Conscious processing and the global neuronal workspace hypothesis. *Neuron*, 105, 776–98.

Mason, M.F., Norton, M.I., Van Horn, J.D., Wegner, D.M., Grafton, S.T., & Macrae, C.N. (2007) Wandering minds: The default network and stimulus-independent thought. *Science*, 315, 393–5.

McCormick, C., Barry, D.N., Jafarian, A., Barnes, G.R., & Maguire, E.A. (2020) vmPFC drives hippocampal processing during autobiographical memory recall regardless of remoteness. *cereb cortex*, 30, 5972–87.

McCormick, C., rosenthal, C.R., Miller, T.D., & Maguire, E.A. (2018) Mind-wandering in people with hippocampal damage. *J Neurosci*, 38, 2745–54.

Mechelli, A., Price, C.J., Friston, K.J., & Ishai, A. (2004) Where bottom-up meets top-down: Neuronal interactions during perception and imagery. *Cereb Cortex*, 14, 1256–1265.

Merten, K. & Nieder, A. (2012) Active encoding of decisions about stimulus absence in primate prefrontal cortex neurons. *Proc Natl Acad Sci USA*, 109, 6289–94.

Middlebrooks, P.G. & Sommer, M.A. (2012) Neuronal correlates of metacognition in primate frontal cortex. *Neuron*, 75, 517–30.

Miele, D.B., Wager, T.D., Mitchell, J.P., & Metcalfe, J. (2011) Dissociating neural correlates of action monitoring and metacognition of agency. *J Cogn Neurosci*, 23, 3620–36.

Miyamoto, K., Osada, T., Setsuie, R., Takeda, M., Tamura, K., Adachi, Y., & Miyashita, Y. (2017) Causal neural network of metamemory for retrospection in primates. *Science*, 355, 188–93.

Morales, J., Lau, H., & Fleming, S.M. (2018) Domain-general and domain-specific patterns of activity supporting metacognition in human prefrontal cortex. *J Neurosci*, 38, 3534–46.

Moris, N., Shrivastava, J., Jeffery, L., Li, J.J., Hayles, J., & Nurse, P. (2016) A genome-wide screen to identify genes controlling the rate of entry into mitosis in fission yeast. *Cell Cycle*, 15, 3121–30.

Moutard, C., Dehaene, S., & Malach, R. (2015) Spontaneous fluctuations and non-linear ignitions: Two dynamic faces of cortical recurrent loops. *Neuron*, 88, 194–206.

Mulcahy, N.J. & Call, J. (2006) Apes save tools for future use. *Science*, 312, 1038–40.

Na, D.G., Ryu, J.W., Byun, H.S., Choi, D.S., Lee, E.J., Chung, W.I., Cho, J.M., & Han, B.K. (2000) Functional MR imaging of working memory in the human brain. *Korean J Radiol*, 1, 19–24.

Nahab, F.B., Kundu, P., Gallea, C., Kakareka, J., Pursley, R., Pohida, T., Miletta, N., Friedman, J., & Hallett, M. (2011) The neural processes underlying self-agency. *Cereb Cortex*, 21, 48–55.

Nee, D.E. & D'Esposito, M. (2018) The representational basis of working memory. *Curr Top Behav Neurosci*, 37, 213–30.

Neubert, F.X., Mars, R.B., Thomas, A.G., Sallet, J., & Rushworth, M.F. (2014) Comparison of human ventral frontal cortex areas for cognitive control and language with areas in monkey frontal cortex. *Neuron*, 81, 700–13.

Noveck, I.A., Goel, V., & Smith, K.W. (2004) The neural basis of conditional reasoning with arbitrary content. *Cortex*, 40, 613–22.

Nunn, J.A., Gregory, L.J., Brammer, M., Williams, S.C., Parslow, D.M., Morgan, M.J., Morris, R.G., Bullmore, E.T., Baron-Cohen, S., & Gray, J.A. (2002) Functional magnetic resonance imaging of synesthesia: Activation of V4/V8 by spoken words. *Nat Neurosci*, 5, 371–5.

Odegaard, B., Knight, R.T., & Lau, H. (2017) Should a few null findings falsify prefrontal theories of conscious perception? *J Neurosci*, 37, 9593–602.

Orr, C. & Hester, R. (2012) Error-related anterior cingulate cortex activity and the prediction of conscious error awareness. *Front Hum Neurosci*, 6, 177.

Osada, T., Adachi, Y., Miyamoto, K., Jimura, K., Setsuie, R., & Miyashita, Y. (2015) Dynamically allocated hub in task-evoked network predicts the vulnerable prefrontal locus for contextual memory retrieval in macaques. *PLoS Biol*, 13, e1002177.

Passingham, R.E. & Lau, H.C. (2019) Acting, Seeing, and Conscious Awareness. *Neuropsychol*, 128, 241–8.

Pearson, J. & Kosslyn, S.M. (2015) The heterogeneity of mental representation: Ending the imagery debate. *Proc Natl Acad Sci USA*, 112, 10089–92.

Pearson, J., Naselaris, T., Holmes, E.A., & Kosslyn, S.M. (2015) Mental imagery: Functional mechanisms and clinical applications. *Trends Cogn Sci*, 19, 590–602.

Persaud, N., Davidson, M., Maniscalco, B., Mobbs, D., Passingham, R.E., Cowey, A., & Lau, H. (2011) Awareness-related activity in prefrontal and parietal cortices in blindsight reflects more than superior visual performance. *Neuroimage*, 58, 605–11.

Philippi, C.L., Tranel, D., Duff, M., & Rudrauf, D. (2014) Damage to the default mode network disrupts autobiographical memory retrieval. *Soc Cogn Affect Neurosci*, 10, 165–75

Postle, B.R., Berger, J.S., & D'Esposito, M. (1999) Functional neuroanatomical double dissociation of mnemonic and executive control processes contributing to working memory performance. *Proc Nat Acad Sci*, 96, 12959–64.

Postle, B.R., Ferrarelli, F., Hamidi, M., Feredoes, E., Massimini, M., Peterson, M., Alexander, A., & Tononi, G. (2006) Repetitive transcranial magnetic stimulation dissociates working memory manipulation from retention functions in the prefrontal, but not posterior parietal, cortex. *J Cogn Neurosci*, 18, 1712–22.

Proffitt, T., Haslam, M., Mercader, J.F., Boesch, C., & Luncz, L.V. (2018) Revisiting Panda 100, the first archaeological chimpanzee nut-cracking site. *J Hum Evol*, 124, 117–39.

Rao, K.V. & Baddeley, A. (2013) Raven's matrices and working memory: A dual-task approach. *Q J Exp Psychol* (Hove), 66, 1881–7.

Raven, J., Raven, J.C., & Court, J.H. (2003) *Manual for Raven's Progressive Matrices and Vocabulary Scales*. Harcourt Assessment, San Antonio.

Rigoni, D., Brass, M., Roger, C., Vidal, F., & Sartori, G. (2013) Top-down modulation of brain activity underlying intentional action and its relationship with awareness of intention: An ERP/Laplacian analysis. *Exp Brain Res*, 229, 347–57.

Rilling, J.K., Barks, S.K., Parr, L.A., Preuss, T.M., Faber, T.L., Pagnoni, G., Bremner, J.D., & Votaw, J.R. (2007) A comparison of resting-state brain activity in humans and chimpanzees. *Proc Natl Acad Sci USA*, 104, 17146–51.

Roca, M., Garcia, M., Torres Ardila, M.J., Gonzalez Gadea, M.L., Torralva, T., Ferrari, J., Ibanez, A., Manes, F., & Duncan, J. (2018) Rule reactivation and capture errors in goal directed behaviour. *Cortex*, 107, 180–7.

Rounis, E., Maniscalco, B., Rothwell, J.C., Passingham, R.E., & Lau, H. (2010) Theta-burst transcranial magnetic stimulation to the prefrontal cortex impairs metacognitive visual awareness. *Cogn Neurosc*, 1, 165–75.

Rowe, J.B., Owen, A.M., Johnsrude, I.S., & Passingham, R.E. (2001) Imaging the mental components of a planning task. *Neuropsychol*, 39, 315–27.

Ruby, E., Maniscalco, B., & Peters, M.A.K. (2018) On a 'failed' attempt to manipulate visual metacognition with transcranial magnetic stimulation to prefrontal cortex. *Conscious Cogn*, 62, 34–41.

Schacter, D.L., Addis, D.R., & Buckner, R.L. (2008) Episodic simulation of future events: Concepts, data, and applications. *Ann N Y Acad Sci*, 1124, 39–60.

Searle, J. (2007) *Freedom and Neurobiology*. Columbia University Press, New York.

Seli, P., Beaty, R.E., Cheyne, J.A., Smilek, D., Oakman, J., & Schacter, D.L. (2018) How pervasive is mind wandering, really? *Conscious Cogn*, 66, 74–8.

Shallice, T. (1982) Specific impairments of planning. *Phil Trans Roy Soc Lond B*, 298, 199–209.

Shallice, T. & Burgess, P.W. (1991) Deficits in strategy application following frontal lobe damage in man. *Brain*, 114 (Pt 2), 727–41.

Shallice, T. & Cooper R, P. (2011) *The Organization of Mind*. Oxford University Press, Oxford.

Shea, N. & Frith, C.D. (2019) The global workspace needs metacognition. *Trends Cogn Sci*, 23, 560–71.

Shulman, G.L., Pope, D.L., Astafiev, S.V., McAvoy, M.P., Snyder, A.Z., & Corbetta, M. (2010) Right hemisphere dominance during spatial selective attention and target detection occurs outside the dorsal frontoparietal network. *J Neurosci*, 30, 3640–51.

Simons, C.J., Tracy, D.K., Sanghera, K.K., O'Daly, O., Gilleen, J., Dominguez, M.D., Krabbendam, L., & Shergill, S.S. (2010) Functional magnetic resonance imaging of inner speech in schizophrenia. *Biol Psychiatry*, 67, 232–7.

Spreng, R.N., Stevens, W.D., Chamberlain, J.P., Gilmore, A.W., & Schacter, D.L. (2010) Default network activity, coupled with the frontoparietal control network, supports goal-directed cognition. *Neuroimage*, 53, 303–17.

Stephan, H., Frahm, H., & Baron, G. (1981) New and revised data on volumes of brain structures in insectivores and primates. *Folia Primatol*, 35, 1–29.

Stevens, N.J., Seiffert, E.R., O'Connor, P.M., Roberts, E.M., Schmitz, M.D., Krause, C., Gorscak, E., Ngasala, S., Hieronymus, T.L., & Temu, J. (2013) Palaeontological evidence for an Oligocene divergence between Old World monkeys and apes. *Nature*, 497, 611–14.

Stoerig, P. & Cowey, A. (1997) Blindsight in man and monkey. *Brain*, 120, 535–59.

Stuphorn, V., Taylor, T.L., & Schall, J.D. (2000) Performance monitoring by the supplementary eye field. *Nature*, 408, 857–60.

Swartz, B.L., Hoffman, M.L., & Evans, S. (2005) Episodic-like memory in a gorilla: A review and new findings. *Learn Motiv*, 36, 226–44.

Tacikowski, P., Berger, C.C., & Ehrsson, H.H. (2017) Dissociating the neural basis of conceptual self-awareness from perceptual awareness and unaware self-processing. *Cereb Cortex*, 27, 3768–81.

Taschereau-Dumouchel, V., Kawato, M., & Lau., H. (2019) Multivoxel pattern analysis reveals dissociations between subjective fear and its physiological correlates. *Mol Psychiatry*, 25, 2342–54.

Thakral, P.P., Madore, K.P., & Schacter, D.L. (2020) The core episodic simulation network dissociates as a function of subjective experience and objective content. *Neuropsychol*, 136, 107263.

Thompson, K.G., Thompson, K.G., & Schall, J.D. (1999) The detection of visual signals by macaque frontal eye field during masking. *Nat Neurosci*, 2, 283–8.

Tranel, D., Hathaway-Nepple, J., & Anderson, S.W. (2007) Impaired behavior on real-world tasks following damage to the ventromedial prefrontal cortex. *J Clin Exp Neuropsychol*, 29, 319–32.

Tulving, E. (2001) Episodic memory and common sense: How far apart? *Philos Trans R Soc Lond B Biol Sci*, 356, 1505–15.

Unterrainer, J.M., Kaller, C.P., Halsband, U., & Rahm, B. (2006) Planning abilities and chess: A comparison of chess and non-chess players on the Tower of London task. *Br J Psychol*, 97, 299–311.

Unterrainer, J.M., Rahm, B., Kaller, C.P., Ruff, C.C., Spreer, J., Krause, B.J., Schwarzwald, R., Hautzel, H., & Halsband, U. (2004) When planning fails: Individual differences and error-related brain activity in problem solving. *Cereb Cortex*, 14, 1390–7.

Van Horn, J.D., Irimia, A., Torgerson, C.M., Chambers, M.C., Kikinis, R., & Toga, A.W. (2012) Mapping connectivity damage in the case of Phineas Gage. *PLoS One*, 7, e37454.
van Vugt, B., Dagnino, B., Vartak, D., Safaai, H., Panzeri, S., Dehaene, S., & Roelfsema, P.R. (2018) The threshold for conscious report: Signal loss and response bias in visual and frontal cortex. *Science*, 360, 537–42.
Webster, M.J., Bachevalier, J., & Ungerleider, L.G. (1994) Connections of inferior temporal areas TEO and TE with parietal and frontal cortex in macaque monkeys. *Cer Cort*, 4, 471–83.
Wechsler, D. (2008) *Wechsler Adult Intelligence Scale IV*. Psychological Corporation, San Antonio, Texas.
Weiskrantz, L. (1997) *Consciousness Lost and Found*. Oxford University Press, Oxford.
Weiskrantz, L. (2009) *Blindsight*. Oxford University Press, Oxford.
Woolgar, A., Duncan, J., Manes, F., & Fedorenko, E. (2018) The multiple-demand system but not the language system supports fluid intelligence. *Nat Hum Behav*, 2, 200–4.
Woolgar, A., Parr, A., Cusack, R., Thompson, R., Nimmo-Smith, I., Torralva, T., Roca, M., Antoun, N., Manes, F., & Duncan, J. (2010) Fluid intelligence loss linked to restricted regions of damage within frontal and parietal cortex. *Proc Natl Acad Sci USA*, 107, 14899–902.
Woolnough, O., Rollo, P.S., Forseth, K.J., Kadipasaoglu, C.M., Ekstrom, A.D., & Tandon, N. (2020) Category selectivity for face and scene recognition in human medial parietal cortex. *Curr Biol*, 30, 2707–15 e2703.
Wyart, V., Dehaene, S., & Tallon-Baudry, C. (2011) Early dissociation between neural signatures of endogenous spatial attention and perceptual awareness during visual masking. *Front Hum Neurosci*, 6, 16.
Wynn, T., Hernandez-Aguilar, R.A., Marchant, L.F., & McGrew, W.C. (2011) 'An ape's view of the Oldowan' revisited. *Evol Anthropol*, 20, 181–97.
Zeidman, P., Mullally, S.L., & Maguire, E.A. (2015) Constructing, perceiving, and maintaining scenes: Hippocampal activity and connectivity. *Cereb Cortex*, 25, 3836–55.

11
Human Prefrontal Cortex
Language, Culture, and Social Rules

Introduction

We do not know for certain when our ancestors first developed the ability to use grammatical language; however, they would have been able to use proto-signs (Arbib, 2005) and proto-speech (Fitch, 2017) before that time. There is no way of knowing whether these developed at the same time or in tandem.

Bonobos have a repertoire of at least thirty-three gestural types, and they share many of these with chimpanzees (Graham et al., 2018). For example, they have different gestures to request an object, to stop a particular behaviour, for the infant to climb onto the mother, and for an adult to accompany them. The gestures can be used singly or in sequence, and chimpanzees often vocalize when the gesture fails to have the desired effect (Hobaiter et al., 2017). Modern human infants of 1 to 2 years old have been reported to have a repertoire of fifty-two gestures, and some of these are the same as ones that are used by bonobos and chimpanzees (Kersken et al., 2019).

However, chimpanzees in the wild rarely point (Hobaiter et al., 2014), perhaps because they move around as a group when they forage and do not need to do so. But chimpanzees in captivity do point when they interact with humans (Halina et al., 2018), and they can use pointing to ask for food whether it is close or more distant (Gonseth et al., 2017). They also understand pointing (Hopkins et al., 2013b), and are sensitive to where the person is looking (Halina et al., 2018).

Handedness

Hopkins and his colleagues have looked to see whether chimpanzees have a preferred hand when they gesture. To do this, they studied the hand used for begging for food in a sample of 227 captive chimpanzees (Hopkins et al., 2005a). They also observed the hand used for throwing, for example during fights in eighty-nine captive chimpanzees (Hopkins et al., 2005b).

The results are shown in Table 11.1 (next page). Hopkins et al. (2013a) have reported data for a bigger sample of 300 chimpanzees in captivity, but they simply divided the animals as being right-handed or left-handed. The figures are not

Table 11.1 Handedness in chimpanzees.

	Right-handed	Left-handed	Either
Gestures	59%	16%	25%
Throwing	56%	26%	18%

therefore included in this table because the paper does not provide figures for the proportion that were as likely to use either hand.

The results in Table 11.1 can be compared with the data from a study by Annett (2004) in which she observed the hand used by a very large sample of children and undergraduates. She included six tests of 'primary actions' such as writing, throwing, striking a match, and using a toothbrush. Table 11.2 presents the results, together with the data for writing and throwing on their own. The results show a very strong shift to right-handedness, and Annett related this to her theory that there is a 'right-shift gene' in the human population.

Comparison of the two tables indicates that the right-shift in chimpanzees is roughly a third of the size of that observed in the human population (Annett, 2006).

Even so, caution is necessary in interpreting the data for chimpanzees. The hand used for gestures such as begging could be influenced by the hand that the carers have used for providing the food while the animals were in captivity. So, data are needed for the hand used when chimpanzees gesture in the wild. Unfortunately, though Roberts et al. (2019) kept a record of whether chimpanzees in the Budongo forest in Uganda gestured with their right or left hand, they did not present any quantitative data in the paper that could be included in Table 11.1.

Nonetheless, the data for throwing are unlikely to be contaminated by feeding. Throwing has been observed by captive chimpanzees not only during fights but also when warding off strangers as they approached. The data for throwing are important because there is a marked right shift for throwing in children and adults (Annett, 2004) (Table 11.2).

To find out whether genetic factors were involved in the right shift in chimpanzees, Hopkins et al. (2015) analysed the effect of familial relationships on

Table 11.2 Handedness in human subjects.

	Right-handed	Left-handed	Either
Six primary actions	78%	5%	17%
Writing and throwing	87%	9%	4%

handedness in captive animals. The test involved using a tool to extract food from a pipe, and it was designed to model the termite fishing observed in the wild. Because infants might imitate the handedness of their mothers, it was critical that there were fifty-three animals in the sample that had been reared in captivity apart from their mother. It was therefore possible to derive estimates of heritability that were not confounded by the opportunity to imitate the mother. The analysis suggested that some of the variance was due to heritability.

Summary

Chimpanzees have a repertoire of gestures in the wild. There is a slight right shift in the hand used by chimpanzees to gesture in captivity. However, it is only about a third of the shift observed in modern human populations.

Cerebral Dominance

When humans point, they are most likely to use their right hand, and this is controlled by the same hemisphere as speech. Milner (1975) used the Wada test to temporarily inactivate one hemisphere to demonstrate the association between handedness and speech. She found that in a sample of 262 patients the left hemisphere was dominant for speech in 96% of right handers, and in 70% of left handers.

Gestures

To find out if chimpanzees showed cerebral dominance for gesturing, Taglialatela et al. (2008) used PET to visualize the areas that were activated while the animals gestured for food. There was an activation in an area in the left hemisphere that Taglialatela et al. described as being the Broca's homologue, that is the inferior part of the caudal PF cortex (area 44).

However, in a follow-up study with four chimpanzees, Taglialatela et al. (2011) were able to compare two chimpanzees that vocalized while they gestured with two chimpanzees that did not. The activation in the inferior caudal PF cortex was only significant in the chimpanzees that vocalized and gestured at the same time.

The Broca's homologue in chimpanzees has been described on the basis of sulcal morphology (Falk et al., 2018). But it is not clear from the horizontal section shown in the paper by Taglialatela et al. (2008) whether the activation was in the dorsal area 44d as opposed to the opercular division area 44v (Amunts et al., 2010). The distinction is important because it is the opercular division that is particularly associated with sequencing human speech (Goucha & Friederici, 2015).

It is possible, though not certain, that proto-speech evolved from the sorts of vocalization that we observe in modern chimpanzees. However, proto-language, whether by sound or gesture, would have differed from the vocalizations that we observe in chimpanzees and other primates in an important respect. This is that it was learned, and that it could be used to refer to things, that is to name. It is true that vervet monkeys (*Cercopithecus aethiops*), for example, have different calls when they spy eagles or snakes (Seyfarth et al., 1980); but the difference can be accounted for by the state of the animal (Fischer & Price, 2017). Words such as names differ in that they retain their meaning, whatever the internal state (Loh et al., 2017).

In the human brain there is an association between the areas that are involved in naming via gestures and naming with words. In a PET study by Emmorey et al. (2007), twenty-nine deaf subjects used American Sign Language (ASL) to name objects. The advantage of using PET is that the hands are outside the scanner and movements of the arms and hands do not cause artefacts in the signal. Emmorey et al. also studied sixty-four hearing speakers while they named objects verbally.

Whether naming occurred via gestures or words, there were activations in the left inferior caudal PF area 45B (Figures 11.1A and B, next page). This area differs from the more posterior area 44 in that it is involved in semantics (Goucha & Friederici, 2015). Thus, the activation was in area 45B because the task was to name the objects.

However, there was a difference between the results for naming with words and signs. Emmorey et al. (2007) only found activation in the left inferior parietal cortex (areas PF and PFG) during signing (Figure 11.1B).

To find out whether these areas were critical for signing, Corina et al. (1999) applied electrical stimulation to disrupt activity during neurosurgery. Stimulation of either the left inferior caudal PF cortex or the inferior parietal cortex caused errors in signing.

Dominance for Skilled Actions

The left hemisphere is also dominant for actions other than gesturing. In a PET study by Schluter et al. (2001), the subjects had to move a different finger depending on the pattern shown on the screen, and to do so as quickly as possible. Choice reaction times were compared with simple reaction times in which the same finger was used whatever the pattern. For choice reaction times, there were activations in the left, but not right, hemisphere irrespective of whether the right or left hand was used. The activated areas included the dorsal PF cortex (area 46) and dorsal premotor cortex, as well as area AIP in the IPS and tissue further posterior in the IPS. These areas form part of an interconnected system (Chapter 6).

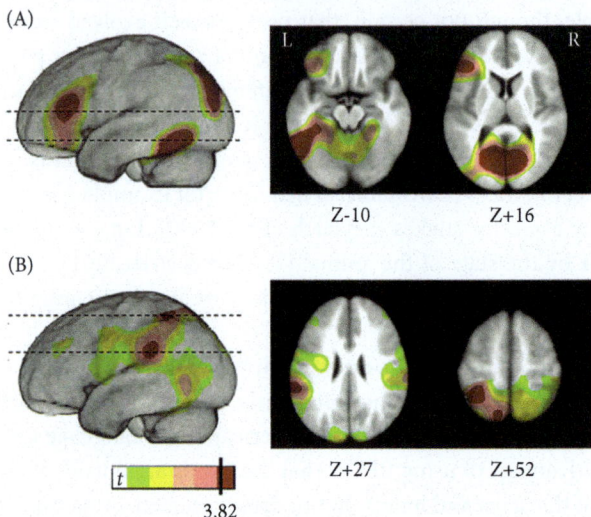

Figure 11.1 (A) Areas that were activated during signing and speaking. (B) Areas that were only activated during signing
The dotted horizontal lines on the lateral views indicate where the horizontal sections shown on the right were taken from.
Reproduced from Emmorey, K., Mehta, S., & Grabowski, T.J. The neural correlates of sign versus word production. *Neuroimage*, 36 (1), 202–8, Figure 1a and b, Doi: 10.1016/j.neuroimage.2007.02.040 Copyright © 2007 Elsevier Inc. All rights reserved.

To find out whether the PF activation was causal, Rounis et al. (2007) applied repetitive transcranial magnetic stimulation (rTMS) at 5 Hz over the left or right dorsal PF cortex before testing subjects on a motor version of the Posner task. A pre-cue told the subjects whether they were going to use a left or right finger, but on 20% of trials this pre-cue proved to be invalid. There was a specific effect: only rTMS over the left dorsal PF cortex slowed reaction times on invalid trials.

The left hemisphere is also dominant for observing actions. In an fMRI study, Biagi et al. (2010) showed subjects a video of a hand playing notes on piano keys. The movements in the videos were performed either with the right or left fingers. Nonetheless, there was an activation in the left, but not right AIP irrespective of the hand used.

Thus, there is an association between dominance for speech and dominance for actions that involve movements of the fingers. The reason may be that humans have a remarkable ability to move all the fingers independently, as in piano playing. And they can perform rapid and accurate movements of the hand and arm, as in striking flakes from stone tools or striking a match. Like the rapid and coordinated laryngeal and oral movements that are used in speaking, these actions with the hands involve skill.

Ringo et al. (1994) suggested a reason why cerebral dominance might be important for skilled performance. There is a time-cost for transmission across the cerebral commissures. Since skilled actions require close coordination and accurate timing, it may be easier to achieve this if they are directed by a single hemisphere.

Anatomical Asymmetry

In the human brain there is a relation between cerebral dominance and anatomical asymmetries. This was first shown by taking measurement along the supratemporal plane (Geschwind & Levitsky, 1968; Galaburda et al., 1978). A similar asymmetry favouring the left hemisphere has been shown to exist in the brains of chimpanzees (Marie et al., 2018).

Heffner and Heffner (1986) have also claimed that there is a functional asymmetry in macaques. They reported that left-sided, but not right-sided, lesions of the superior temporal cortex impaired the ability of Japanese macaques (*Macaca fuscata*) to discriminate between different 'coo' calls. Yet, the lesions were of the same size in the left and right hemisphere.

However, even though there are animals such as home-reared chimpanzees (Savage-Rumbaugh & Lewin, 1994) and dogs (Kaminski et al., 2004) that can understand the meaning of words, they are unable to repeat them. In a study of patients Parker et al. (2014) found that the ability to do this depends on the arcuate fasciculus. In the human brain this connects the middle as well as the superior temporal gyrus with the inferior caudal PF cortex (areas 44 and 45B). Patients with damage to this tract were severely impaired at repeating words.

However, in spite of early claims by Amunts et al. (2003), later reports by the same group (Keller et al., 2009) argued that there is no consistent asymmetry between the left and right inferior caudal PF cortex (areas 44 and 45B) in terms of gross morphology.

Nonetheless, there is an asymmetry in the connections that run in the arcuate fasciculus. Eichert et al. (2019) used diffusion weighted imaging (DWI) to compare the areas that are connected by the arcuate fasciculus in each hemisphere of the human brain. Figure 11.2C (next page) shows in red the areas that are linked by the arcuate fasciculus (Eichert et al., 2018). The histograms in Figures 11.2A and B indicate that the human arcuate fasciculus (AF) and inferior frontal occcipital fasciculus (IFO) are more extensive in the left than right hemisphere. The same Figures indicate that these asymmetries are absent in the macaque monkey brain.

Caspers et al. (2011) have also used DWI to visualize the section of the arcuate fasciculus that links the inferior parietal cortex with the inferior caudal PF cortex. They found that in the human brain the path was asymmetric, being larger on the left.

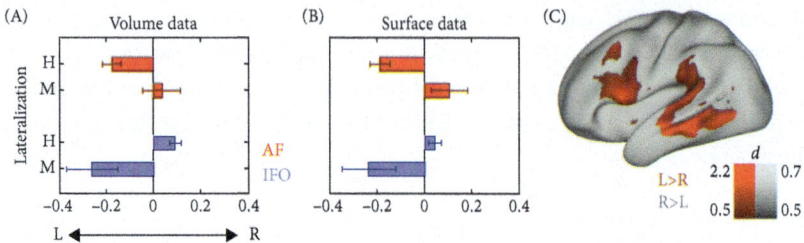

Figure 11.2 The arcuate fasciculus in the macaque monkeys and human brain (A) Surface asymmetries of the arcuate fasciculus (AF) and inferior frontal occipital fasciculus (IFO). (B) Volume asymmetries of the AF and AIFO. (C) Areas marked in red are greater in the left than the right hemisphere of the human brain. M = macaque, H = Human.

Reproduced from Eichert, N., Verhagen, L., Folloni, D., Jbabdi, S., Khrapitchev, A.A., Sibson, N.R., Mantini, D., Sallet, J., & Mars, R.B. What is special about the human arcuate fasciculus? Lateralization, projections, and expansion. *Cortex*, 118, 107–15, Figure 2, Doi: 10.1016/j.cortex.2018.05.005 Copyright © The Authors. Licensed under CC BY 4.0.

Rilling et al. (2012) have claimed that the arcuate fasciculus is also asymmetric in the chimpanzee brain. However, Bryant et al. (2020) have re-analysed the data for the arcuate fasciculus in chimpanzees using a volume measure, and they were unable to demonstrate any such asymmetry in the chimpanzee brain.

It is not clear to what extent the asymmetry in the human brain is the result of experience. It is well established that changes in volume can occur in both the grey matter (Woollett & Maguire, 2011) and white matter (Sampaio-Baptista & Johansen-Berg, 2017) as the result of practice. It has also been found that the arcuate fasciculus as delineated by DWI is more distinct in those who have learned to read than in those who have not (Thiebaut de Schotten et al., 2014). It remains to establish whether there is an initial asymmetry in the human arcuate fasciculus early in life.

Summary

There is cerebral dominance for speech in the human brain. However, the left hemisphere is also dominant for the performance of skilled actions other than speech. It is activated whichever hand is used. Cerebral dominance is also associated with an anatomical asymmetry in the volume of the arcuate fasciculus in the human brain. This asymmetry is not found in chimpanzees or macaque monkeys.

Language

A learned proto-language, whether via gesture or sound, cannot develop unless individuals are able to repeat what they see or hear. Though chimpanzees kept in a zoo have been reported to imitate the actions of visitors, this was in the context of play rather than communication (Persson et al., 2018). In contrast, preschool children of 3 years old spontaneously imitate meaningful gestures, and when they do so they prefer to use their dominant hand (Sebastianutto et al., 2017).

Repeating Gestures and Sounds

Caspers et al. (2010) compared merely observing movements of the hand with imitating them. The study involved a meta-analysis of eighty-seven functional brain imaging experiments in which the subjects were required to observe or imitate actions carried out by the hand, whether these involved gestures or actions performed in relation to objects. Figure 11.3A (next page) shows that there were activations in the inferior parietal cortex both when observing and when imitating actions.

However, when imitation was contrasted with observation. there was an activation in the inferior caudal PF cortex (area 44), shown in green in (Figure 11.3B). This was the opercular division of area 44 (area 44v) (Amunts & Zilles, 2012).

In a later meta-analysis, Kühn et al. (2013) specifically compared the activations in the inferior caudal PF cortex during imitation with the activations during speech, both simple and complex. Figure 11.4 shows sagittal sections through the inferior caudal PF cortex.

It can be seen from Figure 11.4A (page 429) that the activations in the left hemisphere for imitation (red) and for complex and simple speech (green) were very close together, even though there was a tendency for the activation for imitation to be slightly more dorsal than for speech (Figure 11.4B). Figure 11.4C indicates that the activation of tasks requiring motor inhibition were in the inferior frontal junction (IFJ, area 44d), and in the right hemisphere.

The studies included for speech by Kuhn et al. (2013) were not restricted to word repetition: they included fluent speech. Hope et al. (2014) have reviewed the fMRI studies that involved the repetition of words alone. The circuit involved the left middle temporal gyrus and the inferior caudal PF cortex (area 45B). The activation was confined to the superior temporal gyrus when subjects simply listened to words (Price, 2000). The difference between the circuit for imitating hand and finger movements compared with repeating words is consistent with a shift in the origin of the arcuate fasciculus in the human brain. Rather than simply taking its origin from area Tpt (Frey et al., 2014) and the inferior parietal cortex (Catani et al., 2005) as in macaque monkeys, it also connects the superior and middle temporal gyrus with the inferior caudal PF cortex (Eichert et al., 2019).

However, the expansion of the origin of the arcuate fasciculus in the human brain does not tell us whether gestures and sounds took on a communicative meaning at the same time during human evolution. The alternative is that, as Arbib (2005) has proposed, the substrate for proto-signs provided the scaffolding for the later development of proto-speech.

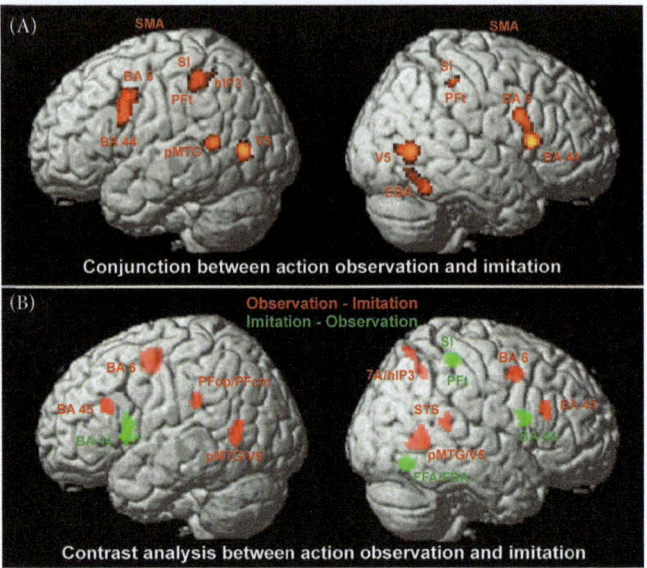

Figure 11.3 (A) The areas that were activated both when observing and when imitating. (B) The areas that were activated when imitating was contrasted with observing alone in green, and when observation was contrasted with imitation in red. The labels on the brains indicate the areas that were activated.

Reproduced from Caspers, S., Zilles, K., Laird, A.R., & Eickhoff, S.B. ALE meta-analysis of action observation and imitation in the human brain. *Neuroimage*, 50 (3), 1148–67, Figure 7, Doi: 10.1016/j.neuroimage.2009.12.112 Copyright © 2009, Elsevier Inc. All rights reserved.

Because chimpanzees lack connections from the superior and middle temporal gyrus to the inferior caudal PF cortex (areas 44 and 45), it is easier to teach them to name by using gestures than sounds. The chimpanzee Washoe was taught a simplified version of ASL by moulding her hands. Once taught, Washoe was able to use gestures to reliably name thirty-two pictures during a blind test (Gardner & Gardner, 1985). However, chimpanzees fail to imitate novel actions or gestures (Tennie et al., 2012), and in this respect they are like new-born babies (Heyes, 2016). The arcuate fasciculus connecting the inferior parietal and temporal lobe to the inferior caudal PF cortex (areas 44 and 45B) is not myelinated at birth and is not fully developed until the age of 7 (Friederici, 2012).

Like the ability to imitate gestures, the ability of human infants to imitate sounds also develops with age (Tchernichovski & Marcus, 2014). It depends in part on the FoxP2 gene. This gene was originally described in members of a four generational family referred to as the KE family. Half the members of the family have a severe speech impairment, and this shows up particularly when they are asked to repeat words (Vargha-Khadem et al., 1995). A genetic analysis of this family pointed to a dominant gene with a locus on chromosome seven, and further work revealed that

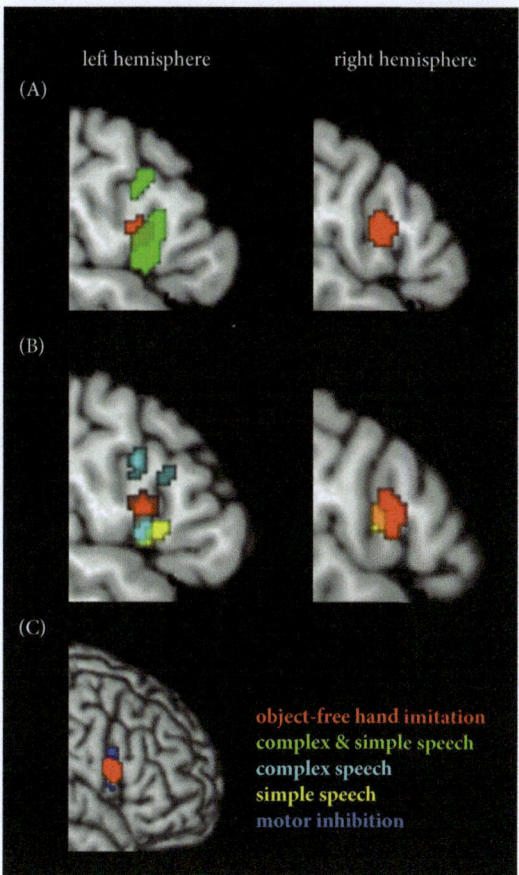

Figure 11.4 Activations for speaking, imitation, and motor inhibition in the left and right hemisphere
(A) Imitation in red, simple and complex speech in green. (B) Imitation in red, simple speech in yellow, complex speech in blue. (C) Surface view of the inferior caudal PF cortex in the right hemisphere. Imitation in red, motor inhibition in dark blue.

Reproduced from Kühn, S., Brass, M., & Gallinat, J. Imitation and speech: Commonalities within Broca's area. *Brain Struct Funct*, 218, 1419–27, Figure 1, https://doi.org/10.1007/s00429-012-0467-5 Copyright © 2012, Springer Nature.

the impairment was caused by a variation in the FoxP2 gene which encodes a transcription factor (Estruch et al., 2016).

In an fMRI study by Liégeois et al. (2003), affected and unaffected members of this family were required either to repeat words or to generate verbs. Figure 11.5 (next page) shows sections through the areas that were under-active in the affected members of the family. These areas were the left inferior caudal PF cortex (area 44), the left ventral premotor cortex, the left inferior parietal cortex (areas PF and PFG), and the right putamen.

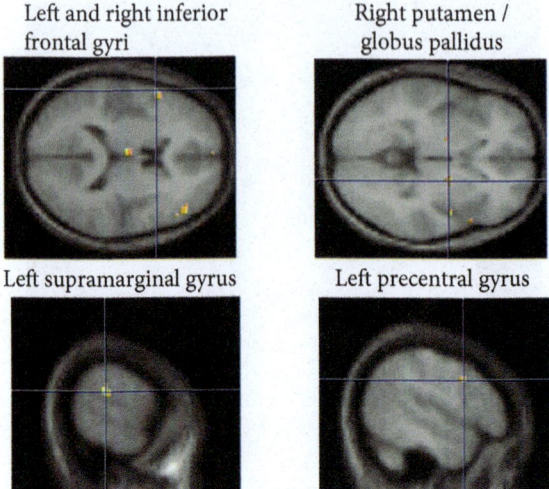

Figure 11.5 Horizontal sections showing the areas that were under-activated in the brains of affected members of the KE family

Reproduced from Liégeois, F., Baldeweg, T., Connelly, A., Gadian, D.G., Mishkin, M., & Vargha-Khadem, F. Language fMRI abnormalities associated with FOXP2 gene mutation. *Nature Neurosci*, 6 (11), 1230–7, Figure 2, Doi: 10.1038/nn1138 Copyright © 2003, Springer Nature.

When unaffected members of the KE family were asked to generate verbs silently, there was strong activation in the left inferior caudal PF cortex. But, remarkably, there was no significant activation in this area in the affected members at all.

Speech

The production of speech involves the control of the laryngeal cartilages and the use of the cheeks and lips to produce stopped consonants such as [p] and [b]. But the inferior caudal PF cortex (areas 44) is neither the cortical larynx area nor the motor area for controlling the mouth. Eichert et al. (2020) have used fMRI to map the activations when human subjects vocalize, and have controlled for breathing so as to identify the motor laryngeal areas. They found two areas, one in the motor cortex and a more ventral on in the premotor cortex.

Flinker et al. (2015) studied the activity of the left inferior caudal PF cortex (areas 44) in patients with electrodes implanted on the cortical surface. Though the inferior caudal PF cortex (area 44) was active before the patients repeated words that they heard, it was silent when they actually did so. The temporal resolution of the recordings enabled Flinker et al. to show that during word repetition the order of events involved activity in the superior temporal gyrus for hearing words, then the inferior caudal PF cortex (area 44) for sequencing the syllables, and finally the mouth area of motor cortex for producing the words.

Syntax

The left inferior caudal PF cortex (area 44) is not simply involved in assembling speech sounds into syllables and words. It is also involved in assembling words into sentences (Goucha & Friederici, 2015).

By definition, sentences are sequences. However, Hauser and Chomsky (2002) have argued that it is a feature of language in the narrow sense that the sequences have a hierarchical structure. They are generated by a tree structure, and this allows the production of an infinite number of sentences via the process of recursion, as in 'I saw the man who patted the cat who meowed at the door that opened for the man [...]'.

One of the requirements in producing a sentence such as this is the ability to generate morphological markers such as tenses and cases. Vargha-Khadem et al. (1995) found that the affected members of the KE family were severely impaired at producing tenses and derivations. Yet there were able to understand grammar when assessed on a test of the reception of grammar, the TROG (Bishop, 1982).

To visualize the areas that support the ability to learn to produce grammatical sentences, Musso et al. (2003) taught German speakers Italian or Japanese. This was compared with teaching pseudo-Italian or pseudo-Japanese which had unreal grammatical rules. An example of Italian was '*La pera è mangiata da Paolo*' (the pear is eaten by Paolo), and an example of pseudo-Italian was '*Perra la mangia Paolo*' (pear the eats Paolo). Musso et al. claimed that there was an activation in the left inferior caudal PF cortex (areas 44 and 45B) only when German speakers learned the natural languages, and not when they learned the pseudo-grammatical rules.

Natural languages all have a hierarchical structure. To find out if this was the critical factor, Pattamadilok et al. (2016) compared sentences with a centre embedded clause with sentences with two adjunctive clauses. The subjects were French, but in translation an example of a centre embedded clause was 'this morning the kids who exhausted their parents slept'; and an example of two adjunctive clauses was 'even if the kids spoke loudly, their parents slept'. The task in this example was to choose which of two probes, 'the kids slept' and 'their parents slept', was appropriate for the sentence.

In an fMRI study, the words of the sentence were presented briefly and serially on a screen, and subjects then had to select between the probes by pressing a button. There was more activation in the left inferior caudal PF cortex (areas 44 and 45B) when the subjects chose between probes when the sentences had an embedded clause as opposed to two adjunct clauses. Pattamadilok et al. also found more activation in the left temporal lobe for the embedded clauses. They claimed that the peak was in the superior temporal sulcus, but according to the atlas of Petrides (2018) it was in the middle temporal gyrus. The assumption was that this activation reflected semantic information, that is the meaning of the words.

It is unclear how the ability to understand and produce embedded sequences evolved. Chomsky and his colleagues (Bolhuis et al., 2014) have argued that there was no precursor whereas others (Progovac, 2016; Jackendoff & Wittenberg, 2017) have suggested that the ability could have developed in stages. For example, Progovac et al. (2018) considered the sentence 'Elena will grow tomatoes'. They pointed out that it consists of a small clause 'grow tomatoes', the specification of the subject 'Elena', and the addition of the verb 'will'. Progovac et al, suggested that proto-language could have consisted of small clauses.

They assumed that if this was the case, there should be activation the left inferior caudal PF cortex (areas 44 and 45B) even for small clauses. In an fMRI experiment, Progovac et al. therefore, compared the activations when subjects read full sentences such as 'the case is closed' with the activations for two-word sentences such as 'fires spread' and small clauses such as 'case-closed'.

As Figure 11.6A (next page) shows, there was significantly more activation in the left inferior caudal PF cortex (area 44) and putamen for full sentences compared with small-clauses. This was not due to the greater number of words since they found the same result for two-word sentences. Nonetheless, the parameter estimates in Figure 11.6B indicate that the left inferior caudal PF cortex (area 44) was significantly activated when the subjects read small clauses, even though to a much lesser extent.

Progovac et al. took these results to be consistent with their account of how sentences might have evolved. But there is, of course, no actual record of the evolutionary development of syntactic language. The best that we can do is test the capacities of other living primates, and to see if they have abilities that could provide a clue.

In order to do this, Jiang et al. (2018) trained two macaque monkeys to reproduce sequences. During the initial training, the monkeys viewed a display with six locations organized in a hexagon. On each trial, two or three locations lit up in turn, and a cue was then presented to tell the monkey whether to touch the same locations in the same order or the mirror order.

The reason for this comparison was that mirror sequences, for example ABC/CBA, involved centre embedding, and in formal linguistics these are described as having a nested structure (Dehaene et al., 2015b). Sequences of this sort have been used in studies of artificial grammar learning with human subjects (Bahlmann et al., 2008), with the difference that consonant vowel syllables were used rather than spatial locations as in the study by Jiang et al. (2018).

The monkeys were able to learn both spatial tasks, but this might simply be a demonstration of rote learning. So, Jiang et al. carried out several transfer tests, analysing the results for the first day on the transfer test. In the first place they found that, once the animals had learned the task with two or three locations, they were able to succeed with four or five locations. Next, once the animals were able to reproduce the locations on a hexagon, they performed above chance on a transfer

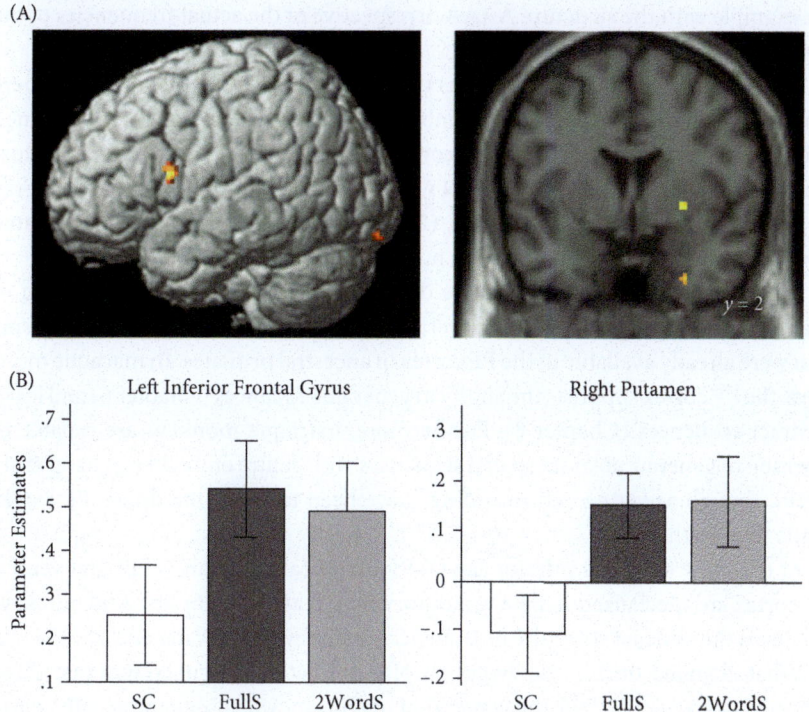

Figure 11.6 Comparing sentences and small clauses
(A) Activations in the left inferior caudal PF cortex (area 44) and the putamen that were greater for full sentences (Full S and 2Word S) than small clauses (Small C).
(B) parameter estimates

Reproduced from Progovac, L., Rakhlin, N., Angell, W., Liddane, R., Tang, L., & Ofen, N. Neural correlates of syntax and proto-syntax: Evolutionary dimension. *Front Psychol*, 9, 2415, Figure 2, Doi: https://doi.org/10.3389/fpsyg.2018.02415 © 2018 Progovac, Rakhlin, Angell, Liddane, Tang and Ofen. Licensed under CC BY 4.0.

test with the locations in different arrangements, for example in a pyramid. Finally, one of the monkeys was trained with clockwise sequences on the hexagonal array. After just one of the locations lit up, the animal could learn to reproduce the rest of the sequence, for example 1–23 321.

The results indicate that macaque monkeys can learn abstract nested rules, and that they show transfer. However, the monkeys needed extensive training to learn the rules. Yet, Jiang et al. (2018) admit that when human infants have been tested on tasks of this sort, they learn them very rapidly.

The sequences in the studies with monkeys were spatial, not sequences of sound. But there is evidence that macaque monkeys can represent abstract sequences of sounds. In an fMRI study by Wang et al. (2015), the monkeys listened to different sequences of sounds. The animals were first habituated to one type of sequence, for example with the structure AAAA, irrespective of the actual frequencies of the sounds used.

If the structure of the sequence was then changed, for example to AAAB, there was an activation in the descending limb of the arcuate sulcus. Wang et al. classified this as being in the ventral premotor cortex. However, it is not clear from the illustration whether in fact it lay in area 44 which lies in the fundus of this sulcus (Frey et al., 2014) (Figure 7.2). Wang et al. (2015) reported an activation for the same condition that clearly lay in area 44 in human subjects.

Taken together, these comparative data suggest that, during the evolution of grammatical language, the human inferior caudal PF cortex co-opted mechanisms that were already available in the PF cortex of ancestral primates. In macaque monkeys, the PF cortex supports the ability to generate sequences (Chapter 6) and learn abstract sequences (Chapter 8). Furthermore, macaque monkeys are capable of learning to generate abstract spatial sequences with centre embedding (Jiang et al., 2018), though as yet no cell recordings have been taken in the dorsal PF cortex while they do this.

At the same time, the inferior caudal PF cortex (areas 44 and 45B) and ventral PF cortex are specialized in having auditory as well as visual inputs. And monkeys can learn the abstract structure of sequences that they hear (Wang et al., 2015).

What changed during the evolution of grammatical language was that these mechanisms were used for the generation of abstract vocal sequences with centre embedding. The activations for generating these sequences occurs either in or near an area that is sensitive to changes in auditory sequences in macaque monkeys and that is activated when chimpanzees vocalize while begging for food (Taglialatela et al., 2008).

It is likely that the ability of the inferior caudal PF cortex to handle tree structures developed in stages, as suggested by Progovac et al. (2018). However, as yet we know nothing of the electrophysiology of the mechanism. The way to find out would be to record from cells in the inferior caudal PF cortex either before or during surgery. And this is feasible. Ojemann (2013) recorded from multi-units in the superior temporal cortex while patients name objects or read words. And Forseth et al. (2018) recorded multi-units from electrodes implanted over the inferior caudal PF cortex (areas 44 and 45B) in epileptic patients before surgery.

Meanings

Learning words for objects or animals involves learning to associate a particular sound with a particular object or animal. As an earlier section points out, names differ from the calls made by vervet monkeys for eagle or snake (Struhsaker, 1980) in that the association is arbitrary. It does not depend on the emotional or motivational state of the individual.

Chapter 7 reviews the evidence that in macaque monkeys the ventral PF cortex is involved in learning new associates, for example between one picture and another (Rainer et al., 1999; Andreau & Funahashi, 2011). The same area is activated when human subjects learn verbal paired-associates. In a PET study by Fletcher et al. (1995), the subjects listened to a list of fifteen paired-associates between words. Each one consisted of a category and exemplar, as in 'poet – Browning'. The task for the subjects was to remember the list. During learning of the associates, there were activations in the left ventral PF cortex as well as the superior temporal cortex bilaterally.

In a later study Nagel et al. (2008) used fMRI instead. They first taught their subjects a list of verbal paired-associates that were semantically unrelated. Then during the scans, the subjects had to produce the target word for each cue word, as in 'think' for 'river'. Two regions of interest were chosen in the PF cortex, one that was centred on the left ventral PF cortex and the other that was centred on the inferior frontal sulcus. There were activations in both areas when the subjects produced the target word for the associates.

There is no reason to suppose that the mechanism for learning names in the human brain is any different than the PF mechanism for learning paired-associates in macaque monkeys. However, this mechanism which presumably existed in ancestral primates has been co-opted for a new use. Nouns are used to *communicate* their associated referent. We mark this difference by saying that the word is a symbol.

To study the areas that support the ability to name, Forseth et al. (2018) used electrocorticography (EcoG) as well as fMRI in sixty-four patients with epilepsy. There were two tasks, one to name pictures and the other to generate a name on the basis of the definition that they heard, as in 'apple' for 'a round red fruit'.

To isolate the semantic content, picture naming was contrasted with saying 'scrambled' when presented with a scrambled visual image. And naming by definition was contrasted with identifying the gender of the voice in scrambled speech.

Figure 11.7 (next page) presents the results. The areas in blue indicate activity that was specific to naming pictures (visual); the areas in red indicate activity that was specific to naming by definition (auditory); and the areas in pink indicate activity that was observed in both conditions (multimodal). The areas that were active in both conditions included the left inferior caudal PF cortex (areas 44 and 45B) and ventral PF cortex, as well as the inferior temporal cortex and the cortex in the IPS.

In order to show that these areas were essential for naming, Forseth et al. also stimulated the cortex electrically. They looked specifically for areas in which the stimulation disrupted naming but not sentence repetition or articulation. The critical areas were the left inferior caudal PF cortex (areas 44 and 45B) and ventral PF cortex (area 45A), as well as the middle and inferior temporal gyri bilaterally.

In an impressive fMRI study, Anderson et al. (2015) investigated the relation between words and their referents. The subjects read fifty-six words in twelve classes,

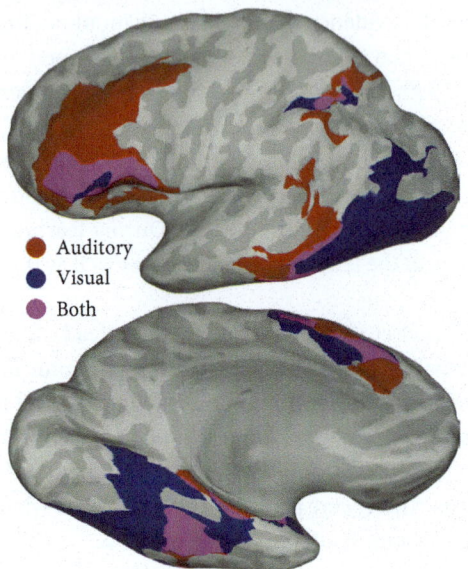

Figure 11.7 Areas of the neocortex which showed activity that was specific to naming pictures (blue), naming by definition (red), or that showed activity for naming in both conditions (pink)

Reproduced from Forseth, K.J., Kadipasaoglu, C.M., Conner, C.R., Hickok, G., Knight, R.T., & Tandon, N. A lexical semantic hub for heteromodal naming in middle fusiform gyrus. *Brain*, 141 (7), 2112–26, Figure 5, Doi: 10.1093/brain/awy120 Copyright © 2018, Oxford University Press.

for example cow and dog for animals, and desk and bed for furniture. Anderson et al. then used representational similarity analysis (RSA) to look for associations between words, for example yellow with banana. They also used RSA to decode pictures of the objects themselves, for example pictures of a dog or bed.

As expected, the areas for which decoding was more successful for decoding pictures of the objects themselves included the occipital lobe and the fusiform gyrus on the ventral surface of the temporal lobe. By contrast, the areas for which decoding was more successful for semantics than pictures included the left superior and middle temporal gyrus and the left inferior caudal PF cortex (areas 44 and 45B) and ventral PF cortex (area 45A). However, decoding was equally successful for semantics and pictures in the lateral part of the inferior temporal cortex.

Anderson et al. took this to be evidence that the words evoked the visual imagery of their referents. It also shows that the system for representing objects (Chapter 7) has been co-opted in the human brain for representing words.

In this study, the inferior parietal cortex was also found to be part of the semantic system, and this might seem surprising. But Huth et al. (2016) looked for areas in which they could decode semantic concepts. The data for their analysis came from subjects who had listened to over 2 hours of stories on the radio. Huth et al. were able to show that the reason why the inferior parietal cortex is part of the semantic system is that it represents concepts such as location and number. In the same study, they also showed that the medial PF cortex represents mental concepts such as 'communal' and 'social'. A later section reviews the evidence that the medial PF cortex is involved in the ability of human subjects to represent the mental states of others.

Summary

The inferior caudal PF area 44 is not the larynx area. It is active when words and sentences are being assembled, but not when they are actually spoken. The left inferior caudal PF cortex (areas 44 and 45B) supports the ability to generate verbs that are appropriate for nouns and area 44 in particular generates sentences according to hierarchical rules. The ventral PF cortex (area 45A) is involved in the learning of paired associates, and names are arbitrarily associated with their referents. The left inferior caudal PF cortex and ventral PF cortex, together with the middle frontal gyrus, are critical for the semantic system. However, activation in the inferior temporal cortex suggests that names can evoke visual imagery of the objects to which they refer.

Teaching

As Chapter 9 mentions, there is a far-reaching consequence of the fact that our ancestors learned to speak. This is that they could teach by using language, rather than by demonstration alone. The advantage of language is that the tutor can describe in detail what the learner must do and can give explicit corrections when errors are made. This avoids the need for trial and error, for example when making complex tools. Even a proto-language would have the advantage that it could direct the attention to the relevant aspects of the task.

Our ancestors survived by cooperating, and as Tomasello (2014) argued this involves 'joint intentionality'. Given a joint enterprise, language has the advantage that the specific role that an individual should take can be discussed. Even a proto-language could be used to assign roles.

Demonstration

Chimpanzees use and modify tools to obtaining food, as in termite fishing and pounding nuts with stones. Whiten et al. (1999) were able to list sixty-five ways in which tools are used or modified, together with the areas in Africa in which the particular use had been observed.

In a more recent review, Whiten and de Waal (2017) listed the evidence for similar traditions in other apes as well as monkeys. For example, cebus monkeys in Brazil use stones for pounding and digging, and pound stones against each other, breaking off flakes (Proffitt et al., 2016). And long-tailed macaques (*Macaca fascicularis* aurea) use stones to open clams on the seashore on one island in a National Park in Burma, but not another (Tan et al., 2015). This is potential evidence for cultural transmission.

However, though infants can be seen near their mothers when they use tools, it could be that the infants learn tool use by themselves because of the availability of tools rather than via imitating their mothers. The way to find out is to experiment.

So, Whiten et al. (1996) compared chimpanzees with human infants of 2 to 4 years of age on a task in which they had to obtain artificial fruit. Each subject saw a demonstration of one of two ways in which this could be achieved. The human infants were more likely to imitate the one that was demonstrated, whereas the chimpanzees were more likely emulate the action rather than strictly imitating. However, when Whiten et al. (2005) taught female chimpanzees in the wild one of two methods for retrieving food from a pipe, most of the chimpanzees in the group adopted that same method.

Thus, it is well possible that the use of twigs, leaves, stones, and branches is passed on from parents to offspring. Nonetheless, there is a stark difference between the traditions observed in chimpanzees and humans. This is that chimpanzee traditions rarely accumulate such that one discovery leads to another (Whiten, 2017). This contrasts with the situation in which human cultures 'ratchet-up' (Tomasello, 1999) or show a 'run-away' effect (Rendell et al., 2011).

The difference is probably due to the fact that chimpanzees, cebus monkeys, and other primates only rely on tools for a proportion of their diet. By contrast, the ancestral hominins were highly dependent on tools for their survival.

It is true that for many thousands of years there was little change in the technology used by the ancestral hominins (Figure 9.1). The Acheulean tradition lasted with little change, with the handaxe serving as an all-purpose tool. The Levallois tradition with flakes struck to form points and spearheads was associated with the presence of modern humans. As they moved out of Africa into Eurasia they were faced with a host of new challenges.

This made demands on creativity and the ability to imagine different solutions. It is the PF cortex that supports the ability to imagine actions, contexts and outcomes (Chapter 10). And there appears to have been an increase in the relative size of the PF cortex in modern humans (Chapter 9). The polar PF cortex in particular is implicated in arbitrating between different alternatives and considering new strategies for solving problems (Donoso *et al.*, 2014).

Once a new solution was discovered, it could be passed on to others by demonstration with or without verbal instructions. The ability to demonstrate has been investigated by Rumiati et al. (2004) by using PET. The subjects were required to demonstrate the use of particular objects by using gestures. In the control condition the subjects simply named the objects. Demonstrating the use of the objects was associated with activations in the ventral PF cortex and the inferior parietal cortex (area PFG).

Demonstration depends on the ability of the observer to copy with high fidelity, and unlike chimpanzees' modern children can do this (Heyes, 2018). As an earlier section mentions, signing chimpanzees have to be taught the gestures of sign language by moulding their hands (Gardner & Gardner, 1969), whereas deaf children spontaneously learn simply by watching the demonstrator.

A previous section mentions a meta-analysis carried out by Caspers et al. (2010) that included studies in which the subjects were required to observe actions. The areas shown in red in Figure 11.3A (page 428) were found to be activated both when subjects observed actions and when they imitated them. They included the inferior caudal PF cortex (IFJ, area 44d) as well as the ventral premotor and inferior parietal cortex in the left hemisphere.

In an fMRI study, Passingham and Vaina (2014) specifically studied the ability of human observers to understand demonstrations. The subjects watched pantomimed gestures, demonstrating for example picking an apple or fitting a lightbulb. The task was to choose which of two objects was being used. The activations were mainly in the left hemisphere. They included the inferior caudal PF cortex (areas 44 and 45B) and ventral PF cortex, as well as the inferior parietal areas identified by Mars et al. (2011) as PFt and PFop.

As expected from these results, left hemisphere lesions that involve the inferior parietal cortex severely impair the ability of patients to understand actions that are performed with objects. In a study by Kalenine et al. (2010) the subjects were shown a word such as 'hammering' and were required to choose between two videos. There were two conditions, one in which one of the videos showed the wrong postures, the other in which it showed the wrong action, for example sawing. Damage to the inferior parietal cortex was associated with errors in discriminating on the basis of posture, whereas damage to the middle temporal cortex was associated with errors in discriminating on the basis of semantics.

The ability to understand demonstrations depends on 'mirror neurones'. In macaque monkeys these have been reported in the inferior parietal area PFG and ventral premotor area F5 (Ferrari et al., 2005). However, there is a difference between the cells in the two areas. The cells in the inferior parietal cortex tend to code for the goals of the actions, whereas the cells in the ventral premotor cortex tend to code for the type of grip that is used (Bonini et al., 2012).

Mirror neurones have been reported in the ventral premotor areas F5p and F5c in macaque monkeys (Rizzolatti & Fogassi, 2014). So Ferri et al. (2015) used fMRI to find the human homologue. The subjects observed people acting or hand gestures. Ferri et al. claimed to find the homologue in the ventral limb of the precentral sulcus in the human brain. However, their figures show that the activation continued anteriorly into the IFJ (area 44d).

There is no mystery concerning mirror neurones. Consider by analogy the cells in the anterior intraparietal area AIP and the ventral premotor cortex that code for the form of an object as well as for the appropriate hand movement (Murata et al., 2000). A comparison of the cells populations in the two areas indicates that there were more cells in AIP that encoded the shape of the objects and more cells in the ventral PF cortex that encoded the action (Murata et al., 2000).

The presence of visuomotor cells in both areas could be explained if the cell activity for action in AIP derives from the ventral premotor cortex. This suggestion could be tested by seeing whether removing the ventral premotor cortex of monkeys abolished the activity encoding action in AIP.

In the same way, it would be possible to see whether a lesion in the ventral premotor cortex in monkeys abolished the mirror activity encoding actions in the inferior parietal area PFG.

It is likely that the properties of mirror neurones develop as the result of experience. Keysers and Perrett (2004) have argued that the ability to understand the actions of others could be learned by associating observation and actions via Hebbian learning. And Heyes (2018) has specifically proposed that ability to imitate could be acquired as the result of infants observing their own hand and arm movements.

There is also evidence from fMRI that mirror neuronal activity results from learning. Calvo-Merino et al. (2005) compared two dance forms, ballet and capoeira, an Afro-Brazilian form of martial art with elements of dance. There was more activation in the dorsal premotor cortex and the IPS when ballet dancers watched the movements of ballet rather than capoeira; but correspondingly, there was more activation in these areas when experts in the martial art watched capoeira rather than ballet. In other words, the activations for observation depended on what the actions that the observers could themselves perform.

The reason why in this experiment there was an activation in the dorsal rather than ventral premotor cortex is that actions involved the body and legs. Whereas there is a representation of the legs in the dorsal premotor cortex (Kurata et al., 1985), the ventral premotor cortex represents the hands alone (He et al., 1993).

Instructions

Whereas macaque monkeys have to be taught tasks over many months, verbal instructions set up the human brain to do the task immediately. In an fMRI study, Hartstra et al. (2011) instructed subjects to perform two tasks. On one, the subjects had to press different fingers depending on which word was presented, and on the other they had to learn the association between objects and particular colours.

There was an activation in the left IFJ (44d) when the instructions were given, even though the subject had not yet carried them out.

The evidence that the instructions set up the specific task comes from an fMRI study by Sakai and Passingham (2006). The subjects were first instructed which task to perform when they were presented with a word after a variable delay. One instruction ('two syllables') told them to decide whether the word had two syllables or not; this was a phonological task. A second instruction ('abstract') told them to decide whether the word was abstract or concrete; this was a semantic task. There was a third instruction ('upper case') that acted as a control; this told them to decide whether the word was in upper or lower case, a visual task.

The advantage of using the two experimental tasks was that the activations while the two tasks were actually being performed were in different cortical areas. Performance of the phonological tasks was associated with activation in the ventral premotor cortex, the reason being that the subjects covertly sounded out the words so as to judge the number of syllables. By contrast, performance of the semantic task was associated with activation in the ventral PF cortex. The delay between presentation of this instruction and the word meant that it was possible to see whether the specific task was set up after the instruction had been given, but before the word had been presented.

Sakai and Passingham observed task-set activation during the delay in the left ventral polar PF cortex, as well as delay-period activation in the task-specific areas. Figure 11.8A (next page) shows the detailed results. On the phonological task there was a significant correlation between the task-set activation in the polar PF cortex and activation in the ventral premotor cortex. On the semantic task there was a significant correlation between activation in the frontal polar cortex and the ventral PF cortex. These results indicate that the specific task was set up before the word was presented.

Although there was also delay-period activation in the dorsal PF cortex (area 46), there were no significant differences between the correlations with the delay-period activations for the three tasks (Figure 11.8B). So, the dorsal PF cortex was not involved in setting up the task.

But the verbal instructions used in this study were just a single word or phrase as in a protolanguage. Full sentences have the advantage that they have subjects and objects. This means that they can distinguish between putting on A on B, or putting B on A. And adverbs add when and how it should be done. And given the infinite number of potential sentences that can be generated, any instruction can be given and, in any detail.

Given the invention of writing, at first with iconic symbols and then arbitrary ones, the generations could also pass on knowledge through written material. Anderson et al. (2017) have used fMRI to decode the words while subjects read sentences. The activation patterns for a training series were estimated using a multiple-regression analysis. Anderson et al. were then able to decode the words

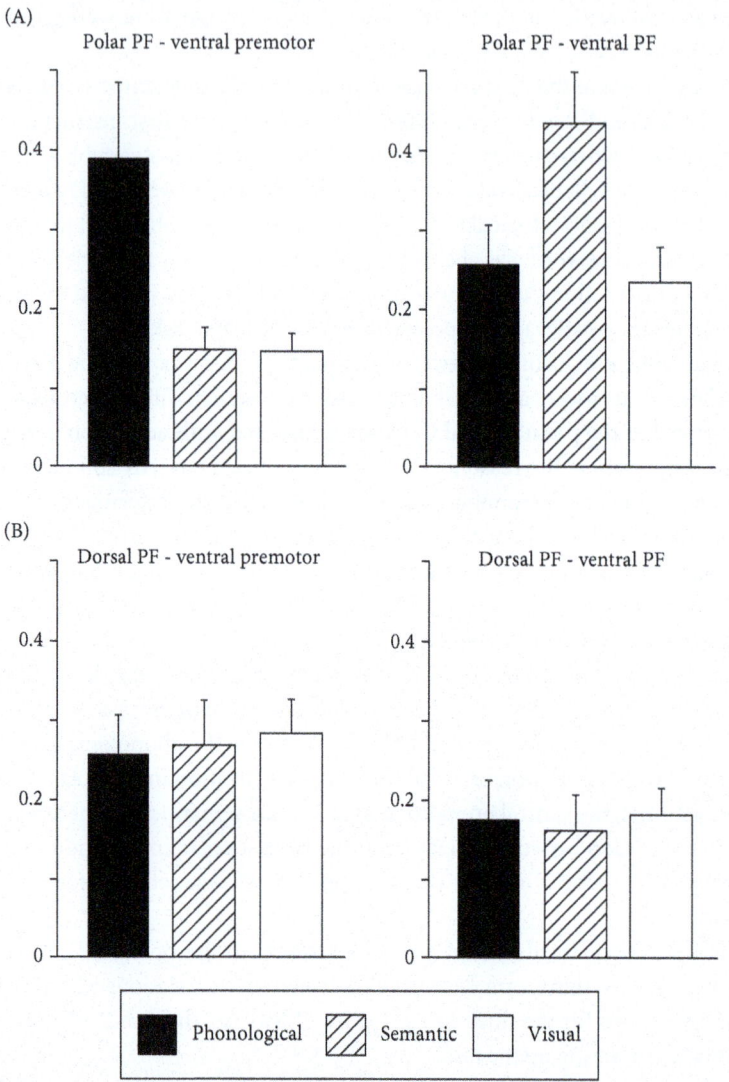

Figure 11.8 Setting up the current task
(A) The correlations between the delay-period activations in the polar PF cortex and the ventral premotor cortex and the ventral PF cortex by the current task. (B) The correlations between the delay-period activations in the dorsal PF cortex (area 46) and the ventral premotor cortex (area 6) and ventral PF cortex (area 45A).

Reproduced from Sakai, K., & Passingham, R.E. Prefrontal set activity predicts rule-specific neural processing during subsequent cognitive performance. *J Neurosci*, 26 (4), 1211–18, Figure 3, Doi: https://doi.org/10.1523/JNEUROSCI.3887-05.2006 Copyright © 2006, The Society for Neuroscience.

in novel sentences, and to do so above the chance level. The areas from which they could do this included the left ventral PF cortex and middle temporal gyrus, areas that are part of the semantic system.

The effect of being taught in sentences is that children can have a formal education. This teaches them knowledge of the world. In an early PET study, Vandenberghe et al. (1996) studied the activations when adults were tested on the semantic knowledge that they had acquired earlier in life. Figure 11.9 presents an example of the display.

The experimental condition is shown in Figure 11.9A. In this example the task was to decide whether the spanner (wrench) or saw is associated with the pliers; the correct answer is the spanner because like pliers it can be used to grip. Figure 11.9B shows the control condition for this example. This was necessary so as to subtract the activations that were related to the presentation of the pictures. Here the subjects had to decide which of the two objects below was nearest in size to the sample object in real life.

The tasks were also presented in verbal form. So, for example, the sample word might be 'pliers' and the choice words 'spanner' and 'saw'. Vandenberghe et al.

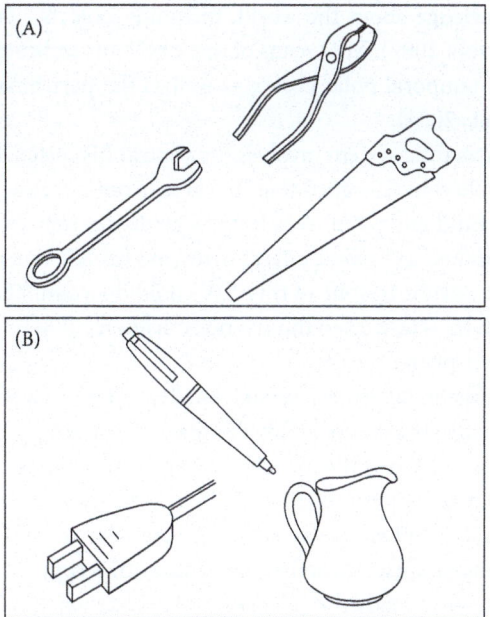

Figure 11.9 Semantic task used by Vandenberghe et al. (1996)
(A) Example of experimental condition using pictures. (B) Example of control condition using pictures.

Reproduced from Passingham, R.E. *Cog Neurosci: A Very Short Introduction*, p. 6, Figure 3 Copyright © 2016, The Author.

looked for activations that were in common whether the problems were set using pictures or words. The finding was that there were activations in the left ventral PF cortex (areas 45A) and inferior caudal PF cortex (area 45B), as well as in the middle and inferior temporal gyrus near the temporal pole. These are part of the semantic system.

Chapter 7 describes populations of cells in the ventral PF cortex and inferior temporal cortex of macaque monkeys that encode categories such as dogs and cats (Freedman et al., 2002). The cells in the PF cortex had activity that was more closely related to the task, in that case matching by category; whereas the cells in the inferior temporal cortex were more likely to code for visual characteristics. The human brain has co-opted this system so as to represent categories in verbal form.

The task shown in Figure 11.9 was based on the 'pyramid and palm trees test', devised to assess semantic knowledge (Howard & Patterson, 1992). On this task a picture is presented, for example of a pyramid, and the subjects have to choose between a picture of a palm tree and a fir tree. In this case, the correct answer is the palm tree since both pyramids and palm trees are found in Egypt. So, this particular test item depends on a knowledge of geography. The fir tree is a foil, chosen because it is more similar perceptually to a pyramid.

Patients who suffer from semantic dementia lose their knowledge of words as well as their knowledge about the world, including objects, concepts and faces (Hodges & Patterson, 2007). MRI scans of these patients indicate that there is degeneration in the temporal polar cortex as well as the perirhinal and entorhinal cortex (Davies et al., 2004).

Although the semantic system involves the ventral PF cortex, Price et al. (Price & Friston, 1999; 1999) have argued that this area is not necessary for performing tasks like the pyramid and palm trees test. As evidence, they cite a PET study of a patient, SW, who had a cerebral infarct involving his left ventral PF cortex and inferior caudal PF cortex. In spite of this, SW could do a simplified version of the pyramid and palm trees task, even though no activation could be detected in these areas in either hemisphere.

But Price and Friston (1999) deliberately made the task easy so that patient SW could do it. The associations were highly familiar, and there is evidence that the PF cortex is not necessary when the associations are easy. Whitney et al. (2012) applied transcranial magnetic stimulation (TMS) at 1 Hz over the left inferior caudal PF cortex (area 45B), and when the subjects were later tested they were able to make easy judgements, such as whether 'salt' goes with 'pepper'.

However, in the same study the subjects made errors when they were tested on more difficult associations, such as whether 'salt' goes with 'grain'. In other words, the inferior caudal PF cortex is involved in controlled, not automatic retrieval (Badre & Wagner, 2002).

Kostopoulos and Petrides (2008) made the same point in an fMRI study of the ventral PF cortex. In the controlled retrieval condition, pairs of words were presented against coloured backgrounds, and the subjects were later tested for their memory of the colour that was associated with one of the words. In a control condition, they simply had to judge whether the word had been presented before. When these conditions were contrasted, there was extensive activation in the ventral PF cortex (area 47/12).

Another way of assessing the ability to judge semantic associations is to present two pairs, and to ask if they are the same. For example, in an fMRI study by Wendelken et al. (2008), the subjects had to judge whether one relation such as 'painter-brush' was the same as another, such as 'writer-pen'. The responses involved pressing one of two keys for 'yes' and 'no'. Comparing these relations led to activations in the right ventral PF cortex.

In a follow-up study, Wendelken et al. (2012) investigated whether the same result would be found, irrespective of the type of association. In one condition, they presented pairs of nonsense drawings, and the subjects had to judge whether they the pairs were similar in their visuo-spatial arrangement. In the other condition, they presented pairs of drawings of actual objects, and the subjects had to judge whether they were similar in their semantic association. There was an activation in the left rostral ventral PF cortex irrespective of the type of judgement made.

Like knowledge of geography, the knowledge required to solve problems such as these is taught at school. In order to investigate new learning, Maguire and Frith (2004) therefore used fMRI to scan adults while they learned new facts. The subjects read sentences that informed them of these facts, such as that 'a grouper is a type of fish'. Acquiring new facts was associated with activations in the left ventral PF cortex (area 47/12) and the middle temporal gyrus. This result is consistent with the finding that there is cell activity in the ventral PF cortex of monkeys when they learn new associations (Andreau & Funahashi, 2011) (Chapter 7).

As well as learning the meanings of words and the semantic associations between objects, children also learn how to use technology and how to play games that involve throwing or kicking a ball. In other words, they learn skills.

One way of studying skills is to use PET to scan subjects while they learn a new skill. Hecht et al. (2015) trained naïve subjects for two years in how to make a stone tool. The subjects were instructed what they had to do to strike off a flake. Hecht et al. were thus able to study the changes in the patterns of activation as the skill was acquired. There were increases in the ventral PF cortex as well as in the left premotor cortex and inferior parietal cortex (area PFG).

In order to study differences in the level of skill, Stout et al. (2008) used PET to scan three expert stone knappers while they made Oldowan and Acheulean tools. The Acheulean hand axes are much more demanding to make than the Oldowan choppers (Stout et al., 2015).

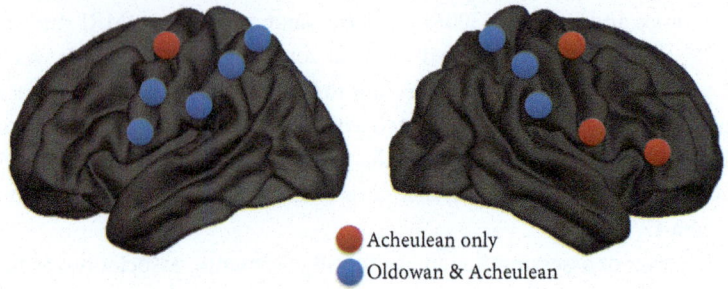

Figure 11.10 Activations while expert stone knappers make Oldowan and Acheulean tools
The blue circles indicate the areas which were activated in common whereas the red circles indicate activations that were specific to making Acheulean hand axes.
Reproduced from Stout, D. & Chaminade, T. Stone tools, language and the brain in human evolution, *Phil Trans RSBB Sci*, 367 (1585), 75–87, Figure 2, Doi: 10.1098/rstb.2011.0099 © 2011, The Royal Society.

Figure 11.10 summarizes the results. The blue circles indicate the activations that were in common for Oldowan or Acheulean tools. The red circles indicate the activations that were specific for making Acheulean tools. There was an activation on the border between the right ventral and dorsal PF cortex that was specific for making the more specialized tool.

One speculative possibility is that it was making stone tools that taught ancestral hominins to understand the relation between cause and effect. Chimpanzees do not understand the properties of objects (Povinelli, 2000). Humans are unique in being able to distinguish between causes and correlations. To establish causality it is necessary to intervene rather than simply observe, and it is the PF cortex that is in a position to represent the consequences given the goal (Chapter 6).

Knowledge of the relation between cause and effect has been passed on through the generations by instruction. It is part of human cultural inheritance. It was the basis for the invention of agriculture, and is the foundation of modern science.

Summary

Children are taught both by demonstration and by verbal instruction. Observing demonstrations is associated with activations in the inferior caudal PF cortex. The advantage of verbal instructions is that they set up the brain for the task immediately, without the need for learning. The instruction leads to set activity in the PF cortex that is associated with activations in the areas that are specifically involved in performance of the tasks.

Learning to Read the Minds of Others

Because our ancestors survived by cooperating, it was important that they be able to infer the intentions of others in the group. For example, hunting big game was only possible because the members of group had a common goal and shared the same intention or 'shared intentionality' (Tomasello et al., 2005).

There have been two main accounts to explain how human subjects can understand the intentions and motives of others. As Chapter 9 mentions, this has been referred to as 'mentalizing' (Frith & Frith, 2006).

One account suggests that human subjects can understand the mental states of others because they can simulate them. This account appeals to the existence of mirror neurones (Rizzolatti & Sinigaglia, 2007), the idea being that when one person observes the actions of another, they simulate the actions themselves, and are thereby able to understand the intentions that accompany them.

The other account has come to be known as 'theory-theory' (Gopnik & Wellman, 2012). The idea is that human subjects learn to understand the intentions and motives of others via a quasi-scientific procedure. The suggestion is that, just as human subjects can learn about causes and their effects, they are learn about the mental states that accompany particular behaviour.

De Lange et al. (2008) used fMRi in an attempt to compare these two accounts. They showed human subjects a series of photos. Some showed ordinary actions, such as drinking from a cup. Others showed extraordinary ways of carrying out the actions, for example drinking from the cup while holding it by its mouth. Yet others showed the person acting in a way that was inexplicable given ordinary intentions, such as holding the cup to their hair.

There were two tasks. In one the subjects were instructed to attend to the intention of the actor; in the other they were instructed to attend to the means used by the actor. When the observers saw the pictures showing extraordinary means, there was activation near the temporo-parietal junction (TPJ). As Chapter 10 mentions, this area is also activated when human subjects observe actions that they find surprising (Grezes et al., 2004) and when the results of their actions as not as expected (Nahab et al., 2011). It presumably reflects the comparison of expectations with outcomes.

When the observers viewed the photos that showed extraordinary intentions, there was an activation bilaterally in the IFJ (area 44d) and the ventral PF cortex (area 45A). De Lange et al. took these activations to suggest covert simulation via a mirror neuronal system.

However, there was also an effect that depended on whether the observers were instructed to attend to the intention of the actor as opposed to the means used. There were activations in the superior temporal sulcus (STS) and the medial PF cortex (prelimbic area 32) when the observers attended to the intention. De Lange et al. interpreted these activations as supporting 'theory—theory', that is an understanding of mental causation.

But the assumption is that there are no mirror neurones in the medial PF cortex, and this is not justified. Caruana et al. (2020) have recorded from cells in the pregenual cingulate cortex in patients, and they found cells that changed their activity both when the patients laughed and when they viewed videos of others laughing. These are mirror neurones.

By using a meta-analysis Vaccaro and Fleming (2018) were also able to show that there is an overlap between the activations for thinking about the mental states of others and thinking about ones' own mental states. The data for mentalizing were extracted by using Neurosynth, an automated platform for synthesizing fMRI data from different studies.

Figure 11.11C (next page) shows that there was overlap between the activations for mentalizing (green) and metacognition (yellow) in the medial PF cortex (dorsal and ventral prelimbic areas 32). However, the activations for mentalizing (green) also extended rostrally into the medial polar PF cortex (Figure 11.11B). On the lateral surface, there was more activation in the STS for mentalizing and in the dorsal PF cortex and IFJ for metacognition (Figure 11.10A).

The ability to mentalize can be assessed in various ways. A standard way has been to present a version of the 'Sally Anne' task. This was developed by Baron-Cohen et al. (1985) to test the ability to understand false belief. The idea is that Sally Anne puts an object down and goes out, but her mother then moves it while Sally is out of the room. The test is to say where Sally Anne will think the object to be when she returns.

Onishi and Baillargeon (2005) developed a version of this test for infants of around 15 months of age. The idea was that, if the infants understood false belief, the infants should be surprised and would therefore look reliably longer when the actor reached for the toy in its new hiding place as opposed to its original location. And this is what Onishi and Baillargeon found. However, the understanding was implicit. Kano et al. (2020) report that macaque monkeys and great apes also show a similar implicit understanding if measures are taken of anticipatory looking.

Yet, children are not able to *say* where Sally Anne would look until around 4 years of age (Baillargeon et al., 2010). The ability to understand irony (Blakemore, 2008) and pretence (Heyes, 2018) is acquired much later, during adolescence (Heyes, 2018).

Heyes and Frith (2014) have reviewed the evidence that the ability to mentalize is acquired in much the same way as reading. In other words, it is taught by parents and peers. As evidence, Taumoepeau and Ruffman (2008) showed that the extent to which the mother talked to their child about their desires predicted the amount of the child's mental-state language at 24 months. Furthermore, the mother's reference to the thoughts of others predicted the child's mental-state language at 33 months.

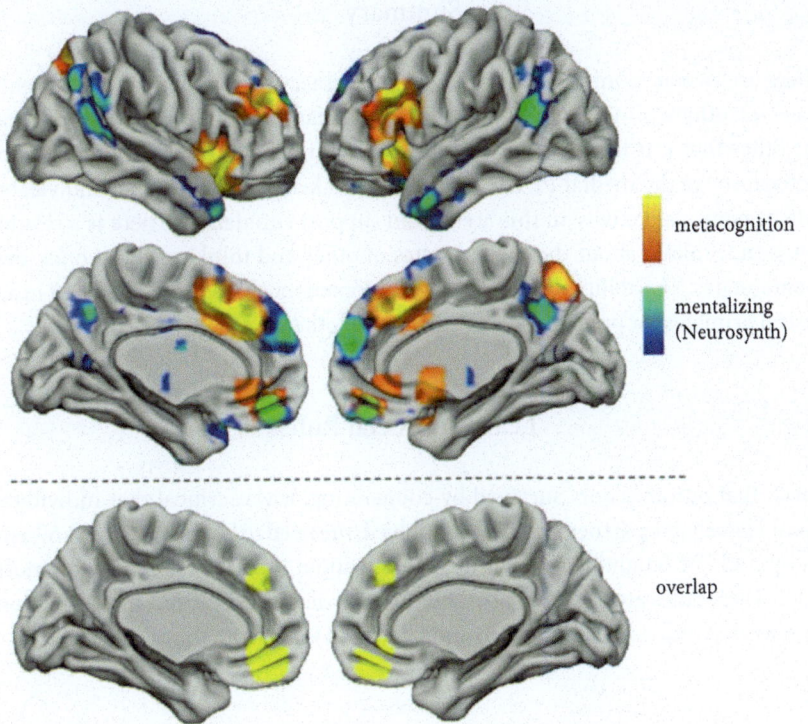

Figure 11.11 The activations resulting from metanalyses of fMRI data for metacognition (yellow and red) and mentalizing (green and blue)

Reproduced from Vaccaro, A.G. & Fleming, S.M. Thinking about thinking: A coordinate-based meta-analysis of neuroimaging studies of metacognitive judgements. *Brain Neurosci Adv*, 2, 1–14, Figure 5, https://doi.org/10.1177/2398212818810591 © The Authors, 2018. Licensed under CC BY 4.0.

But though cultural learning plays a part, it is as important to understand why the activations for mentalizing are located as they are. Chapter 3 describes a study by Sliwa and Freiwald (2017) who used fMRI to visualize the areas of the macaque monkey brain that were specifically activated when they watch videos of other monkeys interacting. Figure 3.5A shows they included the STS and the medial PF cortex.

This suggests that the ability of humans to mentalize co-opted the mechanisms that presumably existed in ancestral primates for predicting the behaviour of conspecifics. Haroush and Williams (2015) reported cell activity in the anterior cingulate cortex (ACC) of macaque monkeys that encoded the choice that another monkey was going to make (Chapter 3). The critical difference is that in the human brain the medial PF cortex is activated when they reflect on the thoughts, and not just the actions, of others.

Summary

There are two accounts of how adult human subjects can understand the mental states of others. One suggests that this understanding depends on simulation, the other that it is learned in the same way as humans learn about other causes. Activations in the medial PF cortex have been taken to favour the second vie, but mirror neuronal activity in this area could support simulation. There is a relation between thinking about the mental states of other and thinking about ones' own mental states. The ability to mentalize has co-opted mechanisms that existed in ancestral primates for predicting the behaviour of others.

Learning Social Rules

Given that our ancestors survived by cooperating, it was critical that individuals should be reliable partners. As speech evolved, the need to be reliable and show empathy could be taught by instructing children about social conventions and moral rules. These rules are backed up by sanctions for disobedience. But these sanctions only work if individuals are able to anticipate and learn from the punishment.

Anticipating Negative Outcomes

The ability to anticipate negative outcomes can be assessed on the Iowa Gambling task. The subjects are presented with four packs of cards, and the aim is to win as much money as possible. But unbeknown to the subject at the start of the task, the packs differ in the probability that the cards will indicate gains or losses. Some packs provide high gains and but also high losses. Over time most subjects come to avoid these packs because they can anticipate the potential losses (Bechara et al., 1994).

Bechara et al. (1999) studied a group of patients with bilateral lesions that included the medial and orbital PF cortex. The medial lesions included the ventral prelimbic area 32, but they also extended into the polar PF cortex. Unlike the control subjects, these patients persisted in choosing from the risky packs.

To try to account for this impairment, Bechara et al. recorded the skin conductance response (SCR) of the subjects as they reached for the high-risk, high-reward pack. As Chapter 10 explains, the SCR is sensitive to slight changes in anxiety because sweat changes the electrical resistance across the skin.

The patients with lesions of the medial and orbital PF cortex still showed an SCR, unlike patients with amygdala lesions in whom the SCR was abolished. However, the patients with lesions of the medial and orbital PF cortex failed to show an anticipatory SCR just before they reached for the risky packs. Bechara et al. interpreted this to mean that the patients persisted in choosing cards from the risky packs because they were poor at anticipating the possibility of a loss.

A subsequent paper from the same research group reported results for a different version of the task, the Cambridge Gambling task (Clark et al., 2008). The subjects had to guess which of the ten red or blue squares hid a token and they had to wager on that guess. The patients with lesions of the medial and orbital PF cortex wagered more. The suggestion was that they were more confident because they failed to anticipate the possibility that they might lose.

However, the patients studied by Bechara (1999) had many different kinds of cerebral lesions. Some has had an anterior cerebral artery haemorrhage and others a surgical resection of a tumour. Abel et al. (2016) therefore collected a larger sample of nineteen patients, all of whom had had surgery to remove a meningioma that affected the medial surface of the hemisphere. The data for sixteen of the same patients were also used in a later study by Barrash et al. (2020). Figure 11.12 shows the lesions, with the colour scale showing the number of patients with lesions that showed overlap with any particular area.

It is important to note that, because these lesions were made by surgical resection, they invaded the white matter; as a result, they would have disconnected other areas. This makes it difficult to pinpoint the area that was critical for the task.

As Chapter 1 mentions, the advantage of using fMRI is that the activations arise from the grey and not the white matter. So, Li et al. (2010) conducted an fMRI study in which healthy subjects were tested on the Iowa gambling task. However, not all of the subjects learned to avoid the risky packs within the limits of testing; and this meant that Li et al. could compare the subjects who learned with those who did not. In the subjects who showed learning from the first to the last block of trials, there was a discrete activation in the medial PF cortex (ventral prelimbic

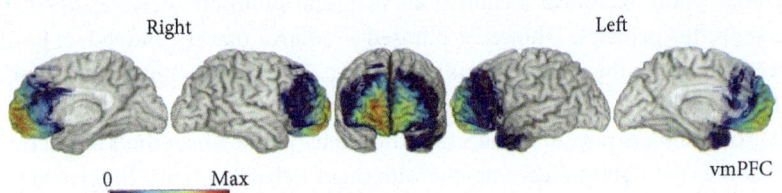

Figure 11.12 The overlap for the lesions that included the medial and orbital PF cortex in 19 patients who had had a medial meningioma removed at surgery (vmpfc) The maximum degree of overlap as shown on the colour scale is for 15 out of 16 patients.

Reproduced from Joseph Barrash, Taylor J. Abel, Katrina L. Okerstrom-Jezewski, Mario Zanaty, Joel E. Bruss, Kenneth Manzel, Matthew Howard, III, & Daniel Tranel, Acquired personality disturbances after meningioma resection are strongly associated with impaired quality of life. *Neurosurgery*, 87 (2), 276–84, Figure 1, Doi: https://doi.org/10.1093/neuros/nyz440 Copyright © 2020, Oxford University Press.

area 32). The reservation is that this analysis was based on too few subjects to be reliable.

This reservation does not apply to another fMRI study by Fukui et al. (2005). They took a different approach. They compared the activations when subjects chose from the risky as opposed to the safe packs, and the sample included 19 subjects. Activation in the medial PF cortex (prelimbic area 32) was associated with the risky choice; and the greater the activation, the greater the success in learning to avoid the risky pack.

The Iowa Gambling Task is meant to model gambling. This means that if it is a valid model, then pathological gamblers should show the same results as the patients with lesions of the medial and orbital PF cortex. Goudriaan et al. (2006) tested this prediction. They reported that indeed gamblers tend to take more risks on the Iowa Gambling task. Furthermore, like the patients, the gamblers showed a reduced SCR before they picked from a risky pack.

Bechara et al. (2000) maintained that the SCR acted as a 'somatic marker', and that it played a causative role in helping subjects to avoid the risky packs. This has proved controversial (Dunn et al., 2006). However, in truth, it does not matter whether the SCR plays a causal role or not. At the very least, the lack of an anticipatory SCR serves as a valuable index of a failure to anticipate a potential loss.

Imagining Outcomes for Others

In most Western societies whether you gamble is up to you. It may be frowned upon, but it is not usually regarded as an issue for morality unless your gambling affects your family or others. Moral rules are imposed when it is the good order of society that is at stake.

The fundamental basis for these rules is 'do as you would be done by'. There are, of course, more technical formulations in moral philosophy (Hare, 1952). But acting on this principle—however phrased—requires that the individual be able to see things from the other person's perspective. This involves imagining the effect that your actions would have on that person.

It is one thing to have an intellectual knowledge about what others might think, but another to be able to feel empathy with them. Sebastian et al. (2012) made this comparison in an fMRI study of boys between the ages of 11 and 16. The boys were presented with a series of cartoons in which there were two people and they were required to pick the one of two possible endings that was appropriate for completing the story.

In the condition assessing empathy, the correct choice depended on understanding the feelings of the people; for example, a mother should comfort her child if it falls off a slide. In the condition assessing knowledge, the correct choice depended on understanding the thoughts of the people; for example, one person should supply a ladder to enable others to reach the apples on the tree. There was also a baseline condition in which the choice depended on understanding physical causation: for instance, the snowman will melt when the sun comes out.

There was an activation in the medial PF cortex (ventral prelimbic area 32) when the comparison assessing empathy was contrasted with the baseline condition. This activation reflected the ability of the boys to appreciate the feelings that others would have in particular emotional situations.

These results would predict that damage to the medial PF cortex should impair the degree to which subjects show empathy. This was tested in a study of patients by Beadle et al. (2018). The degree of empathy was manipulated by using an experimental task. Over an intercom, the subjects heard a competitor in a game discussing the recent death of their son with the research assistant. The degree of empathy so induced was then measured by a series of rating scales. Patients with lesions of the medial and orbital PF cortex were less likely than control subjects to become more empathic.

The degree of empathy that was induced was also measured by requiring the subjects to engage in 20 rounds with the competitor on an economic game. The patients with medial and orbital PF lesions offered less money to the competitor than did the control subjects, so showing less empathy.

The ability to empathize probably depends on mirror neuronal activity. As an earlier section mentions, Caruana et al. (2020) found mirror activity in the pregenual cortex for emotions such as laughter.

But, children also have learn to be empathic as the result of parental guidance. Given this, one would predict that lesions in childhood to interfere with this learning. Anderson et al. (1999) studied two patients who had incurred lesions of the medial and orbital PF cortex in early childhood. The effect was to severely impair their ability to learn and obey social conventions and moral rules. The patients stole, lied, and were impervious to punishment. In general, they showed a lack of guilt concerning their actions.

Darby et al. (2018) found MRI data for seventeen subjects who had behaved in a way that was actually criminal. The lesions included the medial and orbital PF cortex, the temporal polar cortex, and amygdala. Darby et al. then used the platform Neurosynth to carry out a series of meta-analyses using the search words 'value', 'theory of mind', and 'empathy'. This allowed them to visualize the areas that were associated with these concepts in healthy brains. There was an overlap between this map and that the areas that were damaged in these seventeen subjects.

However, there is a limitation in this study. Adolphs and Tranel (2018) have pointed out that most patients with lesions in these areas do not engage in criminal behaviour. This means that there must be an interaction between premorbid individual differences and the effect of the lesions.

Darby et al. (2018) assumed that the lesions led to criminal behaviour because of a decrease in guilt. One way of finding out is to test patients on the ultimatum game. This is an economic game in which one player, the proposer, has the opportunity of splitting a stake of monkey with another player, the responder. Once the proposer communicates their decision, the responder may accept it or reject it. If the responder accepts the offer, the money is split as proposed; if the responder rejects the offer, both players receive nothing.

Krajbich et al. (2009) studied patients with lesions that included the medial and orbital PF cortex as they played the ultimatum game. The patients offered less and felt less guilty that they had done so. Yet, when asked what other people would do, they knew the appropriate behaviour. In other words, they knew the social norms although they failed to abide by them.

Bandura (1997) distinguished between the fear of punishment and guilt by suggesting that guilt involves punishing oneself. Whereas animals can be trained not to perform particular actions through the fear of punishments, humans differ in that they also control their own behaviour according to rules that they have been taught. The ability to do this is acquired by moral education, with the injunction that 'you should feel bad' if you carry out some particular action.

Moral Dilemmas

The ability to exercise self-control can be assessed in a different way, by posing moral dilemmas. Thomas et al. (2011) reported that patients with damage that included the medial and orbital PF cortex were impaired in their ability to reason when faced with moral dilemmas. One such was: 'how appropriate is it for you to flip the switch to kill your daughter to save the five workers?' The patients were more likely than healthy people to make the utilitarian judgment, seeing no reason not to flip the switch.

One reason could be that they were unable to anticipate the feelings of guilt that they would have if they flipped the switch. Alternatively, they may have been poor at empathizing with the consequences for the daughter. Shamay-Tsoory et al. (2003) reported that patients with lesions of the medial and orbital PF cortex failed to recognize what it would be inappropriate to say or do given the feelings of others. Yet, both propriety and moral reasoning depend on this recognition.

Moral reasoning can be assessed by presenting moral dilemmas in words or in pictures, and the dilemmas can be simple or utilitarian. Fede and Kiehl (2020) carried out a meta-analysis of imaging studies on moral reasoning, comparing the activations for moral as opposed to utilitarian reasoning. The areas activated for moral reasoning included medial PF cortex, as well as the amygdala, areas that are interconnected (Beckmann et al., 2009). This is the same system that is involved in actual behaviour, as assessed by the choice of cards from the risky packs on the Iowa Gambling task (Bechara et al., 2000), as well as by criminal behaviour resulting from lesions to these areas (Darby et al., 2018).

The understanding of morality can also be directly assessed by presenting words and requiring the subjects to judge whether the actions that were represented were moral or not. Cameron et al. (2018) used this method to test patients that included the medial and orbital PF cortex. The patients made more errors, judging actions to be moral that others took to be immoral.

As an earlier section mentions, moral behaviour depends on empathy, that is on the ability to feel for others. Of course, the fact that most people can do this cannot be used to prove that a moral principle *must* hold. It is a truism in moral philosophy that it is not possible to base an 'ought' on an 'is' (Hare, 1952), that is a judgement of what ought to happen on something that is the case. In other words, no moral principle follows logically because of some fact. Nonetheless, it is a relevant fact that humans can see things from the perspective of others. Though it may not justify moral principles, it does mean that they are equipped to obey them.

Summary

Human society depends on cooperation, and this means that it has to be regulated by social rules. Moral behaviour depends in part on fear of punishment and the medial and orbital PF cortex are involved in anticipating punishment. Moral behaviour also depends on guilt. The ability to feel guilt is taught during moral education. The neural system that is activated when human subjects fear punishment, feel empathy, or feel guilt involves the medial and orbital PF cortex as well as the amygdala, areas that are anatomically interconnected.

Conclusions

This chapter and the previous one depend heavily on studies that have used functional brain imaging. When interpreting particular activations, it is typical to point to the fact that the same area is activated in other studies. As Poldrack (2006) pointed out this involves the reverse inference that the activations occurred for the same reason. However, Hutzler (2014) has argued that this inference need not be fallacious. The strength of meta-analyses is that they can test the range of tasks that lead to activation in an area, thus ensuring that the comparison between studies of an area is not based on a single task.

Selective Advantage

Pointing to the function only answers one question. It does not tell us why this particular area was activated. More generally, it does not answer the question why the human brain is organized as it is.

As Chapter 1 points out, the answer to this question involves understanding the evolution of the human brain. The reason why there is activation in the human medial PF cortex (prelimbic area 32) and STS when subjects reflect on the thoughts of others is that in ancestral primates this area was involved in predicting the behaviour of conspecifics. Of course, the area has expanded in the human brain (Avants et al., 2006) (Figure 9.8B), and the original function has been modified, partly because of the development of language and thus the ability to form explicit representations of mental states.

The selective pressure to be able to understand the intentions of others was the need for the ancestral hominins to cooperate in order to survive. And the first phrase in the subtitle of this book is 'selective advantage'.

Connectivity

The subtitle also includes the term 'connectivity'. It is not enough for an fMRI study to simply list the areas that were activated. It is necessary to understand why it was this particular set or pattern of areas that was activated. There is no point in calling it a 'system' unless the inter-connections of that system are documented. As Chapter 1 argues, the gold-standard for demonstrating connections is studying the transport of tracers, and Chapters 3–8 have described the connections in the brains of macaque monkeys.

To illustrate how functional systems can be interpreted, consider the pattern of activations shown in Figure 11.10 for making Oldowan choppers and Acheulean handaxes (Stout & Chaminade, 2012). The areas in common for making both types of tools are shown in blue, and the areas that were specific for making handaxes are shown in red.

There were activations for making both types of tool in the cortex of the IPS and the inferior parietal cortex (area PFG) (blue). These areas are interconnected with the ventral premotor cortex (area F5a) (blue) (Gerbella et al., 2010). They are also interconnected with the ventral PF cortex (area 47/12) (red) and the dorsal PF cortex (area 46). In turn, the ventral and dorsal PF cortex project directly to the ventral premotor cortex (area F5a) (Gerbella et al., 2010). In macaque monkeys this is a system for purposeful actions performed with the hand and arm (Borra et al., 2017).

There has been an expansion in all these areas in the human compared with the chimpanzee brain (Avants et al., 2006) (Figure 9.8B). The expansion of the ventral premotor cortex may be responsible in part for the increase in skill. The expansion of the inferior parietal area PFG has been accompanied by an increase in the functional coupling between the inferior parietal and PF cortex during task performance (Sneve et al., 2019).

Neural Operations

The advantage of interpreting the results of experiments on humans in the light of data from macaque monkeys is that it provides a way of accessing information about the neurophysiology of cells in the particular area. That is why the third phrase in the subtitle of the book is 'neural operations'.

It is because accurate localization in PET and fMRI studies provides a link to the neurophysiology that it matters that those using these methods be trained in detailed neuroanatomy. It is not possible to use an atlas such as that provided by Petrides (2018) without such knowledge.

Fortunately, as Chapter 1 describes, there is a way of finding out whether a particular area in the human brain is homologous with an area in the macaque monkey brain. This is to use DWI and resting-state covariance to plot the connections of the area, and to compare them with the connections in the macaque monkey brain. This has been done, for example, by Mars et al. (2011) for the subareas in the inferior parietal cortex in the human and monkey brain. Their parietal areas PFm and PFt correspond to the area PFG as shown in Figure 1.9 for the macaque monkey brain.

It is this allocation that provides an understanding of why the inferior parietal cortex was activated in the study by Stout and Chaminade (2008) (Figure 11.10). There are cells in area PFG and the ventral PF cortex in the macaque monkey that differ in their activity depending on the specific goal of an arm movement (Bonini et al., 2012). Westendorff et al. (2010) have shown that the premotor cells encode this goal before cells do so in the parietal cortex. In turn, the premotor areas derive the representation of the goal from the dorsal and ventral PF cortex (Chapters 6 and 7). It is this connected system (Borra et al., 2017) and mechanism (Bonini et al., 2012) for specifying the target of arm and hand movements that has been co-opted in the human brain for performing rapid, skilled, and targeted movements that are needed when making stone tools or striking a match.

Culture and Education

Chapter 1 suggests that there is one last question to be answered. The scans are typically taken in adult subjects. But this is after a prolonged period of maturation and education. In other words, we need to consider ontogeny.

During child development there are increase in the connectivity in the white matter as shown by DWI. For example, the inferior frontal occipital fasciculus continues to mature during adolescence, and the arcuate fasciculus into adulthood (Asato et al., 2010).

These changes occur in a cultural environment in which children and adolescents are taught. It has been shown that experience and practice influence the connectivity of the brain (Sampaio-Baptista & Johansen-Berg, 2017). And, as an earlier section mentions, the arcuate fasciculus becomes more mature as the result of reading (Dehaene et al., 2015a).

With the development of proto-communication and later grammatical language, our ancestors were brought up in a cultural environment. When considering the environments in which they lived, it is essential to include not only external factors, such as climate, food, and water resources, but also cultural factors such as tool use and tool making, the use of fire for cooking, and language. These also formed part of the environment in which the young were brought up and drove changes in the brain during evolution.

The effect of a change such as stone tool making leads to a further selection pressure to make and use and make tools more effectively; in other words, cultural factors feedback on the development of the brain in just the same way as factors such as a change in the distribution of food and water resources. This interaction and feedback were described by Tomasello (1999) as 'ratchet effect'.

This ratchet effect occurred because individuals that were better adapted to the cultural environment had more access to the resources needed for survival and to mates with similar abilities. The result was that these individuals therefore had more surviving children than others who were less well adapted to the cultural environment. In other words, cultural change drove evolution.

The result was an increase in intelligence associated with changes in the brain. Gentner (2003) asked 'why are we so smart?' and the expansion of the PF cortex was a critical factor. But it is also necessary to take into account the long period of maturation and education. As Tomasello (2019) put it, we learn to be human.

References

Abel, T.J., Manzel, K., Bruss, J., Belfi, A.M., Howard, M.A., 3rd, & Tranel, D. (2016) The cognitive and behavioral effects of meningioma lesions involving the ventromedial prefrontal cortex. *J Neurosurg*, 124, 1568–77.

Adolphs, R., Glascher, J., & Tranel, D. (2018) Searching for the neural causes of criminal behavior. *Proc Natl Acad Sci USA*, 115, 451–2.

Amunts, K., Lenzen, M., Friederici, A.D., Schleicher, A., Morosan, P., Palomero-Gallagher, N., & Zilles, K. (2010) Broca's region: Novel organizational principles and multiple receptor mapping. *PLoS Biol*, 8.

Amunts, K., Schleicher, A., Ditterich, A., & Zilles, K. (2003) Broca's region: Cytoarchitectonic asymmetry and developmental changes. *J Comp Neurol*, 465, 72–89.

Amunts, K. & Zilles, K. (2012) Architecture and organizational principles of Broca's region. *Trends Cogn Sci*, 16, 418–26.

Anderson, A.J., Binder, J.R., Fernandino, L., Humphries, C.J., Conant, L.L., Aguilar, M., Wang, X., Doko, D., & Raizada, R.D.S. (2017) Predicting neural activity patterns associated with sentences using a neurobiologically motivated model of semantic representation. *Cereb Cortex*, 27, 4379–95.

Anderson, A.J., Bruni, E., Lopopolo, A., Poesio, M., & Baroni, M. (2015) Reading visually embodied meaning from the brain: Visually grounded computational models decode visual-object mental imagery induced by written text. *Neuroimage*, 120, 309–22.

Anderson, S.W., Bechara, A., Damasio, H., Tranel, D., & Damasio, A.R. (1999) Impairment of social and moral behavior related to early damage in human prefrontal cortex. *Nat Neurosci*, 2, 1032–7.

Andreau, J.M. & Funahashi, S. (2011) Primate prefrontal neurons encode the association of paired visual stimuli during the pair-association task. *Brain Cogn*, 76, 58–69.

Annett, M. (2004) Hand preference observed in large healthy samples: Classification, norms and interpretations of increased non-right-handedness by the right shift theory. *Br J Psychol*, 95, 339–53.

Annett, M. (2006) The distribution of handedness in chimpanzees: Estimating right shift in Hopkins' sample. *Laterality*, 11, 101–9.

Arbib, M.A. (2005) From monkey-like action recognition to human language: An evolutionary framework for neurolinguistics. *Behav Brain Sci*, 28, 105–24; discussion 125–67.

Asato, M.R., Terwilliger, R., Woo, J., & Luna, B. (2010) White matter development in adolescence: A DTI study. *Cereb Cortex*, 20, 2122–31.

Avants, B.B., Schoenemann, P.T., & Gee, J.C. (2006) Lagrangian frame diffeomorphic image registration: Morphometric comparison of human and chimpanzee cortex. *Med Image Anal*, 10, 397–412.

Badre, D. & Wagner, A.D. (2002) Semantic retrieval, mnemonic control, and prefrontal cortex. *Behav Cogn Neurosci Rev*, 1, 206–18.

Bahlmann, J., Schubotz, R.I., & Friederici, A.D. (2008) Hierarchical artificial grammar processing engages Broca's area. *Neuroimage*, 42, 525–34.

Baillargeon, R., Scott, R.M., & He, Z. (2010) False-belief understanding in infants. *Trends Cogn Sci*, 14, 110–18.

Bandura, A. (1997) *Self-efficacy: The Exercise of Control*. Freeman, New York.

Baron-Cohen, S., Leslie, A.M., & Frith, U. (1985) Does the autistic child have a 'theory of mind'? *Cognition*, 21, 37–46.

Barrash, J., Abel, T.J., Okerstrom-Jezewski, K.L., Zanaty, M., Bruss, J.E., Manzel, K., Howard, M., & Tranel, D. (2020) Acquired personality disturbances after meningioma resection are strongly associated with impaired quality of life. *Neurosurgery*, 87, 276–84.

Beadle, J.N., Paradiso, S., & Tranel, D. (2018) Ventromedial prefrontal cortex is critical for helping others who are suffering. *Front Neurol*, 9, 288.

Bechara, A., Damasio, A.R., Damasio, H., & Anderson, S.W. (1994) Insensitivity to future consequences following damage to human prefrontal cortex. *Cognition*, 50, 7–15.

Bechara, A., Damasio, H., & Damasio, A.R. (2000) Emotion, decision making and the orbitofrontal cortex. *Cer Cort*, 10, 295–307.

Bechara, A., Damasio, H., Damasio, A.R., & Lee, G.P. (1999) Different contributions of the human amygdala and ventromedial prefrontal cortex to decision-making. *J Neurosci*, 19, 5473–81.

Beckmann, M., Johansen-Berg, H., & Rushworth, M.F. (2009) Connectivity-based parcellation of human cingulate cortex and its relation to functional specialization. *J Neurosci*, 29, 1175–90.

Biagi, L., Cioni, G., Fogassi, L., Guzzetta, A., & Tosetti, M. (2010) Anterior intraparietal cortex codes complexity of observed hand movements. *Brain Res Bull*, 81, 434–40.

Bishop, D.V.M. (1982) *Test for Reception of Grammar*. Thomas Leach, Oxford.

Blakemore, S.J. (2008) The social brain in adolescence. *Nat Rev Neurosci*, 9, 267–77.

Bolhuis, J.J., Tattersall, I., Chomsky, N., & Berwick, R.C. (2014) How could language have evolved? *PLoS Biol*, 12, e1001934.

Bonini, L., Ugolotti Serventi, F., Bruni, S., Maranesi, M., Bimbi, M., Simone, L., Rozzi, S., Ferrari, P.F., & Fogassi, L. (2012) Selectivity for grip type and action goal in macaque inferior parietal and ventral premotor grasping neurons. *J Neurophysiol*, 108, 1607–19.

Borra, E., Gerbella, M., Rozzi, S., & Luppino, G. (2017) The macaque lateral grasping network: A neural substrate for generating purposeful hand actions. *Neurosci Biobehav Rev*, 75, 65–90.

Bryant, K.L., Li, L., Eichert, N., & Mars, R.B. (2020) A comprehensive atlas of white matter tracts in the chimpanzee. *Plos Biology*, 18, e3000971.

Calvo-Merino, B., Glaser, D.E., Grezes, J., Passingham, R.E., & Haggard, P. (2005) Action observation and acquired motor skills: An FMRI study with expert dancers. *Cereb Cortex*, 15, 1243–9.

Cameron, C.D., Reber, J., Spring, V.L., & Tranel, D. (2018) Damage to the ventromedial prefrontal cortex is associated with impairments in both spontaneous and deliberative moral judgments. *Neuropsychol*, 111, 261–8.

Caruana, F., Avanzini, P., Pelliccia, V., Mariani, V., Zauli, F., Sartori, I., Del Vecchio, M., Lo Russo, G., & Rizzolatti, G. (2020) Mirroring other's laughter. Cingulate, opercular and temporal contributions to laughter expression and observation. *Cortex*, 128, 35–48.

Caspers, S., Eickhoff, S.B., Rick, T., von Kapri, A., Kuhlen, T., Huang, R., Shah, N.J., & Zilles, K. (2011) Probabilistic fibre tract analysis of cytoarchitectonically defined human inferior parietal lobule areas reveals similarities to macaques. *Neuroimage*, 58, 362–80.

Caspers, S., Zilles, K., Laird, A.R., & Eickhoff, S.B. (2010) ALE meta-analysis of action observation and imitation in the human brain. *Neuroimage*, 50, 1148–67.

Catani, M., Jones, D.K., & ffytche, D.H. (2005) Perisylvian language networks of the human brain. *Ann Neurol*, 57, 8–16.

Clark, L., Bechara, A., Damasio, H., Aitken, M.R., Sahakian, B.J., & Robbins, T.W. (2008) Differential effects of insular and ventromedial prefrontal cortex lesions on risky decision-making. *Brain*, 131, 1311–22.

Corina, D.P., McBurney, S.L., Dodrill, C., Hinshaw, K., Brinkley, J., & Ojemann, G. (1999) Functional roles of Broca's area and SMG: Evidence from cortical stimulation mapping of a deaf signer. *Neuroimage*, 10, 570–81.

Darby, R.R., Horn, A., Cushman, F., & Fox, M.D. (2018) Lesion network localization of criminal behavior. *Proc Natl Acad Sci USA*, 115, 601–6.

Davies, R.R., Graham, K.S., Xuereb, J.H., Williams, G.B., & Hodges, J.R. (2004) The human perirhinal cortex and semantic memory. *Eur J Neurosci*, 20, 2441–6.

de Lange, F.P., Spronk, M., Willems, R.M., Toni, I., & Bekkering, H. (2008) Complementary systems for understanding action intentions. *Curr Biol*, 18, 454–7.

Dehaene, S., Cohen, L., Morais, J., & Kolinsky, R. (2015a) Illiterate to literate: Behavioural and cerebral changes induced by reading acquisition. *Nat Rev Neurosci*, 16, 234–44.

Dehaene, S., Meyniel, F., Wacongne, C., Wang, L., & Pallier, C. (2015b) The neural representation of sequences: From transition probabilities to algebraic patterns and linguistic trees. *Neuron*, 88, 2–19.

Donoso, M., Collins, A.G., & Koechlin, E. (2014) Human cognition. Foundations of human reasoning in the prefrontal cortex. *Science*, 344, 1481–6.

Dunn, B.D., Dalgleish, T., & Lawrence, A.D. (2006) The somatic marker hypothesis: A critical evaluation. *Neurosci Biobehav Rev*, 30, 239–71.

Eichert, N., Papp, D., Mars, R.B., & Watkins, K.E. (2020) Mapping human laryngeal motor cortex during Vocalization. *Cereb Cortex*, 30, 6254–69.

Eichert, N., Verhagen, L., Folloni, D., Jbabdi, S., Khrapitchev, A.A., Sibson, N.R., Mantini, D., Sallet, J., & Mars, R.B. (2019) What is special about the human arcuate fasciculus? Lateralization, projections, and expansion. *Cortex*, 118, 107–15.

Emmorey, K., Mehta, S., & Grabowski, T.J. (2007) The neural correlates of sign versus word production. *Neuroimage*, 36, 202–8.

Estruch, S.B., Graham, S.A., Deriziotis, P., & Fisher, S.E. (2016) The language-related transcription factor FOXP2 is post-translationally modified with small ubiquitin-like modifiers. *Sci Rep*, 6, 20911.

Falk, D., Zollikofer, C.P.E., Ponce de Leon, M., Semendeferi, K., Alatorre Warren, J.L., & Hopkins, W.D. (2018) Identification of in vivo sulci on the external surface of eight adult chimpanzee brains: Implications for interpreting early hominin endocasts. *Brain Behav Evol*, 91, 45–58.

Fede, S.J. & Kiehl, K.A. (2020) Meta-analysis of the moral brain: Patterns of neural engagement assessed using multilevel kernel density analysis. *Brain Imaging Behav*, 14, 534–47.

Ferrari, P.F., Rozzi, S., & Fogassi, L. (2005) Mirror neurons responding to observation of actions made with tools in monkey ventral premotor cortex. *J Cogn Neurosci*, 17, 212–26.

Ferri, S., Peeters, R., Nelissen, K., Vanduffel, W., Rizzolatti, G., & Orban, G.A. (2015) A human homologue of monkey F5c. *Neuroimage*, 111, 251–66.

Fischer, J. & Price, T. (2017) Meaning, intention, and inference in primate vocal communication. *Neurosci Biobehav Rev*, 82, 22–31.

Fitch, W.T. (2017) Empirical approaches to the study of language evolution. *Psychon Bull Rev*, 24, 3–33.

Fletcher, P.C., Frith, C.D., Grasby, P.M., Shallice, T., Frackowiak, R.S., & Dolan, R.J. (1995) Brain systems for encoding and retrieval of auditory-verbal memory. An in vivo study in humans. *Brain*, 118 (Pt 2), 401–16.

Flinker, A., Korzeniewska, A., Shestyuk, A.Y., Franaszczuk, P.J., Dronkers, N.F., Knight, R.T., & Crone, N.E. (2015) Redefining the role of Broca's area in speech. *Proc Natl Acad Sci USA*, 112, 2871–5.

Forseth, K.J., Kadipasaoglu, C.M., Conner, C.R., Hickok, G., Knight, R.T., & Tandon, N. (2018) A lexical semantic hub for heteromodal naming in middle fusiform gyrus. *Brain*, 141, 2112–26.

Freedman, D.J., Riesenhuber, M., Poggio, T., & Miller, E.K. (2002) Visual categorization and the primate prefrontal cortex: Neurophysiology and behavior. *J Neurophysiol*, 88, 929–41.

Frey, S., Mackey, S., & Petrides, M. (2014) Cortico-cortical connections of areas 44 and 45B in the macaque monkey. *Brain Lang*, 131, 36–55.

Friederici, A.D. (2012) Language development and the ontogeny of the dorsal pathway. *Front Evol Neurosci*, 4, 3.

Frith, C.D. & Frith, U. (2006) The neural basis of mentalizing. *Neuron*, 50, 531–4.

Fukui, H., Murai, T., Fukuyama, H., Hayashi, T., & Hanakawa, T. (2005) Functional activity related to risk anticipation during performance of the Iowa Gambling Task. *Neuroimage*, 24, 253–9.

Galaburda, A.M., LeMay, M., Kemper, T.L., & Geschwind, N. (1978) Right-left asymmetries in the brain. *Science*, 199, 852–6.

Gardner, B.T. & Gardner, R.A. (1985) Signs of intelligence in cross-fostered chimpanzees. *Phil Tran Roy Soc Lond B*, 308, 159–76.

Gardner, R.A. & Gardner, B.T. (1969) Teaching sign language to a chimpanzee. *Science*, 187, 752–3.

Gentner, D. (2003) Why we're so smart. In Gentner, D., Goldin-Meadow, S. (eds) *Language in Mind*. MIT Press, Cambridge, pp. 195–236.

Gerbella, M., Belmalih, A., Borra, E., Rozzi, S., & Luppino, G. (2010) Cortical connections of the anterior (F5a) subdivision of the macaque ventral premotor area F5. *Brain Struct Funct*, 216, 43-65.

Geschwind, N. & Levitsky, W. (1968) Human brain: Left-right asymmetries in temporal speech region. *Science*, 161, 186-7.

Gonseth, C., Kawakami, F., Ichino, E., & Tomonaga, M. (2017) The higher the farther: Distance-specific referential gestures in chimpanzees (Pan troglodytes). *Biol Lett*, 13, doi: 10.1098/rsbl.2017.0398.

Gopnik, A. & Wellman, H.M. (2012) Reconstructing constructivism: Causal models, Bayesian learning mechanisms, and the theory theory. *Psychol Bull*, 138, 1085-108.

Goucha, T. & Friederici, A.D. (2015) The language skeleton after dissecting meaning: A functional segregation within Broca's Area. *Neuroimage*, 114, 294-302.

Goudriaan, A.E., Oosterlaan, J., de Beurs, E., & van den Brink, W. (2006) Psychophysiological determinants and concomitants of deficient decision making in pathological gamblers. *Drug Alcohol Depend*, 84, 231-9.

Graham, K.E., Hobaiter, C., Ounsley, J., Furuichi, T., & Byrne, R.W. (2018) Bonobo and chimpanzee gestures overlap extensively in meaning. *PLoS Biol*, 16, e2004825.

Grezes, J., Frith, C., & Passingham, R.E. (2004) Brain mechanisms for inferring deceit in the actions of others. *J Neurosci*, 24, 5500-5.

Halina, M., Liebal, K., & Tomasello, M. (2018) The goal of ape pointing. *PLoS One*, 13, e0195182.

Hare, R.M. (1952) *The Language of Morals*. Oxford University Press, Oxford.

Haroush, K. & Williams, Z.M. (2015) Neuronal prediction of opponent's behavior during cooperative social interchange in primates. *Cell*, 160, 1233-45.

Hartstra, E., Kuhn, S., Verguts, T., & Brass, M. (2011) The implementation of verbal instructions: An fMRI study. *Hum Brain Mapp*, 32, 1811-24.

Hauser, M.D., Chomsky, N., & Fitch, W.T. (2002) The faculty of language: What is it, who has it, and how did it evolve? *Science*, 298, 1569-79.

He, S.-Q., Dum, R.P., & Strick, P.L. (1993) Topographic organization of corticospinal projections from the frontal lobe: Motor areas on the lateral surface of the hemisphere. *J Neurosci*, 13, 952-80.

Hecht, E.E., Gutman, D.A., Khreisheh, N., Taylor, S.V., Kilner, J., Faisal, A.A., Bradley, B.A., Chaminade, T., & Stout, D. (2015) Acquisition of Paleolithic toolmaking abilities involves structural remodeling to inferior frontoparietal regions. *Brain Struct Funct*, 220, 2315-31.

Heffner, H. & Heffner, S. (1986) Effect of unilateral and bilateral auditory cortex lesions on the discrimination of vocalizations by Japanese macaques. *J Neurophysiol*, 56, 683-90.

Heinzle, J., Hepp, K., & Martin, K.A. (2007) A microcircuit model of the frontal eye fields. *J Neurosci*, 27, 9341-53.

Heyes, C. (2016) Imitation: Not in our genes. *Curr Biol*, 26, R412-14.

Heyes, C. (2018) *Cognitive Gadgets*. Harvard University Press, Cambridge, M.A.

Heyes, C.M. & Frith, C.D. (2014) The cultural evolution of mind reading. *Science*, 344, 1243091.

Hobaiter, C., Byrne, R.W., & Zuberbuhler, K. (2017) Wild chimpanzees' use of single and combined vocal and gestural signals. *Behav Ecol Sociobiol*, 71, 96.

Hobaiter, C., Leavens, D.A., & Byrne, R.W. (2014) Deictic gesturing in wild chimpanzees (Pan troglodytes)? Some possible cases. *J Comp Psychol*, 128, 82-7.

Hodges, J.R. & Patterson, K. (2007) Semantic dementia: A unique clinicopathological syndrome. *Lancet Neurol*, 6, 1004-14.

Hope, T.M., Prejawa, S., Parker, J., Oberhuber, M., Seghier, M.L., Green, D.W., & Price, C.J. (2014) Dissecting the functional anatomy of auditory word repetition. *Front Hum Neurosci*, 8, 246.

Hopkins, W.D., Gardner, M., Mingle, M., Reamer, L., & Schapiro, S.J. (2013a) Within- and between-task consistency in hand use as a means of characterizing hand preferences in captive chimpanzees (Pan troglodytes). *J Comp Psychol*, 127, 380–91.

Hopkins, W.D., Reamer, L., Mareno, M.C., & Schapiro, S.J. (2015) Genetic basis in motor skill and hand preference for tool use in chimpanzees (Pan troglodytes). *Proc Biol Sci*, 282, 20141223.

Hopkins, W.D., Russell, J., Freeman, H., Buehler, N., Reynolds, E., & Schapiro, S.J. (2005a) The distribution and development of handedness for manual gestures in captive chimpanzees (Pan troglodytes). *Psychol Sci*, 16, 487–93.

Hopkins, W.D., Russell, J., McIntyre, J., & Leavens, D.A. (2013b) Are chimpanzees really so poor at understanding imperative pointing? Some new data and an alternative view of canine and ape social cognition. *PLoS One*, 8, e79338.

Hopkins, W.D., Russell, J.L., Cantalupo, C., Freeman, H., & Schapiro, S.J. (2005b) Factors influencing the prevalence and handedness for throwing in captive chimpanzees (Pan troglodytes). *J Comp Psychol*, 119, 363–70.

Howard, D. & Patterson, K. (1992) *Pyramids and Palm Trees: A Test of Semantic Access for Pictures and Words*. Thames Valley Tests Company, Bury St Edmunds.

Huth, A.G., de Heer, W.A., Griffiths, T.L., Theunissen, F.E., & Gallant, J.L. (2016) Natural speech reveals the semantic maps that tile human cerebral cortex. *Nature*, 532, 453–8.

Hutzler, F. (2014) Reverse inference is not a fallacy per se: Cognitive processes can be inferred from functional imaging data. *Neuroimage*, 84, 1061–9.

Jackendoff, R. & Wittenberg, E. (2017) Linear grammar as a possible stepping-stone in the evolution of language. *Psychon Bull Rev*, 24, 219–24.

Jiang, X., Long, T., Cao, W., Li, J., Dehaene, S., & Wang, L. (2018) Production of supra-regular spatial sequences by macaque monkeys. *Curr Biol*, 28, 1851–9 e1854.

Kalenine, S., Buxbaum, L.J., & Coslett, H.B. (2010) Critical brain regions for action recognition: Lesion symptom mapping in left hemisphere stroke. *Brain*, 133, 3269–80.

Kaminski, J., Call, J., & Fischer, J. (2004) Word learning in a domestic dog: Evidence for 'fast mapping'. *Science*, 304, 1682–3.

Kano, F., Call, J. & Krupenye, C. (2020) Primates pass dynamically social anticipatory-looking false-belief tests. *Trends Cogn Sci*, 24, 774–8.

Keller, S.S., Crow, T., Foundas, A., Amunts, K., & Roberts, N. (2009) Broca's area: Nomenclature, anatomy, typology and asymmetry. *Brain Lang*, 109, 29–48.

Kersken, V., Gomez, J.C., Liszkowski, U., Soldati, A., & Hobaiter, C. (2019) A gestural repertoire of 1- to 2-year-old human children: In search of the ape gestures. *Anim Cogn*, 22, 577–95.

Keysers, C. & Perrett, D.I. (2004) Demystifying social cognition: A Hebbian perspective. *Trends Cogn Sci*, 8, 501–7.

Kostopoulos, P. & Petrides, M. (2008) Left mid-ventrolateral prefrontal cortex: Underlying principles of function. *Eur J Neurosci*, 27, 1037–49.

Krajbich, I., Adolphs, R., Tranel, D., Denburg, N.L., & Camerer, C.F. (2009) Economic games quantify diminished sense of guilt in patients with damage to the prefrontal cortex. *J Neurosci*, 29, 2188–92.

Kuhn, S., Brass, M., & Gallinat, J. (2013) Imitation and speech: Commonalities within Broca's area. *Brain Struct Funct*, 218, 1419–27.

Kurata, K., Okano, K., & Tanji, J. (1985) Distribution of neurons related to a hindlimb as opposed to forelimb movement in the monkey premotor cortex. *Exp Brain Res*, 60, 188–91.

Li, X., Lu, Z.L., D'Argembeau, A., Ng, M., & Bechara, A. (2010) The Iowa Gambling Task in fMRI images. *Hum Brain Mapp*, 31, 410–23.

Liégeois, F., Baldeweg, T., Connelly, A., Gadian, D.G., Mishkin, M., & Vargha-Khadem, F. (2003) Language fMRI abnormalities associated with FOXP2 gene mutation. *Nat Neurosci*, 6, 1230–7.

Loh, K.K., Petrides, M., Hopkins, W.D., Procyk, E., & Amiez, C. (2017) Cognitive control of vocalizations in the primate ventrolateral-dorsomedial frontal (VLF-DMF) brain network. *Neurosci Biobehav Rev*, 82, 32–44.

Maguire, E.A. & Frith, C.D. (2004) The brain network associated with acquiring semantic knowledge. *Neuroimage*, 22, 171–8.

Marie, D., Roth, M., Lacoste, R., Nazarian, B., Bertello, A., Anton, J.L., Hopkins, W.D., Margiotoudi, K., Love, S.A., & Meguerditchian, A. (2018) Left brain asymmetry of the *planum temporale* in a nonhominid primate: Redefining the origin of brain specialization for language. *Cereb Cortex*, 28, 1808–15.

Mars, R.B., Jbabdi, S., Sallet, J., O'Reilly, J.X., Croxson, P.L., Olivier, E., Noonan, M.P., Bergmann, C., Mitchell, A.S., Baxter, M.G., Behrens, T.E., Johansen-Berg, H., Tomassini, V., Miller, K.L., & Rushworth, M.F. (2011) Diffusion-weighted imaging tractography-based parcellation of the human parietal cortex and comparison with human and macaque resting-state functional connectivity. *J Neurosci*, 31, 4087–100.

Milner, B. (1975) Psychological aspects of focal epilepsy and its neurosurgical management. *Adv Neurol*, 8, 299–321.

Murata, A., Gallese, V., Luppino, G., Kaseda, M., & Sakata, H. (2000) Selectivity for the shape, size and orientation of objects for grasping in neurones of monkey parietal AIP. *J Neurophysiol*, 83, 2580–601.

Musso, M., Moro, A., Glauche, V., Rijntjes, M., Reichenbach, J., Buchel, C., & Weiller, C. (2003) Broca's area and the language instinct. *Nat Neurosci*, 6, 774–81.

Nagel, I.E., Schumacher, E.H., Goebel, R., & D'Esposito, M. (2008) Functional MRI investigation of verbal selection mechanisms in lateral prefrontal cortex. *Neuroimage*, 43, 801–7.

Nahab, F.B., Kundu, P., Gallea, C., Kakareka, J., Pursley, R., Pohida, T., Miletta, N., Friedman, J., & Hallett, M. (2011) The neural processes underlying self-agency. *Cereb Cortex*, 21, 48–55.

Ojemann, G.A. (2013) Human temporal cortical single neuron activity during language: A review. *Brain Sci*, 3, 627–41.

Onishi, K.H. & Baillargeon, R. (2005) Do 15-month-old infants understand false beliefs? *Science*, 308, 255–8.

Parker, J., Prejawa, S., Hope, T.M., Oberhuber, M., Seghier, M.L., Leff, A.P., Green, D.W., & Price, C.J. (2014) Sensory-to-motor integration during auditory repetition: A combined fMRI and lesion study. *Front Hum Neurosci*, 8, 24.

Passingham, R.E., Chung, A., Goparaju, B., Cowey, A., & Vaina, L.M. (2014) Using action understanding to understand the left inferior parietal cortex in the human brain. *Brain Res*, 1582, 64–76.

Pattamadilok, C., Dehaene, S., & Pallier, C. (2016) A role for left inferior frontal and posterior superior temporal cortex in extracting a syntactic tree from a sentence. *Cortex*, 75, 44–55.

Persson, T., Sauciuc, G.A., & Madsen, E.A. (2018) Spontaneous cross-species imitation in interactions between chimpanzees and zoo visitors. *Primates*, 59, 19–29.

Petrides, M. (2018) *Atlas of the Morphology of the Human Cerebral Cortex on the Average MNI Brain*. Academic Press, New York.
Poldrack, R.A. (2006) Can cognitive processes be inferred from neuroimaging data? *Trends Cogn Sci*, 10, 59–63.
Povinelli, D.J. (2000) *Folk Physics for Apes*. Oxford University Press, Oxford.
Price, C.J. (2000) The anatomy of language: Contributions from functional neuroimaging. *J Anat*, 197 Pt 3, 335–59.
Price, C.J. & Friston, K.J. (1999) Scanning patients with tasks they can perform. *Hum Brain Mapp*, 8, 102–8.
Price, C.J., Mummery, C.J., Moore, C.J., Frakowiak, R.S., & Friston, K.J. (1999) Delineating necessary and sufficient neural systems with functional imaging studies of neuropsychological patients. *J Cogn Neurosci*, 11, 371–82.
Proffitt, T., Luncz, L.V., Falotico, T., Ottoni, E.B., de la Torre, I., & Haslam, M. (2016) Wild monkeys flake stone tools. *Nature*, 539, 85–8.
Progovac, L. (2016) A gradualist scenario for language evolution: Precise linguistic reconstruction of early human (and neandertal) grammars. *Front Psychol*, 7, 1714.
Progovac, L., Rakhlin, N., Angell, W., Liddane, R., Tang, L., & Ofen, N. (2018) Neural correlates of syntax and proto-syntax: Evolutionary dimension. *Front Psychol*, 9, 2415.
Rainer, G., Rao, S.C., & Miller, E.K. (1999) Prospective coding for objects in primate prefrontal cortex. *J Neurosci*, 19, 5493–505.
Rendell, L., Fogarty, L., & Laland, K.N. (2011) Runaway cultural niche construction. *Philos Trans R Soc Lond B Biol Sci*, 366, 823–35.
Rilling, J.K., Glasser, M.F., Jbabdi, S., Andersson, J., & Preuss, T.M. (2012) Continuity, divergence, and the evolution of brain language pathways. *Front Evol Neurosci*, 3, 1–6.
Ringo, J.L., Doty, R.W., Demeter, S., & Simard, P.Y. (1994) Time is of the essence: A conjecture that hemispheric specialization arises from interhemispheric conduction delay. *Cer Cort*, 4, 331–343.
Rizzolatti, G. & Fogassi, L. (2014) The mirror mechanism: Recent findings and perspectives. *Philos Trans R Soc Lond B Biol Sci*, 369, 20130420.
Rizzolatti, G. & Sinigaglia, C. (2007) Mirror neurons and motor intentionality. *Funct Neurol*, 22, 205–10.
Roberts, A.I., Murray, L., & Roberts, S.G.B. (2019) Complex sociality of wild chimpanzees can emerge from laterality of manual gestures. *Hum Nat*, 30, 299–325.
Rounis, E., Yarrow, K., & Rothwell, J.C. (2007) Effects of rTMS conditioning over the frontoparietal network on motor versus visual attention. *J Cogn Neurosci*, 19, 513–524.
Rumiati, R.I., Weiss, P.H., Shallice, T., Ottoboni, G., Noth, J., Zilles, K., & Fink, G.R. (2004) Neural basis of pantomiming the use of visually presented objects. *Neuroimage*, 21, 1224–231.
Sakai, K. & Passingham, R.E. (2006) Prefrontal set activity predicts rule-specific neural processing during subsequent cognitive performance. *J Neurosci*, 26, 1211–18.
Sampaio-Baptista, C. & Johansen-Berg, H. (2017) White matter plasticity in the adult brain. *Neuron*, 96, 1239–51.
Savage-Rumbaugh, E.S. & Lewin, R. (1994) *Kanzi: The Ape on the Brink of the Human Mind*. Doubleday, London.
Schluter, N.D., Krams, M., Rushworth, M.F., & Passingham, R.E. (2001) Cerebral dominance for action in the human brain: The selection of actions. *Neuropsychol*, 39, 105–13.
Sebastian, C.L., Fontaine, N.M., Bird, G., Blakemore, S.J., Brito, S.A., McCrory, E.J., & Viding, E. (2012) Neural processing associated with cognitive and affective Theory of Mind in adolescents and adults. *Soc Cogn Affect Neurosci*, 7, 53–63.

Sebastianutto, L., Mengotti, P., Spiezio, C., Rumiati, R.I., & Balaban, E. (2017) Dual-route imitation in preschool children. *Acta Psychol* (Amst), 173, 94–100.

Seyfarth, R.M., Cheney, D.L., & Marler, P. (1980) Monkey responses to three different alarm calls: Evidence of predator classification and semantic communication. *Science*, 210, 801–3.

Shamay-Tsoory, S.G., Tomer, R., Berger, B.D., & Aharon-Peretz, J. (2003) Characterization of empathy deficits following prefrontal brain damage: The role of the right ventromedial prefrontal cortex. *J Cogn Neurosci*, 15, 324–37.

Sliwa, J. & Freiwald, W.A. (2017) A dedicated network for social interaction processing in the primate brain. *Science*, 356, 745–9.

Sneve, M.H., Grydeland, H., Rosa, M.G.P., Paus, T., Chaplin, T., Walhovd, K., & Fjell, A.M. (2019) High-expanding regions in primate cortical brain evolution support supramodal cognitive flexibility. *Cereb Cortex*, 29, 3891–3901.

Stout, D. & Chaminade, T. (2012) Stone tools, language and the brain in human evolution. *Philos Trans R Soc Lond B Biol Sci*, 367, 75–87.

Stout, D., Hecht, E., Khreisheh, N., Bradley, B., & Chaminade, T. (2015) Cognitive demands of lower paleolithic toolmaking. *PLoS One*, 10, e0121804.

Stout, D., Toth, N., Schick, K., & Chaminade, T. (2008) Neural correlates of early stone age toolmaking: Technology, language and cognition in human evolution. *Philos Trans R Soc Lond B Biol Sci*, 363, 1939–49.

Struhsaker, T.T. (1980) Comparison of the behaviour and ecology of red colobus and retail monkeys in the Kibale Forest, Uganda. *African Journal of Ecology*, 18, 33–51.

Taglialatela, J.P., Russell, J.L., Schaeffer, J.A., & Hopkins, W.D. (2008) Communicative signaling activates 'Broca's' homolog in chimpanzees. *Curr Biol*, 18, 343–8.

Taglialatela, J.P., Russell, J.L., Schaeffer, J.A., & Hopkins, W.D. (2011) Chimpanzee vocal signaling points to a multimodal origin of human language. *PLoS One*, 6, e18852.

Tan, A., Tan, S.H., Vyas, D., Malaivijitnond, S., & Gumert, M.D. (2015) There is more than one way to crack an oyster: Identifying variation in Burmese long-tailed macaque (*macaca fascicularis acurea*) stone-tool use. *PLoS One*, 10, e0124733.

Taumoepeau, M. & Ruffman, T. (2008) Stepping stones to others' minds: Maternal talk relates to child mental state language and emotion understanding at 15, 24, and 33 months. *Child Dev*, 79, 284–302.

Tchernichovski, O. & Marcus, G. (2014) Vocal learning beyond imitation: Mechanisms of adaptive vocal development in songbirds and human infants. *Curr Opin Neurobiol*, 28, 42–7.

Tennie, C., Call, J., & Tomasello, M. (2012) Untrained chimpanzees (*Pan troglodytes schweinfurthii*) fail to imitate novel actions. *PLoS One*, 7, e41548.

Thiebaut de Schotten, M., Cohen, L., Amemiya, E., Braga, L.W., & Dehaene, S. (2014) Learning to read improves the structure of the arcuate fasciculus. *Cereb Cortex*, 24, 989–95.

Thomas, B.C., Croft, K.E., & Tranel, D. (2011) Harming kin to save strangers: Further evidence for abnormally utilitarian moral judgments after ventromedial prefrontal damage. *J Cogn Neurosci*, 23, 2186–96.

Tomasello, M. (1999) *The Cultural Origins of Human Cognition*. Harvard University Press, Cambridge.

Tomasello, M. (2014) *A Natural History of Human Thinking*. Harvard University Press, Cambridge, USA.

Tomasello, M. (2019) *Becoming Human*. Harvard University. Press, Cambridge.

Tomasello, M., Carpenter, M., Call, J., Behne, T., & Moll, H. (2005) Understanding and sharing intentions: The origins of cultural cognition. *Behav Brain Sci*, 28, 675–91; discussion 691–735.

Vaccaro, A.G. & Fleming, S.M. (2018) Thinking about thinking: A coordinate-based meta-analysis of neuroimaging studies of metacognitive judgements. *Brain Neurosci Adv*, 2, 2398212818810591.

Vandenberghe, R., Price, C., Wise, R., Josephs, O., & Frackowiak, R.S.J. (1996) Functional anatomy of a common semantic system for words and pictures. *Nature*, 383, 254–6.

Vargha-Khadem, F., Watkins, K., Alcock, K., Fletcher, P., & Passingham, R. (1995) Praxic and nonverbal cognitive deficits in a large family with a genetically transmitted speech and language disorder. *Proc Natl Acad Sci USA*, 92, 930–3.

Wang, L., Uhrig, L., Jarraya, B., & Dehaene, S. (2015) Representation of numerical and sequential patterns in macaque and human brains. *Curr Biol*, 25, 1966–74.

Wendelken, C., Chung, D., & Bunge, S.A. (2012) Rostrolateral prefrontal cortex: Domain-general or domain-sensitive? *Hum Brain Mapp*, 33, 1952–63.

Wendelken, C., Nakhabenko, D., Donohue, S.E., Carter, C.S., & Bunge, S.A. (2008) 'Brain is to thought as stomach is to??': Investigating the role of rostrolateral prefrontal cortex in relational reasoning. *J Cogn Neurosci*, 20, 682–93.

Westendorff, S., Klaes, C., & Gail, A. (2010) The cortical timeline for deciding on reach motor goals. *J Neurosci*, 30, 5426–36.

Whiten, A. (2017) Social learning and culture in child and chimpanzee. *Annu Rev Psychol*, 68, 129–54.

Whiten, A., Custance, D.M., Gomez, J.C., Teixidor, P., & Bard, K.A. (1996) Imitative learning of artificial fruit processing in children (*Homo sapiens*) and chimpanzees (*Pan troglodytes*). *J Comp Psychol*, 110, 3–14.

Whiten, A., Goodall, J., McGrew, W.C., Nishida, T., Reynolds, V., Sugiyama, Y., Tutin, C.E., Wrangham, R.W., & Boesch, C. (1999) Cultures in chimpanzees. *Nature*, 399, 682–5.

Whiten, A., Horner, V., & de Waal, F.B. (2005) Conformity to cultural norms of tool use in chimpanzees. *Nature*, 437, 737–40.

Whiten, A. & van de Waal, E. (2017) Social learning, culture and the 'socio-cultural brain' of human and non-human primates. *Neurosci Biobehav Rev*, 82, 58–75.

Whitney, C., Kirk, M., O'Sullivan, J., Lambon Ralph, M.A., & Jefferies, E. (2012) Executive semantic processing is underpinned by a large-scale neural network: Revealing the contribution of left prefrontal, posterior temporal, and parietal cortex to controlled retrieval and selection using TMS. *J Cogn Neurosci*, 24, 133–47.

Woollett, K. & Maguire, E.A. (2011) Acquiring 'the Knowledge' of London's layout drives structural brain changes. *Curr Biol*, 21, 2109–14.

Index

For the benefit of digital users, indexed terms that span two pages (e.g., 52–53) may, on occasion, appear on only one of those pages.

Notes
Abbreviations used in the index can be found on pages xxvii to xxviii
Tables and figures are indicated by an italic *t* and *f* following the page/paragraph number

abstract categories
  behavioural meaning of cues 271
  conditional tasks 306
  prefrontal-based ganglia interactions 311–12
abstract conditional rules 302
abstractions, invariance 298
abstract matching tasks 299
abstract rules 299–302, 300*f*, 301*f*, 319–20
  conjunctive cells and learning 320–21
  prefrontal-based ganglia interactions 311–12
  sensory input and 301–2
abstract sound sequences
  centre embedding 434
  macaque monkeys 433–34
ACC *see* anterior cingulate cortex (ACC)
accumulation models 161
acetylcholine 356
action reversal tasks 81–82, 93*f*, 135*f*, 135
  feedback dependence 84
actions
  cerebral dominance, speech *vs.* 424
  selections 291
activating responses, changing behaviour 264–68
activation
  activity *vs.* 15–16
  performance *vs.* 384
activity, activation *vs.* 15–16
Aegyptopithecus
  brain–body mass relationship 55, 56*f*
  frontal lobe size 58
Africa, migration from 361
agency awareness *see* awareness of agency
agranular insular cortex (Ia) 119
  connections 120–21
agranular medial prefrontal cortex
  homologies in 73, 74*f*
  timing of actions 107
agranular prefrontal cortex 23–27, 25*t*, 35*f*

AIP *see* anterior intraparietal area (AIP)
allocortex, PF cortex relationship 39
Alouatta (howler monkeys) 48
American Sign Language (ASL) 423
American Standard Code for Information Interchange (ASCII) 253–54
amygdala
  food desirability 147
  food preferences 122, 123*f*
  moral dilemmas 454
  ultrasound in macaques 19–20
amygdala connections
  ACC 76
  mediodorsal nucleus 296
  orbital PF cortex 121, 122, 146
  parietal cortex 290
  prelimbic/infralimbic cortex 76
  ventral PF cortex 275
amygdala lesion studies
  feeding devaluation 130
  food choices 128, 129*f*
  go no-go task 132*f*, 132
  new learning 133
anatomical asymmetry, cerebral dominance 425–26, 426*f*
angiosperms 45–46
  energy-rich foods 50
  food-production history 50
animal foods, climate change and 361
animal phobia testing 407
animate–inanimate object tests 262
A-not-B task 207
anterior cingulate cortex (ACC)
  BrainMap 11
  conflict monitoring 87
  conspecific value encoding 89
  error signal prediction 268
  functional subdivisions 73
  information value judgements 91

anterior cingulate cortex (ACC) (*cont.*)
  reward circuitry 81
  social behaviour 88–92
  social interactions 89–92, 90*f*
  subdivisions in 73
  task performance 374–75
  time to act studies 80
  ultrasound in macaques 19–20
  visual information and 76
anterior cingulate cortex connections 77, 87
  amygdala 76
  area 45A 240
  superior temporal cortex 89
anterior cingulate cortex lesion studies
  choice testing 80
  deterministic reversal tasks 136*f*, 136, 138*f*
anterior cingulate gyrus (gACC) 73
  area 45A connections 240
  social interactions 91
anterior cingulate sulcus (sACC) 73
  choice trials 85
  choice variable encoding 87
  effort costs *vs.* delay costs 88
  expected *vs.* actual outcomes 84
  goal attentive selection 268
  inhibiting & activating responses 267–68
  outcome-related signals 84
  principal sulcus connections 77
  resource decline 86
  self-generated action 88
  time to act studies 80
  value encoding 85
anterior cingulate sulcus lesion studies
  action choices 83
  action reversals 82, 83*f*
anterior insular cortex (Ia)
  BrainMap 11
  task performance 374–75
anterior intraparietal area (AIP) 194–95
  area 44 connections 338
  cerebral dominance in skilled actions 423
  evolution 3–4
anterior middle temporal gyrus 407
anterior parietal area (AIP) evolution 3–4
anthropoid evolution 3
  feeding habits 49
  frontal expansion 38, 59*f*
  PF cortex 54*f*, 55–57, 56*f*, 57*f*, 58*f*
  size increase 48–49
anthropoids
  brain–body mass relationship 56*f*
  definition 34
  foraging strategies 50
  palaeoecology 48–51

PF areas 46*t*
strepsirrhine primates *vs.* 43
Aphelocomoa californica (scrub jay) 98
arcuate fasciculus (AF)
  cerebral dominance and anatomical asymmetry 425
  connections 353, 354*f*
Ardipethecus ramidus 333
  endocranial volume 334
  vocal communication 362
area 8
  area 45A connections 242
  area 45B connections 337–38
  area 47/12 connections 242
  paired-associate tasks 252
  single-pulse TMS 271
area 8A
  functions of 181
  ODR task 175, 177*f*
  order on recall task 213
  recall and re-activation 215–16
  remembered location *vs.* saccadic response 178
  retrospective *vs.* prospective coding 179
  spatial location encoding 173, 174*f*
area 8Ad
  connections 157
  human brain 168*f*, 168
area 8A lesion studies
  food search path trials 199
  return errors 181
  short-term memory 175, 176*f*
area 8Av connections 157
  orbital PF cortex 182
area 8B connections 157
area 8d 270–71
area 8 lesion studies
  bilateral *vs.* unilateral 170
  effects of 169
area 9
  CoCoMac 6*f*, 6
  connections 77, 105, 211–12
  MVPA 406
  retrospinal cortex connectivity 76
area 9/46
  abstract category learning 271
  area 46 division 196–98, 197*f*, 198*f*
  behavioural meaning of cues 270–71
  dorsal PF cortex 191–92
  DR tasks 226
  eye movement sequence tests 221, 222*f*
  new learning 313
  order on recall task 213
  paired-associate tasks 252

planning and automatic sequence
    performance 313
recall and re-activation 215–16
remembered location *vs.* saccadic
    response 178
spatial location encoding 173, 174*f*
area 9/46 connections 228
    LIP 211
area 9/46v
    behavioural meaning of cues 271
    inhibiting & activating responses 267
area 10
    allometry 343–45, 344*f*, 345*f*
    connections 105
    retrospinal cortex connections 76
    spine number according to species 355
area 11
    inactivation effects 130
    orbital PF cortex 118
area 13
    juice–water preference tests 124
    orbital PF cortex 118
area 14
    CoCoMac 6*f*, 6
    orbital PF cortex 118
area 32 406
    imagining outcomes for others 453
    Iowa Gambling task 451–52
    learning the motives & intentions of
        others 447–48
area 44
    cerebral dominance and anatomical
        asymmetry 425
    cerebral dominance in gestures 422
    connections 337
    deductive reasoning tests 388, 389*f*
    demonstration in teaching 439
    event retrieval & future imagining 390–91
    expansion of 409
    inhibiting/activating responses 265
    inner speech 388
    meanings 435
    picture decoding studies 436
    speech 430
    syntax 431, 432
    theta burst stimulation 266
    ventral PF cortex connections 275
area 44d
    area 44v *vs.* 265
    demonstration in teaching 438–39
    learning the motives & intentions of
        others 447
    mirror neurones 439–40
area 44v 265

area 45A 24*f*, 236–37, 239*f*, 239
    abstract category learning 271
    behavioural meaning of cues 270–71
    connections 240–42
    cooling inactivation studies 246
    dorsal PF cortex area 9/46 connections 194
    event retrieval & future imagining 390–91
    inhibiting/activating responses 265
    inhibiting & activating responses 267
    learning categories 256
    learning the motives & intentions of
        others 447
    meanings 435
    picture decoding studies 436
    simultaneous matching 340
    visual–auditory associations 253
    visual–spatial associations 253
    visual–visual associations 250
area 45B 26*f*, 237
    cerebral dominance and anatomical
        asymmetry 425
    cerebral dominance in gestures 423, 424*f*
    connections 337
    deductive reasoning tests 388, 389*f*
    demonstration in teaching 439
    event retrieval & future imagining 390–91
    expansion of 409
    gesture & sound repetition 427
    inner speech 388
    meanings 435
    picture decoding studies 436
    syntax 431
    transcranial magnetic stimulation 444
    ventral PF cortex connections 275
area 46
    area 9/46 division 196–98, 197*f*, 198*f*
    connections 211–12
    dorsal PF cortex 191–92
    DR tasks 226
    high *vs.* low distraction 218
    navigation 100
    new learning 313
    paired-associate tasks 252
    planning and automatic sequence
        performance 313
area 46 lesion studies
    DR tasks 202
    temporal order 209
area 47/12 24*f*, 236, 238, 239*f*, 239
    connections 240–42
    new learning investigations 445
    simultaneous matching 340
    visual–spatial associations 253
area F5c 439–40

area F5p 439–40
area PFG 438–39
arithmetic tasks, fMRI 375
ASCII (American Standard Code for
    Information Interchange) 253–54
ASL (American Sign Language) 423
association automation, cerebellum 318
association via categories, ventral PF cortex 256–64
associative learning
    dorsal PF cortex & dorsal striatum 309
    PF cortex–striatum connections 294
    prefrontal-based ganglia interactions 309–11, 310*f*
    ventral PF cortex 340
asymmetry, cerebral dominance 425–26, 426*f*
Ateles paniscus (red-faced spider monkey) 16
Aterian technology 362
attentional selection 291
attention-guided search and choice task 12
attention, memory coding *vs.* 174*f*
auditory cortex, echolocating bats 37–38
auditory input, orbital PF cortex 121
Australopithecus africanus 336
Australopithecus amensis 334
Australopithecus sediba 336, 337
autobiographical memory retrieval 406
automatic sequence performance
    areas involved 313
    planning and 313
    prefrontal-based ganglia interactions 312–15
    prefrontal–cerebellar interactions 315–16
autonomic nervous system (ANS)
    cortical stimulation 40–41
    granular PF cortex 41
    PF cortex evolution 35*f*, 40–41
awareness of agency 402–4, 403*f*
    testing for 403

basal dendritic spines, layer III pyramidal
    neurones 355
basal forebrain, time to act studies 80
basal ganglia
    associative learning 311
    cortico-subcortical connections 292–94, 293*f*
    output streams 292, 293*f*
    striatum connections 292–94
    structure 297
Bayesian priors, objects reversals 143–45
Beck scale, depression 407
behaviour
    corpus callosum lesion studies 251
    cues, learning of 270–71
    ventral PF cortex 264–69
behavioural ecology 86

beta activity, order on recall task 214
bilateral hippocampal damage, memories 393
binocular rivalry, simulated rivalry *vs.* 400
bipedalism, climate change and 360–61
body mass, brain size *vs.* 53–54, 54*f*, 56*f*
BOLD signal
    dorsal PF cortex 403–4
    fMRI 15–16
    self-representation 406
    temporo-parietal junction 403–4
bonobo *(Pan paniscus)* 16
    relation to humans 333
    spoon test 381
brain evolution 456
brain expansion, self-generated thoughts 393,
    393*t*, 394*f*
BrainMap database
    functional fingerprints 11
    medial cortical areas 92–93
brain size, body mass *vs.* 53–54, 54*f*, 56*f*
Broca's area 13
Broca's homologue, chimpanzees 422
bushbabies *(Galago)*
    granular/agranular PF cortex 35*f*
    saccadic eye movements 42

calculation tasks 375
Californian ground squirrel
    (Otospermophilus) 339
Callithrix (marmoset) 350
Cambridge Gambling task 451
capuchin monkey *(Cebus apella)*
    brain structure 59–60
    connectivity in PF cortex 350
    episodic memory 98
capybara *(Hydrochoerus hydrochaeris)* 339
Carpolestes 42
catarrhines
    evolution 49
    platyrrhines *vs.* 49
category representation, ventral PF
    cortex 261–64, 263*f*
Cattell tests
    Culture Fair Test, Scale 2, Form A 379
    fluid intelligence 377
caudal cingulate motor area (CMAc) 77
caudal prefrontal cortex 27, 29*t*, 153
    areas 24*f*, 153–55, 154*f*
    connections 155–58, 156*f*, 183
    covert visual attention *see* covert visual
        attention
    neglect 169–71
    neural operation 183
    problem solving 390

INDEX   473

saccadic eye movements in bushbabies 42
selective advantage 182–83
short term maintenance 171–81, 172f
short-term memory 175, 225, 247
ventral PF cortex connections 275
visual search 159–63
see also frontal eye field (FEF)
caudal prefrontal cortex lesion studies
   letter identification 180f, 180
   return errors 181
Cavia porcellus (guinea pig) 339
Cebus apella see capuchin monkey (*Cebus apella*)
cell activity
   brain lesions and 20–22
   temporal order 205f, 210–12
central orbital prefrontal cortex 118
centre embedding, abstract sound
      sequences 434
Cercopithecus aethiops (vervet monkeys) 423
Cercopithecus ascanius (red-tail monkey) 50
cerebellum
   automatic sequence performance 313
   cognition and 316–19
   cortico-subcortical connections 293f, 294–95
   dentate nucleus 294–95
   evolution of size increase 341
   eye-blink conditioning 317
   sensorimotor structure as 295
cerebellum lesion studies
   overlearned movement sequences 316
   spatial delayed alternation task 318
cerebral dominance 422–26
   anatomical asymmetry 425–26, 426f
   gestures 422–23, 424f
   skilled actions 423–25
cetaceans (whales) 56
change-shift rule 306, 319–20
change trials, theta burst stimulation of
      area 44 266
cheeks, speech 430
chess 384
Chilecebus
   brain–body mass relationship 55, 56f
   frontal lobe size 58
chimpanzee (*Pan troglodytes*) 16
   anatomical asymmetry 425
   area 10 spine number 355
   cerebral dominance 422, 425
   episodic memory 98
   fMRI of medial networks 94f
   frontal lobe association area comparisons 346
   gestures 422
   gesture & sound repetition 428
   granular PF cortex extent 339–40

handedness 420, 421t
medial network 93
PF cortex as proportion of frontal
      lobe 337, 338f
relation to humans 333
visual cortex 353
visual imagery 381
vocal communication 362
chimpanzee tool use 321–22, 437
   human infants *vs.* 438
choice making
   fMRI studies 383
   medial PF cortex 79–81
   prediction in macaques 91
choices
   definition 78
   reaction times in skilled actions 423
   sensory specific satiety 131
cingulate cortex 13
climate change
   animal foods and 361
   bipedalism and 360–61
   foraging options and 360–61
   Oligocene 48, 106, 360
   PF cortex evolution 360–61
closeness centrality 350–51
CMAc (caudal cingulate motor area) 77
CMAr *see* rostral cingulate motor area (CMAr)
CoCoMac 5, 6f
cognition
   cerebellum and 293f, 316–19
   fMRI 375
commentary stage 396
communication, PF cortex evolution 362–63
comparative psychology 4
conditional tasks
   abstract *vs.* non-abstract strategy 306
   visuo-spatial task & dentate lesion studies 316
conjunctive cells, abstract rule learning 320–21
connectional clusters, PF cortex 6, 7f
connectional fingerprints, macaques 17
connectivity 5–8
   function relation 12–13, 13f
   macaque monkeys 15
   multiple demand system 373–74
   PF cortex evolution 350–54, 351f, 352f, 354f
   prefrontal cortex 456
controlled retrieval condition, ventral PF
      cortex 445
core medial network 391
corpus callosum
   lesion studies and behavioural
      performance 251
   proportion of tissue 342, 343f

474 INDEX

cortex
  autonomic output stimulation 40–41
  striatum connections *see* cortico-striatal connections
  subcortical connections *see* cortico-subcortical connections
  visual stimuli processing 396
corticalization 51–52, 54*f*
cortico-cortical connections 288–92
  areal integration 288, 289*f*
  feedback connections 290–92, 291*f*
  reason for integration 289–90
corticofugal projections, nucleus accumbens 36–37
cortico-striatal connections 40
  PF cortex 39–40
cortico-subcortical connections 292–97
  basal ganglia 292–94, 293*f*
  cerebellum 293*f*, 294–95
  dopaminergic midbrain 296–97
  thalamus 295–96
costs, exploration 87–88
covert trial and error 410
  problem solving 382–83
covert visual attention 163–69
  dorsal systems 168–69
  human testing 164–65, 167
    see also Posner task
  macaque testing 164, 166
  ventral systems 168–69
credit assignment
  fMRI studies 141
  objects reversals 139–41, 140*f*
  meaning of term 140
cryoprobe cooling, short-term memory studies 175
cued go–no-go task 131–32
cues
  behavioural meaning learning 270–71
  ventral PF cortex 269–72, 270*f*, 273*f*
culture 457–58
  environment in development 410
  self-generated thoughts 393, 393*t*, 394*f*
current task rule 272
cytoarchitecture, PF cortex 22–23

DA tasks *see* delayed alternation (DA) tasks
decisions 78
deductive reasoning tests 388
default network 92
delay activity disruption, short term maintenance 175–77, 176*f*, 177*f*
delay costs, effort costs *vs.* 88
delayed alternation (DA) tasks 200–3

dorsal PF cortex lesions 225
  interference 207
delayed matching tasks 250
  ventral PF cortex 341–42
delayed response (DR) tasks 200–3, 201*f*
  brain areas 226
  dorsal PF cortex lesion studies 200, 225, 287
  interference 207
delay-related activity
  perirhinal cortex 248
  working memory and 173
demonstration, teaching 428*f*, 437–40
dentate lesion studies 316
  conditional visuo-spatial task 316
dentate nucleus
  cerebellum 294–95
  pegboard problem 318
  spatial delayed alternation task 318
depression, medial PF cortex 407
descriptions, moral dilemmas 454
designer receptors exclusively activated by designer drugs (DREADDS) 201–2
deterministic reversal tasks 135*f*, 135–36
  objects reversals 136*f*, 136–37, 138*f*
devaluation effect, granular orbital PF cortex/amygdala connections 128
devaluation tasks, human subjects 128–30
devaluation time, sensory specific satiety 130–31
diagnostic traits, PF cortex evolution 39–41
diamidino yellow 15
diffusion weighted imaging (DWI) 5
  borders between areas 12–13
  cerebral dominance and anatomical asymmetry 425, 426*f*
  connectivity in PF cortex 350
  false-positive results 15
  human *vs.* chimpanzee brain connectivity 353
  major fibre tract connectivity 352*f*, 352–53
  tracer methods *vs.* 93, 94*f*
  visual cortex comparisons 353
distance, time *vs.* in navigation 96
distractor array testing 164
distractor-resistant memory, dorsal PF cortex 216–18, 217*f*
distractor trials 217
  high *vs.* low distraction 217–18
DLPFC (dorsolateral prefrontal cortex) 192
DNA structure 5
DO *see* dorsal opercular area (DO)
Dolichocebus 55
dopamine 356
dopamine receptors 296
  visual–spatial associations 256

dopaminergic midbrain
  associative learning 311
  cortico-subcortical connections 296–97
dorsal covert visual attention systems 168–69
dorsal frontal eye field (FEFd)
  connections 155, 159
  human brain 168f, 168
  saccadic eye movements 3
dorsal opercular area (DO) 238
  connections 338
dorsal prefrontal cortex 27, 29t, 191
  area 46 & area 9/46 division 196–98, 197f, 198f
  areal integration 288
  areas 24f, 26f, 191–92, 192f
  associative learning 309
  connections *see* dorsal prefrontal cortex connections
  delay-period activation 206
  distractor-resistant memory 216–18, 217f
  dual task paradigm 314
  events context and order 289–90
  expansion of 409
  interference 207–8
    see also n-back tasks
  lesion studies *see* dorsal prefrontal cortex lesion studies
  matching/non-matching tasks 299, 300f
  mechanisms 212–16
  neural operation 228–29
  neurophysiology 12
  order maintenance on recall tasks 214f
  planning for arrangements 383–84
  recall & re-activation 215–16
  searching 199–200
  selective advantage 227–28
  sequence generation 218–21, 220f
  sequence planning 221–25, 222f, 223f, 224f
  temporal order 212
    see also temporal order
  temporal order maintenance 213–15
  time-resolved fMRI 403–4
  ventral PF cortex *vs.* 274
  visual–spatial association 306, 308f
  see also delayed response (DR) tasks
dorsal prefrontal cortex connections 192–98, 193f, 228
  area 9/46 194
  area 46 194–96
  striatum 294
dorsal prefrontal cortex lesion studies
  delayed response tasks 287
  DR/DA tasks 200, 225
  search paths 199

  short-term memory 175, 176f
  Wisconsin card sorting task 270
dorsal premotor cortex
  area integration 288
  associative learning 309
  automatic sequence performance 313
  demonstration in teaching 440
  dorsal PF cortex area 46 connections 195
  lesion studies in visual–spatial associations 255
  new learning 313
  orbital PF cortex connections 121
  planning and automatic sequence performance 313
  saccades in macaques 154
  ventral PF cortex connections 275
dorsal striatum, associative learning 309
dorsolateral prefrontal cortex (DLPFC) 192
dorsomedial prefrontal cortex 17, 18f
dot display categorization tests 311
DREADDS (designer receptors exclusively activated by designer drugs) 201–2
DR tasks *see* delayed response (DR) tasks
dual-signal trials 266
dual task paradigm 312, 314
  Raven's Progressive Matrices 378
DWI *see* diffusion weighted imaging (DWI)
dynamic causal modelling 394

early primates, palaeoecology 45–48, 46t, 47f
EcoG (electrocorticography) 435
education 457–58
EEG *see* electroencephalography (EEG)
effective connectivity, psychophysiological interaction 402
effort costs, delay costs *vs.* 88
electrocorticography (EcoG) 435
electroencephalography (EEG)
  facial memory 248
  primary somatosensory cortex 14
elementary movements, Purkinje cells & olivary cells 315
emotion, self-representation 407
empathy, child learning 453
encephalization 54f, 56f
  PF cortex evolution 334–36, 335f
encephalization quotient (EQ) 52
  brain size evolution 334, 335f
  mammals 57f
  modern anthropoids 56
  phylogenetic analysis 56–57, 58f
encoding value, medial PF cortex 84
entorhinal cortex lesions 143, 144f
epilepsy 435, 436f

episodic memory
  animals in 98
  hippocampus 99
episodic retrieval, navigation from 97–101
episodic simulation 391
EQ *see* encephalization quotient (EQ)
equitability index 220f, 220–21
ERP event-related potentials 402, 403f
eulaminate areas 23
Euprimates 41
  arboreal life in 47
  grade shifts in PF cortex evolution 52–54, 53f, 54f
  locomotion 47–48
event retrieval, self-generated thoughts 390–95, 392f
everyday life, planning 385–86
evolutionary expansion 357
evolution of brain 456
evolution of frontal lobe 51
  association area comparisons 343f, 347f, 348f, 349t
evolution of hominins 333
evolution of prefrontal cortex 34, 333
  anthropoid grade-shift 54f, 55–57, 56f, 57f, 58f
  association area comparisons 346–49
  autonomic outputs 35f, 40–41
  connectivity 350–54, 351f, 352f, 354f
  cortico-striatal projections 39–40
  diagnostic traits 39–41
  encephalization 334–36, 335f
  euprimate grade shifts 52–54, 53f, 54f
  expansion of 409
  feeding 60
  frontal lobes 336f, 336–41, 338f, 339f, 340t, 341t
  grade shifts & ontogeny 347f, 356–59, 358f
  granular PF cortex *see* granular prefrontal cortex
  homologies 36–38
  microstructure 354–56
  new areas 38
  selection pressures 359–64, 360t
  sensory inputs 40
  spatial arrangement 35f, 39
  upward grade-shifts 51–60
evolution of primates *see* primate evolution
excitotoxic lesions
  orbital PF cortex 137–39, 138f
  surgical lesions *vs.* 137–39, 138f
exploitation
  laboratory models 86–87
  reward rates 87

exploration
  costs in 87–88
  laboratory models 86–87
  reward rates 87
extinction test for neglect 170
eye-blink conditioning, cerebellum 317
eye movements
  cerebellum 318
  sequence tests 221, 222f

faces
  feature tests 262
  matching tasks 247
  memory of 248
feature generalization experiments 259
feedback
  action reversals and 84
  cortico-cortical connections 290–92, 291f
  feedforward signals *vs.* 290, 291f
feedforward signals, feedback signals *vs.* 290, 291f
feeding
  brain evolution 60
  PF cortex evolution 60
  trees, primate evolution 45–46
  see also food; foraging
FEF *see* frontal eye field (FEF)
FEFv *see* ventral frontal eye field (FEFv)
FFA *see* fusiform face area (FFA)
fixed reversals, medial PF cortex 81–82, 90f
flexible choice, medial PF cortex 81–84
fluid intelligence 377–80
  lesion studies and 379f, 379–80
  short-term memory and 378–79, 379f
  tests for 380–81
  see also intelligence
food
  angiosperms 50
  desirability and orbital PF cortex 122–26, 146
  environmental effects 86
  exhaustion of 86
  memory of 106–7
  search paths 199
  shortages in anthropoid evolution 57
  value comparisons 123
food preferences 122–23, 123f, 275
  amygdala 122, 123f
  common currency 123–26
foraging
  climate change and 360–61
  predation in daylight 49
  visual scenes 95
foreign language learning 431
FOXP2 gene 362–63

free choice studies 219
frontal cortex lesion studies, language 380
frontal expansion, anthropoid evolution 38, 59*f*
frontal eye field (FEF) 14, 158–59
   areal integration 288
   cell classification techniques 160
   circuit layout 162*f*
   delay-period 203–4
   dorsal *see* dorsal frontal eye field (FEFd)
   functions of 181
   inactivation in covert visual attention tests 166
   motor activity 160
   non-human primates 155
   planning for arrangements 383–84
   primates foraging studies 158–59
   retrospective *vs.* prospective coding 179
   return errors 181
   saccadic eye movements 3
   task performance 374–75
   ventral *see* ventral frontal eye field (FEFv)
   visual inputs 183
   visual salience encoding 160–61
   visuomotor activity 160
frontal eye field connections
   area 9/46 194
   area 45B 337–38
   medial caudate nucleus 157
   oculomotor nuclei 157
   PF cortex 4
   superior colliculus 157
frontal eye field lesion studies 168*f*, 169
frontal lobectomies
   recency judgements 247
   temporal order 209
frontal lobe lesion studies
   counter intuitive moves on Tower of Hanoi 384
   fluid intelligence assessment 379
frontal lobes, evolution of *see* evolution of frontal lobe
frontal polar cortex connections 77
functional connectivity, resting-state covariance 15
functional fingerprints 8, 9*f*
   fMRI-based 11*f*, 11
   SMA 10
functional magnetic resonance imaging (fMRI)
   ACC in effort costs 88
   agency awareness 402
   area 46 & area 9/46 division 196–97
   automatic sequence performance 313
   BOLD signal 15–16

category representation in ventral PF cortex 261
cerebellum in language 317–18
cerebral dominance in skilled actions 424
choice making 80, 383
cognitive tasks 375
colour & shape matching tasks 270*f*, 270
controlled retrieval condition 445
covert attention testing 164–65
credit assignment studies 141
delay-period activation 203, 246–47
demonstration in teaching 439
dentate nucleus in pegboard problem 318
event retrieval & future imagining 391
facial memory 248
fluid intelligence 377, 378
food preference studies 125
free choice studies 219–20
functional connectivity of medial PF cortex 37
future scene imagining 382
gesture & sound repetition 427, 428*f*, 429
hippocampus 97
imagining outcomes for others 452
Iowa Gambling task 451–52
language meanings 435–36
learning the motives & intentions of others 447
LIP area 262, 263*f*
macaque social interactions studies 90*f*, 90
matching tasks 264
medial networks 94*f*
mirror neurons and 439–40
navigation 97
new category learning 262
new learning 445
no-go *vs.* go trials 265
orbital PF cortex 122, 138*f*, 138–39
recall *vs.* recognition 205*f*, 210–11
right inferior caudal PF cortex 266
risk *vs.* safe choice 452
saccades 154
script generation 386
self-representation 404–5
self-representation vocabulary 406
semantic association studies 445
semantic dementia studies 444
sequence generation tasks 226
semantic studies in epilepsy 435, 436*f*
speech area mapping 430
supplementary eye field 405
synaesthesia patients 400–1
syntax studies 431, 432, 433*f*
temporal order studies 209

functional magnetic resonance imaging
    (fMRI) (cont.)
  TMS combination 272
  Tower of Hanoi 383, 386, 387f
  ventral PF cortex 141–43, 255
  verbal instructions 440–41
  verbal paired-associate studies 435
  visual code for imagination 381
  writing 441–43
functions
  connectivity relation 12–13, 13f
  neural operations 8–12
fusiform face area (FFA) 248–49
  cue changing 271
  visual code for imagination 381
fusiform gyri 436
future imagining
  fMRI studies 382
  self-generated thoughts 390–95, 392f

Gage, Phineas 401
Galago see bushbabies (Galago)
gamma activity, order on recall task 214
gamma aminobutyric acid (GABA)ergic inhibitory interneurons 356
gamma-aminobutyric acid (GABA)ergic projections 292
gelada (Theropithecus gelada) 158–59
general-purpose problem solving 320–21
genetic factors, handedness 421–22
gestures
  cerebral dominance 422–23, 424f
  repetition of 427–30
  verbal naming vs. 423
gibbons (Hylobatidae) 343, 344f
glial cells
  neocortex in 353
  neurones, proportion to 342–43
globus pallidus pars externa (GPe) 292
  associative learning 309
goals
  attentive selection, sACC 268
  dorsal PF cortex 220
  visual maze tests 225
golden lion tamarin (Leontopithecus rosalia) 16
go no-go task 131
  amygdala lesion studies 132f, 132
  inhibiting/activating responses 264
  PF cortex connectivity 374
gorilla (Gorilla gorilla)
  episodic memory 99
  PF cortex as proportion of frontal lobe 337, 338f
GPe see globus pallidus pars externa (GPe)

grade shifts, PF cortex evolution 347f, 356–59, 358f
grammatical language development 363, 410–11, 458
granular frontal opercular (GrFO) connections 338
granular insular cortex (Ia) 121
granular medial prefrontal cortex
  area 9 see area 9
  area 10 see area 10
  connections 75f, 75, 105
granular opercular area (GrFO) 238
granular orbital prefrontal cortex lesions 126–27
granular prefrontal cortex 23–27, 25t, 26f, 35f, 39, 41–45, 72f, 72–73
  autonomic nervous system 41
  evolution of 34, 60
  extent of 339–40, 340t, 341t
  homologies 36
  macaque monkeys 35f, 43
  mammals 36
  new areas 42–43, 44f, 45f
  proportion of frontal lobe as 337–38, 339f
graph theory
  areal integration studies 288
  neocortex connections 288
  PF cortex connectivity 373
great apes (Pongidae), brain–body mass relationship 55
GrFO (granular opercular area) 238
guilt
  anticipation of 454
  punishment fear vs. 454
guinea pig (Cavia porcellus) 339
gustatory information, orbital PF cortex 120–21, 122

hamadryas baboon (Papio hamadryas) 16
  foraging studies 158–59
handedness 420–22, 421t
haplorrhines 34
  evolution of 48
hemineglect 171
heterotypical areas 23
hierarchical organization, PF cortex 224f, 297–99
hippocampus
  episodic memory 99
  fMRI studies in navigation 97
  memories 391–92, 392f
  navigation and 95–96
  object-in-scenes task 105
  paired-associate tasks 252
  personal memory retrieval studies 394
  PF inputs 100

scene immersion 393–94
spatial memory 96–97
temporal order 212
visual replay 95
volume *vs.* PF cortex 393
hippocampus connections 211–12
prelimbic/infralimbic cortex 106
ventral prelimbic cortex 394
hominins
area 10 allometry 343, 344*f*
evolution of 4, 333
PF cortex evolution 333
Homo erectus
frontal lobe evolution 336
Homo heidelbergensis vs. 333
migration from Africa 361
outside Africa 333
striding locomotion 361
Homo ergaster 361
Homo floresiensis 334
Homo habilis 360–61
Homo heidelbergensis
encephalization quotient 334
Europe 333
frontal lobe evolution 336
migration from Africa 361
Homo neanderthalensis 333–34
frontal lobe evolution 336
vocal communication 362
Homo sapiens
encephalization quotient 334
Europe in 333–34
frontal lobe evolution 337
Homo heidelbergensis vs. 333
vocal communication 362
homotypical areas 23
Homunculus
brain–body mass relationship 55, 56*f*
frontal lobe size 58
howler monkeys *(Alouatta)* 48
Human Connectome Project 12, 27
hybrid attractor dynamics and synaptic
    model 215
Hydrochoerus hydrochaeris (Capybara) 339
Hylobatidae (gibbons) 343, 344*f*
hypothalamus connections 77

IFJ inferior frontal junction
IFO (inferior frontal occipital fasciculus) 425
Ignacius 52
imagery, problem solving 380–87
imagination, visual code for 381
imitation, gesture & sound
    repetition 427, 428–29

indifference point, juice–water preference
    tests 124
infants, Sally Anne task 448
inferior caudal prefrontal cortex
abstract vocal sequences 434
connections 242–43
demonstration in teaching 439
gesture & sound repetition 427, 429*f*
short term memory 247
ventral PF cortex connections 275
see also area 44; area 45B
Inferior frontal junction
Posner task 169
supervisory attentional control 375
task performance 374–75
inferior frontal occipital fasciculus (IFO) 425
inferior parietal cortex
area 44 connections 338
demonstration in teaching 438–39
dorsal FEF connections 155
dorsal PF connections 194–95
order on recall task 213
picture decoding studies 436–37
tool-making 456
inferior temporal cortex
behavioural meaning of cues 271
event retrieval & future imagining 390–91
lesions & object discrimination
    problems 304–5, 305*f*
maintenance in 248
MEG studies 398
new category learning tests 262–63
PF cortex size *vs.* 349
visual–spatial associations 254
inferior temporal cortex connections
area 8Av 157
orbital PF cortex 146
perirhinal cortex 99
ventral FEF 155
ventral PF cortex 275
inferior temporal gyri 435
information, value judgements in ACC 91
infralimbic cortex connections
amygdala 76
hippocampus 106
hypothalamus connections 77
nucleus accumbens 76
inhibiting responses, changing
    behaviour 264–68
inner speech 388
instructions, teaching 440–46, 442*f*
intelligence
study of 4
see also fluid intelligence

interference, dorsal PF cortex 207–8
intraparietal area (AIP) 194–95
intraparietal sulcus (IPS)
   connections 294
   delay-period 203–4, 205f
   new learning 313
   order on recall task 213
   planning and automatic sequence
      performance 313
   priority maps in covert attention 165, 177f
   retrospective vs. prospective coding 179
   short-term memory 175
   tool-making 456
   ventral PF cortex connectivity 373–74
invariance, abstractions 298
Iowa Gambling task 450, 451–52
IPS see intraparietal sulcus (IPS)

Japanese macaque (*Macaca fuscata*) 425
joint intentionality 437
   PF cortex evolution 363
judgement, moral dilemmas 454–55
juice–water preference tests 124
   indifference point 124
   value evaluations 125

KE family 428–29
knowledge transfer 411
   evolution of hominins 362

language 426–37
   cerebellum 317–18
   gesture & sound repetition 427–30
   hierarchical structure of 431
   lesion studies 380
   meanings 434–37
   speech 430
   syntax 431–34
   writing 441–43
   see also speech
laryngeal cartilages, speech 430
lateral intraparietal area connections 166
lateral intraparietal area (LIP)
   choices 79
   fixation maintenance training 164
   inactivation in covert visual attention
      tests 166
   microstimulation 79
   short-term memory 247
   temporal order coding 211
lateral intraparietal areas connections 211
   area 9/46 194
   area 46 194
   dorsal FEF 155, 159

lateral orbital prefrontal cortex 118
lateral premotor cortex
   cell activity 10f, 20
   memory sequences 20
   SMA, input from 20
learning
   abstract rules 319–20
   empathy 453
   hierarchical organization and 304
   motives & intentions of others 447–50, 449f
   new subjects see new learning
   ventral PF cortex 256–61, 257f, 258f
learning set 303f
   definition 302
left hemisphere lesions, demonstration in
      teaching 439
left inferior parietal cortex 429
left middle temporal gyrus 427
left premotor cortex 445
left–right discrimination task 374
left ventral prefrontal cortex 441–43
left ventral premotor cortex 429
lesion-mapping analysis, n-back tasks 208
lesion studies 18–22
   amygdala see amygdala lesion studies
   anterior cingulate cortex see anterior cingulate
      cortex lesion studies
   anterior cingulate sulcus see anterior cingulate
      sulcus lesion studies
   area 8A see area 8A lesion studies
   area 46 see area 46 lesion studies
   Cambridge Gambling task 426f, 451
   caudal prefrontal cortex see caudal prefrontal
      cortex lesion studies
   cell activity and 20–22
   cerebellum see cerebellum lesion studies
   dentate see dentate lesion studies
   dorsal PF cortex see dorsal prefrontal cortex
      lesion studies
   entorhinal cortex 143, 144f
   excitotoxic lesions see excitotoxic lesions
   fluid intelligence and 379f, 379–80
   frontal cortex 380
   frontal eye fields see frontal eye field lesion
      studies
   frontal lobe see frontal lobe lesion studies
   granular orbital PF cortex 126–27
   language 380
   left hemisphere 439
   macaque monkeys 19–20
   medial PF cortex see medial prefrontal cortex
      lesions
   orbital PF cortex see orbital prefrontal cortex
      lesion studies

perirhinal cortex *see* perirhinal cortex lesion studies
PF cortex 397
premotor area *see* premotor area lesion studies
presupplementary motor area *see* presupplementary motor area lesion studies
primary visual cortex 396
shopping errands 385
supplementary motor area *see* supplementary motor area lesion studies
total prefrontal lesions 401
uncinate fascicle lesions 251
V1 lesions 396
ventral PF cortex *see* ventral prefrontal cortex lesion studies
letter alphabetization tasks 375
letter identification tests 180*f*, 180
lettering system, brain areas 22
LIP *see* lateral intraparietal area (LIP)
lips, speech 430
load effects, short-term memory 206–7
local field potentials
   learning categories 261
   order on recall task 214
long-tailed macaques *(Macaca fascicularis)* 437
long-term potentiation (LTP), navigation and 96
Lophocebus albigena (mangabeys) 50
lose–shift rules 304
   orbital PF cortex lesion studies 144*f*, 145
LTP (long-term potentiation), navigation and 96

Macaca fascicularis (long-tailed macaques) 437
Macaca fuscata (Japanese macaque) 425
macaque monkeys *(Macaca)* 14–18
  abstract rules 319
  abstract sound sequences 433–34
  activity *vs.* activation 15–16
  agranular PF cortex 35*f*
  brain lesions 19–20
  choice prediction 91
  connectivity 15
  connectivity in PF cortex 350
  cortex, model of 24*f*
  fMRI of medial networks 94*f*
  frontal lobe association area comparisons 346
  granular PF cortex 26*f*, 35*f*, 43, 339–40
  homology 17
  medial PF cortex 72*f*
  models as 16–17
  nested rules 433
  non-representational of monkeys 15–16
  PF areas 25*t*

recognition memory testing 405
visual cortex 353
visual imagery 381
magnetic resonance imaging (MRI)
  criminal behaviour 453
  functional *see* functional magnetic resonance imaging (fMRI)
  macaque social interactions 89
  paramagnetic tracers 15
magneto-encephalography (MEG)
  inferior temporal cortex 398
  saccade-related activity 155
  scene immersion 393–94, 394*f*
  selection entropy 221
  visual replay hippocampus 95
maintenance, posterior areas in 247–49
mammals, PF areas 46*t*
mangabeys *(Lophocebus albigena)* 50
marmot *(Callithrix)* 350
matching tasks 250
  dorsal/ventral PF cortex 299, 300*f*
  medial PF cortex 301
  objects *vs.* faces 247
  rostral cingulate motor area 301
  see also non-matching tasks
matching-to-sample tasks 250
  ventral PF cortex 341, 343*f*
meanings, language 434–37
medial caudate nucleus, FEF connections 157
medial granular prefrontal cortex 101–5, 104*f*
medial lesions, multiple errands test 385
medial networks 92–97, 93*f*, 94*f*
  chimpanzees 93
  connections 93
  fMRI 94*f*
  parietal cortex/parahippocampal cortex connections 98
medial orbital prefrontal cortex 118
medial praecuneus (PCun) 94
medial prefrontal cortex 27, 29*t*, 94
  area 72*f*, 72–73, 74*f*
  choice of when to act 79–81
  choices 78–81
  connections 75*f*, 75–78
  encoding value 84
  expansion of 409
  exploitation & exploration 86–88
    see also exploitation; exploration
  fixed reversals 81–82, 93*f*
  flexible choice 81–84
  functional connectivity 37
  imagining outcomes for others 453
  Iowa Gambling task 451–52

482 INDEX

medial prefrontal cortex (*cont.*)
  learning the motives & intentions of others 447–48
  matching/non-matching tasks 301
  memories 391–92, 392f
  meningioma 386
  metacognition and mentalizing 448, 449f
  moral dilemmas 454
  MVPA 406
  orbital PF cortex connections 76
  peripheral vision inputs 76
  personal memory retrieval studies 394
  probabilistic reversals 82–84
  recognition memory testing 406
  self-representation 407
  time interval generation 101–2
  visual information 107
  see also medial networks
medial prefrontal cortex lesions
  bilateral lesions 450
  imagining outcomes for others 453
  moral dilemmas 454
  ultimatum game 454
mediodorsal (MD) nucleus
  learning and 103f, 296
  magnocellular division 296
  parvocellular division 296
  thalamus 36
mediodorsal nucleus connections
  amygdala 296
  thalamus 295–96
MEG *see* magneto-encephalography (MEG)
memory
  autobiographical memory 406
  coding *vs.* attention 174f
  distractor-resistant memory 216–18, 217f
  episodic *see* episodic memory
  facial memory 248
  perception *vs.* 406
  recognition memory 405
  sequences in lateral premotor cortex 20
  short-term memory *see* short-term memory
  spatial memory 96–97
  spatial sort-term memory tasks 375
  spatial working memory 374
  verbal short-term memory 375
  verbal-working memory 317
  working memory *see* working memory
mentalizing
  metacognition and 448, 449f
  reading *vs.* 448
MEP (motor excitatory potential) 267
metacognition
  accuracy measures 397

  mentalizing and 448, 449f
  perception *vs.* sensory evidence 406
  self-representation measures 405
Microchiroptera 37–38
microstructure, PF cortex evolution 354–56
Microsyops 52
middle superior temporal area (MST), dorsal FEF connections 155
middle temporal (MT) area
  abstract category learning 271
  behavioural meaning of cues 271
  event retrieval & future imagining 390–91
  FEFv connections 159
  visual cortex 37
middle temporal gyrus
  meanings 435
  writing 441–43
"mind-reading" 447–50, 449f
Miocene, anthropoid evolution 49
mirror neurons 439–40, 448
  experience and 440
Montreal Neurological Institute (MNI)
  area 45A 237
  area 46 & area 9/46 division 196
moral dilemmas, social rules 454–55
Morris water maze 96
motor (M) activity, FEF 160
motor excitatory potential (MEP) 267
motor imagery, problem solving 382
motor skill automation 316–17
movement condition (M) 402
movement sequences, overlearned *see* overlearned movement sequences
MRI *see* magnetic resonance imaging (MRI)
MT *see* middle temporal (MT) area
multiple demand system 373–76
  connectivity 373–74
  tasks 374–76, 376f
multiple errands test 385
multivoxel pattern analysis (MVPA) 12
  medial PF cortex 406
  memories 391–92
  new category learning tests 262
  phobia of animals testing 407
  visual representation maintenance 249
muscimol, temporary lesion generation 19
MVPA *see* multivoxel pattern analysis (MVPA)

NA *see* nucleus accumbens (NA)
naïve reinforcement model, reversal learning set 143
navigation
  distance *vs.* time 95–96
  episodic retrieval to 97–101

hippocampal system and 95–96
long-term potentiation and 96
retrieval and 100–1
scenes from memory 101
testing of 96–97
n-back tasks 207–8, 227
  fMRI 375
negative outcome anticipation, social rules 450–52, 451*f*
neglect, caudal PF cortex 169–71
neocortex
  connections 288
  evolution of 51
  frontal lobe proportion 341
  glial cells *vs.* neurones 342–43, 353
  neuronal density 52
  numbers for areas 297
  systems 7
neural operations 8–14, 457
  dorsal PF cortex 228–29
  function 8–12
  function–connectivity relation 12–13
  transformations 13–14
neurocomputational model, stimulus categories 312
neurones
  density in neocortex 52
  glial cells *vs.* 342–43, 353
neuropil space 355
new areas, PF cortex evolution 38
new environments, PF cortex evolution 361–62
new learning
  amygdala lesion studies 133
  area 47/12 445
  areas involved 313
  prefrontal-based ganglia interactions 312–15
nocturnal life, primate evolution 47
no-go trials 266
non-abstract strategy, conditional tasks 306
non-human primates, FEF 155
non-matching tasks 299
  dorsal/ventral PF cortex 299, 300*f*
  medial PF cortex 301
  rostral cingulate motor area 301
  see also matching tasks
non-social choices, social choices *vs.* 91
nucleus accumbens (NA)
  corticofugal projections 36–37
  prelimbic/infralimbic cortex connections 76
  reward circuitry 81

object-in-scenes task 105
object reversal tasks 135*f*, 135

objects
  discrimination problems 304–5
  matching tasks 247
  recognition in perirhinal cortex 298
  reversals in orbital PF cortex 135*f*, 135–46
occipital lobe
  picture decoding studies 436
  temporal area connections 155
oculomotor delayed response task (ODR) 171, 172*f*
  area 8A inactivation studies 175, 177*f*
  cryoprobe cooling studies 175
  parietal lesion studies 176–77
  remembered location *vs.* saccadic response 178
  see also delayed alternation (DA) tasks; delayed response (DR) tasks
oculomotor nuclei, FEF connections 157
ODR *see* oculomotor delayed response task (ODR)
olfaction, orbital PF cortex 120–21, 122
Oligocene, climate change 48, 106, 360
olivary cells, elementary movements 315
ontogeny, PF cortex evolution 347*f*, 356–59, 358*f*
opercular region (op) 9 238
oral inputs, orbital PF cortex 122
orangutan (*Pongo pygmaeus*) 381
orbital lesions, multiple errands test 385
orbital prefrontal cortex 27, 29*t*, 118
  area 118–19, 119*f*
  food desirability 122–26
  lateral sector 147
  negative outcomes 133–34
  neurophysiology 12
  objects reversals 135*f*, 135–46
  outcome encoding 289–90
  sensory specific satiety 126–31, 127*f*, 129*f*
  signs of resources 131–34, 132*f*
orbital prefrontal cortex connections 119–22, 120*f*
  area 8Av 157, 182
  area 45A 240
  area 46 195
  area 47/12 240
  central and lateral sectors 146
  medial PF cortex 76
  ventral PF cortex 275
orbital prefrontal cortex lesion studies
  bilateral lesions 450
  central lesions 123
  central sector 126–27, 127*f*
  excitotoxic lesions 137
  food choices 128, 129*f*
  imagining outcomes for others 453

orbital prefrontal cortex lesion studies (*cont.*)
  moral dilemmas 454
  reversal learning set 142*f*, 144*f*, 145
  simultaneous matching 341
  strip lesions 137–38
  surgical *vs.* excitotoxic lesions 137–39, 138*f*
  three-choice probabilistic reversal tasks 145
  ultimatum game 454
  visual–spatial associations 254*f*, 254
  Wisconsin card sorting task 269
order maintenance on recall tasks, dorsal PF cortex 214*f*
Orrorin 333
Otospermophilus (Californian ground squirrel) 339
outcome imagining
  problem solving 382
  social rules 452–54
overlearned movement sequences
  cerebellum 316
  motor skills as 316

pair-associate learning
  ventral PF cortex 249–56, 254*f*, 257*f*
  visual–auditory associations 252–53
  visual–spatial associations 224*f*, 253–56, 254*f*, 257*f*
pair-coding, temporal area 298
pair-dependent cells, dorsal PF cortex 210
paired-associate tasks 250
  perirhinal cortex 251
paired pulse transcranial magnetic stimulation technique 266, 267
Pan paniscus see bonobo (*Pan paniscus*)
Pan troglodytes see chimpanzee (*Pan troglodytes*)
Papio hamadryas see hamadryas baboon (*Papio hamadryas*)
paracingulate sulcus 73
parahippocampal cortex
  medial network connections 98
  vision and 76
Paranthropus 336
Parapthecus (Simonsius) 55, 56*f*
parcellation, human brain 375, 376*f*
parietal cortex
  inactivation in covert visual attention tests 166
  language lesion studies 380
  ODR task 176–77
parietal cortex connections
  amygdala 290
  area 46 194–95
  areas 8Ad/8Av/8B 157

  medial network 98
parietal lateral intraparietal area
  behavioural meaning of cues 271
  short-term memory 225
parietal lobes
  evolution of 51
  fluid intelligence assessment 379
  order on recall task 213
participation, PF cortex connectivity 373
parvocellular division, mediodorsal nucleus 296
pedunculopontine nucleus 80
pegboard problem, dentate nucleus 318
peptide YY (PYY) 126
perception, memory *vs.* 406
perceptual awareness 376*f*, 395, 399*f*, 401
  intracranial electrode studies 398
  total PF lesions 401
performance, activation *vs.* 384
peripheral vision inputs, medial PF cortex 76
perirhinal cortex
  delay-period activity 248
  paired-associate tasks 251
perirhinal cortex connections
  inferior temporal cortex 99
  orbital PF cortex 146
  ventral PF cortex 275
perirhinal cortex lesion studies
  object recognition 298
  reversal learning 143, 144*f*
personal memory retrieval studies 394
PET *see* positron emission tomography (PET)
phonological tasks 441
picture decoding studies 436
planning
  automatic sequence performance and 313
  everyday life 385–86
  goal arrangements for 383–84
  laboratory tests 383–85
platyrrhines, catarrhines *vs.* 49
plesiadapiforms 41
polar plots 5
polar prefrontal cortex
  expansion of 409
  lesions 102–3, 104*f*
  visual cues 102, 103*f*
Pongidae (great apes), brain–body mass relationship 55
Pongo pygmaeus (orangutan) 381
population bottlenecks 410
population vectors, rotated ODR task 178, 179*f*
position-dependent cells, dorsal PF cortex 210
positron emission tomography (PET)
  abstract categories and rules 313
  cerebral dominance in gestures 422, 423

cerebral dominance in skilled actions 423
chimpanzees 16
demonstration in teaching 438–39
fluid intelligence 377
medial networks 93
n-back task 208
neural operations 457
new learning 445
oculomotor delayed response task 173
Oldowan vs. Acheulean tool making 445, 446f, 456
semantic system 444
simultaneous matching tasks 341
verbal paired-associate studies 435
writing 443f, 443
Posner task 164
inferior frontal junction 169
posterior cingulate cortex (PCC) 94
connections 87
posterior fusiform gyrus
area 46 connections 194–95
cerebral dominance for gestures 423, 424f
posterior parahippocampal gyrus 394
posterior parietal cortex 288
fMRI covert attention testing 164–65
goal representation 290
letter identification 180f, 180
multisensory inputs 290
new category learning tests 263, 264
temporal order 212
PrCo see precentral opercular area (PrCo)
precentral opercular area (PrCo) 238
connections 338
predation
anthropoid evolution 57, 60–61
daylight foraging 49
prefrontal-based ganglia interactions 307–15
abstract categories & rules 311–12
associative learning 309–11, 310f
new learning vs. automatic performance 312–15
prefrontal–cerebellar interactions 315–19
automatic performance 315–16
prefrontal cortex 22–27
abstract category learning 271
agranular areas 23–27
allocortex relationship 39
connectional clusters 6, 7f
cytoarchitecture 22–23
decision-making 78
dorsal see dorsal prefrontal cortex
evolution of see evolution of prefrontal cortex
granular areas see granular prefrontal cortex
hierarchical organization 224f, 297–99

lesion studies 37
medial see medial prefrontal cortex
social interactions 89
subareas of 27, 28f, 29t
transfer see transfer
ventral see ventral prefrontal cortex
visual stimuli categorization 21
pregenual prelimbic cortex 90
resource cooperation/competition 90–91
prelimbic cortex connections 77
amygdala 76
hippocampus 106
hypothalamus 77
nucleus accumbens 76
premotor area lesion studies
object discrimination problems 305f, 305
premotor area, species comparisons 347f
preSMA see presupplementary motor area (preSMA)
prestriate cortex, species comparisons 347f
presubiculum connections 211–12
presupplementary motor area (preSMA)
awareness of agency 402
dual task paradigm 314
MVPA 406
new learning 313
paired pulse TMS technique 267
sequence learning tasks 314
presupplementary motor area connections 77
area 46 195
presupplementary motor area lesion studies
choice testing 80
navigation 100
sequence learning tasks 314–15
primary actions, handedness 421, 421t
primary somatosensory cortex 14
primary visual cortex lesion studies 396
primate evolution
social groups 46, 47f
varied diet advantages 146–47
principal sulcus, sACC connections 77
probabilistic 3-arm bandit tasks 141, 142f
probabilistic reversal tasks 135f, 135–36
medial PF cortex 82–84
objects reversals 139–41, 140f
problem solving 390
cerebellum 317
imagery and 380–87
symbolic thought 387–90, 389f
Proconsul 55, 56f
propositional code 387–88
prospective coding
retrospective coding vs. 177–81, 179f, 180f
short-term memory 204f, 205f, 205–6

proto-communication development 458
proto-speech evolution 423
psychophysical interactions, effective connectivity 402
psychometrics 4
punishment fear, guilt vs. 454
Purkinje cells 315
putamen
  automatic sequence performance 313
  reward circuitry 81
pyramidal neurones, basal dendritic spines 355
pyramid and palm tree test 443f, 444

rats
  abstract rules 319
  granular/agranular PF cortex 35f
Raven's Progressive Matrices 377f, 377, 378, 379f
  dual task paradigm 378
re-activation, dorsal PF cortex 215–16
reading 441–43
  mentalizing vs. 448
reasoning, spatial codes and 390
recall
  dorsal PF cortex 215–16
  recognition vs. 210–11
receiver operating characteristic (ROC) curves 397
recency judgements, frontal lobectomies 247
recognition memory testing 405
recognition, recall vs. 210–11
recurrent neural networks, computational models 215
red-faced spider monkey (*Ateles paniscus*) 16
red-tail monkeys (*Cercopithecus ascanius*) 50
reinforcement skill 321
relational reasoning tasks, fMRI 375
remapping factors 340, 340t, 341t
repeat-stay rules 306, 319–20
repetitive transcranial magnetic stimulation (rTMS)
  cerebral dominance in skilled actions 424
  letter alphabetization tasks 375
representational similarity analysis (RSA) 12, 85, 248
  language meanings 435–36
resources
  cooperation & competition for 90–91
  orbital PF cortex 131–34
  volatility 135
resting-state coactivation, STS 92–93, 93f
resting-state covariance 89
  area 46 & area 9/46 division 196, 197f
  functional connectivity 15
  PF cortex connectivity 350
  ventral PF cortex 239

retrieval, navigation and 100–1
retrograde amnesia 99
retrospective coding
  prospective coding vs. 177–81, 179f, 180f
  short-term memory 203–5, 204f, 205f
retrosplenial cortex 94
  area 9 & 10 connections 76
  area 46 connections 195
  connections 211–12
reversal learning set 143–45, 144f
reward circuitry 81
rewards
  exploitation vs. exploration 87
  magnitude vs. cell activity 85
  probabilistic rewards 82, 132f
ROC (receiver operating characteristic) curves 397
rodents, brain area proportions 339
Rooneyia 52–53, 53f
rostral anterior cingulate sulcus
  choices 80
  resource cooperation/competition 90–91
rostral cingulate motor area (CMAr)
  connections 77
  dorsal PF cortex area 46 connections 195
  matching/non-matching tasks 301
  navigation 100
RSA *see* representational similarity analysis (RSA)
rTMS *see* repetitive transcranial magnetic stimulation (rTMS)

saccades
  bushbabies 42
  caudal PF cortex 3
  macaque dorsal premotor cortex 154
  targeting & visual salience encoding 160–61
  visual-visual associations 309
Sahelanthropus 333
  endocranial volume 334
Saimiri sciureus (squirrel monkeys) 339
Sally Anne task 447
  infants 448
scene immersion, magneto-encephalography 393–94, 394f
scenes from memory, navigation 101
SCR *see* skin conductance response (SCR)
script generation, fMRI 386
scrub jay (*Aphelocomoa californica*) 98
second somatosensory area (S2), area 44 connections 338
SEF (supplementary eye field) 155, 405
selection pressures, PF cortex evolution 359–64, 360t

selective advantage 3–5, 321–22, 410–11
self-generated (voluntary) actions 71–72
  sACC 88
self-generated thoughts 390–95
  brain expansion & culture interplay 393, 393t, 394f
  event retrieval 390–95, 392f
  future imagining 390–95, 392f
self-generation mechanisms 395
self-representation 404–7, 408f
  emotion 407
semantic dementia 444
semantic processing, cerebellum 317
semantic tasks 441
sensory inputs
  abstract rules and 301–2
  PF cortex evolution 40
sensory specific satiety, orbital PF cortex 126–31, 127f, 129f
sequences
  generation in dorsal PF cortex 218–21, 220f
  learning tasks 314
  planning in dorsal PF cortex 221–25, 222f, 223f, 224f
serotonin 356
set shifting, cerebellum 317
shopping errands, PF lesion tests 385
short term maintenance
  caudal PF cortex 171–81, 172f
  delay activity disruption 175–77, 176f, 177f
  retrospective vs. prospective coding 177–81, 179f, 180f
short-term memory 203–7, 225
  caudal PF cortex 175
  cryoprobe cooling 175
  dorsal PF cortex 274
  fluid intelligence and 378–79, 379f
  IPS 175
  load effects 206–7
  prospective coding 204f, 205f, 205–6
  retrospective coding 203–5, 204f, 205f
signed prediction error 84
signs of resources, orbital PF cortex 132f
Simonsius (Parapthecus) 55, 56f
simulated rivalry, binocular rivalry vs. 400
simultaneous matching, ventral PF cortex 341–46, 343f
single-pulse transcranial magnetic stimulation 271
skilled actions, cerebral dominance 423–25
skilled performance, cerebral dominance 425
skin conductance response (SCR)
  negative outcome anticipation 450
  phobia of animals 407

SMA see supplementary motor area (SMA)
SOA (stimulus onset synchrony) 396
social choices, non-social choices vs. 91
social factors
  ACC 88–92
  PF cortex evolution 363–64
social groups
  ACC 89–92, 90f
  brain evolution effects 89–90
  development 49
  fMRI studies 90f, 90
  primate evolution 46, 47f
social rules 363, 450–55
  imagining outcomes for others 452–54
  moral dilemmas 454–55
  negative outcome anticipation 450–52, 451f
sound repetition, language 427–30
spatial arrangement, PF cortex evolution 35f, 39
spatial codes, reasoning and 390
spatial delayed alternation task 318
spatial memory, hippocampus 96–97
spatial sort-term memory 375
spatial working memory 374
specific conditional rules 299
speech 430
  cerebral dominance, actions vs. 424
  see also language
spoon test 381
squirrel monkeys (Saimiri sciureus) 339
stimulus categories, neurocomputational model 312
stimulus onset synchrony (SOA) 396
STN (subthalamic nucleus) 267
stop-signal task 265
STP (superior temporal polysensory area) 77, 288
strategy scoring
  computation 305f, 306
  food search path trials 199
strepsirrhines 34
  anthropoids vs. 43
  brain–body mass relationship 56f
  evolution of 48
  PF areas 46t
striatum
  time to act studies 80
striatum connections
  basal ganglia 292–94
  dorsal PF cortex 294
striding locomotion 361
strip lesions, orbital PF cortex 137–38
strokes 19
  neglect 170
Stroop task 374

structural equation modelling 313–14
STS *see* superior temporal sulcus (STS)
subareas, PF cortex 28f, 29t
subgoal calculation, visual maze tests 225
subjective experience 410
substantia nigra
 reward circuitry 81
 time to act studies 80
subthalamic nucleus (STN) 267
superior colliculus connections
 areas 8Ad/8Av/8B 157
 FEF 157
 SEF 160
superior temporal cortex connections
 ACC 89
 area 47/12 240
 ventral PF cortex 275
superior temporal polysensory area (STP) 77, 288
superior temporal sulcus (STS)
 deductive reasoning tests 388
 learning the motives & intentions of others 447–48
 resting-state coactivation 92–93, 93f
 social interactions 89
superior temporal sulcus connections 77, 228
 area 9/46 194
 area 45A 240
 areas 8Ad/8Av/8B 157
supervisory attentional control, inferior frontal junction 375
supplementary eye field (SEF) 155, 405
 superior colliculus connections 160
supplementary motor area (SMA) 8
 automatic sequence performance 313
 awareness of agency 402
 cell activity 10f, 20
 connections 77
 functional fingerprints 10
 lateral premotor cortex, input to 20
 movement from memory *vs.* visual cues 10f, 10–11
 sequence learning tasks 314
supplementary motor area lesion studies
 choice testing 80
 sequence learning tasks 314–15
surgical resections 19
symbolic thought, problem solving 387–90, 389f
synaesthesia, fMRI 400–1
syntax, language 431–34
systems 7

tactile inputs, orbital PF cortex 122
transcranial magnetic stimulation (TMS) 249
tarsiers *(Tarsiidae)* 34
 brain–body mass relationship 56f
 nocturnal activity 48
 trichromatic vision 48
tasks
 multiple demand system 374–76, 376f
 switching mechanisms 272
TE *see* temporal area (TE)
teaching 437–46
 demonstration 428f, 437–40
 instructions 440–46, 442f
temporal area (TE)
 areas 8Av connection 157
 pair-coding 298
temporal lobe
 evolution of 51
 lesion studies & fluid intelligence assessment 379
 picture decoding studies 436
 size comparisons 348
 ventral PF cortex connections 251, 275
temporal order 209–12
 cell activity encoding order 205f, 210–11
 cell activity encoding time 211–12
temporal order maintenance, dorsal PF cortex 213–15
temporal pole connections 77
temporo-parietal junction (TPJ)
 BOLD 403–4
 learning the motives & intentions of others 447
 stokes and neglect 170
 time-resolved fMRI 403–4
temporo-parietal transition area (Tpt) 427
thalamus
 mediodorsal nucleus 36
 new learning 313
thalamus connections
 cortico-subcortex 295–96
 mediodorsal nucleus 295–96
theory-theory 447
Theropithecus gelada (gelada) 158–59
theta-burst transcranial magnetic stimulation
 activation *vs.* performance 384
 area 44 266
thoughts, self-generated *see* self-generated thoughts
3-arm bandit task 143
three-choice probabilistic reversal tasks 145
time
 distance *vs.* in navigation 96
 interval generation in medial PF cortex 101–2

time-resolved functional magnetic resonance
    imaging
  dorsal PF cortex 403–4
  TPJ 403–4
tool use
  climate change and 361
  Oldowan vs. Acheulean 445, 446f, 456
Torrejonia 42
total prefrontal lesions, perceptual
    awareness 401
Tower of Cambridge test 383, 384f
Tower of Hanoi test 318, 383
  fMRI 386, 387f
Tower of London test 384f, 384
TPJ *see* temporo-parietal junction (TPJ)
Tpt (temporo-parietal transition area) 427
tracer methods
  anatomical connections 288, 289f
  DWI vs. 93, 94f
tractography, PF cortex connectivity 350
transcranial magnetic stimulation (TMS) 19
  area 45B 444
  covert visual attention test in humans 167
  fMRI combination 272
  repetitive *see* repetitive transcranial magnetic
      stimulation (rTMS)
  retrospective vs. prospective coding 179
  single-pulse 271
transcription factors, PF cortex evolution 359
transfer 302–7, 303f
  definition 302
  syntax tests 432–33
transformations, neural operations 13–14
transgenic mice, hippocampus in navigation 96
tree shrews
  granular PF area 43, 44f
  lesion studies 43, 45f
Tremacebus 55
trichromatic vision evolution 48
tumours 19

ultimatum game 454
ultrasound
  macaques 19–20, 84
  time to act studies 80
uncinate fascicle lesions, visual–visual
    associations 251
upright walking evolution 334
  climate change and 360–61
upward grade-shifts, PF cortex evolution 51–60

V1 bilateral lesions 396
V2, FEFv connections 155, 159
V3, FEFv connections 155, 159

V4
  abstract category learning 271
  behavioural meaning of cues 271
  FEFv connections 155, 159
  timing in covert visual attention 166
value evaluations, juice–water preference
    tests 125
ventral attention system 168–69
ventral frontal eye field (FEFv)
  connections 155, 159
  saccadic eye movements 3
  timing in covert visual attention 166
ventral intraparietal areas (VIP)
  dorsal FEF connections 155
  multisensory inputs 290
  recall and re-activation 215–16
ventral prefrontal cortex 27, 29t, 141–43,
    142f, 236
  areas 236–39, 237f, 238f, 239f
    see also area 45A; area 45B; area 47/12
  association via categories 256–64
  changing behaviour 264–69
  controlled retrieval condition 445
  cue changing 269–72, 270f, 273f
  current contexts 289–90
  demonstration in teaching 438–39
  dorsal PF cortex vs. 274
  event retrieval & future imagining 390–91
  maintenance in 248
  matching/non-matching tasks 299, 300f
  matching-to-sample 341
  neural operation 275–76
  new learning 445
  pair-associate learning 249–56, 254f, 257f
  selective advantage 272–76
  semantic tasks 441, 444
  visual code for imagination 381–82
  visual–spatial association 306, 308f
ventral prefrontal cortex connections 240–43,
    241f, 275
  area 8Av 157
  intraparietal sulcus 373–74
  temporal lobe 251
  ventral premotor cortex 146
ventral prefrontal cortex lesion studies
  delayed matching 245f, 342
  delayed non-matching tasks 246
  simultaneous matching 341
  visual–spatial associations 254f, 254
  Wisconsin card sorting task 269–70
ventral prelimbic cortex, hippocampus
    connections 394
ventral premotor cortex 3–4, 8
  areal integration 288

ventral premotor cortex (cont.)
  demonstration in teaching 440
  evolution of 3–4
  orbital PF cortex connections 121
  phonological tasks 441
  ventral PF cortex connections 146, 275
ventral visual system, spines in 298
ventrolateral polar prefrontal cortex 383
verbal naming, gestures vs. 423
verbal paired-associate studies 435
verbal short-term memory tasks 375
verbal-working memory, cerebellum 317
verb generation, cerebellum 317
vervet monkeys *(Cercopithecus aethiops)* 423
Victoriapithecus
  brain–body mass relationship 55, 56*f*
  frontal lobe size 58, 59*f*
VIP *see* ventral intraparietal areas (VIP)
visceral information, orbital PF cortex 120–21, 122
vision
  ACC and 76
  control over feedback 403
  foraging fields 95
  obstacles in visual maze tests 223, 224*f*
  parahippocampal cortex and 76
  search in caudal PF cortex 159–63
visual areas, vFEF connections 155
visual attention, covert *see* covert visual attention
visual-auditory association
  pair-associate learning 252–53
  ventral PF cortex 252–53
visual code, imagination for 381
visual cortex
  diffusion weighted imaging 353
  evolution of 37, 59
  visual stimuli categorization 21
visual cues
  food preferences 123–24
  polar PF cortex 102, 103*f*
visual imagery, problem solving 381–82

visual inputs
  orbital PF cortex 121, 122
  ventral PF cortex 275
visual maze tests 223*f*, 223
  visual obstacles 223, 224*f*
visual salience encoding
  FEF 160–61
  saccade targeting 160–61
visual–spatial association
  dorsal/ventral PF cortex 306, 308*f*
  pair-associate learning 224*f*, 253–56, 254*f*, 257*f*
  ventral PF cortex 253–56, 254*f*, 257*f*
visual stimuli
  categorization in brain areas 21
  cortical processing 396
visual–visual association
  pair-associate learning 250–52
  saccades 309
  ventral PF cortex 250–52
visuomotor activity (VM), FEF 160
visuomotor cells 183, 440
visuospatial information 373–74
VM (visuomotor activity), FEF 160
voices, matching tasks 247
voluntary actions *see* self-generated (voluntary) actions

Washoe 428
whales (cetaceans) 56
win–shift errors 139
win–stay strategy 304
  orbital PF cortex lesion studies 144*f*, 145
Wisconsin Card Sorting Task 269
  cerebellum 318
  monkeys 269
  PF cortex connectivity 374
working memory
  delay-related activity and 173
  misuse of term 226
writing 441–43
WWW (what, where, when) memories 98

The manufacturer's authorised representative in the EU for product
safety is Oxford University Press España S.A. of El Parque Empresarial
San Fernando de Henares, Avenida de Castilla, 2 - 28830 Madrid
(www.oup.es/en or product.safety@oup.com). OUP España S.A. also acts
as importer into Spain of products made by the manufacturer.
Printed and bound by CPI Group (UK) Ltd, Croydon, CR0 4YY

15/12/2025
02019516-0013